Nonconventional and Vernacular Construction Materials

Related titles

Eco-efficient Masonry Bricks and Blocks: Design, Properties and Durability
(ISBN: 978-1-78242-305-8)

Eco-efficient Construction and Building Materials
(ISBN: 978-0-85709-767-5)

Modern Earth Buildings
(ISBN: 978-0-85709-026-3)

Woodhead Publishing Series in Civil and Structural Engineering: Number 58

Nonconventional and Vernacular Construction Materials

Characterisation, Properties and Applications

Edited by

K.A. Harries, B. Sharma

AMSTERDAM • BOSTON • CAMBRIDGE • HEIDELBERG
LONDON • NEW YORK • OXFORD • PARIS • SAN DIEGO
SAN FRANCISCO • SINGAPORE • SYDNEY • TOKYO
Woodhead Publishing is an imprint of Elsevier

Woodhead Publishing is an imprint of Elsevier
The Officers' Mess Business Centre, Royston Road, Duxford, CB22 4QH, UK
50 Hampshire Street, 5th Floor, Cambridge, MA 02139, USA
Langford Lane, Kidlington, OX5 1GB, UK

ISBN: 978-0-08-100871-3 (print)
ISBN: 978-0-08-100055-7 (online)

British Library Cataloguing-in-Publication Data
A catalogue record for this book is available from the British Library

Library of Congress Cataloging-in-Publication Data
A catalog record for this book is available from the Library of Congress

For information on all Woodhead Publishing publications
visit our website at http://store.elsevier.com/

Contents

List of contributors   xi
Woodhead Publishing Series in Civil and Structural Engineering   xiii
Preface   xvii

Part One   Nonconventional materials and vernacular
            construction   1

1   What we learn from vernacular construction   3
    Randolph Langenbach
    1.1   Bam   3
    1.2   Vernacular   7
    1.3   Vernacular architecture   9
    1.4   The "vernacular" of industrial architecture   10
    1.5   Srinagar, Kashmir, India   13
    1.6   The great 1906 San Francisco earthquake and fire   19
    1.7   Skeleton-frame construction   22
    1.8   Frames and solid walls   26
    1.9   Conclusion: the "ecology" of the vernacular   28
          Sources of further information   33
          References   33

2   Introduction to nonconventional materials and an historic
    retrospective of the field   37
    Khosrow Ghavami
    2.1   Introduction   37
    2.2   Natural materials in historic construction   39
    2.3   Reinforced adobes as energy-saving construction materials   44
    2.4   Bamboo material for the 21st century   48
    2.5   Application of natural and alternative materials in modern bridge
          construction   57
    2.6   Concluding remarks   58
          Acknowledgements   59
          References   59

3   **Future directions for nonconventional and vernacular material
    research and applications**   **63**
    S. Suffian, R. Dzombak, K. Mehta
    3.1   Introduction   **63**
    3.2   Analyzing potential for sustainable scale   **66**
    3.3   Deciding between materials   **75**
    3.4   Conclusion   **77**
          Acknowledgments   **78**
          References   **78**

Part Two   Natural fibres   **81**

4   **Characterization of vegetable fibers and their application in
    cementitious composites**   **83**
    V.C. Correia, S.F. Santos, G.H.D. Tonoli, H. Savastano, Jr.
    4.1   Introduction   **83**
    4.2   Characterization of vegetable fibers for engineering applications **85**
    4.3   Characterization of vegetable fiber composites   **95**
    4.4   Final remarks and perspective for future research needs   **100**
          Acknowledgments   **102**
          References   **103**

5   **Natural fibre-reinforced noncementitious composites (biocomposites)** **111**
    S.J. Christian
    5.1   Introduction   **111**
    5.2   Description of biocomposites   **111**
    5.3   Durability concerns   **117**
    5.4   Managing durability   **121**
    5.5   Conclusions and future research   **122**
          References   **123**

6   **Straw bale construction**   **127**
    P. Walker, A. Thomson, D. Maskell
    6.1   Introduction   **127**
    6.2   Material properties   **130**
    6.3   Characterisation of straw bale construction materials   **137**
    6.4   Applications   **142**
    6.5   Future of straw bale construction   **152**
          References   **153**

Part Three   Concrete and masonry   157

**7   Utilization of industrial by-products and natural ashes in mortar
     and concrete: development of sustainable construction materials   159**
R. Siddique, Kunal
7.1   Introduction   159
7.2   Cement kiln dust   160
7.3   Silica fume   170
7.4   Waste foundry sand   173
7.5   Rice husk ash   182
7.6   Palm oil fuel ash   190
7.7   Conclusions   195
      References   196

8   Dry-stack and compressed stabilised earth-block construction   205
H.C. Uzoegbo
8.1   Introduction   205
8.2   Materials selection and block production   209
8.3   Block production   223
8.4   Strength evaluation of block units and masonry   231
8.5   Strength evaluation of block units and masonry walls   236
8.6   Durability of CSEBs   245
8.7   Conclusions   247
      References   248

9   Unfired clay materials and construction   251
J.M. Kinuthia
9.1   Introduction to clay-based materials   251
9.2   Structure and properties of clay soils   252
9.3   Unfired clay material systems   255
9.4   Clay materials without additives   256
9.5   Industrial additives   258
9.6   Agricultural additives   263
9.7   Construction using clay-based systems   264
9.8   Future trends   268
      References   269

10   Earthen materials and constructions   273
A. Fabbri, J.C. Morel
10.1 Earth as a building material   273
10.2 Particularities of earthen materials and constructions   275

10.3   Assessing the performance of earthen material for construction   **290**
10.4   Concluding remarks on earthen building sustainability   **294**
       Acknowledgments   **296**
       References   **296**

11   Ancient stone masonry constructions   **301**
       L. Dipasquale, L. Rovero, F. Fratini
11.1   Introduction   **301**
11.2   Overview of ancient applications   **302**
11.3   Stone masonry materials   **303**
11.4   Masonry constructions   **314**
11.5   Mechanical performance   **322**
11.6   Applications in modern architecture and innovative uses   **325**
11.7   Role of masonry in sustainable construction   **327**
       References   **330**

Part Four   Timber, bamboo and paper   **333**

12   Nonconventional timber construction   **335**
       K.I. Crews
12.1   Natural forms of timber materials   **335**
12.2   Traditional and historic applications (round and sawn timber)   **337**
12.3   Material properties   **342**
12.4   Modern engineering applications   **349**
12.5   Significant case studies   **354**
12.6   Codes and standards   **359**
12.7   Conclusions   **361**
       References   **362**

13   Bamboo material characterisation   **365**
       D. Trujillo, L.F. López
13.1   Introduction   **365**
13.2   Bamboo as a plant   **365**
13.3   Material properties of bamboos   **367**
13.4   Tests for material and physical property determination   **374**
13.5   Derivation of design values   **385**
13.6   Further work and future developments   **386**
13.7   Concluding remarks   **388**
       Acknowledgements   **389**
       References   **389**

14   Bamboo design and construction   **393**
       J.F. Correal
14.1   Introduction   **393**
14.2   Bamboo as a material   **399**

14.3   Design considerations   404
14.4   Bending members   408
14.5   Axial force members   414
14.6   Combined loading elements   416
14.7   Shear walls   417
14.8   Connections   419
14.9   Fabrication and construction   427
14.10   Concluding remarks and future research   428
        References   429

15 Engineered Bamboo   **433**
Y. Xiao
15.1   Introduction   433
15.2   Production of glubam   434
15.3   Material properties of engineered bamboo   437
15.4   Structural components   439
15.5   Applications: buildings   441
15.6   Applications: bridges   444
15.7   Analysis of carbon footprints   447
15.8   Future research needs   448
        References   448
Engineered bamboo for structural applications   **450**
B. Sharma
        References   452

16 Paperboard tubes in structural and construction engineering   **453**
L.C. Bank, T.D. Gerhardt
16.1   Introduction   453
16.2   Paper tube manufacturing and primary uses   455
16.3   Mechanics and properties of paperboard and tube materials   459
16.4   Structural systems   466
16.5   Structural elements   466
16.6   Structural analysis and design   473
16.7   Conclusion   477
        References   477

Index   **481**

List of contributors

L.C. Bank The City College of New York, New York, NY, United States

S.J. Christian Carnegie Mellon University, Pittsburgh, PA, United States

J.F. Correal Universidad de los Andes, Bogotá, Colombia

V.C. Correia University of São Paulo, São Paulo, Brazil

K.I. Crews University of Technology Sydney, Ultimo, NSW, Australia

L. Dipasquale DIDA (Department of Architecture), University of Florence, Florence, Italy

R. Dzombak University of California, Berkeley, CA, United States

A. Fabbri University of Lyon, Vaulx-en-Velin, France

F. Fratini CNR ICVBC (National Council of Researches-Institute for Conservation and Promotion of Cultural Heritage), Sesto Fiorentino, Florence, Italy

T.D. Gerhardt Sonoco Products Company, Hartsville, SC, United States

Khosrow Ghavami Pontifícia Universidade Católica, Rio de Janeiro, Brazil

J.M. Kinuthia University of South Wales, Cardiff, United Kingdom

Kunal Thapar University, Patiala, Punjab, India

Randolph Langenbach Conservationtech Consulting Oakland, California, United States

L.F. López Independent Consultant, Bogotá, Colombia

D. Maskell University of Bath, Bath, United Kingdom

K. Mehta The Pennsylvania State University, University Park, PA, United States

J.C. Morel Coventry University, United Kingdom

L. Rovero DIDA (Department of Architecture), University of Florence, Florence, Italy

S.F. Santos São Paulo State University, São Paulo, Brazil

H. Savastano, Jr. University of São Paulo, São Paulo, Brazil

B. Sharma   University of Bath, Bath, United Kingdom

R. Siddique   Thapar University, Patiala, Punjab, India

S. Suffian   Villanova University, Villanova, PA, United States

A. Thomson   University of Bath, Bath, United Kingdom

G.H.D. Tonoli   Federal University of Lavras, Lavras, Minas Gerais, Brazil

D. Trujillo   Coventry University, Coventry, United Kingdom

H.C. Uzoegbo   University of the Witwatersrand, Johannesburg, South Africa

P. Walker   University of Bath, Bath, United Kingdom

Y. Xiao   College of Civil Engineering, Nanjing Tech University, Nanjing, China

Woodhead Publishing Series in Civil and Structural Engineering

1   **Finite element techniques in structural mechanics**
    C. T. F. Ross
2   **Finite element programs in structural engineering and continuum mechanics**
    C. T. F. Ross
3   **Macro-engineering**
    F. P. Davidson, E. G. Frankl and C. L. Meador
4   **Macro-engineering and the earth**
    U. W. Kitzinger and E. G. Frankel
5   **Strengthening of reinforced concrete structures**
    Edited by L. C. Hollaway and M. Leeming
6   **Analysis of engineering structures**
    B. Bedenik and C. B. Besant
7   **Mechanics of solids**
    C. T. F. Ross
8   **Plasticity for engineers**
    C. R. Calladine
9   **Elastic beams and frames**
    J. D. Renton
10 **Introduction to structures**
    W. R. Spillers
11 **Applied elasticity**
    J. D. Renton
12 **Durability of engineering structures**
    J. Bijen
13 **Advanced polymer composites for structural applications in construction**
    Edited by L. C. Hollaway
14 **Corrosion in reinforced concrete structures**
    Edited by H. Böhni
15 **The deformation and processing of structural materials**
    Edited by Z. X. Guo
16 **Inspection and monitoring techniques for bridges and civil structures**
    Edited by G. Fu
17 **Advanced civil infrastructure materials**
    Edited by H. Wu
18 **Analysis and design of plated structures Volume 1: Stability**
    Edited by E. Shanmugam and C. M. Wang
19 **Analysis and design of plated structures Volume 2: Dynamics**
    Edited by E. Shanmugam and C. M. Wang

20 **Multiscale materials modelling**
    Edited by Z. X. Guo
21 **Durability of concrete and cement composites**
    Edited by C. L. Page and M. M. Page
22 **Durability of composites for civil structural applications**
    Edited by V. M. Karbhari
23 **Design and optimization of metal structures**
    J. Farkas and K. Jarmai
24 **Developments in the formulation and reinforcement of concrete**
    Edited by S. Mindess
25 **Strengthening and rehabilitation of civil infrastructures using fibre-reinforced polymer (FRP) composites**
    Edited by L. C. Hollaway and J. C. Teng
26 **Condition assessment of aged structures**
    Edited by J. K. Paik and R. M. Melchers
27 **Sustainability of construction materials**
    J. Khatib
28 **Structural dynamics of earthquake engineering**
    S. Rajasekaran
29 **Geopolymers: Structures, processing, properties and industrial applications**
    Edited by J. L. Provis and J. S. J. van Deventer
30 **Structural health monitoring of civil infrastructure systems**
    Edited by V. M. Karbhari and F. Ansari
31 **Architectural glass to resist seismic and extreme climatic events**
    Edited by R. A. Behr
32 **Failure, distress and repair of concrete structures**
    Edited by N. Delatte
33 **Blast protection of civil infrastructures and vehicles using composites**
    Edited by N. Uddin
34 **Non-destructive evaluation of reinforced concrete structures Volume 1: Deterioration processes**
    Edited by C. Maierhofer, H.-W. Reinhardt and G. Dobmann
35 **Non-destructive evaluation of reinforced concrete structures Volume 2: Non-destructive testing methods**
    Edited by C. Maierhofer, H.-W. Reinhardt and G. Dobmann
36 **Service life estimation and extension of civil engineering structures**
    Edited by V. M. Karbhari and L. S. Lee
37 **Building decorative materials**
    Edited by Y. Li and S. Ren
38 **Building materials in civil engineering**
    Edited by H. Zhang
39 **Polymer modified bitumen**
    Edited by T. McNally
40 **Understanding the rheology of concrete**
    Edited by N. Roussel
41 **Toxicity of building materials**
    Edited by F. Pacheco-Torgal, S. Jalali and A. Fucic
42 **Eco-efficient concrete**
    Edited by F. Pacheco-Torgal, S. Jalali, J. Labrincha and V. M. John

43   **Nanotechnology in eco-efficient construction**
     Edited by F. Pacheco-Torgal, M. V. Diamanti, A. Nazari and C. Goran-Granqvist

44   **Handbook of seismic risk analysis and management of civil infrastructure systems**
     Edited by F. Tesfamariam and K. Goda

45   **Developments in fiber-reinforced polymer (FRP) composites for civil engineering**
     Edited by N. Uddin

46   **Advanced fibre-reinforced polymer (FRP) composites for structural applications**
     Edited by J. Bai

47   **Handbook of recycled concrete and demolition waste**
     Edited by F. Pacheco-Torgal, V. W. Y. Tam, J. A. Labrincha, Y. Ding and J. de Brito

48   **Understanding the tensile properties of concrete**
     Edited by J. Weerheijm

49   **Eco-efficient construction and building materials: Life cycle assessment (LCA), eco-labelling and case studies**
     Edited by F. Pacheco-Torgal, L. F. Cabeza, J. Labrincha and A. de Magalhães

50   **Advanced composites in bridge construction and repair**
     Edited by Y. J. Kim

51   **Rehabilitation of metallic civil infrastructure using fiber-reinforced polymer (FRP) composites**
     Edited by V. Karbhari

52   **Rehabilitation of pipelines using fiber-reinforced polymer (FRP) composites**
     Edited by V. Karbhari

53   **Transport properties of concrete: Measurement and applications**
     P. A. Claisse

54   **Handbook of alkali-activated cements, mortars and concretes**
     F. Pacheco-Torgal, J. A. Labrincha, C. Leonelli, A. Palomo and P. Chindaprasirt

55   **Eco-efficient masonry bricks and blocks: Design, properties and durability**
     F. Pacheco-Torgal, P.B. Lourenço, J.A. Labrincha, S. Kumar and P. Chindaprasirt

56   **Advances in asphalt materials: Road and pavement construction**
     Edited by S.-C. Huang and H. Di Benedetto

57   **Acoustic emission (AE) and related non-destructive evaluation (NDE) techniques in the fracture mechanics of concrete: Fundamentals and applications**
     Edited by M. Ohtsu

58   **Nonconventional and vernacular construction materials: Characterisation, properties and applications**
     Edited by K. A. Harries and B. Sharma

59   **Science and technology of concrete admixtures**
     Edited by P.-C. Aïtcin and R. J. Flatt

60   **Textile fibre composites in civil engineering**
     Edited by T. Triantafillou

61   **Corrosion of steel in concrete structures**
     Edited by A. Poursaee

62   **Innovative developments of advanced multifunctional nanocomposites in civil and structural engineering**
     Edited by K. J. Loh and S. Nagarajaiah

63   **Biopolymers and biotech admixtures for eco-efficient construction materials**
     Edited by F. Pacheco-Torgal, V. Ivanov, N. Karak and H. Jonkers

Preface

Safety in the built environment is a fundamental right.[1] While adequate shelter is perhaps the most critical aspect of a community infrastructure, the lack of stable infrastructure is often cited as the primary barrier to sustainable development. Indeed, it is suggested that "sustainable development" may be largely unattainable to all but the wealthiest and most advanced global communities.[2] Sustainable development is most commonly defined as that which "meets the needs of the present without compromising the ability of future generations to meet their own needs..."[3] What is often omitted from this definition, however, are the lines that immediately follow: "...It contains within it two key concepts: (1) the concept of needs, in particular the essential needs of the world's poor, to which overriding priority should be given; and (2) the idea of limitations imposed by the state of technology and social organization on the environment's ability to meet present and future needs." It is the challenge of coupling both global and regional sustainability practices with social equity that lies at the heart of the grand engineering and social challenges of the 21st century. Nonconventional and vernacular construction materials are one aspect of addressing these challenges and bringing sustainable development within reach of the entire global community.

There is considerable interest—if judged by the proliferation of journal articles in recent years—in the modern engineering application of nonconventional materials (NOCMAT). Often this interest is driven by the interests of sustainable engineering and/or the newer trends toward "engineering for humanity" and "social entrepreneurship." Nonetheless, there are few repositories of such information. In English, a few international conferences (the NOCMAT series being one of the oldest) and regionally focused journals dominate the literature. Some work makes its way into archival journals, but typically such articles are at the periphery of the journal focus and are often included based on their "sustainability" credentials. This volume aims to fill this significant gap in the available literature and should serve as an important resource for engineers, architects, academics, and students interested in nonconventional materials.

---

[1] 1948 United Nations *Universal Declaration of Human Rights* reiterated in the 1994 United Nations *Special Rapporteur's Report to the United Nations Commission on Human Rights*.

[2] Silberglitt, R. et al. (2006) "The Global Technology Revolution 2020, In-Depth Analyses Bio/Nano/Materials/Information Trends, Drivers, Barriers, and Social Implications." *Report TR-303-NIC*, Rand Corporation, 314 pp.

[3] World Commission on Environment and Development (1987) *Our Common Future*. Oxford University Press. [so-called Brundtland report.]

The focus of this volume is the materials science and modern structural engineering (and some architectural) applications of ancient, vernacular, and nonconventional building materials. Such materials are defined to include relatively unprocessed naturally occurring materials and processed or engineered materials developed from the natural material. Chapters focus on individual construction materials and address both material characterization and structural applications. Each chapter reflects the current state of the art in terms of the modern use and engineering of the material. For ease of reference and use, the volume is divided into four parts.

Part I provides an introduction to nonconventional and vernacular materials, exploring their role in both social and engineering contexts. Chapter 1 places vernacular construction methods in a modern social context. Chapter 2 considers the historic context of these materials and describes the development of their formal study, while Chapter 3 looks to future needs and applications.

Part II focuses on natural fibers. Chapter 4 describes natural fiber-reinforced cementitious composites, while Chapter 5 considers non-cementitious "bio-composites." Chapter 6 describes straw bale construction, which has been gaining traction in diverse communities around the globe.

Part III focuses on nonconventional concrete and masonry materials. Chapter 7 describes the state of the art in natural ash— and waste-based concrete. Chapters 8 through 11 describe various masonry practices: dry stack and pressed brick (Chapter 8), unfired clay (Chapter 9), and adobe (Chapter 10). Nonconventional aspects of stone masonry construction are described in Chapter 11.

Part IV covers nonconventional forest products. Chapter 12 considers nonconventional timber materials in construction. Chapter 13 introduces full-culm (pole) bamboo materials, while Chapter 14 describes full-culm bamboo construction. Engineered bamboo materials are described in Chapter 15. Finally, Chapter 16 introduces a thoroughly modern nonconventional method of construction: paper tubes.

It is hoped that this volume will prove instructive and educational while serving as an important resource for engineers, architects, academics, and students interested in nonconventional materials.

The editors would like to thank all contributing authors for their outstanding work and for maintaining our ambitious schedule for this volume. Thanks are due all at Elsevier, in particular Project Managers Alex White, Kate Hardcastle, and Charlotte Cockle who kept the editors in line throughout this process.

Finally, the editors would like to dedicate this book to the memory of our friend, colleague, and collaborator, Ms. Gayatri Kharel, who provided the opportunity for us both to delve deeper into this field than we may have imagined.

Kent A. Harries, Pittsburgh
Bhavna Sharma, Bath
January 2016

Part One

Nonconventional materials and vernacular construction

What we learn from vernacular construction

1

Randolph Langenbach
Conservationtech Consulting Oakland, California, United States

Photographs by © Randolph Langenbach, except as marked.

1.1 Bam

On December 26, 2003, while people were still sleeping, an earthquake with a magnitude of 6.6 (USGS 2004) struck the ancient city of Bam in Iran. Despite the relatively modest magnitude, the surface shaking was intense enough to devastate the city, killing approximately 30,000 people, one-third of the population. Many of the news photographs that circled the globe were of the devastated ruins of the city's most iconic heritage site, the ancient citadel recognized as the largest earthen structure in the world, the "Arg-e-Bam" (Fig. 1.1). Because of these dramatic views of the pulverized parts of the Arg in the news, many people assumed that most of the 30,000 people who were killed died in the ancient earthen buildings of the Arg. In fact, they did not. The death toll was almost exclusively caused by the collapse of buildings in the modern city, almost all of which had been constructed during the past 30 years.

   The "Arg" was an ancient walled city that became an archeological site and museum of the history of the city. After being continuously occupied from as early as the 6th century BCE, it had fallen into disuse in the early 19th century, when people

Figure 1.1 The Arg-e-Bam archeological site in Bam, Iran, before and after the 2003 Bam earthquake.

Nonconventional and Vernacular Construction Materials. http://dx.doi.org/10.1016/B978-0-08-100038-0.00001-9

Figure 1.2 Earthquake destruction of the restored parts of the Arg-e-Bam photographed four months after the earthquake.

felt safe enough to abandon their houses the walls and move into their date palm orchards that were planted around the Arg. By the middle of the 20th century, well after the Arg buildings had fallen into roofless ruins, a major restoration project was begun to turn the Arg into a tourist attraction. Over the course of the half-century preceding the 2003 earthquake, a wide swath of buildings from the main gate to the Governor's House on the hill were restored (Figs. 1.2 and 1.3). These included the main market street, two caravanserais, the main mosque, the Governor's House, and a large part of the fortifications. By the time of the earthquake the restored site, with its undulating earthen walls, had successfully become a world-famous attraction.

The 2003 earthquake rendered the Arg almost unrecognizable. Many of the formless rubble piles actually stood higher than the walls that were left still standing. The site quickly became a symbol for the earthquake in the same way as did the National

Figure 1.3 View of an archeological site in Bam with high walls that came through the 2003 earthquake with very little damage.

Palace in Port-au-Prince, following the 2010 Haiti Earthquake. When I arrived to inspect the site as a delegate in a UNESCO-organized international reconnaissance and conference held four months after the earthquake, the conventional wisdom was that the damage could be explained by the simple fact that the Arg was constructed of unfired clay.

For those of us who arrived in Bam shortly after the earthquake, the argument that the Arg collapsed simply because it was unfired clay did not immediately seem to be unreasonable. This impression was reinforced by what could be witnessed in the surrounding settlement, where many modern steel-frame buildings had also collapsed. In fact, prior to my arrival in Bam I had been told by a seismic engineer in Tehran that earthen construction should be banned, despite its continued use and practicality in the desert climate and a shortage of timber in Iran and the rest of the Middle East.

At first glance, after entering the Arg-e-Bam there was little to be seen to disabuse one of such an opinion. The devastation was vast. However, on traversing the site more than one time, there were increasing visual signals that being made of unfired clay could not alone explain the nature and extent of the damage. First, there was almost no evidence of diagonal tension cracks in the collapsed or still-standing walls. Instead, almost everywhere the walls looked as if they collapsed vertically — like a slump test with too much water in a concrete mix [1].

Looking further, it became evident that some walls and structures had appeared to survive almost entirely intact, while others were only a pile of rubble. As I continued to look a pattern began to emerge. This was particularly evident when I came upon an area that had *not* been restored. The standing walls of unrestored structures, abandoned for more than a century and a half, had survived with little damage, while those that collapsed were almost always the ones that had been restored in recent decades. In addition, when examining the military fortifications, counterintuitively, the thicker the wall the *more* likely it was to have been collapsed by the earthquake.

As I continued to look, I also discovered that in a random sampling of the cracked open sections of the collapsed or broken walls, there was frass (insect excrement) in practically every single one. The Iranian archeologists told me this was from a local species of termites. *Termites?* In an archeological site with walls dating back 2000 to 8000 years? This comment launched me on a study that in the end had more to do with materials science than with structural engineering (Fig. 1.4).

The first question was: Why did the unrestored roofless buildings survive the earthquake better than the restored structures? And did the termites have something to do with this? To search for an answer to this question, I turned to a new subject of inquiry — the science of the cohesion of unfired earthen construction. The question was how it may have been compromised by the 20th-century restoration work. A chance meeting with an engineering doctoral student from the United Kingdom, Paul Jaquin, provided the first steps toward an answer. He had been doing detailed research on rammed earth construction, including its internal cohesion [2]. From him, and later also from colleagues at the French research and teaching institute CRA-terre, I learned that the most significant contributor to cohesion in unfired clay is the meniscus forces of water [3]. *Water?* In ancient structures in a hot dry desert?

Figure 1.4 Evidence of termites in a wall in the Arg-e-Bam.

Jaquin and CRAterre's research had shown that the cohesive force from water is so effective that the use of stabilizers can on occasion reduce, rather than increase, the cohesion of the clay. Even in very dry climates such as that at Bam, there is a residual level of moisture that normally remains in intact earthen walls. This can be enough to hold the walls together, sometimes for centuries. So a plausible hypothesis of what had happened in Bam is that the residual moisture had disappeared prior to the earthquake. But *why?*

While in Bam I learned that the modern restoration work was reinforced with straw, while historically the clay was either without fiber reinforcement or reinforced with shredded date palm bark. Thus, the modern-day restorers had, in effect, been feeding the termites a banquet, whereas historically the builders had used a material, date palm bark, which was resistant to termites. There had been, in effect, a loss of traditional knowledge between the time when the Arg was last inhabited and when it was restored a century and a half later.

To understand how this may have made a difference, I could see that the termites had perforated the walls by consuming the straw reinforcement. This then likely contributed to a desiccation of the walls, resulting in a loss of suction provided by the residual water that had given the walls a large part of their cohesive strength.

This investigation then led to the evidence of other causes of a loss of clay cohesion that may have been more significant than the termite infestations alone. Perhaps the most important pathology was that over centuries of occupancy, with erosion and repair, the ancient walls had evolved from having horizontally bedded mud layers to walls with vertical construction joints. The later restoration work, which included the reconstruction of the missing upper parts of the walls and roofs, added overburden weight to the surviving lower parts of the walls. The high-frequency earthquake vibrations thus caused this new work to simply crush the older parts of the walls. This explained why the walls had collapsed in place rather than falling over, as evidenced by the location of the rubble piles with little still standing that remained recognizable [4].

1.2 Vernacular

So, what is the significance of this story in a book about "nonconventional and vernacular" construction materials? Since unfired clay may still be the most commonly used building material worldwide, how can it also be classified as being "nonconventional?"

Of course, one must first understand that the use of the terms "vernacular" and "nonconventional" in the context of architecture and construction technology are not meant to connote "primitive" or lacking in sophistication or scientific basis. For example, Roman concrete, which has demonstrated a remarkable resilience and longevity, continues to defy efforts to answer all questions about its original prehydration and carbonation constituents because of the complexity of its polymerization over many years into a sequence of different minerals. Like the labeling of a human language, from which the term "vernacular" originally has been derived, vernacular construction is founded on the duality alluded to in the previous sentence, namely that it has evolved from human observation and experience over hundreds, if not thousands, of years (Fig. 1.5), while at the same time is scientifically complex enough to justify many dissertations and books. This is one of those books.

The story of the earthquake-induced collapse of the Arg-e-Bam is a story of a shift from an empirically based traditional knowledge handed down over centuries and even millennia, to our own era in which that traditional knowledge has been lost — both because of the introduction of new materials and construction systems, changes in education, and increasingly disused construction systems and technologies forgotten through neglect. This time-honored empirically based knowledge has been replaced by some very sophisticated scientific research in the fields of materials science and structural engineering, but this sophisticated knowledge is not often communicated to masons, or even known by people where it is needed most: those on the construction site. While modern science has often confirmed the time-honored empirical knowledge, it has not and cannot entirely replace the kind of knowledge passed from generation to generation that made structures such as the Arg last as long as they have, based on trial and error experiences over centuries and even millennia.

Figure 1.5 Continuation of a vernacular building tradition in Luong Prabang, Laos.

1.2.1 "Vernacular" defined

"Vernacular" has its origin in the Latin *vernaculus*: "domestic," "native" from *verna*: "home-born slave," a word of Etruscan origin. From the beginning of the 18th century, it came to mean the "native speech or language of a place" [5].

"Vernacular architecture" according to the Oxford English Dictionary is "Architecture concerned with domestic and functional rather than monumental buildings." It is interesting to find that it is described in part as "contrasted against polite architecture which is characterized by stylistic elements of design intentionally incorporated for aesthetic purposes which go beyond a building's functional requirements" [6]. This raises the question of aesthetic intentions within the domain of vernacular architecture. One can argue that vernacular architecture — particularly that which remains universally loved by people from most walks of life — shows evidence of many deliberate aesthetic intentions, such as are found in an English Cotswold cottage, a traditional Turkish Ottoman house, a farmstead in Kerala, India, or a shop house in Pingyao, China. The differences with "polite" architecture are not the presence or absence of aesthetic intentions but rather the predominant manifestation of time-honored locally based cultures (Fig. 1.6).

The same applies to the quarrying and harvesting of building materials. Vernacular architecture is most often created from materials obtained or manufactured locally. For example, while brick and lime mortar are manufactured products, brick and lime kilns were often erected in villages and towns where the raw materials, clay and limestone (or in Rome, sadly, the marble from the abandoned ancient Roman temples [7]) were available (Fig. 1.7). Where limestone was not available, mud mortar would serve almost as well. Where pozzolan or shale was mixed (at first by nature) with the limestone, natural hydraulic cement would result such as is found in the vernacular construction in Santorini, Greece, and in Italy close to Mount Vesuvius and the village of Pozzouli, from which the term "pozzolan" originated.

Romans turned this into a business by marketing pozzolan for the making of hydraulic cement mortars around the Mediterranean. One may thus ask if the resulting

Figure 1.6 New construction in reinforced concrete next to a traditional earthen house near Thimphu, Bhutan, 2010.

Figure 1.7 Traditional brick kilns near Kathmandu, Nepal, 2000.

constructions are "vernacular?" I will leave that question unanswered, using it only to point out that the definition that is best may be the one that allows for exceptions, in the same way that a vernacular language is a language that is not only local in its origins, but fluid and ever-changing.

1.3 Vernacular architecture

Perhaps the best distinction between vernacular and "polite" architecture and construction may be the fact that in vernacular buildings, the basic structural system is more often exposed or, at the most, covered with a contiguous layer of mud or lime plaster. This is in contrast to buildings with a veneer of nonstructural dressed stone or brick on the exterior and dropped ceilings and ornate finishes on the interior walls.

Therein lays the essence of the aesthetics of vernacular architecture. Not only are there simple design rules or "patterns" based on human-scaled proportions, as have been documented and described by Christopher Alexander in his influential book *A Pattern Language*, but also the aesthetics of the craftsmanship of the principal

Figure 1.8 Contemporary design of a resort (Playa Viva) constructed with traditional materials near Zihuatanejo, Mexico.

structural elements, such as the masonry, timber frames, or earthen walls [8]. In a sense, the demands on the quality of the craftsmanship of the original builders is even greater than in "polite" architecture, where the finish work hides the structural walls or frames. For example, the tied bamboo joints or the undulating mud stucco surface laid over the adobe blocks on the interiors and exteriors is the architectural aesthetic of vernacular buildings (Fig. 1.8). This difference is apparent when one compares these buildings with many modern concrete-frame buildings, where the many rock pockets (gaps in the concrete where the unconsolidated aggregate forms a dam) and other construction faults must be covered by finishes.

1.4   The "vernacular" of industrial architecture

My research on the subject of traditional building construction did not begin in the rural countryside of a distant land, but rather in the heart of the first industrial cities of the United States. In the mid-1960s I began a documentary study of the early architecture of the industrial revolution — the first generation of masonry and timber textile mills and planned industrial cities in early 19th-century New England [9]. These brick and stone buildings were constructed with timber plank and beam floors. They were

Figure 1.9 Nineteenth-century mill buildings with water power canal, Amoskeag Millyard, Manchester, New Hampshire, USA, in 1966. The buildings on the right have since been destroyed and the canal filled in for roads and parking as part of a U.S. government funded "urban renewal" program in the late 1960s and 1970s, before their heritage value was recognized.

usually five or six stories in height, but in some rare cases rose to as high as nine stories, making them the largest and tallest multistory buildings of their time — further complicated in that they were heavily loaded with cast iron machinery and wares.

Over the course of the 19th century, these water-powered factory complexes and mill towns evolved into strikingly impressive examples of architecture and urban design. However, unfortunately, at the time of this research and documentation in the middle of the 20th century, these historic factories in New England were largely abandoned and many were being demolished — and just as with the ancient marbles in Rome — they were quarried for their brick, stone, and timber for new construction, which was often only used for suburban houses for the wealthy (Fig. 1.9).

It is interesting to consider whether these early American industrial buildings and their counterparts in Great Britain, where the Industrial Revolution gained its first foothold, meet the definition of "vernacular construction." When one studies the evolution of building construction from the pre-industrial era to the present, one must recognize one important fact which separates most building construction from industrial products: for the most part, even in the present, buildings are still constructed by hand. True, more of the materials and larger elements are manufactured and preassembled, but the process of building a structure is still usually unique and thus dependent on on-site craftsmanship as much as on off-site manufacturing. In the case of 19th-century factory buildings the stone, brick, and timber were obtained as locally as possible. Limestone and granite quarries were opened and brick kilns constructed very close to the sites.

In Britain, where coal was locally plentiful but structural timber largely depleted, cast and wrought iron were more often used for the interior post and beam framework. Interestingly, in the United States cast iron was more commonly used for columns in the early part of the 19th century than it was later, because in 1855 in Lawrence, Massachusetts a large mill collapsed in the middle of a workday as a result of flaws in its cast columns. From that day forward well into the 20th century, textile mills were constructed with what one would think would be the less "industrial" material — timber. However, wood had proved to be more resilient for the heavy loads and strenuous vibrations of a mill filled with oscillating power looms than had cast iron.

Today, one can only reflect on the fact that if the *Rana* Plaza, an eight-story reinforced concrete frame industrial building near Dhaka, Bangladesh, had instead been supported on timber columns, the 1,129 people who lost their lives in the 13th of May 2013 collapse of the structure would still be alive. Thus, one can also say that an important additional definition of vernacular construction is that it is more easily understood by local owners and builders than is construction that is dependent for structural reliability on complex scientific processes and theoretical analysis with differential equations, as is reinforced concrete (Figs. 1.10 and 1.11).

Princeton Professor Robert Mark in 1994 touched on this issue when he made the following counterintuitive yet prescient observation: One of the reasons for the success of the builders of monumental masonry structures over the past two millennia *"is the forgiving nature of typical masonry construction as compared with the relatively 'high strung' nature of slender modern structural elements of reinforced concrete and steel"* [10].

Figure 1.10 Front façade of the *Rana* Plaza, Bangladesh before collapse. Elevation view of front façade of *Rana* Plaza near Dhaka, Bangladesh. This rectified image was constructed by author from a combination of news media TV video images to show the façade prior to collapse. This photo reconstruction of the building serves to bring into focus a fact that, in addition to its structural failings, it was unremittingly ugly as a work of architecture. I would say that these two aspects of building design are actually closely related — the developer Sohel Rana's grotesque failure even to grasp the truth of his own engineers' warnings of imminent collapse are of a piece with his ignorance and even contempt for architectural expression that is in any way responsive to the local and national culture (see TEDxAmoskeagMillyard talk [11]).

Figure 1.11 *Rana* Plaza after 2013 pancake collapse that killed 1,129 people. Photo by Rijans, Wikimedia CC.

1.5   Srinagar, Kashmir, India

While it can be argued that the construction technology found in 19th-century factories is consistent with the definition of "vernacular *construction*," few would argue that all but the most remote rural factories are examples of "vernacular *architecture.*" While I have always enjoyed vernacular architecture, my more in-depth engagement with it as a topic of research followed a visit in 1981 to Srinagar, Kashmir, in northern India [12]. At that time, more than a third of a century ago, entering Srinagar was like going back a half a millennium in time. The houses appeared to be ancient and timeless, with much evidence of wear and tear. V.S. Naipaul made the observation in 1964:

> *It was a medieval town, and it might have been medieval Europe, It was a town of smells: of bodies and picturesque costumes... a town...of disregarded beauty... a town of narrow lanes and dark shops and choked courtyards [13].*

Srinagar was in 1981 a densely packed city full of houses (Fig. 1.12). Unfortunately, over the last 30 years much has disappeared to be replaced by roads widened for military vehicles and "modern" reinforced concrete buildings. In the oldest portions

Figure 1.12 Traditional houses along the Raniwari Canal, Srinagar, Kashmir, 2005.

of the city in 1981, the residences were mixed together with shops and even small manufacturing industries, such as carpet weaving and metalworking. The view of this dense development and teeming activity was like a scene out of the pages of Dickens or a canvas by the 16th-century Dutch painter Bruegel.

The houses themselves were (and many continue to be) what formed the backdrop to this remarkable scene. They appeared rickety and insubstantial, almost as if they were deliberately built only as a stage set for the human pageant which took place around them. These buildings provided an opportunity to study the linkage between traditional construction technology, vernacular design, and a traditional way of life (Fig. 1.13 and 1.14).

When embarking on historical research of Srinagar, I came across remarkable mid- and late 19th-century descriptions of the buildings by British writers who described their observations of two separate earthquakes that occurred in 1868 and 1885. These observations were of particular interest to me, as I was in the process of moving to the earthquake-prone area of California in 1984. To find masonry buildings recognized in the mid-19th century as earthquake resistant provided an interesting point of

Figure 1.13 Narrow lane with timber-laced masonry bearing wall buildings in central Srinagar, Kashmir, 2005.

Figure 1.14 View of four- and five-story buildings of traditional *taq* (on the left) and *Dhajji dewari* (on the right) construction in the city center of Srinagar, Kashmir, 2005.

Figure 1.15 Contemporary continuation of preindustrial technique of sawing timbers into boards in Kashmir.
Photo by Abbid Hussain Khan, INTACH, 2006.

comparison to the United States at a time when historic unreinforced masonry buildings were broadly being condemned and either torn down or retrofitted with their masonry walls entombed behind steel and concrete.

Wood in the Himalayan region was comparatively plentiful before the modern era, but it had to be sawn by hand as there were no saw mills. Also, nails were hand forged rather than purchased by the box. Thus, the primary locally available building material in the Vale of Kashmir was clay, which was fired into bricks or used for mortar. Timber, however, was used not only to support the floors and roof but also to reinforce the masonry (Fig. 1.15).

Most of the traditional buildings in Srinagar can be divided into two basic systems of construction. The first system, referred to as *taq* in Kashmiri and **bhatar** in Pashtun, consists of thick load-bearing masonry piers with thinner masonry walls in between

Figure 1.16 Canalside building of *taq* construction in central Srinagar, showing the timber-laced masonry construction, photographed in 1981. Notice that the windows do not line up vertically and that the masonry is divided into panels and piers, which is only possible structurally because of the timber ring beams in the masonry walls.

(Fig. 1.16). These separate sections of masonry walls and piers are then laced together with timbers that form ring beams around the exterior walls and are held together with wooden ties that pass through the walls at each floor level. This timber lacing is configured like ladders laid horizontally in the wall. They are laid below and above the joists at each floor level, which then tie the walls together with the floors because of the friction provided by the overburden weight of the masonry above.

What is important to note regarding this system is that there are no vertical timbers or any other form of vertical tension reinforcement. This is something that seems counterintuitive to most engineers today, including the engineers in India who produced the otherwise remarkably forward-thinking Indian Building Code IS 13828 for "non-engineered" construction that is based on this system [14]. As a result of their insecurity with having only horizontal reinforcement: this code recommends that a cement grout-encased steel reinforcing bar be embedded in the masonry in each exterior corner. My concern is that this column of steel-reinforced concrete over time is likely to prevent the natural changes in dimension of the surrounding load-bearing masonry laid in soft mud mortar. In the event of an earthquake, it could disrupt the masonry in such a way that it will be much more difficult or impossible to repair [15].

The second system, known as "*dhajji dewari*" construction, consists of a braced timber frame with masonry infill. It is brick-nogged timber-frame construction that in Britain is known as "half-timber." *Dhajji-dewari* comes from the Persian words for "patch quilt wall" (Figs. 1.17 and 1.18). It is similar to what in Turkey is called *hımış* [16].

Figure 1.17 *Dhajji-dewari* building showing the single-leaf construction of the walls. The building was under demolition for a road widening, 2005.

Figure 1.18 New construction in *dhajji-dewari* photographed in 1989. More recently, this form of construction has largely gone out of use, in favor of reinforced concrete.

After the October 2005 Kashmir earthquake, structural engineering Professors Durgesh Rai and Challa Murty of the Indian Institute of Technology Kanpur reported that

> *"In Kashmir traditional timber-brick masonry (dhajji-dewari) construction consists of burnt clay bricks filling in a framework of timber to create a patchwork of masonry, which is confined in small panels by the surrounding timber elements. The resulting masonry is quite different from typical brick masonry and its performance in this earthquake has once again been shown to be superior with no or very little damage."*

They cited the fact that *"the timber studs resist progressive destruction of the. wall and prevent propagation of diagonal shear crack. and out-of-plane failure."* They went on to suggest that: *"there is an urgent need to revive these traditional masonry practices which have proven their ability to resist earthquake loads"* [17].

This contemporary observation is not without precedent. In the 19th century, British geologist, Frederic Drew, said in a book published in 1875 that the Srinagar houses have *"mixed modes of construction [that] are said to be better as against earthquakes (which in this country occur with severity) than more solid masonry, which would crack"* [18]. He thus notes not only that the houses demonstrated a level of earthquake resistance but also that this attribute was recognized by the local population.

Later, the British physician and award-winning writer Arthur Neve, in a book describing his experiences during 30 years in Kashmir during the late 19th and early 20th centuries, described at length the resilience demonstrated by the traditional Srinagar houses of both systems during the earthquake of 1885. After describing these Srinagar houses as *"tumbledown and dilapidated to a degree"* (which was not unlike my first impression a century later), he went on to say that their construction was *"suitable for an earthquake country."* His explanations of why the traditional construction worked so well are particularly significant. After observing that *"Part of the Palace and some other massive old buildings collapsed... [but] it was remarkable how few houses fell,"* he goes on to explain that *"wood is freely used and well jointed,"* but then he made the counterintuitive claim that *"clay is employed instead of mortar, and gives a somewhat elastic bonding to the bricks."*

In describing the *taq* construction, he says that the bricks are constructed *"in thick square pillars, with thinner filling in,"* by which he describes the bearing wall masonry construction that is constructed as a series of wall sections with unbonded construction joints between the sections, as seen in Fig. 1.16. This construction, he says *"If well built in this style the whole house, even if three or four stories high, sways together, whereas more heavy rigid buildings would split and fall"* [19].

Both of these 19th-century commentaries seem as remote from modern engineering theory and practice as the buildings they describe are picturesque. Neither Drew nor Neve was an architect, engineer, or mason. In addition, the comparative sizes and characteristics of the earthquakes are only vaguely documented. What is interesting, though, is that these quotes can be used to shed light on what has proved to be revolutionary changes that have occurred over the past century and a half in the design and construction of buildings. These changes effectively have divorced the building technical community, including architects, engineers, and builders, from the kind of tactile understanding of traditional materials and construction systems that these 19th-century quotes demonstrated — a problem that has only recently begun to be recognized and addressed, as shown by the previous quote by Professors Rai and Murty.

1.6   The great 1906 San Francisco earthquake and fire

Turning to the United States, consider the engineering study describing the damage from an earthquake half-way around the planet from the one in Kashmir — the 1906 San Francisco earthquake and fire. Six years prior to the publication of Arthur Neve's book about Srinagar, the American Society of Civil Engineers (ASCE) in their *Transactions* published the *"The Effects of the San Francisco Earthquake of April 18th, 1906 on Engineering Construction."* In their committee report on the damage to buildings one finds the following statement:

> *It may be stated, as one of the most obvious lessons of the earthquake, that brick walls, or walls of brick faced with stone, when without an interior frame of steel, are hopelessly inadequate. As a method of building in earthquake countries, such types are completely discredited [20].*

Later in the report they repeatedly emphasize this point, as follows:

> *The writers simply reiterate the statement that, speaking generally, buildings of brick walls and wooden interiors cannot be built which will not be wrecked in a severe shock, it being a fault of design and not of materials or workmanship [21].*

This was followed by some heated rebuttals in the "Discussion" portion of the publication, in which 25 ASCE member engineers contributed their own reports. While seven offered strong rebuttals to the Committee's position, two expressed unequivocal support, adding their endorsement to the evidence of the transition from empirical observation to theoretical analysis in engineering. (The others were either neutral or did not discuss masonry building performance.) One supporting member, C. Derleth, wrote: *"It is not right to argue that brick buildings are adequate for earthquake countries because a hundred buildings stood in certain places"* [22], while another, J.D. Galloway, said: *"It is strange that engineers will still champion this material which, analytically and by evidence, proves itself to be 'hopelessly inadequate"* [23].

In defense of masonry, one engineer, Edwin Duryea, made the following apt observation: *"This conclusion is so severe that, if it were true, it would entirely exclude the safe use of ordinary brick buildings, for any purpose, on the San Francisco peninsula"* [24]. All of this flowed from the repeated observations that only some, but by no means all unreinforced masonry buildings were damaged by the earthquake. Another engineer, W.W. Harts, made the even more directed rebuttal of the absolutist nature of the Committee's position, basing his position as a defense of empirical methodology and evidence:

> *It is very dangerous, in any scientific discussion, to formulate sweeping general rules....The explanation has been made that the test of engineering structures is not experience, and that the majorities [of engineering professionals] cannot be relied upon...to discover the truth as to their stability. Can it be maintained that the science of engineering is not the resultant of years of experience? Can it be said that successes are not guides and that failures are not warnings? [25].*

His colleague, Bernard Bienenfeld, then made the more specific observation:

> *Regarding the contention of the Committee that the ordinary brick construction is inelastic, observation of the action of the brick walls of wrecked buildings that were being torn down subsequent to the earthquake and fire would indicate that there is indeed a considerable and surprising internal elasticity in this form of construction [26].*

This quote is the one that comes closest to the 1885 observations by Arthur Neve cited above, when he said that weak clay mortar gives a somewhat elastic bonding to the bricks, which allows the masonry buildings to *"sway together."* [19]. It is interesting in light of this debate over the behavior of masonry to also take note of the famous 19th-century French architect Violet-le-Duc's description of masonry behavior as showing *elasticité,* by which he meant that masonry has the ability to structurally adapt to changes in loading and even allow for cracking, without losing its stability. (He was describing it in general, not specific to earthquakes.) Thus, we are confronted with the use of the term *"elastic"* to describe phenomena that includes inelastic behavior of a structural system. As described next, this seeming anomaly may in fact lie at the core of what makes masonry construction difficult for engineers to analyze and accept when confronting the mitigation of earthquake risk.

While these quotes concern the engineering analysis of the unreinforced masonry buildings in San Francisco, most of these buildings had probably not been designed by engineers, except for the simple static load calculations and detailing of internal framing elements. , However, the most historic watershed is that this is the first earthquake to affect steel skeleton-frame buildings with curtain walls of masonry, rather than load-bearing exterior masonry walls with internal post and beam framing like that of the British and American 18th and 19th-century industrial buildings (see Fig. 1.9). This system of construction, now so common as to seem unremarkable, had only first been used 20 years earlier in Chicago.

From an engineering perspective, mid- and high-rise skeleton-frame buildings represent much more than the simple replacement of the exterior bearing walls with an extra bay of posts and beams. As Donald Friedman states in his book *Historical Building Construction*, the few short decades before the 1906 earthquake also mark the *"change in structural theory from masonry-based compression forms such as bearing walls and arches to flexure-based designs such as knee-braced portal frames"* [27]. At this same time, the industrial production of large quantities of steel made possible after the invention and proliferation of the Bessemer converter and open hearth furnace, replaced wrought and cast iron within less than a decade after 1980. These two changes made the skeleton frame possible, which inevitably led to the construction of much taller buildings that became known as 'skyscrapers', a word for which came from the top sails on ocean-going sailing ships. Only then did the skeleton frame conform to what the engineering term-of-art means by the word "frame" — that is, a braced or moment frame with rigid beam/column connections.

The invention of the skeleton-frame skyscraper fits well with the other remarkable inventive engineering feats of the age, such as bridges and towers. The most remarkable of these towers at the time was most certainly the Eiffel Tower. However, there was a profound difference: unlike bare steel bridges and towers, buildings require interior and exterior walls. Also, as a consequence of the flammable nature of their contents, the steel frame itself must be protected against collapse from the heat of a raging fire inside a building (Figs. 1.19—1.21). For more on 1906 Fire, see Ref. [28].

Figure 1.19 View of San Francisco showing three of the steel skeleton frame buildings after the fire had burned them out. These included the 315 foot high steel skeleton frame Call Building constructed in 1890. All of the surrounding buildings, except for the skeleton frame structures, were completely destroyed. These three skeleton frame structures, as well as most of the others in the city from that time, were repaired and about half of them still exist today.

Figure 1.20 View of the 1904 steel skeleton frame with infill masonry Flood building after the 1906 San Francisco earthquake and fire.

Figure 1.21 The same view shown in Fig. 1.20 but 100 years later. Most of the steel skeleton-frame buildings were repaired after the earthquake despite the fact that all were burnt out by the fire.

At that time, the walls and the fireproofing around the steel framework were still of unreinforced masonry resting on the steel frame, although these walls were no longer load bearing. Of course, such enclosure walls not only added weight but also complicated the engineering analysis of the frame. Because of the walls, this was a problem unique to buildings, rather than to bridges or the Eiffel Tower. This difference, as we shall see later, is more profound than was perhaps realized at the time.

1.7   Skeleton-frame construction

The steel skeleton frame skyscrapers in San Francisco which survived both an earthquake and uncontrolled fire stand at the threshold of a sea change in engineering and building construction. The conventional descriptions of the trajectory of architectural and engineering history of this time often described the changes of the late 19th and early 20th centuries as "progress," as they led to the elimination of heavy masonry facades and interior walls and the opening up of floor plans. This was idealized by Le Corbusier in his model Domino House, an idealized drawing of a reinforced concrete post and slab structure, which was followed by the adoption of curtain walls of glass and lightweight panels to enclose such structures. The consequence of this in earthquake areas was only later to be confronted as successive earthquakes around the globe caused increasing numbers of collapses with many victims.

Interestingly, at the time of the 1906 earthquake, engineers had little idea how the masonry-clad skeleton-frame buildings would perform. As reported by A.L.A. Himmelwright, head of the Roebling Construction Company [29]:

The successful manner in which the tall, steel skeleton-frame buildings withstood the effects of the earthquake and the fire is most reassuring, in fact wonderful, and proves conclusively that the best modern practice is directed along correct and efficacious lines.. These buildings had never before been subjected to violent earthquake shocks, and many architects and engineers doubted their ability to withstand such surface movements without injury…. In all cases when the structural details were designed in accordance with the best modern practice and executed with skill and workmanship of only fair quality, the buildings passed through the earthquake without structural injury.

Their remarkably good performance in 1906 San Francisco firmly established them in the ASCE Transactions Report as the structural system acclaimed to have the most resilience — despite the many earthquake-induced cracks in the masonry interior and exterior walls. The 1907 ASCE Committee Report observed that *"The damage to steel frames was almost negligible"* and *"the writers are of the opinion that the steel frame offers the best solution of the problem"* [30] after making the observation that *"buildings of this type were those most exposed to earthquake damage, the type including most of the tall buildings in the city"* [31].

This, of course, is where the *masonry* comes into the discussion. The report states that the exterior brick *"wall adds little, if any, to the bracing of a steel frame. Many of such walls were cracked badly, and moved on the supporting girder. No reliance should be placed upon them, as they are open to all the objections stated in connection with brick walls in general. The well-designed steel frame offers the best solution of the question of an earthquake-proof building, as all the stresses can be cared for"* [30]. With this and other observations, the committee makes it clear that they saw the masonry as simply a load on the frame, rather than working (in the engineering meaning of the term) in partnership with the frame to resist the lateral forces and dampen the destructive vibrations of the earthquake.

In contrast to this analysis, the author of what may be the first engineer's text book on the subject of skyscrapers, Joseph Kendall Freitag expressed a diametrically opposing view in both the 1895 and 1901 editions of his textbook *Architectural Engineering With Special Reference to High Building Construction*. In his 1901 edition, he says:

> *'Skeleton Construction'… suggests a skeleton or simple framework of beams and columns, dependent largely for its efficiency upon the exterior and interior [masonry] walls and partitions which serve to brace the structure, and which render the skeleton efficient, much as the muscles and covering of the human skeleton (to borrow a comparison used by various writers) make possible the effective service of the component bones" [32].*

In a historically important way, Freitag's quotes reveal the dialectic between that which is empirically obvious versus what is still an engineer's reluctance to rely on that which cannot be calculated, described as follows:

> *A building with a well-constructed iron frame should be safe if provided with brick partitions and if the base is a large proportion of, or equal to the height, or if the exterior of the iron framework is covered with well-built masonry walls of sufficient*

thickness; for the rigidity of the solid walls would exceed that of a braced frame to such an extent that were the building to sway sufficiently to bring the bracing rods into play, the walls would be damaged before the rods could be brought into action [33].

This was in fact a phenomenon that was confirmed years later by tests carried out on the Empire State Building in New York City [34]. Freitag, however, is critical of the total reliance on the infill masonry for lateral support:

This method of filling in the rectangles of the frame by light partitions may be efficient wind bracing, but the best practice would certainly indicate that it cannot be relied upon, or even vaguely estimated [35].... While the steel frame is more or less reinforced by the weight and stiffening effects of the other materials, still no definite or even approximate values can be given to such items, except their purely static resistance or weight.... Any dependence placed on curtain walls [36] and partitions for lateral strength is opened to very grave question [37].

Despite his *"muscles and skeleton frame"* metaphor describing the structural integration and dependency between the masonry walls and steel frames, Freitag himself reveals the inconsistency in his own views of the role of the masonry. In so doing he recognizes the very problem that continues to exist for engineers today − the lack of reliable ways to quantitatively analyze and calculate its role in a structural engineering design for multistory frame buildings. He thus anticipates the elimination of the masonry for lateral support when, on the very next page, he describes what he calls "cage" construction:

The steel framework, originally introduced to carry vertical loads only, has been gradually developed and systematized as increased attention has been bestowed upon the questions of lateral strength and stiffness against wind or other external forces. The use of a well-braced frame now permits the substitution of curtain or veneer walls for the solid masonry construction formerly required, and the reduction in thickness of such walls to 12" or 16" protective veneer walls only, makes it possible to obtain much larger window areas ... [and] also possible to omit heavy interior walls [38].

However, 12 to16 inches of masonry is still a considerable amount of masonry infilling and cladding, and if called upon in an earthquake to resist the lateral forces, as it was in San Francisco in 1906, it could provide added resistance and damping from its cracking that serves to prevent the underlying steel building frame from deforming beyond its elastic range. This is empirically true, even if the original designers had disregarded it, as is shown by the fact that the ASCE committee in San Francisco concluded that it was of no value. Local regulations in Chicago and New York City at the turn of the century mandated thick exterior masonry walls, even for skeleton frame structures, but during the decades to follow such requirements were soon lifted.

While most of the historical focus is on the transition to the use of frames for taller buildings, the watershed event in this transition is not so broadly known − it is, within the field of structural engineering, the "invention" of a way of doing a portal frame analysis using the contraflexure methodology for isolating moments." In lay language, this method allowed the calculation of the bending stresses on multi-story frames by mathematically separating the frame into parts at each neutral point of bending reversal of the columns and beams. This then allows for the forces to be calculated using the three equations of equilibrium. Prior to that, the forces inherent in "frame action" could not be accurately calculated. As long as, in effect, columns were pin connected at the floor levels, this was not a problem (Fig. 1.23), but as structures grew taller, this was structurally inefficient. Modern skyscrapers in every practical sense date their origin to this change in engineering analysis methodology.

The contraflexure methodology of portal frame analysis [39], eliminates what Freitag described as the "table leg" principle — the pin connection of columns at each floor level which is required to avoid the complication of calculating an indeterminate (hyperstatic) structure that would exist if all connections were rigid (Fig. 1.22). Ironically, contraflexure-based analysis allowed for much greater precision and thus economy in the amount of steel needed for the frame, but in so doing, it reduced the structural redundancy that the earlier methods required and thus reduced the de facto safety margin in the event of earthquakes which exceed the elastic capacity of the frame structures. Since this refinement in the engineering calculations coincided with the reduction in the thickness of the masonry walls, the skeleton frames became significantly lighter and more flexible. Elwin Robinson in 1989 remarks on this when he says: *"The exterior curtain was continually reduced in mass until it eventually became a thin glass envelope, unable to contribute to the structural stability of the system in any way"*. This would come to have profound and even tragic consequences over the next century up to and beyond the present day [40].

To fully understand this historical phenomenon, we must jump ahead 100 years to the centenary of the 1906 earthquake and a report written by two of the most highly regarded seismic engineers in San Francisco today, Ronald O. Hamburger, and John D. Meyer.

Figure 1.22 An office interior remodeling project in the south wing of the 1891−1893 steel-frame Monadnock Building in Chicago briefly revealed this view, an example of what Freitag defines as the "table leg" portal arch. The photograph was taken in 2005.

They address this topic from the perspective of the current day, and in so doing, they find some interesting counterintuitive facts:

> *The outstanding performance of the infill steel-frame buildings in the 1906 earthquake remains an issue of considerable technical curiosity. Though many of these buildings had design deficiencies known to have caused very poor performance in more recent earthquakes, including soft and weak stories, torsional configuration, large seismic mass, and low stiffness, none collapsed, and essentially all were repaired and restored to service, many remaining in use to this day.... Certainly, this superior performance is not predicted when standard procedures for seismic evaluation and upgrade, such as ASCE-31 ASCE 2003 and FEMA 356 are employed.*
>
> *[A] possible explanation lies in the accuracy, or lack of accuracy, with which our current evaluation techniques model buildings. Although some analyses of buildings of this type have been performed, accounting for the masonry and frame interaction effects in exterior walls..., we are not aware of any such models that also incorporated the effects of the hollow clay tile partitions [41].*

These comments are particularly interesting in light of what happened during the 1989 Loma Prieta earthquake in Oakland, across the bay from San Francisco. In that earthquake the epicenter was 60 miles (97 km) distant from the Bay Area, and it resonated with the soil under Oakland City Center at a frequency that affected early 20th-century skeleton-frame mid-rise buildings with masonry curtain walls. These included the 320 feet (98 m) high 18-story City Hall constructed in 1914, and a number of other buildings within the downtown area. None of these buildings came close to collapse during the earthquake, but, as expected, many manifested cracks in their masonry exterior walls. One such building, the former Hotel Oakland, an eight-story block-sized building recently converted to housing, sustained considerably more damage to its façade than the others. After the earthquake, it was surprising to find out that of all of the buildings of this type, this had been the only one that recently had been seismically upgraded. As part of that upgrade, all of the interior hollow clay tile masonry walls had been demolished and replaced with gypsum plasterboard on light weight steel studs.

The questions raised by Hamburger and Meyer point to the fact that the masonry in general, and the hollow clay tile interior walls in particular, fell outside of modern seismic analysis models, and that in a retrofit project, the typical decision was simply to eliminate them. However, their absence left the exterior walls during the earthquake to do all of the work. The better performance of the dozens of surrounding unretrofitted buildings, by their comparison with the Hotel Oakland, serves to demonstrate the positive contribution that the infill masonry does have in an earthquake — the very thing that was under-appreciated by the engineers, both in the post-1906 reconnaissance quoted above, and in the modern day [42].

1.8   Frames and solid walls

One may wonder why one focuses on the evolution of skeleton-frame construction. What does this have to do with "vernacular construction" and nonconventional materials? The answer is that the development of the analytics and mathematics that accompanied the invention of the skeleton frame changed the practice of structural engineering, in such a profound way that it has served to build a conceptual barrier between the design of solid wall masonry structures in past centuries, and modern skeleton-frame structures.

This increasingly opaque barrier within the structural engineering profession between the empirical past and quantitative frame analysis of the present can be seen in Freitag's textbook when he says that: *"the stability of a building must depend entirely either upon the masonry, that is, the inertia or dead weight of the structure, or upon the steel framework"* [43]. *"In veneer buildings [masonry curtain wall buildings], which are being considered here in particular, the system of bracing the metal-work [with steel] must be used, with the [masonry] walls as light [meaning as thin] as possible"* [44].

More than any other, the process that led to these changes in the quantitative analytical methods now so commonly accepted in structural engineering is also what has created the barrier between the modern era and historical approaches to building design and construction. At the time this evolved, engineering education had largely shifted from the building sites to the university, with its attendant specializations and concentration on mathematical theory rather than empirical approaches to building construction combined with direct hands-on apprenticeship at the building sites. Frame analysis thus became the core of the curriculum, such that now both engineers and architects think and model most structures in their minds and on paper as frames to be analyzed mathematically. Shear walls, when included, are modeled as deep beams. Masonry walls, where they exist, most often have been treated simply as dead weight and given lateral resistance factors that only recognize their limited elastic capacities, rather than their ultimate strength, resilience, and energy dissipation.

Princeton engineering Professor Robert Mark touched on a significant part of the changes that have occurred when he states: *"it was only in the nineteenth century that the merger of science and the art of building. finally took hold, abetted by the introduction of new construction materials to which the old rules of building no longer applied and by the establishment of architecture and engineering as university subjects"* [45].

What is interesting is that now, a century later, there is still a struggle within the fields of architecture and engineering to find a balance between an empirical process founded on observation and experience, and quantitative approaches that are now a seamless part of engineering education and largely dictated by the building codes. In 2004, California-based engineer, Sigmund Freeman wrote a criticism of recent changes to the seismic building code, as follows:

> *Making codes more restrictive and more complicated does not necessarily make for better buildings…. The analytical methods used are generally precise, but are they accurate? Materials can have many variables that are not accounted for in the analytical models…. Continued studies and interaction between engineers, as well as with other design professionals, is required. Engineers need to think more about how buildings perform than how to satisfy building codes [46].*

Freeman's observations are reminiscent of the comments a century earlier in defense of brick buildings by W.W. Harts, quoted above, in the ASCE Report on the 1906 San Francisco earthquake when he wrote:

> *It is very dangerous, in any scientific discussion, to formulate sweeping general rules…. Can it be maintained that the science of engineering is not the resultant of years of experience? Can it be said that successes are not guides and that failures are not warnings? [47].*

1.9   Conclusion: the "ecology" of the vernacular

The evolution of modern skeleton-frame construction, and changes to the engineering profession in response to it, have without a doubt made possible the many extraordinary structures and structurally safe buildings of the modern era. These disciplines have become specialized, and the responsibilities for the safety of structures continue to be extensive and comprehensive.

It is perhaps hard to appreciate this today, but where the difference between the premodern and the modern era became most apparent to me was on a visit to Kabul, Afghanistan, on an assignment in 2006 for The Turquoise Mountain Foundation, a UK-based nongovernment organization. This foundation had undertaken the restoration of a number of historic structures of traditional timber and masonry construction in a district of Kabul. The Afghan professional who was supervising and directing the teams of workmen was called by the honorific "Engineer" by his workmen and others in the community; his full name thus being "Engineer Hedayatullah" in all of the communications by the foundation. No drawings were used for the work, yet all work was carefully guided to meet the stringent restoration requirements of the foundation. Engineer Hedayatullah did not have a formal university-based education but rather had learned his discipline through the trade guild system that still exists there, just as it had existed throughout Europe before the modern era (Figs. 1.22—1.24).

Figure 1.23 The Peacock House, Murad Khane, Kabul, Afghanistan, before and after restoration.
Photograph by Randolph Langenbach.

Figure 1.24 The same house shown in Fig. 1.23 after restoration by the Turquoise Mountain Foundation.
Photo by Andre Ullal for the Turquoise Mountain Foundation.

Despite this anachronistic example, it cannot be a surprise that a Renaissance man approach to design in the premodern era is no longer seen as appropriate or even possible in more developed contemporary urban environments. With its disappearance, however, there has been a loss of knowledge about traditional materials and technologies that do not fall within the parameters of the current building codes, and the subjects taught in universities.

This evolution is parallel to the ongoing phenomenon of the disappearance of what may be termed "vernacular" languages — regional languages that are now disappearing with the spread of English and other major languages around the world. A recent PBS video included a report by David Crystal, a linguist, who said: *"Half of the languages of the world are so in danger that they are going to die out in the present century. That means one language dying out somewhere every two weeks"* [48]. This same program closed with a quote that in my opinion speaks also to the importance of traditional forms of construction within the fields of architecture and engineering (Fig. 1.25):

Just as the physical ecology of the earth depends on the healthy interaction among plants and animals. there is an ecology of consciousness and interdependence of knowledge, culture and wisdom we find in and through our languages. Language is a lens through which we see the world.

Vernacular architecture and construction belong in that same "ecology" and they both need to exist in the world seen through the "lens" of idiomatic local language and culture. It is no coincidence that the emergence of movements to save dying languages, including Welsh and Hawaiian, parallels a growing interest in alternative "vernacular" forms of construction that are more than simply architectural styles - but carry the substance of deeply historic construction technologies.

Thus, while it may seem a stretch to compare the establishment of frame theory and the displacement of traditional forms of construction with reinforced concrete moment frames worldwide to the loss of languages, when one sees how extraordinarily

Figure 1.25 Srinagar, Kashmir, photographed in 1981 showing the collision of scale, architectural language, and construction technologies in the historic city after the construction of a large reinforced concrete building. In the three decades that have followed, many more buildings of similarly disruptive scale and design have been constructed in what had been a remarkably well-preserved historic city center.

dominant reinforced concrete construction has become, and the alien scale and visual language of the resulting architecture in many parts of the world, it is clear that something has been lost and not replaced (Fig. 1.25). What has disappeared has everything to do with the *"knowledge, culture and wisdom"* that had been unique to the particular regions and communities.

What is encouraging is that this disappearance has stimulated a growing interest around the world in nonconventional materials and vernacular construction. This is not because these materials and construction methods are new or, in the converse, have entirely ceased to be used, but because what are now conventional materials and construction typologies have become so distant from these previously common materials and systems as to be on the other side of the barrier that has emerged between what are now different disciplines.

Earthquakes are the ultimate test of buildings, and thus, the growing interest among engineering students in technologies that were spurned and totally absent from university curricula less than a decade ago gives a chance to revisit some of those same questions that confronted the engineers who climbed over the ruins of San Francisco in 1906, just as their predecessors had in 1755 in Lisbon, Portugal, and 1783 in Calabria, Italy. More recently, it is also just as the itinerant builders and residents have observed in Turkey after the 1999 Kocaeli and Duzce earthquakes (Figs. 1.26–1.29), and in Pakistan after the 2005 Kashmir earthquake (Fig. 1.30).

Following the example in Fig. 1.31 and other spontaneous examples, the government of Pakistan approved *dhajji* 1 year after the earthquake and *bhatar* 2 years after the earthquake for government assistance. By 2009, a mere four years after the earthquake, there are *"at least 150,000 new homes"* constructed in northern Pakistan in either of these two traditional typologies [49].

Figure 1.26 View after the 1999 Duzce, Turkey, earthquake showing collapsed reinforced concrete building immediately adjacent to a traditional *hımış* residence with only one panel of brick infill missing.
Photographer unknown, image provided courtesy of Adem Dogangun.

Figures 1.27 and 28 This family were living in Düzce, Turkey, in the house in Fig. 1.27 at the time of the 1999 Düzce earthquake. At that time he was constructing a new house of reinforced concrete. After the earthquake, when he saw that so many RC buildings had collapsed, he stopped the construction and began instead to build a house in *hımış* construction shown in Fig. 1.28.

Figure 1.29 The interior of the new house being constructed by this family shown in Figs. 1.27 and 1.28 being built of traditional Turkish *hımış* construction.

Figure 1.30 *Dhajji-dewari* building in rural Pakistan, Kashmir, after the 2005 earthquake, showing the few missing wall panels of infill masonry.

Figure 1.31 The owner and carpenter of a new house under construction in rural Pakistan near to the building shown in Fig. 1.29 after the earthquake destroyed the owner's original rubble stone home, the ruins of which are on the right. The decision to reconstruct in *dhajji* construction was because the one home in the village that did not collapse was of that construction and the reinforced concrete village store collapsed.

Now, over the course of our new 21st century, there have been increasing numbers of students and professors of architecture and engineering turning their attention and interest toward examples of vernacular construction. Both Indian and Pakistani *dhajji-dewari* and Turkish *hımış* have begun to be explored for their seismic resistance in India and Pakistan, as well as in Turkey, Italy, and Haiti. Perhaps soon, the lessons to be learned from these premodern and preindustrial forms of construction that date back to ancient Rome and even to ancient Minoan civilization can become an inspiration not only for heritage preservation but also even for methods to introduce an increased resilience into modern buildings of steel and reinforced concrete.

Sources of further information

1. Langenbach Randolph. Soil dynamics and the earthquake destruction of the Arg-e Bam. Iranian Journal of Seismology and Earthquake Engineering 2004. Tehran, Iran, Special Issue on 26 December 2003 Bam Earthquake, vol. 5:#4 & vol. 6:#1. Available at: www.conservationtech.com.
2. Langenbach Randolph. Performance of the earthen Arg-e-Bam (Bam Citadel) during the 2003 Bam, Iran, earthquake. EERI Earthquake Spectra December 2005. Special Issue on Bam, Iran Earthquake, #1, vol. 21. Available at: www.conservationtech.com.
3. Langenbach Randolph. Don't tear it down! Preserving the earthquake resistant vernacular architecture of Kashmir. UNESCO; 2009. p. 60. For copies of this book, go to, www.traditional-is-modern.net.
4. Langenbach Randolph, Kelley Stephen, Sparks Patrick, Rowell Kevin, Hammer Martin, Co-authors. Preserving Haiti's Gingerbread Houses. World Monuments Fund; December 2010. 85 pages.

References

[1] Langenbach Randolph. Soil Dynamics and the Earthquake Destruction of the Arg-e-Bam. Iranian Journal of Seismology and Earthquake Engineering; 2004. Tehran, Iran, Special Issue on 26 December 2003 Bam Earthquake, vol. 5:#4 & vol. 6:#1. Available at: www.conservationtech.com.
[2] Information on Paul Jaquin can be found at: http://www.historicrammedearth.co.uk/.
[3] Henri Van Damme, Mokhtar Zabat, Jean-Paul Laurent, Patrick Dudoignon, Anne Pantet, David Gélard, Hugo Houben. Nature and Distribution of Cohesion Forces in Earthen Building Materials. In: Agnew N, editor. Conservation of Ancient Sites on the Silk Road: Proceedings of the Second International Conference on the Conservation of Grotto Sites, Symposium Papers. Los Angeles: Getty Publications; 2010.
[4] Langenbach Randolph. Performance of the Earthen Arg-e-Bam (Bam Citadel) During the 2003 Bam, Iran, Earthquake. Earthquake Spectra; December 2005. EERI, Special Issue on Bam, Iran Earthquake, #1, vol. 21. Available at: www.conservationtech.com.
[5] www.etymonline.com.
[6] www.wikipedia.com.

[7] For 1500 years, the temples and monuments of Imperial Rome were used as quarries and lime kilns were often constructed next to them where the marble was simply burned. See www.piranesian.com [for a movie trailer of a documentary by the author on Rome].

[8] Alexander Christopher, Silverstein Murray, Ishikawa Sara. A Pattern Language, Towns, Buildings, Construction. Oxford University Press; 1977.

[9] Langenbach Randolph. To see publications, exhibitions and photographs from this documentary work, go to: www.conservationtech.com "Publications" + "Exhibitions" + "Large Format Photographs (1966−81)."

[10] Mark Robert. Light Wind and Structure, The Mystery of Master Builders. Cambridge, MA, USA: MIT Press; 1994. p. 170.

[11] For more details about the Rana Plaza collapse, and about the urban design of the Amoskeag Millyard, see 2014 TEDx Amoskeag Millyard talk *"Reconsidering Sustainable Architecture"* by Randolph Langenbach (www.conservationtech.com)

[12] Langenbach Randolph. India in Conflict: Urban Renewal Moves East. Historic Preservation. National Trust for Historic Preservation May−June 1982;34(3):46−51. Available at: www.conservationtech.com.

[13] Naipaul VS. An Area of Darkness. London: André Deutsch; 1964. p. 123.

[14] The Indian Building Codes for non-engineered construction, as well as other codes and references are available at: http://www.traditional-is-modern.net/LIBRARY.html.

[15] See explanation for this critique in Randolph Langenbach Don't Tear it Down! Preserving the Earthquake Resistant Vernacular Architecture of Kashmir, 2009. UNESCO, p. 60. For copies of this book, go to: www.traditional-is-modern.net.

[16] For detailed descriptions, see: Don't Tear it Down! Preserving the Earthquake Resistant Vernacular Architecture of Kashmir, UNESCO, 2009. Available at: www.traditional-is-modern.net and also see www.conservationtech.com [Publications].

[17] Rai Durgesh C, Murty CVR. Preliminary report on the 2005 North Kashmir earthquake of October 8, 2005. Kanpur: Indian Institute of Technology; 2005. www.EERI.org.

[18] Drew Frederic. The Jummoo and Kashmir Territories. London: Edward Stanford; 1875. p. 184.

[19] Neve Arthur. Thirty Years in Kashmir. London: Arnold; 1913. p. 38.

[20] ASCE. Transactions of the American Society of Civil Engineers December 1907;LIX. p. 234.

[21] ASCE. Transactions of the American Society of Civil Engineers December 1907;LIX. p. 250.

[22] ASCE. Transactions of the American Society of Civil Engineers December 1907;LIX. p. 313.

[23] ASCE. Transactions of the American Society of Civil Engineers December 1907;LIX. p. 326.

[24] ASCE. Transactions of the American Society of Civil Engineers December 1907;LIX. p. 264.

[25] ASCE. Transactions of the American Society of Civil Engineers December 1907;LIX. p. 383.

[26] ASCE. Transactions of the American Society of Civil Engineers December 1907;LIX. p. 8.

[27] Donald Friedman 1995, 2010. Historical Building Construction. W. W. Norton & Company, Inc., New York City. p. 50.

[28] For more about the effects of fire on early skeleton frame buildings and comparisons with the World Trade Center towers from the 9/11 attacks, see: Langenbach Randolph. The Integrity of Structure: The Armature of Our Architectural Heritage. In: CHS Newsletter, No. 67, December 2003, London, Construction History Society. Edited and Republished from Proceedings of the 2002 ICOMOS General Assembly, Madrid; 2002. Available at: www.conservationtech.com.

[29] Abraham Lincoln Artman Himmelwright. The San Francisco Earthquake and Fire: A Presentation of Facts and Resulting Phenomena, with Special Reference to the Efficiency of Building Materials Lessons of the Disaster. New York: The Roebling Construction Company; 1907. pp. 7 & 242.

[30] ASCE. Transactions of the American Society of Civil Engineers December 1907; LIX. p. 235.

[31] ASCE. Transactions of the American Society of Civil Engineers December 1907; LIX. p. 232.

[32] Freitag Joseph Kendall. Architectural Engineering with Special Reference to High Building Construction, Including Many Examples of Prominent Office Buildings. 2nd ed. NYC: John Wiley & Sons; 1901. p. 9.

[33] Freitag Joseph Kendall. Architectural Engineering with Special Reference to High Building Construction, Including Many Examples of Chicago Office Buildings. NYC: John Wiley & Sons; 1895. p. 137 & 250.

[34] Langenbach Randolph. Saga of the Half-timbered Skyscraper: What Does Half-timbered Construction Have to do with the Chicago Frame? Proceedings of the Second International Congress on Construction History. Cambridge University; 2006. Available on: www. conservationtech.com.

[35] Freitag Joseph Kendall, Architectural Engineering with Special Reference to High Building Construction, Including Many Examples of Chicago Office Buildings, NYC: John Wiley & Sons; 1895, p. 137, and Freitag Joseph Kendall, Architectural Engineering with Special Reference to High Building Construction, Including Many Examples of Prominent Office Buildings, 2nd ed. NYC: John Wiley & Sons; 1901, p. 249.

[36] It is interesting to note that the term "curtain wall" was already in common use before the turn of the century, although with a slightly different meaning than today. Today, after half a century of construction of curtain walls of glass and lightweight materials, the brick and terra cotta—clad tall buildings of the period from 1884 until the 1940s are not often identified as "curtain wall" buildings. It is also interesting to note that the word in Turkish used for "shear wall" translates literally to "curtain wall."

[37] Freitag Joseph Kendall. Architectural Engineering with Special Reference to High Building Construction, Including Many Examples of Prominent Office Buildings. 2nd ed. NYC: John Wiley & Sons; 1901. p. 256.

[38] Freitag Joseph Kendall. Architectural Engineering with Special Reference to High Building Construction, Including Many Examples of Prominent Office Buildings. 2nd ed. NYC: John Wiley & Sons; 1901. p. 10.

[39] Langenbach Randolph. The Great Counterintuitive: Re-evaluating Historic and Contemporary Building Construction for Earthquake Collapse Prevention. Proceedings, International Conference on Structures and Architecture 2013. Guimarães, Portugal.

[40] Robison Elwin C. Windbracing: Portal Arch Frames and the Portal Analysis Method. unpublished paper. Kent State University; 1989. p. 9.

[41] Hamburger Ronald O, Meyer John D. The Performance of Steel-frame Buildings with Infill Masonry Walls in the 1906 San Francisco Earthquake. Earthquake Spectra 2006;4. p. 64. EERI, Oakland, CA.

[42] Langenbach Randolph. Earthquakes: A New Look at Cracked Masonry. Civil Engineering Magazine November 1992. p. 56—58. Available at: www.conservationtech.com.

[43] Freitag Joseph Kendall. Architectural Engineering with Special Reference to High Building Construction, Including Many Examples of Prominent Office Buildings. 2nd ed. NYC: John Wiley & Sons; 1901. p. 254.

[44] Freitag Joseph Kendall. Architectural Engineering with Special Reference to High Building Construction, Including Many Examples of Prominent Office Buildings. 2nd ed. NYC: John Wiley & Sons; 1901. p. 250.

[45] Mark Robert. Light Wind and Structure, The Mystery of Master Builders. Cambridge, MA, USA: MIT Press; 1994. p. 139.

[46] Freeman Sigmund A. Why Properly Code Designed and Constructed Buildings Have Survived Major Earthquakes. In: Paper #1689, Proceedings, World Conf. on Earthquake Engineering, Vancouver, BC, Canada; 2004. p. 14.

[47] ASCE. Transactions of the American Society of Civil Engineers December 1907;LIX. p. 283.

[48] Crystal David. Linguist, quoted in Public Broadcasting System (PBS) TV documentary "Language Matters," produced by David Grubin. 2015. Available at: http://video.pbs.org/video/2365391566/.

[49] International Federation of Red Cross and Red Crescent Societies. World Disasters Report 2014, focus on culture and risk, chapter 5 "Culture, Risk and the Built Environment". 2014. p. 132. Available at: www.ifrc.org.

Introduction to nonconventional materials and an historic retrospective of the field

Khosrow Ghavami

Pontifícia Universidade Católica, Rio de Janeiro, Brazil

2.1 Introduction

Since the Industrial Revolution, human activity has put an increasing strain on the environment — consuming both renewable and nonrenewable resources at rates previously unseen in the natural ecology of the planet Earth. The mechanization of processes, the creation of factories and the use of coal as a fuel source beginning towards the end of the 18th century improved the quality of life for people at the time by enabling a period of economic growth but also created a dependency on energy needed to fuel these new, mechanical technologies. In addition, new synthetic materials began to be used in place of natural materials. In construction, the increased use of steel required the mining and intensive processing of iron ore—consuming large amounts of energy and resulting in the emission of pollutants into the environment. In the 250 years since the Industrial Revolution, about 337 billion metric tons of greenhouse gases have been added to the atmosphere, with an increasing percentage attributed to industrial activities (Fig. 2.1), especially cement manufacturing, which currently makes up about 0.8% of global emissions (Fig. 2.2) according to Boden et al. (2010). The production of

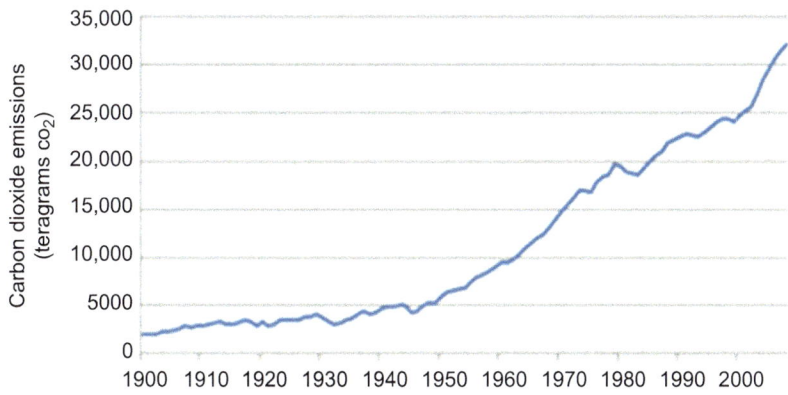

Figure 2.1 Trends in global carbon dioxide emissions (Boden et al., 2010).

Nonconventional and Vernacular Construction Materials. http://dx.doi.org/10.1016/B978-0-08-100038-0.00002-0

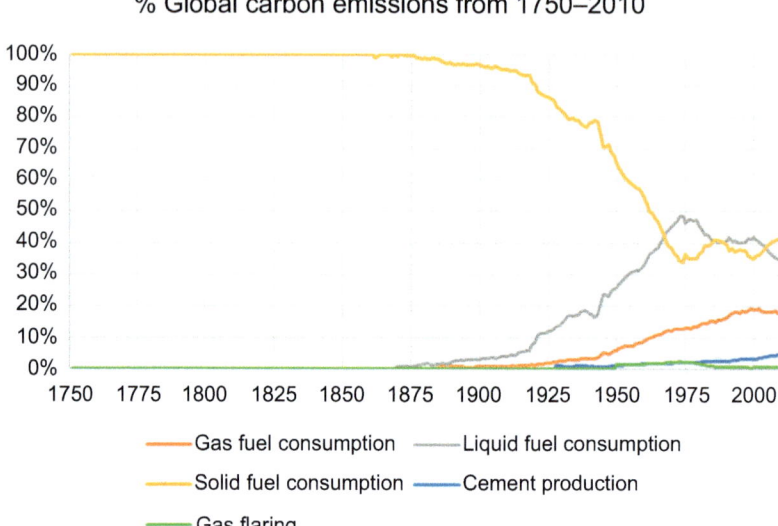

Figure 2.2 Sources of global carbon emissions from 1750 to 2010, showing the increasing contribution of cement production (Boden et al., 2010).

greenhouse gases at this rate is not sustainable and action must be taken to reduce the burden of anthropogenic activities on the environment, particularly for the mitigation of climate change.

Current concerns about dwindling reserves of nonrenewable resources and the increasing importance of global climate change are driving the search for more sustainable practices. This includes the use of locally produced materials, with a preference for using renewable resources and waste products that require minimal processing. This would lessen the environmental impact of the construction sector by reducing energy and greenhouse gas emissions related to manufacturing and transportation. In recent years, there has been an increase in research into the use of natural and nonconventional construction materials, which have suitable physical and mechanical properties for structural applications, and are desirable as ecological alternatives to more commonly used industrialized materials. The study of nonconventional materials and technologies (NOCMAT) for construction began in the 1970s. This field of research has continued to grow in the decades since. The increased use of locally available natural and waste materials in construction can promote environmental sustainability and aid in the eradication of extreme poverty.

Many of the world's most poverty-stricken people live in remote locations, typically rich in natural resources. In many of these areas, the use of steel and reinforced concrete in construction has become a symbol of economic status. These materials are imported for use when local, natural materials can also be used to create structures that can successfully meet the intended need at a lower cost and environmental impact (considering precautions are taken to prevent deforestation or depletion of other resources). Organizations like ABMTENC (Brazilian Society of the use of

NOCMAT 1996) and the International Network for Bamboo and Rattan (INBAR) strive to achieve sustainable development and poverty alleviation by promoting the use of these two natural resources: bamboo and rattan (INBAR, 2014).

This chapter presents a few historic and modern examples of buildings and bridges constructed with natural and alternative materials, demonstrating the suitability of using these nonconventional materials in structural applications, as well as their potential for use in modern civil and bridge construction.

2.2 Natural materials in historic construction

The earliest written record of bridge construction was by Herodotus, a 5th-century Greek historian who described a bridge built across the Euphrates River around 600 BC (Bennett, 2008). The bridge connected ancient Babylonian palaces located on opposite sides of the river and was 10 m wide by 200 m in length with 100 stone piers, which supported the wooden beams that formed the roadway. In his writings, Herodotus also described a bridge built by Persian ruler Xerxes to cross the Hellespont (now the Dardanelles). This bridge consisted of two parallel pontoon bridges each made up of 314 and 360 boats, which were tied to the riverbank and anchored to the riverbed to support the weight of the Persian (Iranian) army (with 2 million men and horses), which crossed the river to meet the Greeks for the battle at Thermophalae in 480 BC (Fig. 2.3). In his description, the boats were lashed together, side-by-side with ropes made of white flax and papyrus, covered with wooden boards, then paved with a mixture of brushwood

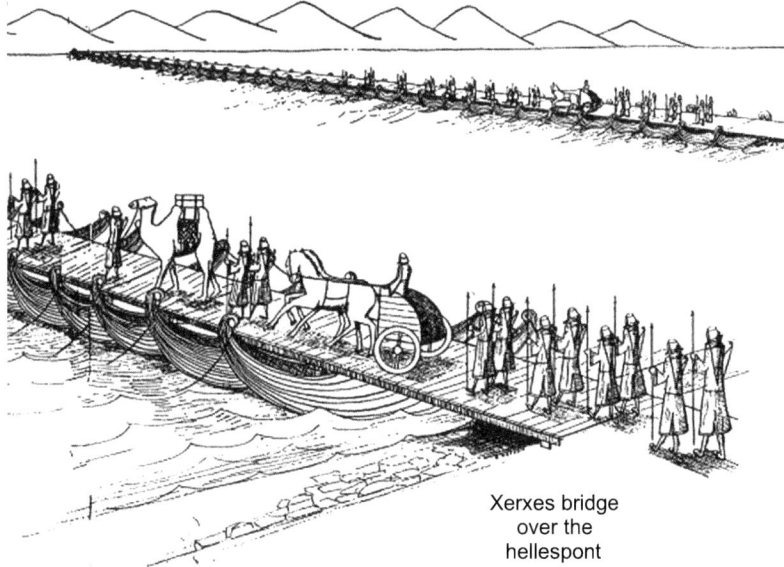

Xerxes bridge
over the
hellespont

Figure 2.3 A depiction of Xerxes' Pontoon Bridge, 480 BC (Johnson, 2000).

and compacted earth to make a solid roadway. It remains that this bridge was constructed entirely with natural materials.

Along with the writings of Herodotus, there are also records from the time of Emperor Yoa of bridge building in China around 2300 BC. The earliest Chinese bridges were also pontoon bridges, where sampan boats about 10 m long were fastened side-by-side from bank to bank then covered with a walkway to provide a river crossing. Later bridges incorporated the arch form, starting around AD 100 (Bennett, 2008)

While some natural materials have mechanical properties comparable to those of materials more commonly used in modern construction, they do have disadvantages of low durability when used in their natural state. Despite this, there are many bridges made from natural materials, centuries ago, that remain standing today. These structures provide some insight into historic construction methods and show that it is possible to work around the weaknesses of specific materials to create strong bridges with nonindustrial materials. Four bridges from different parts of the world are discussed — the Zhao-Zhou Bridge, the Inca Suspension Bridge at Huinchiri, the Kintaikyo Bridge and the Living Root Bridge of Cherrapunji. These bridges were made using regional construction techniques, utilizing local natural resources as building materials and are well designed to serve the designated function.

Zhao-Zhou Bridge The Zhao-Zhou (also Zhaozhou or Anji) Bridge is one of the world's oldest stone segmental arch bridges, constructed between AD 595 and 605 (Fig. 2.4). Built by Li Chun during the Sui Dynasty, the Zhao-Zhou Bridge was one of many construction projects undertaken during the reign of Emperor Wen as part of his desire to create a unified China (Bingham, 1941). Located in the Hebei Province in China, the bridge is composed of limestone slabs joined with iron dovetails and stretches about 50 m, with the largest arch spanning about 37 m. The Zhao-Zhou Bridge has survived 1400 years through floods, earthquakes and fires without significant damage as recorded by Ling-Xi (1987).

Ling-Xi conducted a numerical analysis to assess the load-carrying capacity of the Zhao-Zhou Bridge. This assessment used structural limit states analysis to determine the minimum thickness of the arch required to support the self-weight of the bridge and a 10-ton point load representing the weight of a truck crossing the bridge. Additionally, the analysis calculated the thickness of the bridge arch required to support the 10-ton live load when applied in different positions along the bridge. Once the minimum thicknesses from the entire load cases were calculated, these are compared to the actual

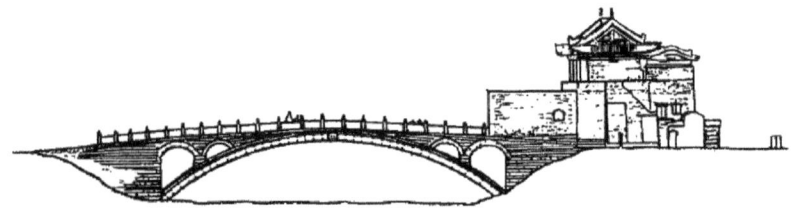

Figure 2.4 Zhao-Zhou Stone Bridge—AD 590—600. Span 37.02 m, rise 7.05 m, radius 27.82 m and ring thickness 1.0 m (Ling-Xi, 1987).

thickness of the stone arch bridge to determine the factor of safety inherent in the bridge design. For the assumed loading conditions, the bridge was overdesigned by a factor of three in the extreme condition considered. Furthermore, the maximum required compressive strength calculated is at least 1/10th of the compressive strength of the limestone blocks used to build the bridge, and the maximum required shearing stress is even less significant. Ling-Xi's numerical analysis shows that the design of the Zhao-Zhou Bridge is very conservative, explaining how this bridge has survived 14 centuries with minimal damage.

Inca Suspension Bridge The Inca Empire was one of the greatest in the Americas, extending from Colombia into southern South America, and over a period of about 100 years, incorporating parts of Peru, Chile, Bolivia and Argentina into its vast civilization. To maintain this expansive empire, a large network of roads and bridges were constructed to facilitate the distribution of information throughout the territory. The Inca, having mastered the making of textiles, used this knowledge to weave crude suspension bridges from the local grass to span the deep gorges that crisscross the mountainous Andean region. The technique of making these bridges has survived until today, with a good example of an Inca Bridge remaining in Huinchiri, Peru (Fig. 2.5).

Similar suspension bridges made from woven plant fibres had already been constructed in mountainous regions of Asia centuries before the Inca created their grass bridges. These bridges include the Anlan Bridge in China (AD 300) and the Vine Bridges of the Iya Valley, Japan (AD 1100) (Fig. 2.6). This suggests that ancient societies found the suspension bridge form to be ideal for spanning long distances, requiring minimal material and resulting in lightweight structures. Each bridge is made from intertwining plant fibres to create cables, which serve as the main structural components of the bridge. The Anlan Bridge is made with bamboo rope, the Iya Valley

Figure 2.5 Inca suspension bridge in Huinchiri, Peru.
Photo credit: Rutahsa Adventures.

Figure 2.6 Ancient suspension bridges made from plant-fibre ropes: Anlan Bridge, Chengdu, China *(Photo credit: Prince Roy)* and Iya Valley Bridge, Kazurabashi, Japan *(Photo credit: Osaki)*.

Bridges are made with Wisteria vine rope and the Inca Suspension bridge is made with local grass rope. While the Anlan Bridge and some of the Bridges of the Iya Valley have now been reinforced with modern materials, the grass bridge at Huinchiri remains as it was built in Inca times (ChinaTourOnline, 2014; Atlas Obscura, 2014, accessed 27.07.14). The Inca Bridge at Huinchiri has a 33-m span, but what makes this bridge especially remarkable is the stone abutments supporting it. The stone abutments penetrated about 60 m of rock without the technologies available today (Ochsendorf, 1996). Although these grass bridges have a short lifespan of only about 2 years, this characteristic of the bridge has resulted in the knowledge of this construction technique surviving until today. The bridge is rebuilt every year in a 3-day-long festival with full participation of the approximately 500 members of surrounding communities. Each family is responsible for making rope to bring to the festival, where these chords are woven into larger cables, which will form the main structure of the bridge.

A study carried out by Ochsendorf (1996) calculated the load capacity of the bridge at Huinchiri. This study involved laboratory testing and field measurements, first conducting laboratory tests on ropes and cables collected from a previous bridge (one that had been taken down to be replaced), then by conducting field measurements of an actual bridge to calculate the in situ capacity of the bridge. The field measurements were taken when the bridge was the same age as the rope samples tested in the laboratory so that tensile resistance of the cables determined from testing could be used in the determination of the capacity of the bridge to the greatest degree of accuracy. It was determined that the bridge could support a maximum of 56 people assuming each to weigh about 75 kg, creating the maximum tensile force in each of the four main cables supporting the bridge structure.

During the height of the Inca Empire, hundreds of these bridges crisscrossed the Andean Region in a network of roads covering nearly 10,000 km (Hyslop, 1984). The larger of these bridges easily spanned 50 m — free spans that could not be achieved with the available 16th-century Spanish technology, and spans longer than that of any masonry bridge that had been built up to that point in history (Ochsendorf,

2006). A plan to replace the Inca suspension bridge over the Apurímac river with a masonry arch bridge during 1588−1595 was abandoned due to high costs and high mortality rates associated with the project (Harth-Terre, 1961). The Inca bridge technology was not surpassed until the 19th century, where only the advances of the Industrial Revolution provided other means of crossing the deep gorges of the Andes (Ochsendorf, 2006).

Root Bridge of Cherrapunji The region of Cherrapunji (also Cherrapunjee), India is one of the wettest places in the world, receiving up to 9.3 m of rainfall in one calendar month (July 1861). Besides this, Cherrapunji is also home to bridges unlike any other in the world. These bridges are made from the roots of living trees (Fig. 2.7). The trees used to create these bridges are a type of rubber tree, *Ficus elastica*, native to the region. These trees have aerial roots, which locals guide from one side of an area they would like to cross to another (using the hollowed-out trunks of the betel palm) where they take hold and continue to grow, eventually forming the strong mesh of roots that form the bridge structure. The initial process of forming the bridge can take 10−15 years before it is ready for use. Many of these bridges present in Cherrapunji today have existed for centuries and continue to gain strength as they grow.

The roots of the *F. elastica* serve the added function of stabilizing the soil and preventing soil erosion due to the massive amounts of rainfall regularly experienced in the Cherrapunji region (Atlas Obscura, 2014, accessed 28.05.14). There have been no engineering analyses of the Cherrapunji root bridges to date, however, making a few assumptions the maximum load capacity of these bridges can be estimated. The main assumptions used for this approximation are that the bridge acts as a cable subjected to a uniform applied load and that the aerial roots of *F. elastica* have the same mechanical properties as the wood of *Hevea brasiliensis*, the more common species of rubber tree. Using the geometry of the bridge shown in Fig. 2.7, the sag was determined and equilibrium equations were subsequently used to solve for the maximum load that the bridge can support, producing the maximum tensile stress in the bridge material.

Figure 2.7 Cherrapunji root bridge in India.
Photo credit: Kumar.

The maximum distributed load that can be applied to the bridge is approximately 1545 kg/m. This corresponds to a total load of 46,350 kg, which is the equivalent of about 660 people standing on the bridge (at 70 kg per person). The results of this calculation would need to be updated with the density and tensile strength of *F. elastica*, but this initial analysis gives a close approximation to the actual load capacity of this tree root bridge in Cherrapunji, India. The results of this analysis show that the bridge is well designed to safely carry the expected loads in its application as a pedestrian walkway.

This brief overview of historic constructions shows the importance of continued research and development of different materials for use in construction. Since the development of industrialized materials, including steel and concrete, natural materials have been abandoned for their structural use. While the mechanical properties of natural materials are less predictable than those of the industrial materials, they are readily available in many rural locations and can be used with significantly less processing. The availability of steel and cement, on the other hand, depend on the presence of factories and an abundant feedstock supply to create these materials, as well as adequate infrastructure to permit transportation to the locations for their use. Now, with energy consumption and sustainability becoming increasingly important issues in the construction industry, industrialized materials are not always optimal when there are many natural materials that have excellent structural properties, are renewable and require minimal energy input to be construction-ready. Therefore, as the progress of the civil engineering industry relies on the continued development of different materials in construction, NOCMAT should be considered in the search for the low-energy structural materials of the future.

2.3 Reinforced adobes as energy-saving construction materials

Clay is the main material for adobe production. It is a natural, earthy, fine-grained material. It is composed essentially of silica, alumina and water. Adobe is a type of earthen building material consisting of sandy clay, water, straw and other organic materials. This natural building material is traditionally produced in handmade moulded or compressed shapes and is air-dried in a shaded area. Adobes, in general, are used in hot and dry climates and are extremely durable according to studies carried out by Ghobadian (1994). Adobe or earthen constructions are one of the oldest building materials in the human society; there are many known old buildings around the world with nearly 10,000 years of history. As an example, the Bam citadel located in Kerman−Iran (Persia) is a fortified town completely built of adobes. Ghrishnan (1978) studied the Bam citadel, which was constructed more than 2500 years ago, and was inhabited up to end of 19th century by nearly 10,000 people. Later in the 20th century, it was used as army barracks until 1932. Now designated as a world heritage site by UNESCO, this town made of adobe reinforced with wheat straw included a fortress, more than 60 towers, caravanserais, a bazaar and hundreds of houses. On

December 26, 2003, an earthquake destroyed the citadel (see chapter 'What We Learn from Vernacular Construction', Fig. 1).

One of the most common types of reinforced adobe in the Bam citadel is the composition of sand, clay, straw and water, known as Kah-gel in the Persian language. In order to make Kah-gel, the ingredients are mixed very well with a combination of stirring and beating, up to a homogenized appearance. The traditional way to achieve this combination is mixing with feet and tarps. The first step is throwing some crushed soil on a tarp; next, sand is added and the two are then well mixed. The function of sand particles in the Kah-gel is as the aggregate in concrete; each particle usually has rough angled sides preventing the high shrinkage of clay during the drying process. The straw acts as reinforcement similar to steel or polymer fibres in a cementitious matrix. The volume fractions of each component of the soil composites depend on their application. Traditionally, before adding straw the weight fraction of sand is between 50% and 85% and that for clay between 15% and 50%. The weight fraction of the straw is based on the sand–soil mixture: the more clay in the mix, the more straw is used. As a rule of thumb, the amount of straw will not be greater than 30%. One of the main objectives of using natural fibres as reinforcing elements with soil matrices is to prevent cracking of the soil resulting from shrinkage. Tensile shrinkage cracks in the soil are mainly due to rapid and non-uniform drying. Reinforcing fibres in the soil matrices prevent cracking by adhesion or bonding (Ghavami et al., 1999; Miraki et al., 2006).

The large application of soil as a construction material in the 21st century is built on the knowledge of the chemical, physical and mechanical properties of adobe. For example, by the 1920s, X-ray diffraction techniques allowed determination of the structure of the clay minerals while its chemical composition was obtained using infrared spectral analysis. In general, it is agreed in Pineda-Piñon et al. (2007), Ramirez et al. (2007) and Ghobadian (1994) that the atomic lattices of most of the clay minerals consist of a unit in which a silicon atom is at the centre of, and equidistant from, four oxygen or hydroxyls arranged to form a tetrahedron. Another structural unit consists of two sheets of closely packed oxygen atoms, hydroxyls have aluminium, iron or magnesium atoms sandwiched between them. Each metal atom is in the centre of, and equidistant from, six oxygen atoms arranged in an octahedron.

2.3.1 Adobe attributes and properties

One of the most significant characteristics of adobe is the return to its original natural state without any waste. For example, when adobe loses its performance it is returned to the environment as agricultural soil. From an economical point of view, adobe is categorized as the least expensive building material, since clay, as its primary raw material, is widely found in different areas around the world. Additionally the production process of adobe does not need high-energy consumption or complicated equipment. Moreover using materials like wheat straw for reinforcing the soil is actually converting a minor agricultural product into a valuable building material. Table 2.1 presents the amount of energy required for adobe production in comparison with other building materials (Ghavami et al., 1999).

Table 2.1 Comparative primary energy requirements of building materials (Ghavami et al., 1999)

Very high energy	(MJ/kg)	High energy	Per (MJ/kg)
Aluminium	200–250	Steel	20–60
Copper	100+	Lead, zinc	25+
Stainless steel	100	Glass	12–25
Plastics	50–100	Cement	5–8
Medium energy		Low energy	
Lime	3–5	Sand	<0.5
Clay bricks and tiles	2–7	Fly ash	<0.5
Concrete		Soil	<0.5
Blocks	0.8–3.5	Adobe	<0.2
Precast	1.5–8.0		

Adobe is one of the best energy saving materials used in building construction. The adobe is fabricated manually, air-dried in a covered open space with voids left in between the adobe blocks. The voids cause a reduction in the heat transfer rate between the inside and outside of the building. As a comparison, the thermal conductivity coefficient for adobe ranges from 0.5 to 0.7 W/m k while that for concrete and burned brick ranges from 1.4 to 1.6 W/m k, respectively (Miraki et al., 2006). Adobe tensile strength is low in comparison to that of other building materials. Adobe compression strength varies between 1 and 7 MPa (Badillo and Rico, 1997), whereas the compression strength of concrete is 20 to 60 MPa.

Adobe is highly permeable; air moisture, rain and snow are absorbed quickly into the material causing soil particles to expand; therefore, adobe does not possess good stability in humid regions. Crack growth is a considerable weakness that reduces the reliability of adobe for use in modern buildings. Use of natural fibres can provide reinforcement and bridge cracks in adobe.

2.3.2   Improvement of physical and mechanical properties (fabrication stage)

The dimensional changes of natural fibres due to moisture and temperature variation influence three adhesion characteristics. During mixing and drying of the soil, the fibres absorb water and expand. The swelling of the fibres pushes away the soil, at least at the micro-level. Then at the end of the drying process, the fibres lose their moisture and shrink back almost to their original dimensions leaving very fine voids around themselves. The swelling and shrinkage of a natural fibre in drying soil, as shown

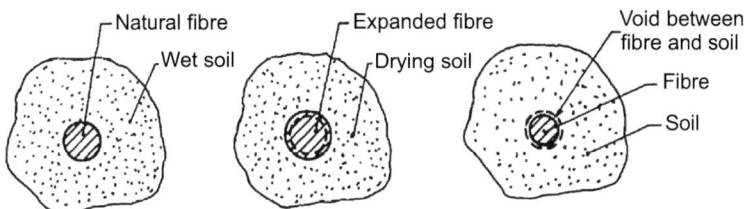

Figure 2.8 Interaction of natural reinforcing fibre and drying soil.

in Fig. 2.8, create limitations in the use of natural fibres with soil matrices (Ghavami et al., 1999).

Another effect of natural fibres is increasing the ductility of the adobe. For example, the influence of an optimum 4% of natural fibres (coconut and sisal) on the ductility of the adobe is illustrated in Fig. 2.9. It can be seen that for the natural soil the final failure occurs immediately after the ultimate load. However, in tests on soil with 4% natural fibres work softening occurs. This can be explained by considering the redistribution of internal forces from the soil matrix to the reinforcing fibres. After final failure, the soil–fibre composite does not disintegrate completely in contrast to natural soil specimens. In addition, the fibres hold the soil matrix together and no rupture of fibres occurs, although Ghavami et al. (1999) observed an insignificant loss of fibre bond.

In addition to natural fibres, different types of supplementary materials can be added to the mixture as described next, resulting in a spectrum of adobe materials, each having specific advantages (Zemorshedi, 1998).

- Ordinary adobe composed of homogeneous sandy clay and water shaped into different dimensions of bricks by means of wooden frames.
- Mineral clay, powdered stone and water are mixed and an alternative form of adobe is produced.

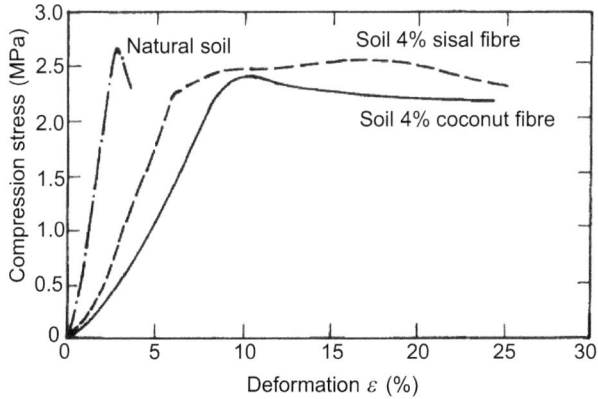

Figure 2.9 Influence of 4% fibres on the stress–strain relationship of the adobe.

- Composition of clay, small size sands, straw and water, which is described previously as Kah-gel. Using straw causes a significant increase in adobe strength, also binding adobe ingredients together. Further, small sands increase adobe stability and compressive strength.
- Mixture of clay and ash. There is a little fat in ash that causes a reduction in water absorption of adobe. The colour of this type of adobes is grey and is useful for humid regions.
- Adding dung to clay causes significant reduction in adobe moisture diffusivity. This type of adobe has a greater resistance to separation of particles affected by water absorption, so it is usable for humid climates.
- Adding goat hair to clay. The crack growth in the adobe produced by this composition is very low and in some cases disappears completely.
- Rice husks are added to clay causing a reduction of crack growth.
- Mixing some plant roots and clay will result in the increase of adobe strength.
- Mixture of clay and date fibre. The strength of this adobe is better than ordinary adobe.
- In place of organic materials, it is possible to use conventional building materials such as cement, lime and bitumen to improve the stability of adobes.

The supplementary materials should be mixed with clay homogeneously; otherwise, shrinkage and expansion of adobe will be different in different locations and adobe performance decreases.

2.3.3 Effective use of adobes in building construction (building construction stage)

Before using adobes in building construction, factors such as climate type, yearly amount of rain, time of the maximum rain in year, the amount of reduction of temperature in rainy days, day and night temperature difference and the amount of air moisture should be considered (Zemorshedi, 1998). Adobe is used mostly in hot and dry climates, and because humidity is not considerable in such areas, the building structures are extremely durable. In order to construct adobe buildings in humid regions walls must be constructed on a stone plinth of more than 25 cm in height in order to prevent moisture diffusion from the ground. In addition, as adobe absorbs the humidity quickly, the walls should be covered by plaster of clay and straw. This makes a homogeneous composition with the adobe, as soil, plaster of clay and straw belong to the renewable earthen building materials group.

2.4 Bamboo material for the 21st century

To facilitate bamboo becoming a commonly used engineering material, substituting conventional materials, certain steps such as plantation, harvesting, curing, treatment and posttreatment process should be considered. Additionally, a thorough statistical analysis of physical, mechanical and meso-, micro- and nano-structural properties of the whole bamboo culm should be carried out. In the following, some results are presented establishing such properties. Such an approach helps the user to obtain bamboo with the desired properties for his or her project.

2.4.1   Efficiency of bamboo compared with other materials

The structural efficiency of bamboo compared to that of other engineering materials, considering the relation between their elastic modulus, E, and density, ρ, using material selection method, was developed at Cambridge University by Ashby (1992) and Wegst et al. (1993) and is shown in Fig. 2.10. The line presenting the efficiency factor $C = E^{1/2}/\rho$ applies to the properties of bamboo. Materials that have a better performance than bamboo are situated above the line and those having a poorer structural efficiency are below the line. It is noted that only palm trees are in the same range as bamboo while steel, concrete and aluminium are situated far below the line.

2.4.2   Scientific research of the application of bamboo in engineering

Isolated and sporadic scientific research on engineering applications of bamboo dates back to 1914 in China and the United States. In years that are more recent, research has extended to Germany, Japan, India and the Philippines, among other countries. In Brazil, the first scientific studies in relation to bamboo were initiated in 1979 by Ghavami and Hombeeck (1981), Culzoni (1986), Ghavami and Zielinski (1988) and Ghavami and Moreira (2002). Since then, various research programmes were developed into the use of bamboo and natural fibres (sisal, coconut fibre, piassava, curaua and bamboo pulp) as low-cost and energy-saving materials for civil construction, particularly focussing on the use of bamboo as an alternative for steel reinforcement in concrete structures and for spatial structures (Ghavami, 1990, 1986, 1989, 1995).

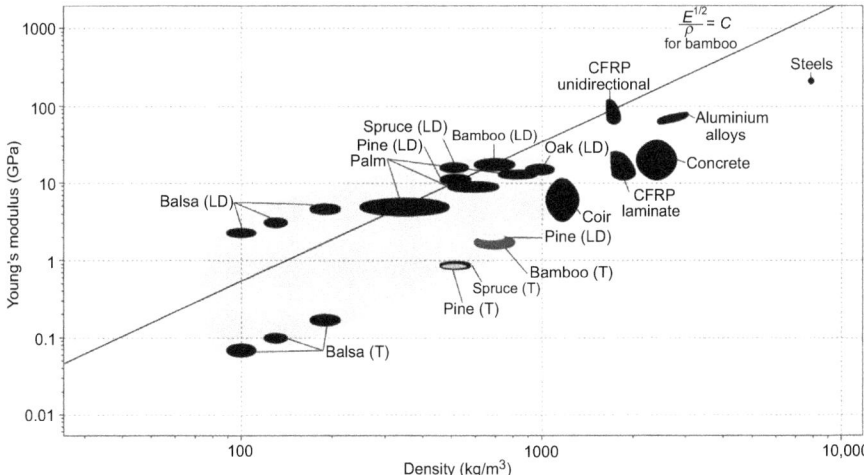

Figure 2.10 Performance of bamboo and other materials, in relation to the Young modulus and their density.

2.4.3 Bamboo as a functionally graded composite material

Bamboo is a composite material, consisting of long and aligned cellulose fibres immersed in a matrix of lignin. The distribution of fibres through the culm-wall thickness varies. This presents a functionally graded material (FGM) Amada (1996), Ghavami et al. (2003a,b), Ghavami and Marinho (2005). The FGM develops as in accordance with the stress distribution in the culm as it grows in its natural environment. As shown in Fig. 2.11, the fibres are concentrated towards the outer culm-wall in a manner that helps the culm to resist wind forces (by increasing the effective section moment of inertia) to which they are constantly subjected.

In establishing the mechanical properties of bamboo in the elastic range, the rule of mixtures for composite materials, which accounts for the properties of the fibres and matrix constituents and their volumetric fractions, is used. Eq. [2.1] presents the calculation of the elasticity modulus of a composite, E_c, where E_f and E_m are the elasticity modulus of the fibres and matrix, respectively. V_f and V_m are the volumetric fractions of the fibres and matrix respectively.

$$E_c = E_f V_f + E_m V_m \quad \text{or} \quad E_c = E_f V_f + E_m(1 - V_f) \qquad [2.1]$$

The use of Eq. [2.1] assumes long and aligned fibres distributed uniformly within the matrix and having perfect bond with the matrix. However, as seen in Fig. 2.11, the fibre distribution, as well as the size of the fibres, varies through the thickness of the bamboo culm-wall. In order to use Eq. [2.1] for the analysis of bamboo, the fibre variation and volume fraction through the culm-wall thickness should be taken into account. In Eq. [2.1], this is taken into account by defining the fibre volumetric fraction as a function of x, the distance through the culm-wall, measured from the inner culm-wall (ie, $V_f(x)$ in which x varies from 0 to t, the culm-wall thickness):

$$E_c = f(x) = E_f V_f(x) + E_m(1 - V_f(x)) \qquad [2.2]$$

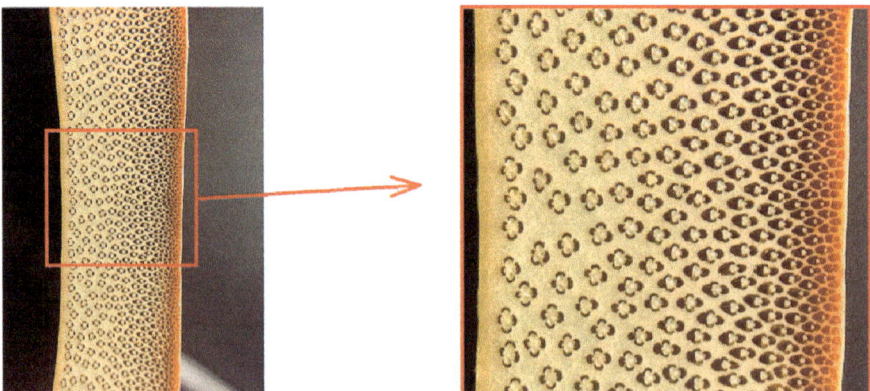

Figure 2.11 Fibres distribution on cross section of bamboo.

The variation of the fibres through the thickness of the culm-wall, $V_f(x)$, may be determined using digital image processing (DIP), for instance. This procedure was applied to two types of bamboo: Moso (*Phyllostachys heterocycla pubescens*) and *Dendrocalamus giganteus*. Moso is of a cylindrical shape, as are most bamboo species. However, in this study a square bamboo, produced artificially by growing the culms in restraining metal plates, was analysed to investigate the nonuniform distribution of the fibres not only through the culm-wall but also around the perimeter of the culm. A slice was cut from this 'quadratic bamboo' culm from which two samples were taken, one from a straight part and one from a curved corner, as shown in Fig. 2.12. The two samples were compared for the difference in fibre distribution through their thickness (Ghavami et al., 2003a,b).

As seen in Fig. 2.12, the sample from the curved part of the quadratic bamboo has a uniform distribution of fibres through the cross section in comparison to the sample from the straight part. Using data presented in Fig. 2.13, the mean volume fraction variation of fibres was found to be that given by Eq. [2.3], giving a value of V_f that varies from 0.12 at the inner culm-wall to 0.61 at the outer:

$$V_f(x) = 49.83x^2 - 0.49x + 12.01 \qquad [2.3]$$

The DIP approach was used to study the variation of fibre volume fraction along the length of bamboo *D. giganteus*. For this purpose, three samples were taken from the base, middle and the top part of bamboo culm, as shown in Fig. 2.14. The variation

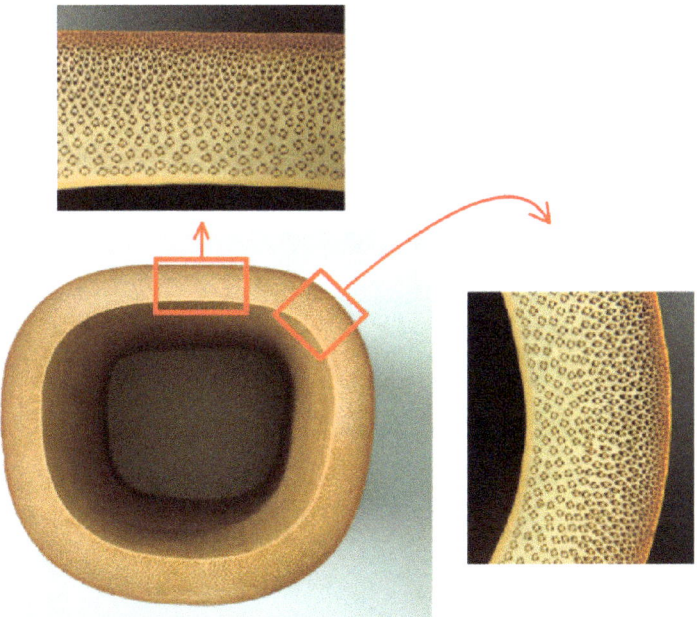

Figure 2.12 Samples of bamboo *Moso (Phyllostachys heterocycla pubescens).*

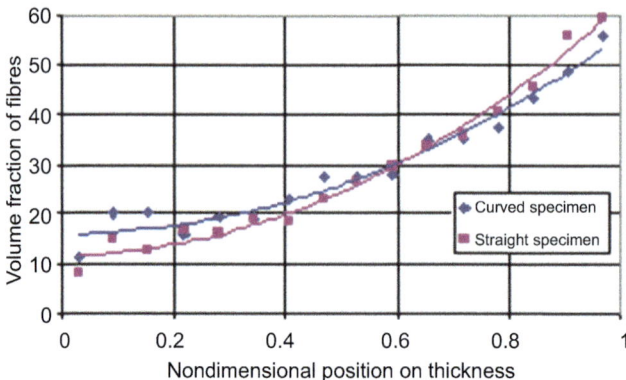

Figure 2.13 Comparative graphs of different samples of *Phyllostachys heterocycla pubescens.*

Figure 2.14 Original location of samples in the culm.

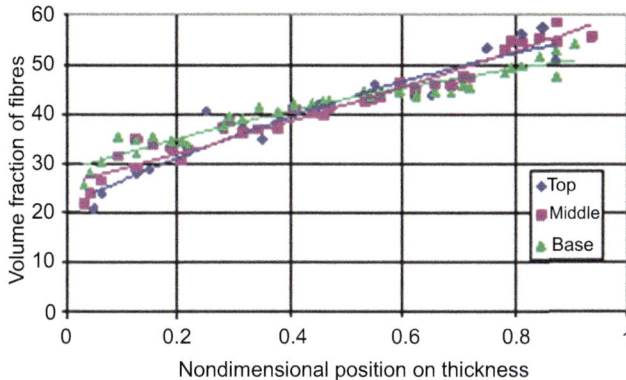

Figure 2.15 Variation of fibre distribution across the bamboo thickness at base, middle and top part.

of fibre volume fraction across the thickness of bamboo, $V_f(x)$, at each of these sections is given in Fig. 2.15. The fibre distribution is uniform at the base than at the top and middle parts. This difference can be explained considering that at the base, the bamboo is subjected to maximum bending due to wind and is supporting a greater self-weight (Ghavami and Marinho, 2005).

There is no simple and specific scientific method yet for categorizing different species of bamboo. Liese (1992) observed that the morphology of the vascular bundles of various types of bamboo differ one from another. A research programme is under way to classify bamboo according to the morphology of their vascular bundles, using DIP.

2.4.4   Mapping of physical and mechanical properties of full-culm bamboo

This section presents the mapping of the physical properties of the full-culm bamboo. The bending strength of culms of the following bamboo species is presented: *D. giganteus*, *Guadua angustifolia*, *Guadua tagoara*, *Phyllostachys heterocycla pubscens* (*Moso*) and *Phyllostachys bambusoides* (*Matake*) (Ghavami et al. (2003a,b) and Ghavami and Marinho (2005)).

This information permits the civil, agricultural and forestry engineer, the architect and, in general, people working with bamboo to easily choose an appropriate bamboo for their projects and analyse these with the correct dimensions and properties.

2.4.4.1   Dimensions of the bamboo culm

Fig. 2.16a identifies culm dimensions and coordinates along the culm. The equivalent external diameter (*D*) of each internode having length, *l*, was determined from measured values of the circumference. To determine the wall thickness (*t*), three borings of 5-mm diameter were made in each internode, as can be seen in Fig. 2.16(b),

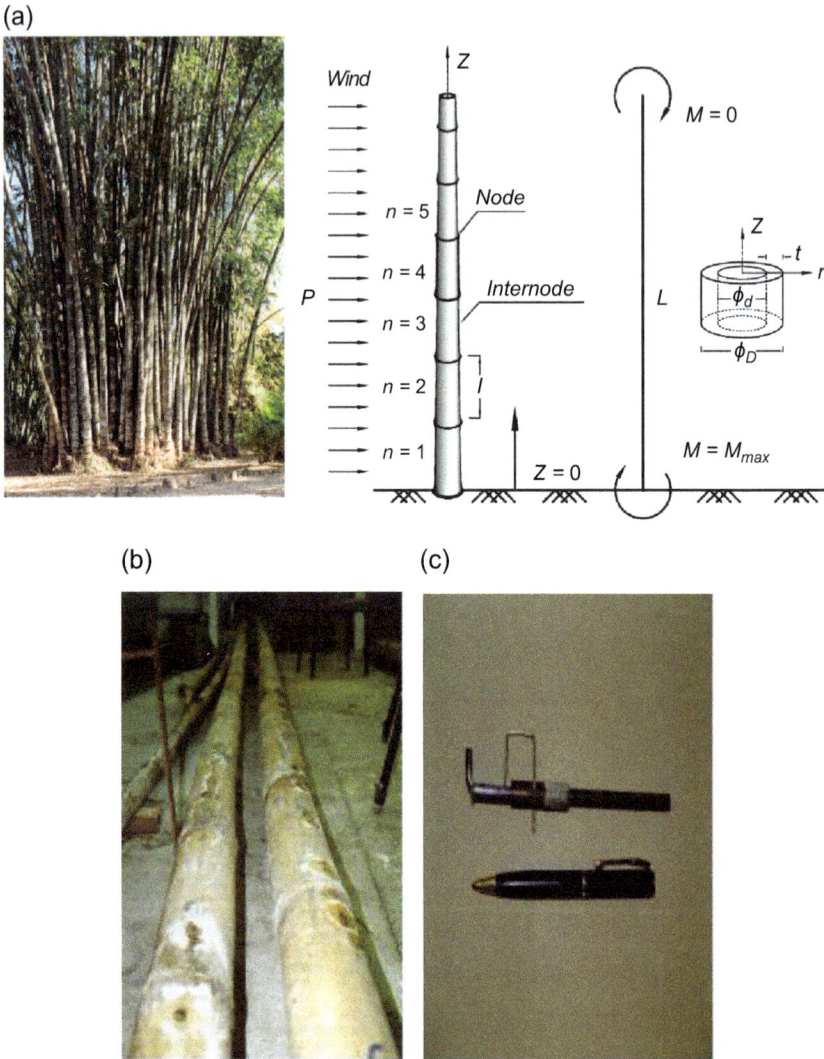

Figure 2.16 Coordination, positions of borings and device for measuring the culm wall.
(a) Axis of the coordinates. (b) Borings along the bamboo. (c) Developed devise for measuring
the thickness.

through which a special designed device was introduced and the thickness measured
(Fig. 2.16(c)). Variations of these dimensions along the culm height are shown in
Fig. 2.17(a) and best-fit equations for diameter and wall thickness are given in
Table 2.2. Based on these geometry values, the value of the elastic section modulus
(Z) along the culm height may be calculated and, from fundamental mechanics, the
extreme-fibre bending-induced stresses (σ_b) determined using an estimate of the
bending moment due to wind loading (M_m).

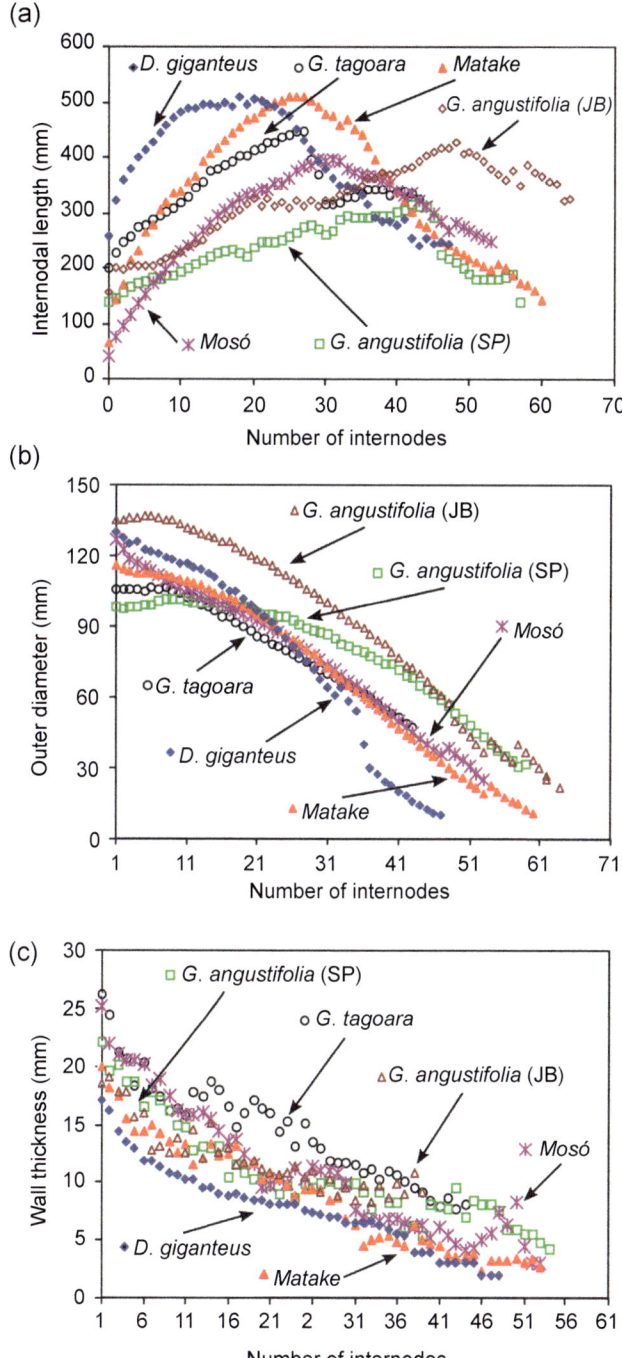

Figure 2.17 (a) Internodal length in relation to the internodes along the bamboo culm. (b) External diameter in relation to number of internode along the bamboo culm. (c) Wall thickness in relation to number of internode along the bamboo culm.

Table 2.2 Mathematical equations determining the wall thickness and external diameter in relation to the culm length

Bamboo species	Equation of thickness $t = f(z)$ (cm)	Equation of diameter $D = f(z)$ (cm)
Phyllostachys heterocycla pubenscens − *Moso*	$t = -0.0092z^3 + 0.297z^2 - 3.607z + 22.75$	$D = -0.0475z^2 - 5.061z + 118.95$
Phyllostachys bambusoides − *Matake*	$t = -0.0014z^3 + 0.056z^2 - 1.341z + 17.44$	$D = -0.193z^2 - 1.072z + 114.4$
Dendrocalamus giganteus	$t = -3 \times 10^5\, z^2 + 0.013z - 0.872z + 15.303$	$D = -0.019z^3 + 0.169z^2 - 3.176z + 130.19$
Guadua angustifolia (SP)	$t = -3.832 \ln(z) + 16.475$	$D = -0.0004z^4 + 0.041z^3 - 1.308z^2 + 7.425z + 87.63$
Guadua angustifolia (RJ)	$t = 15.93z^{-0.2257}$	$D = -0.654z^2 - 4.596z - 142.62$
Guadua tagoara	$t = -0.0002z^3 + 0.030z^2 - 1.348z + 22.96$	$D = -0.137z^2 - 1.944z + 108.96$

2.4.4.2   Data for the selection of bamboo for engineering projects

The variation in internode length, l, is shown in Fig. 2.17(a). Near the base of the culm, the internode length is smaller; it reaches a maximum value in the central part and decreases again in the upper part of the culm. This geometry occurs for all studied bamboo species. The variation of external diameter and culm-wall thickness along the height of the culm are presented in Fig. 2.17(b) and (c), respectively. The external diameter, D, shows a near linear behaviour, with the values decreasing from the base to top. By contrast, the wall thickness, t, shows a great variation along the culm and from species to species. It can also be seen in Fig. 2.17(a) that differences occur among culms of the same species, grown in different places. An example of this is the *G. angustifolia* from São Paulo and the same species planted in the Botanical Gardens of Rio de Janeiro, which presented differences in values of the internodal length, wall thickness and external diameter. This difference could occur in relation to the variation of climate, surrounding vegetation and altitude at each location. Table 2.3 shows approximate nondimensional values of wind-induced moment $(\overline{M_m})$, elastic section modulus $(\overline{Z_m})$ and bending-induced flexure stress bending stress $(\overline{\sigma_b})$, at heights of 2 m (base), 7 m (intermediate) and 12 m (top) along the culms. The nondimensional values are normalized by the greatest value of each parameter (at the base of the culm) in each case. The results indicate that despite being cultivated in different countries and are harvested at the ages between 2 and 3 years,

Table 2.3 **Nondimensional deflection moment, resistance modulus and bending stress in three lengths along the culm (z), in relation to analysed bamboo species**

Bamboo species	Deflection moment $(\overline{M}_m)$			Resistance modulus $(\overline{Z}_m)$			Bending stress $(\overline{\sigma}_b)$		
z (m)	2	7	12	2	7	12	2	7	12
Phyllostachys heterocycla pubescens	0.76	0.30	0.06	0.44	0.20	0.05	0.91	0.82	0.68
Phyllostachys edulis Riv (Amada, 1996)	0.70	0.22	0.02	0.61	0.20	0.02	0.94	0.80	0.48
Phyllostachys bambusoides	0.82	0.44	0.17	0.71	0.41	0.15	0.86	0.79	0.86
Guadua angustifolia – SP	0.82	0.44	0.02	0.73	0.24	0.10	0.83	0.53	0.15
G. angustifolia – JB	0.82	0.58	0.18	0.65	0.47	0.13	0.68	0.67	0.75
Guadua tagoara	0.78	0.29	0.04	0.78	0.46	0.20	0.90	0.60	0.22
Dendrocalamus giganteus	0.78	0.31	0.04	0.70	0.45	0.18	0.99	0.62	0.21

bamboo maintains mechanical characteristics similar to each other. In Table 2.3, one can also observe how *G. angustifolia* varies from culms grown in the State of São Paulo to the ones from Rio de Janeiro.

2.5 Application of natural and alternative materials in modern bridge construction

Despite the vast pool of research demonstrating the benefits of natural and alternative building materials, they have yet to be widely adopted in common construction practice. Nonetheless, a few innovative structures have been built incorporating nonconventional materials to produce some remarkable works of engineering.

In 2007, Shigeru Ban designed a temporary bridge made of recycled paper tubes, which was assembled across the Gardon River in southern France (see chapter 'Paperboard Tubes in Structural and Construction Engineering', Fig. 16.13(e)). The bridge, located just half a mile from the Pont du Gard, provides a material contrast with this ancient Roman stone arch bridge. The 7.5-ton structure was constructed of 281 paper tubes 10 cm in diameter having 1.3-cm wall thickness. Cardboard is lightweight and is easy to dismantle and to collect locally. Twenty-seven architecture students from France and Japan erected the structure over the course of one month and tested it using balloons filled with 1.5 tons of water to ensure that it would support the intended 20-person design load (Dilworth, 2007).

Notable bamboo bridges include the Crosswaters Ecolodge, Jenny Garzón, Coquiyo and Cúcuta bridges. The main structure of the Crosswaters Ecolodge Bridge in Huizhou, China, and the Jenny Garzón Bridge (see chapter 'Bamboo Design and Construction', Fig. 14.2) in Bogotá, Colombia — which are of similar design — is a trussed arch made up of thick bamboo bundles with a concrete abutment to maximize the load-carrying capacity of the bridge. Additionally, the bridges are covered with a ceramic-tiled roof to protect the structure from the elements and to further increase the longevity of the structures.

The bamboo bridge in Coquiyo, Colombia, has a span of 52 m. Its main structural components include two arches each made up of bundles of 12 bamboo poles. The main structure of the bridge in Cúcuta, Colombia, also includes two arches — one on each side of the pedestrian walkway — to support bridge loads and also includes two bamboo masts that support a tensioned fabric roof (Rottke, 2002; Wright, 2010).

Although little technical information is available for many of the aforementioned structures, they demonstrate natural and nonconventional materials used successfully in structural applications. It will take much work, however, for the use of these materials to gain full acceptance within the realm of modern engineering practice.

2.6   Concluding remarks

There are instances in which promising research was misinterpreted or methodologies and/or materials were misused resulting in the failure of products, which had already and could still greatly benefit humanity. Hemp was such a product that fell victim to defamation in the press and lobbying from industrialists who then wanted to introduce synthetic fibres such as Nylon, produced by DuPont, whose main shareholder was at the time the secretary of the treasury. Hemp was declared as an illicit plant and as being linked to drug abuse and crime, (Hounshell and Smith, 1988). Crops were destroyed on the order of government agencies despite existing publications by Dewey and Merrill (1916), chief scientists at the Department of Agriculture, pointing out the potentials and superior qualities of hemp for the fabrication of paper and rope. The use of hemp in an engineering context was mostly abandoned because of conflicting business interests.

A new era began. There is no doubt that new synthetic materials have found their way into our daily lives but we should always consider that natural materials, in use for thousands of years, might be equal or superior to equivalent industrialized materials. As described in this chapter, these natural materials, when locally available, are sustainable and generally do not pollute our environment, indeed, in some cases, they may even filter pollution caused by the production of industrial materials. The scientific knowledge collected by researchers on natural materials and technologies needs to be taught as a mainstream topic at colleges and universities. It is of utmost importance that research findings are published and easily available facilitating the search for new solutions.

Acknowledgements

The author would like to acknowledge the financial support granted by FAPERJ, CNPq and CAPES. Thanks are also due to all the technicians helping in the realization of the laboratory work. Special thanks are also due to all my students and colleagues who contributed in realizing the research programmes. Finally, I thank Ursula Ghavami, who checked the English text. Thanks also to the PhD student Arah Azadeh for the photos of the chapter.

References

Amada, S., 1996. The mechanical structures of bamboo in viewpoint of functionally gradient and composites materials. Journal of Composites Materials 30 (7).

Ashby, M.F., 1992. Materials Selection in Mechanical Design. Pergamon Press, Oxford.

Atlas Obscura. The Vine Bridges of Iya Valley. http://www.atlasobscura.com/places/vine-bridges-japan (accessed 27.07.14.).

Atlas Obscura. The Root Bridges of Cherrapunji—Centuries-Old Bridges, Grown from Tangled Roots. http://www.atlasobscura.com/places/root-bridges-cherrapungee (accessed 28.05.14.).

Badillo, J., Rico, R., 1997. Mecánica de Suelos. In: Tomo, I. (Ed.), Fundamentos de mecánicas de suelos. Editorial Limusa. Tercera Edición.

Bennett, D., 2008. The History and Aesthetic Development of Bridges. Institution of Civil Engineers. http://dx.doi.org/10.1680/mobe.34525.0001. ICE Manual of Bridge Engineering. http://www.icemanuals.com.

Bingham, W., 1941. The Founding of the T'ang the Sui Dynasty: The Unification of China. A.D. 581−617. Alfred A. Knopf, New York (pbk) ISBN:0-394-49187-4; 0-394-32332-7.

Boden, T.A., Marland, G., Andres, R.J., 2010. Global, Regional, and National Fossil-Fuel CO_2 Emissions. Carbon Dioxide Information Analysis Center, Oak Ridge National Laboratory, U.S. Department of Energy, Oak Ridge, Tenn., U.S.A. http://dx.doi.org/10.3334/CDIAC/00001_V2010.

ChinaTourOnline.com. Brief Information on Anlan Suspension Bridge. http://www.chinatouronline.com/china-travel/chengdu/chengdu-attractions/Anlan-Suspension-Bridge_230.html (accessed July 27.07.14.).

Culzoni, R.A.M., 1986. Características dos Bambus e sua Utilização como Material Alternativo no Concreto (Characterization of Bamboo and its Utilization as alternative material in Concrete). Master thesis. Pontifícia Universidade Católica do Rio de Janeiro/PUC-Rio, Rio de Janeiro.

Dewey, L.H., Merrill, J.L., 1916. Hemp Hurds as Paper-making Material USDA. Bulletin No. 404, Washington, DC, p. 25.

Dilworth, D., August 14, 2007. Shigeru Ban Bridges Stone and Cardboard. http://archrecord.construction.com/news/daily/archives/070814ban.asp.

Ghavami, K. (Ed.), 1986. Low Cost and Energy saving Construction Materials, vol. 2. EXPED-Expressão e Cultura, Rio de Janeiro, 300 pp.

Ghavami, K., 1989. Application of bamboo as a low-cost energy material in civil engineering. In: Third CIB/RILEM Symposium Materials for Low Income Housing, Mexico City, 1989. Proceedings. CIB/RILEM, Mexico City, pp. 526−536.

Ghavami, K., 1990. Application of bamboo as a low-cost construction material. Bamboo Current Research, Kerala Forest Research Institute & IDRC, Singapore, pp. 270−279.

Ghavami, K., 1995. Ultimate load behaviour of bamboo reinforced lightweight concrete beams. International Journal of Cement and Concrete Composites, Elsevier Science/England 17 (4), 281−288.

Ghavami, K., Toledo, R.D.F., Barbosa, N.P., 1999. Behaviour of composite soil reinforced with natural fibers. Cement and Concrete Composites 21, 39−48.

Ghavami, K., Hombeeck, R.V., 1981. Application of bamboo as a construction material. Part I -Mechanical properties and water repellent treatment of bamboo. Part II - Bamboo reinforced concrete beams, Latin American Symposium, Rational Organization of Building Applied to Low-cost Housing, IPT/CIB, Sao Paulo vol. 1, pp. 49−66.

Ghavami, K., Zielinski, Z.A., 1988. Permanent Shutter Bamboo Reinforced Concrete Slab. Department of Civil Engineering, Concordia University, Montreal/Canada (Report BRCS1).

Ghavami, K., Moreira, L.E., 2002. The influence of initial imperfections on the buckling of bamboo columns. Asian Journal of Civil Engineering, Tehran, Iran 3 (3 & 4), 01−16.

Ghavami, K., Rodrigues, C., Paciornic, S., 2003a. Bamboo: functionally graded composite material. Asian Journal of Civil Engineering, Iran, Tehran 4 (1), 1−10.

Ghavami, K., Marinho, A.B., Cruz, M.I.S., 2003b. Analisis de las propriedades mecanicas de tallos interos de bambu. Ingeniería Civil (Madrid), Madrid−Espanha 132, 119−126 (In Spanish).

Ghavami, K., Marinho, A.B., 2005. Propriedades Físicas e Mecânicas do Colmo Inteiro do Bambu da Espécie *Guadua angustifolia*. Revista Brasileira de Engenharia Agrícola e Ambiental, Campina Grande PB 9 (1), 107−114 (In Português).

Ghrishman, R., 1978. Iran from the Earliest Time to the Islamic Conquest. Penguin Books, New York.

Ghobadian, V., September 1994. Climate Analysis of the Traditional Iranian Buildings. Tehran University Publications.

Harth-Terre, E., 1961. El historico puente sobre el río Apurímac, vol. I. Revista del Archivo Nacional del Perú, Book XXV. Lima.

Hounshell, D.A., Smith Jr., J.K., Fall 1988. The Nylon Drama, Invention and Technology. Cambridge University Press, pp. 40−55.

Hyslop, J., 1984. The Inka Road System. Academic Press, New York.

INBAR. International Network for Bamboo and Rattan. http://www.inbar.int (accessed 03.11.14.).

Johnson, W., 2000. Seeking classical manuscripts at Mount Athos and seiches at ancient chalkis. International Journal of Mechanical Sciences 42 (12), 2469−2487.

Kumar, A. https://commons.wikimedia.org/wiki/File%3ADouble_Decker_Living_Root_Bridge2. jpg and https://commons.wikimedia.org/wiki/File%3ADouble_Decker_Living_Root_ Bridge4.jpg (Flickr: Double Decker Living Root Bridge) [CC BY-SA 2.0 (http://creative commons.org/licenses/by-sa/2.0)], via Wikimedia Commons from Wikimedia Commons.

Liese, W., 1992. The structure of bamboo in relation to its properties and utilization. In: Proceedings of the International Symposium on Industrial Use of Bamboo, Beijing, China, pp. 95−100.

Ling-Xi, Q., 1987. New insight into an ancient stone arch bridge—The Zhao-Zhou bridge of 1400 years old. International Journal of Mechanical Sciences 29 (12), 831−843.

Miraki, M.K., Ghavami, K., Daneshmand, F., 2006. Reinforced adobes as low-cost, energy saving building materials. In: Brasil-nocmat 2006-Salvador,conference in Bahia, Proc. in CD, pp. 1−12. ISBN:85-98073-07-05.

Ochsendorf, J., 1996. An engineering study of the last standing Inca suspension bridge. In: International Bridge Conference.

Ochsendorf, J., April 2006. Keynote lecture: engineering analysis for construction history: opportunities and Perils. In: Dunkeld, M., Campbell, J., Louw, H., Tutton, M., Addis, B., Powell, C., Thorne, R. (Eds.), Proceedings of the Second International Congress on Construction History, vol. 1. Construction History Society, Cambridge, UK, pp. 89−107.

Osaki, J., December 13, 2011. Seeking Mountains Around the World: Kazurabashi, Iya Valley. http://johnblog.mountainhikingholidays.com/2011/12/kazurabashi-iya-valley.html.

Pineda-Piñón, J., Vega-Durán, J.T., Manzano-Ramírez, A., Pérez-Robles, F., Balmori-Ramírez, H.y, Hernández-Landaverde, M.A., 2007. Enhancement of mechanical and hydrophobic properties of Adobes for building industry by the addition of polymeric agents. Building and Environment 42 (2), 877−883.

Prince Roy. https://commons.wikimedia.org/wiki/File%3AAnlan_Suspension_Bridge-1.jpg By Prince Roy (Flickr: Chengdu Day 5) [CC BY 2.0 (http://creativecommons.org/licenses/by/2.0)], via Wikimedia Commons from Wikimedia Commons.

Ramirez, A.M., Pinon, J.P., Ghavami, K., 2007. Characterization of clay for human habitat. In: International Symposium on Earthen Structures (ISES-2007). Characterization of Clay for Human Habitat, Bangalore, pp. 22−24.

Rottke, E., October 27, 2002. Jörg Stamm. http://bambus.rwth-aachen.de/eng/reports/joerg_stamm/referatstamm.html.

Rutahsa Adventures. https://commons.wikimedia.org/wiki/File%3AIRB-8-NewBridgeSide View-KC603-11.jpg Photo courtesy of Rutahsa Adventures mailto:www.rutahsa.com−uploaded with permission by User:Leonard G. at en.wikipedia [CC BY-SA 1.0 (http://creativecommons.org/licenses/by-sa/1.0)], via Wikimedia Commons from Wikimedia Commons.

Wegst, U.G.K., Shercliff, H.R., Ashby, M.F., 1993. The Structure and Properties of Bamboo as an Engineering Material. University of Cambridge, Cambridge, UK.

Wright, B., January 2010. Unusual Bamboo and Fabric Structure Offers Design Advantages. Fabric Architecture. http://fabricarchitecturemag.com/articles/0110_cs_bamboo.html.

Zemorshedi, H., 1998. The Architecture of Iran, Introduction to Traditional Building Materials. Zomorod Publications, Tehran.

Future directions for nonconventional and vernacular material research and applications

3

S. Suffian
Villanova University, Villanova, PA, United States

R. Dzombak
University of California, Berkeley, CA, United States

K. Mehta
The Pennsylvania State University, University Park, PA, United States

3.1 Introduction

The global middle class includes over 1.8 billion people and is expected to nearly triple to 4.9 billion people by 2030 (Kharas, 2010). Growth of population and economies strains scarce natural resources and impacts local and regional stability. Sustainability is now a central tenet of design theory that serves to mitigate these impacts. Sustainable design seeks to minimize the environmental and social impacts of a given product and its material components as well as its journey from cradle to grave. Achieving true sustainability is not always easy; it requires one to navigate complex decisions and weigh the costs and benefits of different alternatives. Impending resource constraints and the looming effects of climate change mandate that sustainable practices be adopted to reduce waste and ensure that global needs are satisfied in the short and long terms.

Shelter is a basic human need, providing safety, security, and a sense of home. Construction materials for shelter vary widely around the world depending on regional availability and material cost as shown in Fig. 3.1. In some areas, common resources for construction like lumber and clay are limited, leading to the development of innovative, low-cost, and renewable alternatives. Nonconventional and vernacular construction leverages indigenous knowledge, use of local building materials, and sustainable design principles to optimize resources for construction. Emerging developments in design combine historical context with innovative technologies resulting in more efficient and resilient buildings.

For the purposes of this chapter, a conventional material is characterized by widespread use, industrialized processing, and minimal unit costs. Conventional materials are well integrated into global supply chains and are used pervasively without consideration of local needs. Low production costs through mass-manufacturing and a broad

Nonconventional and Vernacular Construction Materials. http://dx.doi.org/10.1016/B978-0-08-100038-0.00003-2

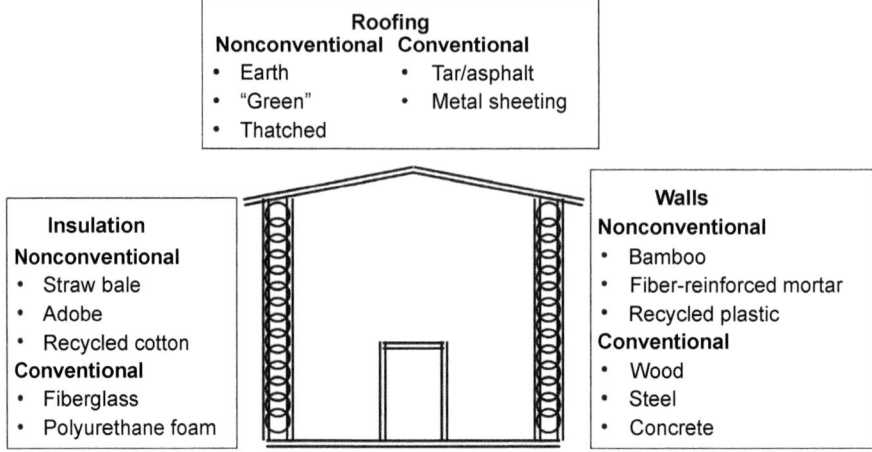

Figure 3.1 Shelter with nonconventional and conventional material examples.

customer base, achieved by leveraging global distribution, drives the use of conventional materials. Conventional materials are processed to ensure standardization and quality output. New technologies are developed to optimize manufacturing productivity to minimize cost. Steel, concrete, and wood are common conventional materials for buildings, with concrete being the most abundant construction material worldwide (Aitcin, 2000).

In contrast to materials found throughout global supply chain networks, "nonconventional" materials are used for their cultural, environmental, or technological benefits. Vernacular materials symbolize and can be derived from traditional customs and cultural activities. Adobe, rammed earth, and thatched roofing are examples of vernacular materials in use today (Miqueleiz et al., 2012; Pacheco-Torgala and Jalali, 2012; Kramer, 2013). Sustainable materials can also seek to mitigate potential negative environmental and social impacts. Some examples include renewable resources such as bamboo (Gatóo et al., 2014), straw bales (Veerbeck and Ponet, 2012), and biomass or recycled waste products such as plastic and used cotton. A third subset of nonconventional materials includes materials recently developed through research and technological advancements. Many newly developed materials exhibit improved material resilience to environmental shocks such as earthquakes and floods. For example, natural fiber-reinforced bricks increase tensile strength necessary for earthquake protection (Binici et al., 2005).

Truly sustainable construction practices require the consideration of material externalities, such as cultural relevance, environmental degradation, and social impacts. However, designing for sustainability can compromise material affordability and mass-availability. Further, without rigorous supply chain analysis, seemingly sustainable or technologically appropriate materials can actually lead to worse impacts than their conventional alternatives. We suggest new criteria for analyzing material choices, called "sustainable-scale," which combines the attributes of nonconventional materials and the scale of conventional materials as seen in Fig. 3.2.

	Nonconventional	Conventional
Sustainable-scale	• Local availability • Low cost (for repairs/maintenance) • Knowledge of how to work with material as well as its strengths and limitations • Environmental benefits • Localized resources • Cultural significance (eg, roofs as meeting places) • Contextual appropriate • Naturally processed materials	• Large-scale production leads to availability • Standardization yields replicability • Industrialized processing can yield longer material lifetime or more effective decommissioning • New technologies used to optimize material effectiveness
Failed design	• Not designed for dynamic conditions • Small-scale production	• Negative externalities such as environmental damage, social system disruption

Figure 3.2 Material sourcing and potential for scale.

A truly sustainable material should have minimal environmental impact and reduced material consumption. Extraction and use of these materials should also be conducted in a socially responsible manner. It is important that materials are analyzed for their sustainability within different contexts. The use of a material that is considered environmentally benign in one region may be a threat to the ecosystem in another region. For example, bamboo, with its high natural tensile strength and fast growth rate, may seem like a global solution to reducing deforestation by replacing timber as a primary construction material source. However, excessive demand of this material can lead to the destruction of forests to make room for bamboo plantations, as is currently taking place in China (Gallagher, 2011). Similarly, locally sourced stone can cause more environmental damage than imported wood or bricks depending on labor practices associated with quarrying and timber extraction. If local resource extraction relies on worker exploitation and dangerous conditions, imported materials can provide a more sustainable solution. Security and reliability issues often accompany imported materials. Resource-rich regions must create measures to protect against material overuse. Further, some nonconventional materials may satisfy the criteria for sustainability but lack the characteristics for scalability. For example, shredded currency notes can serve as building insulation, as indicated by an artist who used 1.4 billion shredded Euros to construct his house in Dublin, Ireland (Carbone, 2012). However, the use of recycled money as a construction material is not a scalable or replicable solution. A material of sustainable-scale optimizes the sustainability constraints while also being replicable, low-cost, and accessible to the masses.

While these constraints may seem overwhelming, there are several tools to help engineers make appropriate decisions when using nonconventional materials for construction. In the next section, we step through life cycle analysis, which can be used for rigorously comparing the impacts of different materials for a construction project. This approach identifies the limitations for materials to be sustainable within the context of a specific project. We subsequently describe several technical and systemic design factors that are necessary for new materials to achieve sustainable scale. This chapter provides a set of tools and considerations for design practitioners, engineers, and materials researchers using materials that are sustainable, affordable, scalable, and accessible.

3.2 Analyzing potential for sustainable scale

There are several decisions that must be made when choosing sustainable and scalable materials. Material selections decisions incorporate the specific social, political, and environmental context of the region to determine where, when, and how to source a material. A life cycle approach addresses difficult decisions by taking a systems-level view to conduct a cost, energy, and resource balance across the entire supply chain of a construction project.

3.2.1 Life cycle approach

A common method of conducting a holistic comparison between different methods of construction is life cycle assessment (LCA). A traditional LCA involves conducting a material and energy balance for each stage of a product or process, from resource extraction through decommissioning. The per-unit impacts of each step, such as water use, carbon emissions, and various polluting gases like nitrogen oxides and sulfer oxides are quantified. LCA enables intelligent decision making by normalizing overall environmental impacts for comparison. The procedures for conducting an LCA are composed of four main phases, officially standardized by the International Organization for Standards as ISO 14040.

The first phase develops the goal and scope of the assessment. The goal of the assessment helps to narrow the details and context of what exactly is being compared between two materials. For example, a goal may be to compare the relative environmental impacts of bamboo construction against modern construction in a certain village in China. The scope of an LCA defines the boundaries of the analysis. "Cradle-to-grave" and "cradle-to-gate" are common scopes used for various analyses. The "cradle" in these cases refers to the earliest point of the supply chain that can be reached, including the impacts of raw material extraction. "Gate" means that the analysis will end at the conclusion of manufacturing, whereas "grave" would follow the analysis through use and decommission. For construction materials, "gate" would imply an LCA that focuses only on the manufacturing of the building material, whereas "grave" would include impacts from the construction and eventual degradation and demolition. A different scope, used for materials that are designed for recycling, is "cradle-to-cradle." Cradle-to-cradle analysis includes analysis of the resources necessary to transform a material for reuse in a second application. In cradle-to-cradle design, all outputs of a given cycle are transformed or transferred into inputs to be used in a new process or product. In our example, a decision would be made to analyze bamboo from its growth until the house is no longer in use, or a "cradle-to-grave" analysis.

The second phase of an LCA is the life cycle inventory (LCI). This phase establishes the material and energy balance of each life cycle stage through the collection of supply chain data. It can be difficult to obtain sufficient data for each stage, leading to assumptions and generalizations being made. These assumptions must be transparent among decision-makers, with a qualitative explanation of how they might impact the end results. An important consideration is which global unit should be used to normalize the data. This may seem like an easily convertible step after an LCA is established, but that is not always the case. A global unit may be a certain quantity of material, or could be a single unit of the end product (such as whole home). Material properties such as density, physical size, and structural strength will lead to different quantities depending on which unit is used. For the bamboo example, data collection would take place regarding the inputs of bamboo planting, growing, harvesting, and transport. A good functional unit might be a house of standardized size, and the necessary bamboo to construct such a building.

After conducting an LCI, the Life Cycle Impact Assessment takes place. In this phase, previously collected data are analyzed across certain predefined metrics. One of the more common comparisons is relative carbon emissions. This metric allows two materials to be compared relative to their impacts on global climate change. Contribution to acid rain, water use, and toxicity can also be used as measurements. This analysis can be conducted through a combination of contextual data collection and more generalized public datasets on the environmental impacts of certain materials.

The final phase of assessment is the interpretation of results. The simplicity of this step is dependent on the scope of the LCA's original goal. If the goal is to compare material impacts as measured by quantifiable numbers, then the interpretation is largely a look at what stages and inputs lead to the greatest impacts, and overall which has the lower impact. There is conversation among researchers about the expansion of LCA to also incorporate socioeconomic metrics, thus creating an even more comprehensive comparative report on the selected materials (Norris, 2006). Interpretation in these cases becomes more complex due to multiple simultaneous metrics. Environmental impacts such as carbon emissions, toxicity, and water use must be compared and valued against other impacts like resource availability, land-use equity, and harvesting complexity. For example, while bamboo is considered a strong, lightweight material that has been shown to withstand earthquake forces (Jayanetti and Follett, 2008), locally produced bamboo can yield some negative impacts. Bamboo growth is difficult to contain, and when wild it acts as an invasive species. Due to its invasiveness, bamboo requires dedicated land and active management to prevent its spread to other areas. Regions with already limited arable land, or those that would require deforestation to make room for bamboo cultivation, must decide whether that land would be best suited to grow construction materials or should instead be allocated toward food-based agriculture. If timber is already available, sustainable forestry practices may reduce damage to the local ecosystem, even if its per-unit carbon footprint proves greater. An LCA of bamboo and timber construction provides the information necessary to navigate such a trade-off. More importantly, it reveals opportunities for significant improvement of the supply chain by indicating what stages use the most energy and resources.

Measuring impacts of coir binderless board

Coconuts are an ample resource in many coastal regions such as Thailand, Indonesia, the Philippines, and French Polynesia. When coconuts are harvested, the husks are often left as a waste product. The husk of a coconut, called coir, is 70% pith and 30% fiber by weight (van Dam et al., 2004). The high lignin content of the pith, combined with the tensile strength of the fiber enable the materials to form a strong binderless insulating building material. The coir binderless board has been shown to have similar levels of thermal conductivity compared to more traditional insulation materials such as fiberglass, mineral wool, and polyurethane foams. However, coir board has a significantly lower density than commercially available alternatives (Panyakaew and Fotios, 2011), which means greater quantities of coir are required to produce the same insulating effect as the mineral wool or foams.

An LCA completed by Payakaew and Fotios compared the embodied energy of coir binderless board, fiberglass, and mineral wool. The **functional unit** for the analysis was defined as "the mass (kg) of insulating board that produces a thermal resistance of $1.0 \text{ m}^2 \text{ K/W}$." The authors chose to use a cradle-to-gate **scope**, including material extraction, transportation, and manufacturing, providing only the pre-use energy requirements of the material. Subsequently, the **inventory** provided data on energy costs associated with transforming the material from a raw material to a final product as well as information on energy costs for comparative materials. For example, the majority of building materials found in Thailand and French Polynesia must be imported, and the necessary material transportation significantly raises the embedded energy of the product. **Impact assessment** revealed that insulating boards made from agricultural products all had higher energy impacts because of the high board density; more agricultural material is needed to produce the same insulating effect as traditional materials. While agricultural boards are a local resource, they do not yield environmental benefits over importing more traditional construction materials. Payakaew and Fotios looked only at pre-use energy requirements and the total life-cycle energy costs may shift if they included end-of-life and recyclability potential to their analysis.

Decisions around material selection are of increasing importance within resource-constrained environments with the exacerbation of climate change. LCA is one of many frameworks that can guide projects and estimate externalized costs. The strength of LCA is that it forces one to enumerate impacts that otherwise would not have been recorded. Often, many construction projects only focus on the financial bottom line, as its effects are the most acute. However, the construction industry is a major source of waste worldwide (Kofoworola and Gheewala, 2009; Formoso et al., 2002; Docksai, 2014). The building and deconstruction processes lead to massive resource utilization and the accumulation of waste, both toxic and benign. The use of LCA illuminates the challenges to sustainable-scale construction. What characteristics do new materials and construction methods require to overcome these drawbacks? How can we leverage improved design processes to reduce the resource intensity of construction?

3.2.2 Factors for sustainable-scale design

Nonconventional and vernacular materials are not sufficient for meeting global construction needs by themselves. Improved design processes, supply chains, and end-of-life management must accompany natural materials selection to ensure that resources are utilized to their full potential. Before we can advocate for the increased usage of a material, we must ensure that agricultural and resource management systems are equipped to handle the increased strain. Projects outspend their budgets or cause unintended consequences when teams fail to view all project aspects systemically. At the onset of an initiative, one must consider resilience against various ecosystem shocks, the presence of different failure modes, as well as when and how a building will eventually become obsolete. Subsequently, design decisions should account for these complex factors and attempt to create ways by which challenges to sustainability can be overcome. The following section presents a series of design factors that influence the environmental, social, and economic costs of construction projects.

3.2.2.1 Technical factors

Disassembly

To design for disassembly or deconstruction is to create products with the intention of minimizing value loss at the end of life. In the context of an LCA, this design factor would lead to materials that are "cradle-to-cradle" in scope. Waste generation is inherent to almost every process or service and leads to overall degradation of both raw material utility and monetary value. For example, minerals used in the electronics industry are limited in availability; however, disassembly of electronics and material recovery proves difficult due to components' composite structure. Furthermore, current processes by which materials are recovered from electronics pose a significant threat to human health. In the construction industry, many materials used on a large scale can be recovered, reused, and repurposed. Steel, wood, concrete, and asphalt can contribute to a more sustainable built environment by being reused. A material's capability for reuse depends on its installation and formulation. Composite materials, or materials that are amalgamations of various natural resources and additives, are increasingly becoming incorporated into infrastructure, particularly for roofing and building exteriors. However, the material complexity of composites, such as fiber-reinforced polymer, makes disassembly difficult and highly energy intensive (Horvath, 2004). Design for disassembly and reuse facilitates cradle-to-cradle design and must be championed wherever possible.

Closed-loop manufacturing

All materials require some form of processing or manufacturing prior to installation and use for construction. Closed-loop manufacturing involves redesigning products so that any waste generated can later be used as an input into either the same cycle or in the creation of a different product or service. The carbon cycle provides an example of a closed loop system. Carbon moves between terrestrial and aquatic eco-systems serving as both a nutrient and a waste product at different times. In juxtaposition, most human-driven processes degrade natural resources without replenishing their source. One can leverage nonconventional and vernacular materials to develop systems that encourage material recovery and recyclability.

Dried sludge generated from wastewater can be used as an alternative brick material (Weng et al., 2003). Though sludge has a lower density than clay, it has other nonmaterial benefits, such as the destruction of otherwise dangerous pathogens during the firing process. Improper storage of wastewater sludge can lead to leaching of heavy metals contained within the sludge. Leaching metals cause pathogen transmission as well as contamination of surface and ground water. Currently, 1.3 billion tons of solid waste is produced every year. With increased wealth and urbanization, it is estimated that this number will reach 2.2 billion tons per year by 2025, thereby putting additional strain on local municipalities that are typically responsible for their own waste management (Hoornweg and Bhada-Tata, 2012). Using sludge for bricks represents a way to close a loop, as a current waste product can be manufactured to meet a growing need. At end-of-life, sludge bricks can be broken up and either used again as an industrial input or can be used in agricultural systems. However, this process must be carefully monitored to ensure that the toxic components of the sludge bricks are properly managed to prevent contamination.

Supply chains

Today's global economy has made it commonplace for products to be manufactured in one geographic location and used thousands of miles away. The environmental costs associated with transport can often offset the benefits of using natural materials. Further, some materials require multiple stages of processing with each intermediate step occurring in a different geographic region. Disparate production can yield a large spectrum of potential emissions, depending on factory locations and available energy sources. Emissions are not uniform by location; depending on the energy sources (renewable, coal power plant, etc.) the emissions profile of a single process can vary greatly if done in Pittsburgh, USA, or Nairobi, Kenya. Thus, it is critical that one considers the impacts of all steps from resource extraction to end-of-life.

Local production, processing, and usage have both costs and benefits. Shortening a supply chain can yield significant environmental benefits. For one, reducing transportation needs can decrease emissions from fossil fuels. Additionally, shorter supply chains often are accompanied by increased transparency. This is extremely important when it comes to natural resource extraction. Suppliers can ensure that labor rights are upheld, that materials are replenished at an appropriate rate, and that over-utilization does not lead to the eradication of a given material. However, lower manufacturing costs accompany increased scale. If an existing industry, or cluster of industries, does not exist, costs will generally be higher. Creating industries around natural materials can lead to increased employment opportunities. Suppliers as well as other supply chain stakeholders must be able to see the value in the benefits and accept sacrifices along at least one bottom line.

Resilience

One result of climate change is more frequent extreme weather events (Easterling et al., 2000). Longer droughts, heavier winters, and stronger storms can adversely affect the built environment. Environmental ecosystem shocks are expected to alter global energy and water supply, as well as shift our material resources. The ability of infrastructure and its ability to withstand sudden system changes, such as a change in demand or political instability, will be repeatedly tested. In the forthcoming dynamic age, engineers and design practitioners must consider time scales over which they expect buildings to last. Should large amounts of energy and resources be put into buildings to ensure that they are ruggedized and can withstand shocks? Are the construction materials strong, or will they necessitate continuous maintenance and rebuilding? Knowledge of whether one is designing for either a short or long time span can change material selection and construction processes.

Heavy usage of timber in a region facing desertification in the next century would prove detrimental. For one, cutting down trees to source the timber accelerates the rate of desertification. Also, if timber became unavailable due to a resource shortage, houses needing repair would have difficulty finding a replacement component. Instead of replacing a single piece, a more intensive solution may be required. The construction industry must anticipate change and determine mechanisms of adaptation. Lessons can be taken from the people of Fiji, who have adapted their shelters to become more resilient to cyclones. They tie banana leaf veins with green coconut leaves to secure thatched houses and place heavy cement bricks or sand bags on top to secure iron roofs (McNamara and Prasad, 2013).

In some cases, the use of traditional materials, leveraged by modern research, can improve the resilience of construction. For example, in rural Turkey, most buildings are constructed with limestone, traditional concrete masonry block, and adobe. This region of the world is tectonically active and has had several major earthquakes in the last decade alone. New research is being conducted into improving the tensile strength of these traditional building materials. Using new geometric shapes and incorporating certain fibrous materials bring these materials up to modern earthquake-resistant standards (Binici et al., 2005). The focus of modern research on traditional materials has the opportunity to advance vernacular architecture in our dynamic world.

Cost

It is estimated that more than one-third of the world's urban population lives in inadequate housing (U.N. Habitat, 2003; Keivani and Werna, 2001). Designs are needed that provide low-cost housing solutions without sacrificing shelter quality. Often, cost is seen as the only significant factor when creating products for individuals living in resource-constrained environments. However, if a building is characterized by regular maintenance and a short lifespan, the total structure cost could end up significantly higher than if a higher grade of materials was used during the nascent stages of construction. For example, many houses found in India use either corrugated cement or metal sheets for roofing materials. In addition to causing high indoor temperatures, cement can crack easily, leading to water leaks. Concrete, a more desirable and durable roofing material, is financially out of reach for the majority of residents. As a result, roofs must be replaced often or patched with temporary solutions. ReMaterials, a social enterprise based in Ahmedabad, created a modular low-cost roofing material from agricultural and packaging waste (ReMaterials, 2014). By using a material that would otherwise be discarded, ReMaterials is able to keep costs low for the consumer. Additionally, the modular design means that if damage occurs, a resident only replaces a single tile. Thus, roof durability is increased through innovative design without increasing the price point and not through a significantly higher price point.

As stated previously, in constructing the built environment, one must consider not only the materials but also the processes by which materials are fabricated and altered. Each phase of a material's life can influence costs for a consumer in both acute and distal ways. Inefficient or unnecessarily high levels of processing can manifest as an increased price point for a given product. Unsafe resource extraction or energy intensive machining can lead to social and environmental costs that last for decades.

3.2.2.2   Systemic factors

Cultural significance

Shelters, and vernacular materials used in their construction, hold value beyond their structural integrity. In many cases, the same materials have been used for generations and are woven into the cultural fabric of a community. In other cases, there are certain social functions within a shelter that require additional design constraints for the composed construction materials. For example, vernacular roofs in Iran are used as living spaces, for public gatherings, and as cultural interaction centers (Maghsoudi et al., 2013). Materials that may be more environmentally sustainable but do not allow for the load requirements of multiple people would not succeed in this context.

For example, in 1985, pucca or "strong" houses were introduced into Indian culture as part of a social welfare initiative. Pucca houses are defined as industrially produced houses, whereas their counterpart, kachcha or "raw" houses are built with thatched roofs and natural materials. However, the meaning behind these labels has strong emotional undertones. "The Hindi work kachcha generally has a negative connotation and conversely, its opposite—pucca—holds positive connotations. The terms 'kachcha' and 'pucca' are far from neutral—with 'kachcha' being associated with poverty and backwardness and 'pucca' with progress and modernity"

(Barenstein and Iyengar, 2010). Pucca houses have garnered significant aspirational value, leading many to reject the potentially more sustainable and resilient kachcha houses. Even though vernacular materials are more affordable and sustainable, the lower middle class sees them negatively, instead desiring pucca houses made of reinforced concrete and modern materials. The problem is primarily a marketing challenge; consumer education and campaigns are needed to sway public opinion and expose individuals to the benefits of vernacular materials.

Localization of resources, materials, and processes

Shelter design is often context specific. If an area is accustomed to high temperatures, pervasive pests, or heavy rain, the vernacular shelter form will change in accordance to these driving influences. It is critical that shelter users and their contextual knowledge be used throughout the design process. This is particularly true for individuals living in vernacular homes. As stated by Carl Mitcham, "The vernacular house is human dwelling made visible, enriched with material traces from its lived past" (Mitcham, 2005). Vernacular homes have been built over centuries by optimizing the materials to which one has access. Lessons can be taken from examples of indigenous populations who have been forced to move due to regional conflicts or changing climates. Peasants recently relocated by the government to La Quetzal in Guatemala used their knowledge of forest resources for construction materials. While 50% of this population used tin metal roofing, many used thatched palm leaves of *Sabal mauritiiformis* (botán) for their kitchen roof to keep their homes cooler in the summer (Nesheim et al., 2006). Construction with local materials and resources can satisfy community demand with minimal environmental and social impacts. How can knowledge of vernacular construction be used and shared to address global needs?

While most vernacular designs are localized, standardization can yield benefits. For one, standardized designs and construction processes can yield lower costs for users. IKEA has used this model of design for the masses to distribute aesthetically appealing and low-cost furniture to millions of people around the world. If one can establish replicable construction processes, vernacular construction may be more easily disseminated.

Localization versus standardization in policy can also influence construction processes. Often building standards will be set forth by governments in an attempt to force standardization throughout a given country. In low-resource communities, such standards can be unrealistic because of the costs associated with achieving them. For example, governments set regulations in the upgrade of an informal urban settlement without taking into consideration all of the economic, social, and environmental constraints on the project. The lack of alignment between building directives and available resources leads to the project failures (Das and Takahashi, 2009).

Sophistication of processing

While technology has the power to increase the efficiency of many production systems associated with vernacular materials, such options are not always pursued. This can sometimes lead to decreased lifespans of these materials. For instance, some bamboo and timbers are vulnerable to insects and other forms of degradation if not properly treated. Labor-intensive processing employs large populations in low-income regions. Mechanized processes can lead to increased resource destruction, carbon emissions, and job disruption. However, sometimes mechanization can help to preserve the use of vernacular materials in a given context. For example, in French Polynesia, vegetal roofs made from pandanus leaves strongly signify Polynesian culture and history. French Polynesia is known to many as a major tourist destination, though much of the indigenous population lives in poverty. Low-income housing options are available to Polynesian citizens; however, the homes use conventional materials instead of traditional thatch because of high processing costs. The pandanus vegetal thatched roofs are comprised of prefabricated tiles that require highly specialized, labor-intensive production by trained artisans. Though historically production was common throughout the Polynesian islands, now a cohort of producers on a single island, Maiao, monopolizes the industry. This monopoly raises the price of pandanus, rendering it unaffordable for the average Polynesian. Thatched roofs have their own benefits and detriments. While their thermal properties allow Polynesian houses to stay comfortable during high heat, the lack of durability leads to roofs replacement being necessary every 5 to 6 years. Ultimately, though, the true value of thatched roofs is their cultural significance. Mechanized thatch production could lead to market diversification and significantly lower costs. Industrialization, in this case, presents an opportunity to actually make the roofs and their installation process more affordable and sustainable.

Industrialization and quality control

Industrialized materials have undergone processing or are produced in a factory setting where economies of scale lead to lower product cost. Industrialization has the benefits of quality control and replicability, two factors that can increase the efficiency of a construction process. For example, brick presses and molds enable fast creation of standardized bricks. However, variability in the bricks can still occur due to the person operating the press. Proper training, appropriate salary, and a corporate culture that values quality workers and their products help to improve standardization. It is important to implement systems to test the quality of the final product. These problems become even more complex for nonconventional materials where natural processing, local extraction, or small-scale manufacturing can make standardization difficult. Some vernacular materials require indigenous knowledge to overcome these barriers during construction. For example, the construction of thatched roofs requires expertise in choosing materials, weaving the thatches, and stitching to construct a lasting roof (Rustemeyer et al., 2012). While these roofs may not be standardized, quality control is maintained through the transfer of indigenous knowledge using traditional oral traditions. Melding indigenous knowledge and processes with western knowledge and industrialization mechanisms can lead to efficiencies that result in more sustainable products and processes that align with cultural sensitivities and ways of living and doing.

3.3 Deciding between materials

The previously discussed design criteria yield a variety of trade-offs that engineers, architects, and designers must navigate. Should a building be as low cost as possible, or should funds be spent on aesthetics to ensure popularity? Should a building be constructed for permanence or should it be transportable to a sparse rural environment? Should a building be ruggedized for a harsh climate or should infrastructure drive development in more tepid regions? Contextual considerations drive design and can lead to different outcomes depending on what is most important to the customers or community involved. However, it is important to note that designs are never "finalized." These complex considerations must be addressed as individuals and projects undergo a constantly evolving design process. Therefore, flexible and innovative design allows for interconnectivity and switching out materials as needs and resource availability evolves. One way to address trade-offs is to consider each option as a variable within a larger system, and calculate the Pareto optimality of that system.

3.3.1 Pareto optimality

Pareto optimality is the state at which resources in a given system are optimized in a way that one dimension cannot improve without a second worsening. Mapping optimality, as shown in Fig. 3.3, enables decisions between design choices. Using Pareto optimality, one can assess how engineered systems can best meet multiple criteria. In this context, it can be used to understand how a construction project balances environmental, social, and economic factors. For example, a farmer in Pakistan might be deciding on the building materials for a new home. She wants a solution that is inexpensive but also is resistant to the increasing frequency of floods in her area. In a more complex example, Pareto efficiency would be calculated using several dimensions, but in this case we will limit them to flood vulnerability and cost. She has a few different

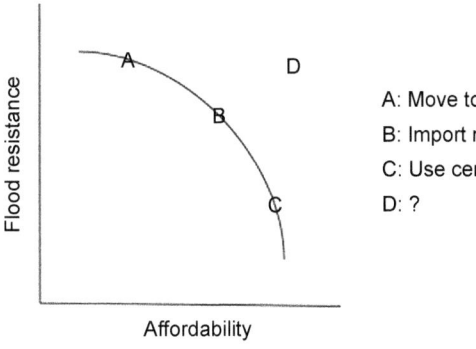

Figure 3.3 Example of Pareto optimality graph.

A: Move to drier land
B: Import materials
C: Use cement
D: ?

options regarding building materials. She can use concrete and construct her house in a conventional manner, as many of her neighbors do, she can move to a less flood prone area, or she can import stone and other materials from the nearest city.

In order to make an informed decision, she decides to rate her options on a scale of 1−10 in terms of both flood vulnerability and cost. For each option, a function will be defined that will lead to a numeric normalized value. We will call the $v(X)$ the function that provides the flood vulnerability of option X on a scale of 1−10, and $c(X)$ the function that provides the cost of option X on a scale of 1−10. Both functions will include various variables regarding available options. For example, $c(X)$ could include material costs, transportation costs, and increased cost of living.

Her most expensive option, moving to a new location, will be costly but provide protection against flooding. She calls this option A, and using her previously defined functions v and c finds that $v(A) = 2$ and $c(A) = 8$. Another option, called option B, is to import new materials from the nearest manufacturer. In this case, $v(B) = 4$ and $c(B) = 4$. Her third option, option C, is to use concrete to build her home. This option leaves her vulnerable to flooding ($v(C) = 8$) but saves her the most money ($c(C) = 2$).

To find the best option, the farmer multiplies these two values together ($v(x)*c(x)$) and realizes that they all equal the same value (24) and are therefore all on the existing Pareto optimum curve. External factors, such as simplicity of construction and proximity to family, would lead to the farmer choosing to use concrete and build a conventional home. This would be called the weak Pareto optimum. The farmer is looking for an option D where $v(x)*c(x) < 24$, which would provide a strong Pareto optimum, and thus be the best choice for her.

A possibility for option D is the inclusion of limestone in the mortar for an otherwise conventionally constructed house. The limestone provides greater flood resistance at minimum additional cost. Using her previously defined functions, she finds $v(D) = 3$, and $c(D) = 5$. These values multiply to 15, providing a new strong Pareto optimum, and therefore the best choice for the construction of her home.

This example illustrates the use of Pareto optimality not only for making decisions but also for discovering opportunities for new technologies and nonconventional materials. For the farmer, options A, B, and C are equivalent when considering both cost and flood vulnerability. The farmer seeks an option D that is cheap and resistant to floods. In this case, modern research combined with indigenous knowledge offers a ready solution. As early as 2600 BC, Pakistan used lime as part of their mortar for bricks in construction. With the availability of cement, lime use in mortar has decreased. However, lime has inherent waterproofing properties while its breathability allows water that is already trapped to escape. While this indigenous knowledge may not have survived in many of the communities in the region, historical studies and materials research have shown again its importance in flood-resistant construction (ACTED News, 2014). The farmer could purchase some locally extracted limestone for use in the mortar of an otherwise traditional house and achieve greater flood resistance without having to import raw materials from far away. Through the use of Pareto optimality, better design decisions can be made and opportunities for new innovations identified. Hartikainen et al. (2011) provide more detailed information on mathematically modeling Pareto optimality and decision-making.

3.3.2   Intersectional innovation

The use of limestone in mortar to improve material resilience to flooding is just one example of the opportunities available for improving construction materials around the world. A major advantage of living in the 21st century is our ability to connect across time zones, cultures, and geographies. This leads to cross-pollination of ideas, which can serve as the impetus and building block for innovation. Recently, researchers and engineers have used cardiac stents to inspire technologies for drain repairs and modeled LEDs from a lens belonging to a firefly. How can we encourage the expansion and proliferation of innovation? Nations and industries can borrow innovations from one another and determine how to build upon ideas to develop solutions applicable to their own contexts and challenges. One example from the materials sector comes from Ecovative Design, a company that leverages one of nature's key recycling agents, fungi, to create building materials. Ecovative cures fungi, effectively transforming it into a rigid material that can be used for building insulation and product packaging, among other applications (Ecovative, 2014).

Ecovative represents a merging of disciplines to produce a new vernacular material. Their original use of mushroom material for packaging products provides a sustainable option for a growing industry. Within the United States, containers and packaging materials represented more than 30% of municipal solid waste in 2012, accounting for nearly 75 million tons of waste (Levis et al., 2014). Just over one-third of all disposed packaging waste is recovered, while the rest remains permanently in landfills. The waste stream of packaging is only set to grow in the coming decades. Increased consumption of fast-moving consumer goods by developing economies such as China and India will increase the demand for packaging materials (Zaman and Lehmann, 2013). As climate patterns become increasingly erratic and the number of people in the global middle class increases, a need and desire for housing insulation will also grow. From work in the packaging industry, Ecovative determined that mycelium and agriculture waste could be used to create insulated sheathing called MycoFoam. Though not yet commercially available, this technology would serve the same purpose as conventional spray foam insulation (Ecovative, 2014). Ecovative's insulating product has low embodied energy and is composed of natural materials as opposed to potentially harmful chemicals as similarly rated products.

3.4   Conclusion

Climate change will necessitate shifts in living styles. Many architecture forms will prove irrelevant because of excessive heat, prolonged droughts, or heavy rains. The effects of climate change will be exacerbated in developing nations where housing crises are already severe. It is imperative for engineers, designers, and entrepreneurs to share context-specific knowledge and design resources to create scalable and sustainable housing solutions. Cross-pollination of ideas can yield answers to critical questions including: What natural resources are currently underutilized and can become value-added products if processed? What innovations from one region can be adapted and translated to other places with similar challenges? Practitioners working to improve global housing can also use analytical tools to navigate building trade-offs and inform policy decisions. A life cycle perspective of construction materials informs evaluative questions and includes consideration of both technical and design factors. Pareto optimality focuses broad analyses on specific contexts and ensures that a given solution satisfies local needs. Current research in the advancement of vernacular architecture can benefit from using analytical tools and design processes when creating more resilient, sustainable, and culturally contextual buildings.

Acknowledgments

This material is based on work supported by the National Science Foundation Graduate Research Fellowship Program. Any opinion, findings, and conclusions or recommendations expressed in this material are those of the authors(s) and do not necessarily reflect the views of the National Science Foundation.

References

ACTED News, June 16, 2014. Re-inventing Shelter: Lime for Safer, Stronger Homes for Flood-affected People (Pakistan).

Aitcin, P.-C., September 2000. Cements of yesterday and today: concrete of tomorrow. Cement and Concrete Research 39 (9), 1349−1359.

Barenstein, J.D., Iyengar, S., 2010. India: from a culture of housing to a philosophy of reconstruction. In: Lyons, M., Schilderman, T., Boano, C. (Eds.), Building Back Better, p. 163.

Binici, H., Aksogan, O., Shah, T., 2005. Investigation of fibre reinforced mud brick as a building material. Construction and Building Materials 19 (4), 313−318.

Carbone, N., January 28, 2012. Broke Irishman Builds Billion-Euro House—Out of Shredded Bills. Retrieved from TIME: http://newsfeed.time.com/2012/01/28/broke-irishman-builds-billion-euro-house-out-of-shredded-bills/.

Das, A.K., Takahashi, L.M., 2009. Evolving institutional arrangements, scaling up, and sustainability: emerging issues in participatory slum upgrading in Ahmedabad, India. Journal of Planning Education and Research 29 (2), 213−232.

Docksai, R., March−April, 2014. A world without waste? The Futurist 48 (2).

Easterling, D.R., Meehl, G.A., Parmesan, C., Changnon, S.A., Karl, T.R., Meams, L.O., 2000. Climate extremes: observations, modeling, and impacts. Science 289 (5487), 2068−2074.

Ecovative, 2014. Retrieved December 1, 2014, from Ecovative Design: http://www.ecovativedesign.com/.

Formoso, C.T., Soibleman, L., Cesare, C.D., Isatto, E.L., 2002. Material waste in building industry: main causes and prevention. Journal of Construction Engineering and Management 128 (4), 316−325.

Gallagher, S., August 12, 2011. China's Appetite for Bamboo Is Damaging Forests (Pullitzer Center for Crisis Reporting) Retrieved from pullitzercenter.org: http://pullitzercenter.org/reporting/china%E2%80%99s-appetite-bamboo-damaging-forests.

Gatóo, A., Sharma, B., Bock, M., Mulligan, H., Ramage, M.H., 2014. Sustainable structures: bamboo standards and building codes. Proceedings of the ICE—Engineering Sustainability 165 (5), 189—196.

Hartikainen, M., Miettinen, K., Wiecek, M.M., 2011. Constructing a Pareto front approximation for decision making. Mathematical Methods of Operations Research 73 (2), 209—234.

Hoornweg, D., Bhada-Tata, P., 2012. What a Waste: A Global Review of Solid Waste Management. World Bank, Washington, DC.

Horvath, A., November 2004. Construction materials and the environment. Annual Review of Environment and Resources 29, 181—204.

Jayanetti, D.L., Follett, P.R., 2008. Bamboo in construction. In: Xiao, Y., Inoue, M., Paudel, S.K. (Eds.), Modern Bamboo Structures. CRC Press, London, UK, pp. 23—32.

Keivani, R., Werna, E., February 2001. Modes of housing provision in developing countries. Progress in Planning 55 (2), 65—118.

Kharas, H., 2010. The Emerging Middle Class in Developing Countries. OECD Development Centre, Paris.

Kofoworola, O.F., Gheewala, S.H., 2009. Estimation of construction waste generation and management in Thailand. Waste Management 29 (2), 731—738.

Kramer, K., 2013. Applying the lessons of Indian vernacular architecture. In: Weber, W., Yannas, S. (Eds.), Lessons from Vernacular Architecture, p. 129.

Levis, J.W., Barlaz, M.A., DeCarolis, J.F., Ranjithan, S.R., 2014. Systematic exploration of efficient strategies to manage solid waste in US municipalities: perspectives from the solid waste optimization life-cycle framework (SWOLF). Environmental Science and Technology 48 (7), 3625—3631.

Maghsoudi, E., Rahman, N.N., Yazid, M.Y., 2013. Sustainable roofs in Iranian vernacular residential buildings. In: International Sustainable Tropical Environmental Desing Conference. Serdang, Selangor, Malaysia.

McNamara, K.E., Prasad, S.S., 2013. Valuing indigenous knowledge for climate change adaptation planning in Fiji and Vanuatu. UNU-IAS TKI. http://www.unutki.org/news.php?news_id=182&doc_id=39. Accessed on 20.11.2014.

Miqueleiz, L., Ramírez, F., Seco, A., Nidzam, R.M., Kinuthia, J.M., Tair, A.A., et al., 2012. The use of stabilised Spanish clay soil for sustainable construction materials. Engineering Geology 133, 9—15.

Mitcham, C., 2005. Thinking re-vernacular building. Design Issues 21 (1), 32—40.

Nesheim, I., Dhillion, S.S., Stølen, K.A., February 2006. What happens to traditional knowledge and use of natural resources when people migrate. Human Ecology 34 (1), 99—131.

Norris, G.A., 2006. Social impacts in product life cycles towards life cycle attribute assessment. The International Journal of Life Cycle Assessment 11 (1), 97—104.

Pacheco-Torgala, F., Jalali, S., 2012. Earth construction: lessons from the past for future eco-efficient construction. Construction and Building Materials 29, 512—519.

Panyakaew, S., Fotios, S., 2011. New thermal insulation boards made from coconut husk and bagasse. Energy and Buildings 43 (7), 1732—1739.

ReMaterials, 2014. In: Ganatra, H., Gosain, S., Rabenau, L.V. (Eds.), ModRoof. Retrieved December 1, 2014, from ReMaterials: re-materials.com/product.

Rustemeyer, T., Elsner, P., John, S.-M., Maibach, H.I., 2012. Roofers. In: Belsito, D.V. (Ed.), Kanerva's Occupational Dermatology. Springer Berlin Heidelberg, Berlin, pp. 1681—1688.

U.N. Habitat, 2003. The Challenge of Slums: Global Report on Human Settlements 2003. Earthscan, London.

van Dam, J.E., van den Oever, M.J., Teunissen, W., Keijsers, E.R., Peralta, A.G., 2004. Process for production of high density/high performance binderless boards from whole coconut husk. Industrial Crops and Products 20 (1), 97−101.

Veerbeck, G., Ponet, J., 2012. Straw bale houses in a moderate climate: adaptable to meet future energy performance requirements?. In: 28th International PLEA Conference, Opportunities, Limits & Needs Towards an Environmentally Responsible Architecture. Lima, Peru.

Weng, C.-H., Lin, D.-F., Chiang, P.-C., 2003. Utilization of sludge as bridge materials. Advances in Environmental Research 7 (3), 679−685.

Zaman, A.U., Lehmann, S., 2013. The zero waste index: a performance measurement tool for waste management systems in a 'zero waste city'. Journal of Cleaner Production 50, 123−132.

Part Two

Natural fibres

Characterization of vegetable fibers and their application in cementitious composites

V.C. Correia
University of São Paulo, São Paulo, Brazil

S.F. Santos
São Paulo State University, São Paulo, Brazil

G.H.D. Tonoli
Federal University of Lavras, Lavras, Minas Gerais, Brazil

H. Savastano, Jr.
University of São Paulo, São Paulo, Brazil

4.1 Introduction

4.1.1 Availability and potential of vegetable fibers

Sustainability is a concept that is being adopted all over the world in an effort to address the shortage of natural resources and energy, the generation of several types of solid wastes, and gas emissions from various sources. The rational use of vegetable fiber can be an alternative solution for the production of durable and more sustainable goods (Savastano et al., 2009a). Several vegetable fibers have potential for use in the manufacture of different products including paper, paperboard, tissues, polymeric or inorganic composite materials, etc. The performance of the vegetable fiber application depends on the vegetable source and the method used to obtain or harvest the fiber. Knowledge of the vegetable microstructure is important to understand the ways in which the fibers are obtained and why they are used in so many products. Further research is still needed to characterize the fibrous raw materials and for the development of new products and production methods. Tropical countries in particular present great potential for the production of cellulosic fibers, especially if they are available as residues from other agriculture activities (eg, bagasses, straws, or husks). To advance the use of cellulosic pulp fibers for different materials, continued research on the fiber characteristics, properties, and methods of obtaining the fibers is essential.

Fibers are widely available in tropical and equatorial countries, where they are primarily used in the cordage, textile, and papermaking sectors. The heterogeneity and restricted market in these industry sectors have led to intense generation of residues with high pollution potential. For example, each ton of commercially used sisal fibers

Nonconventional and Vernacular Construction Materials. http://dx.doi.org/10.1016/B978-0-08-100038-0.00004-4

(for rope) yields three tons of residual fibers, whose disposal results in environmental hazards (Garcia-Santos, 2004).

4.1.2 Vegetable fibers as reinforcement for brittle matrices

The purpose of fiber reinforcement is to improve mechanical properties of a given brittle matrix, which would be otherwise unsuitable for several practical applications (Agopyan et al., 2005). Examples of brittle matrices with practical applications include (1) Portland cement (Tonoli et al., 2010, 2011); (2) gypsum (Mármol et al., 2013); and (3) clay (Millogo et al., 2014). A major advantage of fiber reinforcement of a brittle material is the resulting composite behavior after cracking. Postcracking toughness produced by low modulus of elasticity fibers in the matrix may permit the use in large-scale construction of such composites (Agopyan et al., 2005). Reinforcement is distributed into the composite, reinforcing the matrix and bridging cracks during bending or tensile tests.

Alkali-resistant synthetic fibers, such as polypropylene (PP) and polyvinyl alcohol (PVA), are commonly used as the main reinforcement in cement-based composites. However, the production of such synthetic fibers consumes a large amount of energy, chemicals, and petrochemical raw materials. For example, as the demand for propylene (raw material for PP fibers) continues to outpace ethylene demand, there is increasing interest and need for finding or developing alternative sources of propylene without adversely affecting ethylene availability. In addition to PP, fiber manufacture has a high embodied energy, about 38 kJ/kg, and it produces effluents (process water, caustic steams), emissions (airborne polymer powder, fumes containing carbon monoxide, formaldehyde, and acrolein), and wastes (contaminated polymer scraps, spent catalyst) (Nexant, 2000).

Within the context of a sustainable economy and innovative construction, vegetable fibers, from wood and nonwood sources, are widely available in most developing countries as suitable reinforcement materials for brittle matrices. Accounting for the mechanical properties of the fibers as well as their broad variation, one may develop building materials with suitable properties by means of adequate mix design (Agopyan et al., 1989; Agopyan and John, 1989).

The concept of vegetable fiber reinforcement in cement-based materials was developed in the 1970s, when vegetable fibers were evaluated as substitutes for synthetic and asbestos fibers. Natural vegetable fibers, including sisal, bamboo, fique, hemp, flax, jute, and ramie, for example, are used in regions where these materials are readily available. Motivations for their use include reducing raw materials costs and contributing to sustainability of the sector. As reported in Coutts (1992), vegetable fibers contain cellulose (which is a natural polymer), providing the main reinforcement of these materials. The chains of cellulose form microfibrils, which are held together by hemicellulose and lignin to form fibrils. The latter are then assembled in various layers to build up the fiber structure. Fibers or cells are cemented together in the plant by lignin (Gram, 1983). Vegetable lignocellulosic macrofibers present several interesting advantages, including their low real ($1.3-1.5$ g/cm^3) and apparent density ($0.4-1.5$ g/cm^3), high specific stiffness ($1.1-80.0$ GPa) and strength ($0.1-3.0$ GPa), biodegradability, renewable character, low processing energy in the case of chopped natural fibers, and availability everywhere at modest cost in a variety of morphologies and dimensions (Jarabo et al., 2012; Tomczak et al., 2007; Satyanarayana et al., 2007).

There are two main approaches for the development of new composites in fiber-reinforced cement (Agopyan, 1988). The first is based on the production of thin sheets and other non-asbestos-like components. The components are similar to asbestos-cement-based materials and are produced by well-known industrial-scale processes such as Hatschek and Magnani methods (see Box 4.1) and commercially used with high acceptance for building purposes (Coutts, 1986).

The use of cellulosic fibers as reinforcement in cement-based composite materials has been studied to partially replace synthetic fibers, especially glass and polymeric fibers in construction materials (Tonoli et al., 2010; Pacheco-Torgal et al., 2011a,b; Tan et al., 2012). Nevertheless, significant losses in long-term mechanical performance have been observed in vegetable fiber—cement composites after natural or accelerated aging (eg, wet/dry cycling), due to the degradation of cellulose fibers in the cementitious environment (Mohr et al., 2007; Tonoli et al., 2011; Melo Filho et al., 2013).

Several efforts have been made to enable new air-cured hybrid composites reinforced with vegetable fibers (Agopyan, 1988). These strategies include blended cement as the matrix, which aims to reduce the free alkalis by developing low-alkaline binders based on industrial and/or agricultural by-products (Jarabo et al., 2012, 2013; Pereira et al., 2013; Mármol et al., 2013; Teixeira et al., 2013). Other approaches are the use of cellulose nanofibrils or "whiskers" from vegetable fibers as nanoreinforcement (Claramunt et al., 2010, 2011; Hoyos et al., 2013) and the use of carbonation curing (Almeida et al., 2013; Pizzol et al., 2014a,b; Santos et al., 2015).

4.2   Characterization of vegetable fibers for engineering applications

4.2.1   Macrofibers

Considerable research effort has gone into the study of fast-growing, inexpensive agricultural crops and crop residues, especially for those countries with limited forest resources. Table 4.1 presents primary material properties of some macrofibers and those of comparable commercially available synthetic fibers.

Nonwood pulps constitute abundant and low-cost raw materials in many developing countries. Savastano et al. (2003) and Joaquim et al. (2009) presented attempts to produce viable fiber—cement materials using the Hatschek method and nonconventional pulps of sisal (*Agave sisalana*) as reinforcement. In those studies the performance as reinforcement of the fibers obtained from commercial and by-product sisal (*Agave sisalana*) by thermomechanical pulping, chemithermomechanical pulping, and organosolv processes were investigated (Savastano et al., 2003; Joaquim et al., 2009).

Box 4.1 Industrial-scale processes for producing components of fiber-reinforced cement

Hatscheck process

The first method of fiber—cement manufacturing process was invented by Ludwig Hatschek in the 1890s. He combined slurry with cellulose, reinforcing fibers, and Portland cement in water. This fed into a paper-making machine in which a cylindrical sieve or sieves rotate through the slurry. The solids are deposited on the sieve, which on each rotation transfers the layer of solids onto a continuous belt. The layers are built up to the desired thickness and then removed and, if necessary, compressed by stack pressing.

Some stages in the Hatschek process are numbered in Fig. 4.1. The first step is to prepare the slurry (1), which consists of mixing an adequate proportion of solid materials, with water in a low solid concentration (approximately 20% of total mass). Portland cement (2), reinforcement fibers (3), cellulose fibers, limestone filler (4), and water (5) are the most commonly used materials by the fiber—cement technology. The slurry is then transported to the vats (6) with sieve cylinders (7) where wet solid material is deposited. Sequentially, the running felt (8) removes the material from the sieve cylinders, thus forming a green lamina. Vacuum (9) is applied to remove water from the lamina before it is transferred to the formation cylinder (10) where the stacking is performed. Finally, the green sheet (F) is cut (11), shaped (eg, corrugated sheets and accessories) (12), and submitted to curing (Dias et al., 2010).

Magnani process

The industrial manufacturing process of cylindrical water tanks is based on the method known as the modified Magnani process. The slurry used in this process is thicker, with concentration of solids to water of about 1:1. The raw materials typically used in this process are Portland cement, limestone materials, recycled cellulose pulp, and polymeric fibers. The mass is applied in a single layer on the mold (Fig. 4.2(a)). The tank is formed by rotating the mold. Excess water is

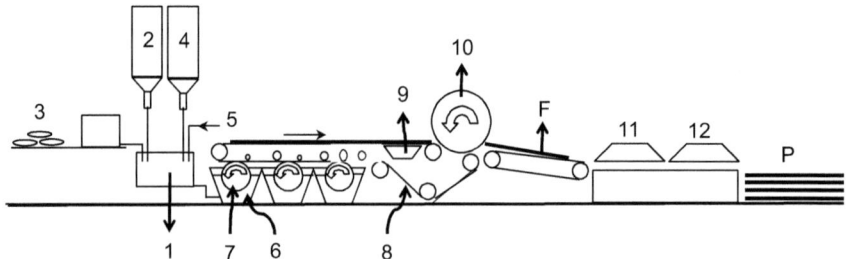

Figure 4.1 Schematic drawing of a typical Hatschek process. In this figure, F is the green sheet fiber—cement, and P is a final product.
Adapted from Dias, C.M.R., Savastano Jr., H., John, V.M., 2010. Exploring the potential of functionally graded materials concept for the development of fiber cement. Construction and Building Materials 24, 140—146.

Box 4.1 Industrial-scale processes for producing components of fiber-reinforced cement—cont'd

(a)   (b)

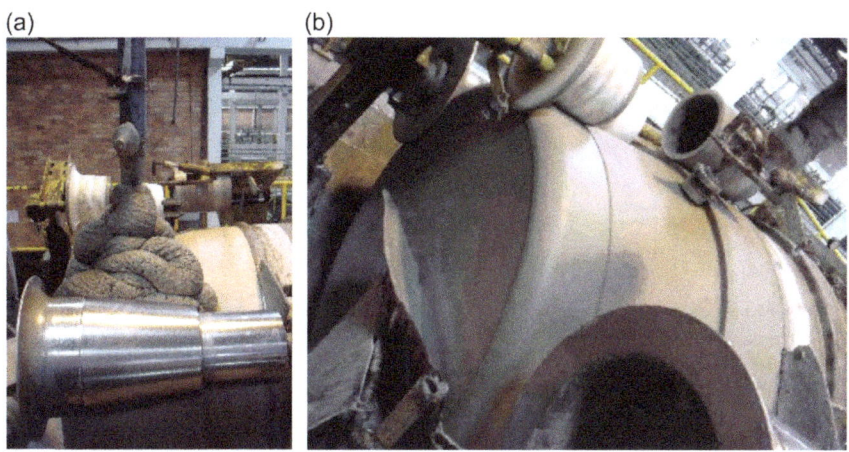

Figure 4.2 (a) Magnani method for industrial production of water tanks. (b) External surface finish of the water tank by rollers.
Courtesy of Prof. Vanderley Moacyr John, University of São Paulo, Brazil.

removed by vacuum suction, and the tank sides and bottom are finished with rollers (Fig. 4.2(b)).

A second approach consists of producing composites for different types of building components such as hollow load-bearing walls, roofing tiles, and ceiling plates, which are similar to components commercially produced with asbestos—cement.

Estimated as several millions of tons per year (Heinricks et al., 2000), consumption of fiber-reinforced cement building components is rapidly increasing, especially in developing countries. This is because such materials allow one to produce lightweight building components with good mechanical performance and acceptable thermal-acoustic insulation properties, while being economically attractive.

Correia et al. (2014a,b,c) produced bamboo cellulosic pulp by the organosolv process, evaluating different temperatures and reaction times, and the pulps were analyzed aiming their future application for the reinforcement of composites. Bamboo presents advantages such as rapid growth, short-term renewal, and broad availability, making it a promising potential raw material for cellulosic pulp production.

Banana (*Musa cavendishii*) pseudo-stem strand fiber subjected to chemithermomechanical pulping and kraft methods was also studied as reinforcement in fiber—cement (Savastano et al., 2005, 2009b). The researchers presented the results of an

Table 4.1 **Physical and mechanical properties of some macrofibers**

Fiber	Property of interest				
	Density (g/cm³)	Tensile strength (MPa)[a]	Modulus of elasticity (GPa)	Elongation at failure (%)	Water absorption (%)
Jute (*Corchorus capsularis*)[b]	1.36	400−500	17.4	1.1	250
Coir (*Cocos nucifera*)[c]	1.17	95−118	2.8[d]	15−51	93.8
Sisal (*Agave sisalana*)[c]	1.27	458	15.2	4	239
Banana (*Musa cavendishii*)[b]	1.30	110−130	−	1.8−3.5	400
Bamboo (*Bambusa vulgaris*)[c]	1.16	575	28.8	3.2	145
Curauá (*Ananas erectifolius* L. B. Smith)[e]	1.10	900 ± 200	36 ± 10	2.5 ± 1	−
E-glass[f]	2.50	2500	74	2−5	−
Polypropylene[f]	0.91	350−500	5−8	8−20	−

[a]Tensile strength strongly depends on the type of the fiber tested: a bundle or a single filament.
[b]Rehsi (1988).
[c]Agopyan (1988).
[d]Guimarães (1984).
[e]Spinacé et al. (2009).
[f]Fordos (1988).

experimental study of resistance-curve (R-curve) behavior and fatigue crack growth in cementitious matrices reinforced with eco-friendly natural fibers.

Tonoli et al. (2010) evaluated the effects of 14 years of weathering exposition in Cali, Colombia, on the microstructure and mineral composition of cementitious roofing tiles, still in service, reinforced with fique fibers (*Furcraea andina*). The kinetics of sisal fiber degradation and the mechanisms responsible for deterioration of continuous sisal fiber−cement composites are also presented in this paper (Tonoli et al., 2010).

Jarabo et al. (2012, 2013) have studied the potential of both corn stalk (*Zea mays* L.) and industrial hemp core (*Cannabis sativa* L.) fibers as renewable sources of cellulose fibers in the production of fiber−cement. For each source of fiber, several chemical processing treatments were studied. The results indicated that both types of pulp can be used for the production of fiber−cement, due to the morphological similarities with the pine fibers that are currently used.

Amazonic fibers such as those originated from curauá plant and jacitara palm have also been studied as reinforcement for fiber−cement composites (Fonseca et al., 2013).

Curauá plant (*Ananas erectifloius*) has been recognized since pre-Columbian days for its valuable fibers, which are among the unique lignocellulosic fibers of Brazil. D'Almeida et al. (2010) produced cement-based composites reinforced with 2, 4, and 6% short curauá fibers. The results obtained were comparable with those found for cement composites reinforced with sisal fibers. The jacitara palm (*Desmoncus polyacanthos* Mart.) is also widely used by the artisans of the Amazon Basin region of Rio Negro, Brazil, and it is known to provide excellent fiber characteristics and appearance. However, there remains a lack of technical/scientific information about these important Amazonic fibers. To fill this gap, the researchers evaluated the main properties of jacitara fibers for their future technological application as reinforcement in composites (Fonseca et al., 2013).

Recently, Sena Neto et al. (2015) presented a comparative study of the mechanical and thermal characteristics of 12 pineapple leaf fiber varieties, which gives valuable information for selection of those vegetable plants when the goal is to provide fibers for a specific application.

4.2.2   Pulp fibers

Pulp fibers are normally obtained from either softwood or hardwood and are composed of three main chemical components: cellulose, hemicelluloses, and lignin. Both hardwood and softwood unitary cells are tubular structures and can be pulped. The chemistry and fiber morphology of hardwood and softwood pulps, however, differ significantly. The average lengths of hardwood and softwood fibers are approximately 1 and 3 mm, respectively. The width of the fibers varies between 10 and 50 μm and the wall thickness between approximately 1 and 5 μm. Pulp fibers of nonwood materials (macrofibers) may also be obtained. Table 4.2 depicts the average dimensions of some pulp fibers from wood and nonwood materials.

During the production of pulp fibers in the paper industry, the pulping process affects each component differently. For example, cellulose pulps can be produced from residual crop wood species, by chemical pulping (eg, kraft pulping) with alkaline liquors (eg, sodium hydroxide reaction), fractions, or organosolv pulping with organic solvents (eg, ethanol) (Higgins, 1996; Savastano et al., 2003; Aziz and Sarkanen, 1989). Strand fibers that undergo low-temperature chemithermomechanical pulping result in modification of the chemical composition and reduction in the lignin and hemicellulose contents (Savastano et al., 2003).

Pulps have been evaluated as processing fibers in the Hastchek method of fabrication. The fiber–cement industry has used mainly *Pinus* and *Eucalyptus* pulps in the Hatschek process. Pulped fibers are also preferred for composites production at a laboratory scale using the slurry vacuum de-watering technique, which is a crude simulation of the Hatschek process. During the de-watering stage, the pulp forms a net that retains cement grains. Advances have been proposed with new methods for flocculant selection (Blanco et al., 2010) and to evaluate the best refining degree (Coutts, 2005; Tonoli et al., 2007, 2009, 2013) required to improve fiber bonding, processing, and strength of the fiber–cement composites produced. Moreover, there are particular aspects of the bleached and unbleached pulps that have been discussed in the literature.

Table 4.2 **Average dimensions of wood pulp fibers in comparison with nonwood (macrofibers) pulp fibers (Hurter, 1991; Atchison, 1993)**

Source of fibers	Average length (mm)	Average width (μm)
Leaf		
Abaca	6.0	20
Sisal	3.0	17
Stem		
Sugar cane bagasse	1.7	20
Bamboo	1.4–4.0	8–30
Cereal straw	1.5	13
Rice straw	1.4	8
Corn	1.3	16
Cotton	0.9	19
Sorghum	1.6	47
Grass		
Esparto	1.1	9
Sabai	2.1	9
Hemp	20	22
Woods		
Softwood	3.0	30
Hardwood	1.3	25

Bleached (ie, low-lignin) pulp exhibited accelerated progression of fiber mineralization in a cement matrix compared to unbleached fibers (Mohr et al., 2005).

The cellulosic pulp can also be considered as microreinforcement in cement-based matrices (Mármol et al., 2013; Almeida et al., 2013). An interesting waste product is the paper packaging used for binders (eg, ordinary Portland cement). It is obtained from civil construction waste, as recycled Portland cement kraft bags. The recycling of this paper-based residue is not trivial as it is for other types of paper. A consequence is the residual content of cement (or powdered binder, in general) in the bags which can be highly abrasive and, consequently, undesirable for the cycling processing. In many countries (eg, Brazil) the amount of Portland cement commercialized in kraft bags is significant, and the destination of the used bags is considered a key issue for the cement producers in these countries. Therefore, the approach of recycling Portland cement packaging is of interest as the reinforcement in fiber−cement and requires simple techniques for processing and consequent application for the manufacturing of these composites (Mármol et al., 2013).

4.2.3   Organosolv pulp production and properties

Pulp is the most important raw material for processing of cellulose. It is industrially used for paper production and is regarded as a promising material for use as reinforcement of polymeric, ceramic, and cement materials (Alemdar and Sain, 2008; Sundar et al., 2010). Organosolv is an unconventional pulping method, and an alternative process to that used in industry, such kraft and sulfite pulping. In the organosolv process, organic solvents are usually associated with water in a volume ratio from 10 to 50%. Most organosolv processes are carried out at high temperatures (185−210°C), and there is no need for the addition of acid since organic acids released from the biomass act as catalysts for the rupture of the lignin−carbohydrate complex bonds (Duff and Murray, 1996).

The organosolv process has the advantage of using substances which are less harmful to human health and the environment compared to conventional processes (Ruiz et al., 2011). This process also has economic and environmental advantages, since the use of organic solvents (eg, ethanol) enables the operation of smaller and more compact plants, eliminates the need for the recovery of inorganic reagents, and eliminates sulfur emission. Additionally, the lignin removed can be recovered in the recycling operations of the organic solvent, which involves stages of distillation and precipitation (Sridach, 2010; Ruiz et al., 2011; Li et al., 2012).

González et al. (2008) compared the properties of non-woody pulp (palm oil fibers) produced using the organosolv and kraft methods. The organosolv process was more effective, yielding more pulp having higher cellulose content and lower lignin content than the kraft method. The zero-span index of the resulting pulp is 204 N m/g. The zero-span test was carried out according to Tappi T 273 cm-95 and T 231 cm-96 Standards. This value is high compared to commercial microfiber used to produce paper such as eucalyptus pulp, which has values between 90 and 160 N m/g (Foelkel, 2007). Silva et al. (2010) used eucalyptus pulp with a zero-span index of 90 N m/g as reinforcement of a polymeric matrix, and they showed an improvement of the mechanical performance of the rigid polyurethane foam reinforced with eucalyptus pulp. The zero-span index of bamboo pulp also indicates its potential for use as reinforcement in polymer matrices and for the production of paper (Correia et al., 2014b).

4.2.4   Nanofibrillated cellulose

Initial efforts to explore nanotechnology and/or nanoscience in the development of the construction industry have focused on understanding nanoscale phenomena and improving the performance of commercial materials and products. There is a focus on using nanoscience to increase the strength and durability of cementitious composites using nanofibers and nanoparticles. The nanofibers act as reinforcement in nanoscale dimensions. This approach makes possible the manufacture of more resistant cement products, preventing the formation and propagation of micro- or nanocracks (IRC, 2002; Konsta-Gdoutos et al., 2010a). Nano- and microscale fiber reinforcement systems have been found to be particularly effective in cement composites. Introduction of fibers and, more recently, nanomaterials offers effective means of enhancing toughness, for instance (Peyvandi et al., 2013). Development of nanocomposites based on nanocellulosic materials is a new but rapidly evolving research area. Cellulose is abundant in nature, biodegradable, and relatively inexpensive; it is a promising source of nanoscale reinforcement material for cement-based composites.

Cellulose nanocrystals (also reported in the literature as cellulose whiskers), nano-fibers, cellulose crystallites, or crystals, are the crystalline forms of cellulosic fibers, and they can be isolated using several methods (Nishiyama, 2009; Tonoli et al., 2013). Chemical methods use an aggressive acid (normally sulfuric acid) to break down fibril aggregates to cellulose rods (whiskers) which form a weak physical network by hydrogen bonds. In the mechanical methods, microfibrillated cellulose is prepared by mechanical disintegration (refining or sonication, for example) with application of high shear forces and cavitation, which leads to highly entangled and inherently connected fibrils and fibril aggregates and mechanically strong networks (Bufalino et al., 2014; Guimarães et al., 2015a). Enzymatic hydrolysis is another method that yields a mixture of dominantly cellulose I fibrils (about 5 nm thickness) and fibril aggregates (about 10 to 20 nm thickness) (Pääkkö et al., 2007). Various definitions have been given to the fibrillated materials, eg, nanofibrillated cellulose, nanofibers, nanofibrils, microfibrils, and nanocellulose (Chinga-Carrasco, 2011). Fig. 4.3 depicts the morphology of cellulose nanocrystals (whiskers obtained by acid hydrolysis) and nanofibrils (obtained by sonication) from eucalyptus pulp as presented in Tonoli et al. (2013).

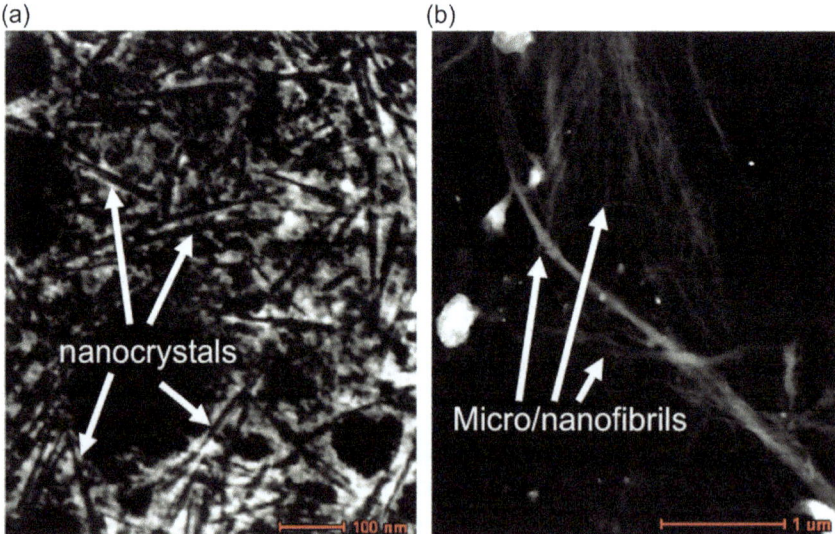

Figure 4.3 Transmission electron microscopy images of nanofibers obtained from eucalyptus pulp: (a) cellulose nanocrystals (whiskers obtained after 60-min acid hydrolysis) and (b) cellulose micro/nanofibrils pulled out after sonication. (Note that the magnification of (b) is 50x greater than that of (a).)

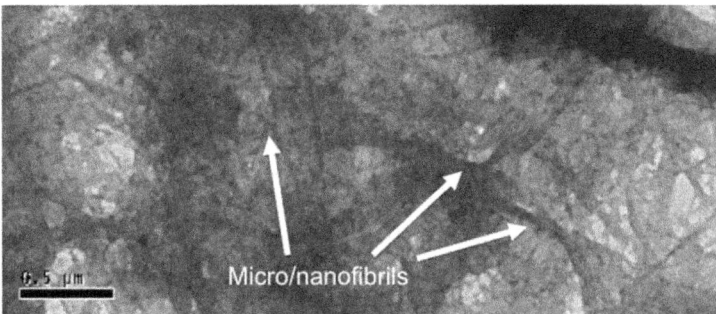

Figure 4.4 Scanning electron microscopy micrographs of micro/nanofibrils network obtained by mechanical defibrillation of jute fibers.

Although wood cellulose fibers have constituted the main source of microfibrillated cellulose production, the utilization of nonwood (macrofibers and culms/canes) pulp fibers, agricultural crops, and vegetable by-products have also been explored as natural sources (Guimarães et al., 2015a,b). Fig. 4.4 shows micro/nanofibrils obtained by mechanical defibrillation of jute fibers.

Hoyos et al. (2013) evaluated the influence of interactions between cellulose microcrystalline particles (MCC) and cement hydration products on rheology, hydration kinetics, microstructure, and mechanical properties of cement-based products. Some results showed that interactions between MCC, cement particles, hydration products, and water decreased the workability and delayed the hydration reaction. However, the results from thermogravimetric analyses showed that both accelerated curing and MCC addition (0 and 3 wt%) increased the hydration degree of cement materials due to increases in temperature during the curing process and the MCC releasing its water content, respectively, both contributing to the hydration process. Thus, the potential of nanofibrils as reinforcement in brittle materials must be further explored to seek new processing techniques, modeling techniques, and designs utilizing the high mechanical performance and durability.

4.2.5   Nanofibrillated cellulose obtained by the grinding method

The production of nanofibrillated cellulose to obtain nanofibers requires intensive mechanical fibrillation treatment, in which most of the amorphous phase is maintained, since there are no chemicals used. This mechanical treatment causes irreversible changes in the fibers, increasing their bonding potential by modifying the morphology and size of the fibers (Kamel, 2007; Gardner et al., 2008). Nanofibrillated cellulose is already commercialized and available internationally (Siró and Plackett, 2010). The typical production of commercial nanofibrillated cellulose uses a mechanical method consisting of refining and high-pressure homogenization steps.

The *grinding* process has the advantage of not requiring chemical pretreatment of the fibers before the fibrillation, which is necessary in other processes (eg, acid hydrolysis) to obtain nanofibers (Nakagaito and Yano, 2004; Pääkko et al., 2007;

Stenstad et al., 2008). *Grinding* is the process of fibrillation using an ultrafine *grinder* having one disc which rotates against another stationary disc with an adjustable gap between the discs (Abe et al., 2007).

Fibrillation by the *grinding* method occurs in the cell wall structure, which is formed by multilayer structured nanofibers and hydrogen bonds that are broken by application of shear force generated by the discs of the *grinder* over repeated cycles, producing nanoscale fibers from the pulp (Iwamoto et al., 2007; Guimarães et al., 2015a). The main advantage of the *grinding* is that mechanical pretreatment to fiber shortening, required in the other nanofibrillation techniques, is not required, and unlike the homogenizer process, the *grinding* process requires fewer cycles to obtain nanofibrillated cellulose (Spence et al., 2011; Lavoine et al., 2012).

4.2.6 Properties of nanofibrillated cellulose

The production of nanofibrillated cellulose and its application in composite materials as reinforcement has gained increased attention due to its biodegradability and renewability, high strength, and stiffness combined with nanoscale nature, resulting in a very high surface area, high aspect ratio, and low weight (Turbak et al., 1983; Nakagaito and Yano, 2004; Siró and Plackett, 2010; Hassan et al., 2012).

Nanofibrillated cellulose are bundles of elementary fibrils separated by less ordered regions. Depending on the plant species, the diameter of the fibers ranges 10–100 nm, and the length is in the micrometer range. Depending on the raw material and fibrillation technique, the degree of polymerization, morphology, and aspect ratio of the nanofibers can vary (Svagan et al., 2008; Alila et al., 2013).

The morphology of nanofibrillated cellulose (diameter and length) changes with mechanical treatment and pretreatment. Microscopy techniques, such as scanning electron microscopy, transmission electron microscopy, and atomic force microscopy, are used to observe the morphology changes and make the measurement of the fiber sizes.

Fig. 4.5(a) and (b) compares the nanofibrillated surface of bleached bamboo organosolv pulp and fibers after 15 cycles of nanofibrillation by the *grinding* method, respectively. The images show the difference in the surface of the fibers. The average width of 50 fibers was measured using image analysis techniques and was found to be 7.4 μm and 32.9 nm for pulp and nanofibrillated cellulose, respectively (Correia et al., 2014c). These results are in agreement with those announced by Iwamoto et al. (2007), who reported nanofibrillated cellulose having 20–50 nm diameter and more than 1 μm length following 15 nanofibrillation cycles through the *grinder*.

Nanofibrillated cellulose is composed of crystalline and amorphous regions, and measurement of the degree of crystallinity explains the behavior and the properties of the material. In general, an increase in crystallinity brings about an increase in tensile strength and stiffness and a decrease in chemical reactivity (Chen et al., 2012; Yuan et al., 2013). Another parameter that has been affected by crystallinity is swelling: an increase in the crystallinity results in a decrease in swelling of fiber (Wan et al., 2011).

Iwamoto et al. (2007) determined the degree of crystallinity of nanofibrillated cellulose obtained following different numbers of cycles through a *grinder*, and

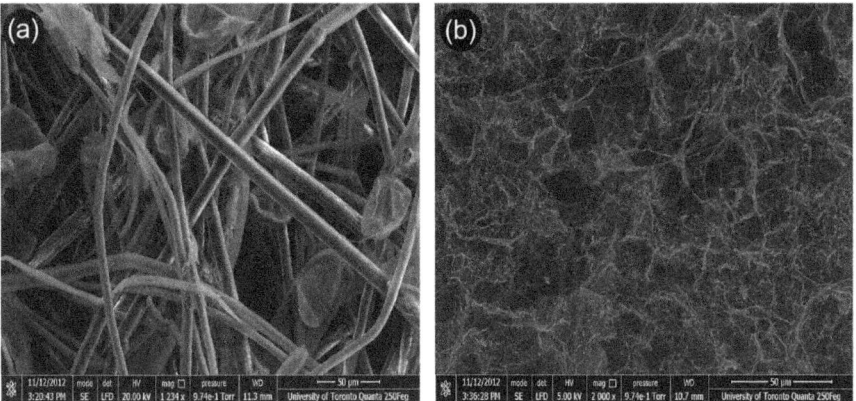

Figure 4.5 Bleached bamboo organosolv pulp (a) and fibers after 15 times of nanofibrillation by the *grinding* method (b).
Reproduced from Correia, V.C., Sain, M., Leão, A. L., Santos, S.F., Savastano Jr., H., Nanofibrillated cellulose by *grinding* method from bamboo organosolv pul as nanoreinforcement in composites.

according to their results, the degree of crystallinity decreases with an increase in the number of cycles up to about 30 cycles. Iwamoto et al. (2005) report that 10 nanofibrillation cycles are enough to produce nanofibrillated cellulose having uniform width of 50–100 nm, and no significant changes occur in the fibers beyond 10 cycles.

Another property of nanofibrillated cellulose is the reinforcement capacity of the nanofibers, considering their high contact area and aspect ratio. Mechanical tensile test of the nanofibrillated film shows that its strength is related to the high interfibril adhesion properties, great fibril strength, and fewer defects resulting from homogeneous distribution (Henriksson et al., 2008).

4.3 Characterization of vegetable fiber composites

4.3.1 Composites reinforced with bamboo cellulose fiber

Bamboo is a potential vegetable fiber for use as a raw material in cellulosic pulp production, with a pulping yield around 50%, similar physical and mechanical characteristics compared to the other fibers used as reinforcement. Bamboo fibers have been used as reinforcement for organic and inorganic matrices, such as polymers and cement composites. The growing interest in these fibers is due to their characteristics as a fast-growing (rapid growth species can reach up to 25 cm/day) (Escamilla and Habert, 2014) renewable resource, worldwide abundance, relatively low density, and good mechanical properties. Mechanical and physical properties of bamboo fibers are similar to those of synthetic fibers, such as E-glass fibers, used in industry for reinforcement in composite materials (see Table 4.3). Bamboo fibers are often called "natural glass fibers" (Zakikhani et al., 2014).

Table 4.3 **Physical and mechanical properties (by tensile test) of bamboo and E-glass fibers**

Fiber	Density (g/cm³)	Diameter (μm)	Elongation at failure (%)	Tensile strength (MPa)	Modulus of elasticity (GPa)
Bamboo	0.88−1.1	100−200	−	391−713	18−55
E-Glass	2.5	9−15	2.5	2500	70

Adapted from Osorio, L., Trujillo, E., Van Vuure, A.W., Verpoest, I., 2011. Morphological aspects and mechanical properties of single bamboo fibers and flexural characterization of bamboo/epoxy composites. Journal of Reinforced Plastics and Composites 30, 396−408.

Polymeric matrices used for bamboo-based composites include polyethylene, PP, hybrid composites such as PP and polylactic acid (Han et al., 2008; Ying-Chen et al., 2010; Xu et al., 2011), and biodegradable composites (Siro and Plackett, 2010; Satyanarayana et al., 2009). Bamboo fibers are also used as reinforcement of construction materials such cement and concrete (Coutts and Ni, 1995; Lima et al., 2008; Correia et al., 2014a).

Mandal et al. (2010) produced hybrid polymeric composites, replacing up to 25% of glass fibers with bamboo fibers and no reduction in mechanical properties of the composites. Thwe and Liao (2002) replaced glass fiber with bamboo fiber in a polymeric matrix and reported an increase in tensile modulus and strength. According to Abdul Khalil et al. (2012), there is ample opportunity for bamboo fibers to replace or reduce the utilization of synthetic fibers, such as glass, in polymeric composites, and to do so commercially on an industrial scale.

Bamboo fibers have been used in inorganic matrices as reinforcement of concrete and fiber−cement. Coutts and Ni (1995) and Correia et al. (2014a) used bamboo pulp as reinforcement of fiber−cement. The mechanical behavior of the composites reinforced with bamboo pulp was compatible with that of the commercial fiber−cement reinforced with *Pinus* pulp (Coutts and Ni, 1995). The strength of the fiber−cement reinforced with 8% of bamboo pulp increased after accelerated aging tests and showed promising behavior concerning the durability of the fiber−cement (Correia et al., 2014a).

4.3.2 Pulp-reinforced cement-based composites

The use of cellulosic pulp has been present in construction building components since the 1980s. Cellulosic pulps, which are fibers in the microscale, contribute to improve the fiber-matrix bonding in cement materials, and therefore, they have a greater efficiency for use as reinforcement compared to macrofibers. The use of microfibers in cement-based materials also results in better packing in the matrix; it is possible to use higher fiber contents (of around 10% by mass), resulting in improvements in the mechanical properties (Coutts, 1986; Savastano et al., 2003). Another advantage of pulped cellulose fibers is that they require less energy in

Table 4.4 **Mechanical properties of the cement-based composites reinforced with 6%, 8%, 10%, and 12% bamboo organosolv pulp**

Bamboo pulp content (%)	Modulus of rupture (MPa)	Limit of proportionality (MPa)	Modulus of elasticity (GPa)	Specific energy (kJ/m^2)
6	6.4 ± 0.9	3.8 ± 1.5	10.6 ± 0.5	0.8 ± 0.1
8	7.5 ± 0.1	5.4 ± 0.6	9.6 ± 0.8	1.2 ± 0.1
10	6.8 ± 1.4	3.2 ± 1.2	6.2 ± 1.3	1.9 ± 0.3
12	5.8 ± 1.5	2.3 ± 1.0	4.5 ± 0.6	2.1 ± 1.0

Reproduced from Correia, V.C., Santos, S.F., Mármol, G., Curvelo, A.A.S., Savastano Jr., H., 2014a. Potential of bamboo organosolv pulp as reinforcement element in fiber-cement. Construction and Building Materials 72, 65−71.

their preparation than most common synthetic reinforcing fibers, such as PVA and PP, and can be considered "green" products (Silva et al., 2010).

Correia et al. (2014a) have produced cement composites reinforced with bamboo organosolv pulp. The authors tested four different levels of pulp (Table 4.4 and Fig. 4.6), intending to evaluate the content of bamboo pulp to produce a material

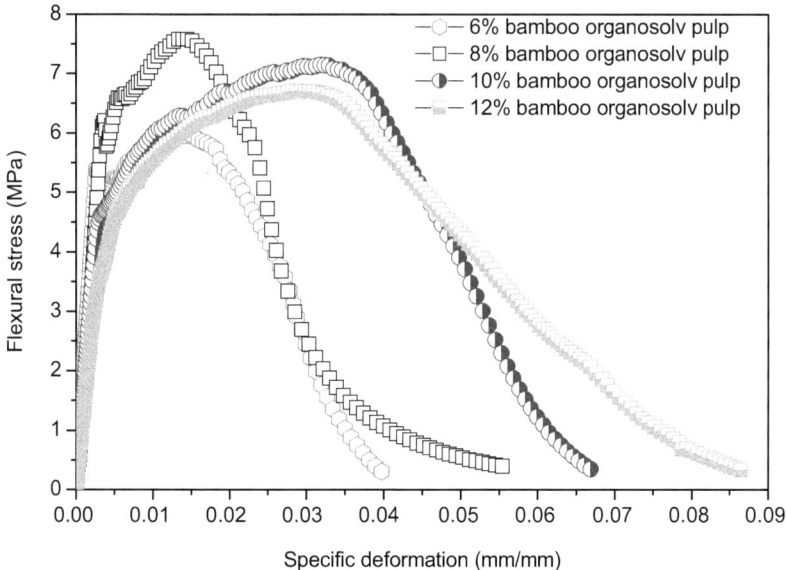

Figure 4.6 Typical stress−strain curves of the composites with 6%, 8%, 10%, and 12% of bamboo organosolv pulp at under flexure tests.
Reproduced from Correia, V.C., Santos, S.F., Mármol, G., Curvelo, A.A.S., Savastano Jr., H., 2014a. Potential of bamboo organosolv pulp as reinforcement element in fiber-cement. Construction and Building Materials 72, 65−71.

with the improved mechanical properties. The flexural test was performed with a four-point bending configuration to evaluate the modulus of rupture (MOR), limit of proportionality (LOP), modulus of elasticity (MOE), and specific energy (SE) of the composites.

Bamboo pulp content of 8% was sufficient to assure the reinforcement before and after initial crack propagation, as indicated by MOR and LOP. The mechanical performance of the composites reinforced with 10% and 12% pulp was lower due the porosity added by the pulp, and the toughness in postcracking conditions was higher.

Coutts and Ni (1995) tested the levels of bamboo Kraft pulp ranging from 2% to 14%, and the maximum MOR was at 10% of pulp. Khorami and Ganjian (2013) have produced fiber—cement composites reinforced with waste kraft pulp to test the optimum content of reinforcement. The authors tested the levels from 1% to 14% of pulp, and they concluded that the optimum content was 8%, for which the MOR value was higher compared to the other levels. Pulp content higher than 8% led to the reduction of MOR.

4.3.3   Engineered composites with hybrid reinforcement

Embedded nanofibers in a cement matrix provide a partial solution to cracking since they act as bridges to transfer stress in nanocracks. Many studies have been reported on the use of carbon nanofibers and nanotubes as reinforcement of cementitious materials at the nanoscale (Makar et al., 2005; Li et al., 2005; Yakovlev et al., 2006; Konsta-Gdoutos et al., 2010b; Galao et al., 2014).

In hybrid composites, two or more different types of fibers are combined to produce a composite that benefits from each of the individual fibers exhibiting a synergistic response. In hybrid composites with fibers of different sizes, these are combined to achieve a dense packing and dimensional stability (Banthia and Nandakumar, 2003). The combination of micro- and nanofibers potentially serves as multiple reinforcement in the scales corresponding to the fiber sizes.

Additionally, some works reported the use of cellulose nanofibers as reinforcement of cement materials. Ardunay et al. (2012) used 3.3% (by weight) sisal nanofibrillated cellulose as mortar reinforcement, compared to mortar reinforced with some level of sisal pulp at 28 days. The nanofibrillated cellulose increased the flexural strength (26.4%) and the flexural modulus (41.5%).

Fig. 4.7 shows the typical stress—strain curves of the composites subjected to the flexural test, comparing the composites only with bamboo pulp (8% by weight) and hybrid composites (5% bamboo pulp + 3% nanofibrillated cellulose). These composites were produced by the slurry vacuum de-watering and pressing method (using pressure of 3.2 MPa during 5 min) and subjected to thermal curing at 45°C for 8 days.

The curves show that 3% by weight of nanofibrillated cellulose did not contribute to the improvement of the mechanical performance of the composites. The composite with 8% of pulp showed the highest ability to absorb energy during the fracture process. The water/cement (w/c) ratios corresponding to composites reinforced only with pulp and to hybrid composites were 0.46 and 0.54, respectively. The higher w/c ratio of the hybrid composites is attributed to the high specific surface area of the

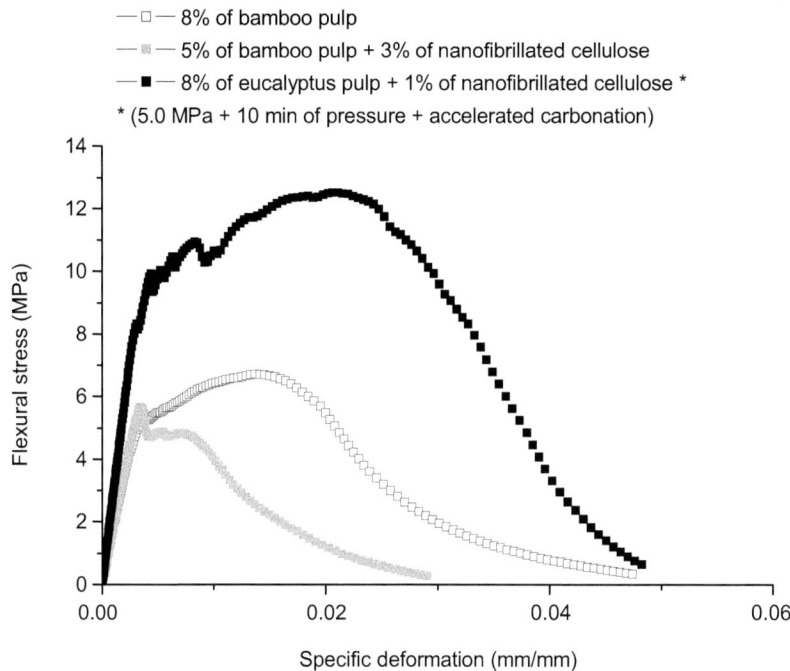

Figure 4.7 Typical stress–strain curves of the composites reinforced with 8% of bamboo organosolv pulp compared to the hybrid composites with bamboo pulp + nanofibrillated cellulose and eucalyptus pulp + nanofibrillated cellulose.

nanofibrillated cellulose, which absorbs more water and consequently has contributed to increase the porosity and to decrease the mechanical properties of the composites with nanofibrillated cellulose.

Fig. 4.7 also compares the hybrid composites with 8% of eucalyptus pulp + 1% of nanofibrillated cellulose, produced by modification of the slurry vacuum de-watering and pressing method. The pressure of 3.2 MPa during 5.0 min was increased to 5.0 MPa during 10 min, and such composites were subjected to supercritical carbonation cure according to Santos et al. (2015).

The curves show that the increase of pressure has promoted the mechanical performance of the composite with 8% eucalyptus pulp + 1% of nanofibrillated cellulose. The MOR and the SE of composites with 8% eucalyptus pulp + 1% of nanofibrillated cellulose was, respectively, 57.2% and 78.3% higher than the composite with 5% bamboo pulp + 3% of nannofibrillated cellulose. From the modification of the production method, the matrix was densified, which can be observed in the higher MOE.

The cure by accelerated carbonation also contributed significantly to improvements in mechanical performance of composites with eucalyptus pulp. This behavior is attributed to the reactions that occur during carbonation, from which occurs the precipitation of calcium carbonate ($CaCO_3$) to the pores of the matrix and favors its densification.

The reactions that occur during accelerated carbonation (filling the pores with $CaCO_3$) result in reduction of porosity and water absorption, and they make the cement matrix denser, which improves the mechanical properties of the resulting composite (Akers and Studinka, 1989; Macvicar et al., 1999; Lesti et al., 2013).

The aforementioned results indicate that the type of material and its production method and the cure procedure affect the nanoreinforcement in the cement-based materials.

4.3.4 Durability

Carbonation is the reaction of cement hydration products with carbon dioxide (CO_2). Accelerated carbonation of cement materials can be used to improve the durability of cellulose fiber−cement composites because it reduces the alkalinity of the cement matrix, making it less aggressive to the cellulose fibers (Macvicar et al., 1999; Savastano et al., 2003; Toledo Filho et al., 2005). During carbonation, CO_2 is diffused through unsaturated pores of the cementitious matrix, and it is dissolved in the aqueous phase in the pores and transformed into carbonic acid (H_2CO_3). This is dissociated into HCO^{-3} and CO^{-3} ions, along with the dissolution of $Ca(OH)_2$, that releases Ca^{2+} and OH^- ions, which precipitate and form calcium carbonate ($CaCO_3$) (Fernández et al., 2004; Peter et al., 2008; Almeida et al., 2013; Santos et al., 2015).

The durability of the cement-based composites reinforced with vegetable fibers can be assessed by the alternating of soak and dry cycles. Accelerated aging testing has the advantage of providing results in the shortest time interval and simulates the natural aging of the material.

Fig. 4.8 shows the typical curves of the hybrid composites reinforced with 8% of eucalyptus pulp + 1% of nanofibrillated cellulose, and composites reinforced only with eucalyptus pulp (9%). The curves present the effect of the accelerated carbonation in the mechanical performance of both composites. There is an increase of SE and MOR in the composites after accelerated carbonation, as a consequence of the matrix densification.

Fig. 4.9 present the typical curves of the composites reinforced with 8% of bamboo organosolv pulp before and after 200 wet and dry cycles. The results show a decrease of the SE after accelerated aging, which indicates their embrittlement; however, the MOR increased. This performance is due to the higher fiber-matrix bonding caused by the matrix densification, which occurs on account of filling of the pores by the calcium carbonate produced during the carbonation (Bentur and Akers, 1989). The increase of the MOR of the composites after accelerated aging indicates that despite the embrittlement of the composite, the fibers and the composite have not impaired their durability.

4.4 Final remarks and perspective for future research needs

Nonconventional building materials and their application have been extensively investigated as an alternative for cost-effective housing in developing countries.

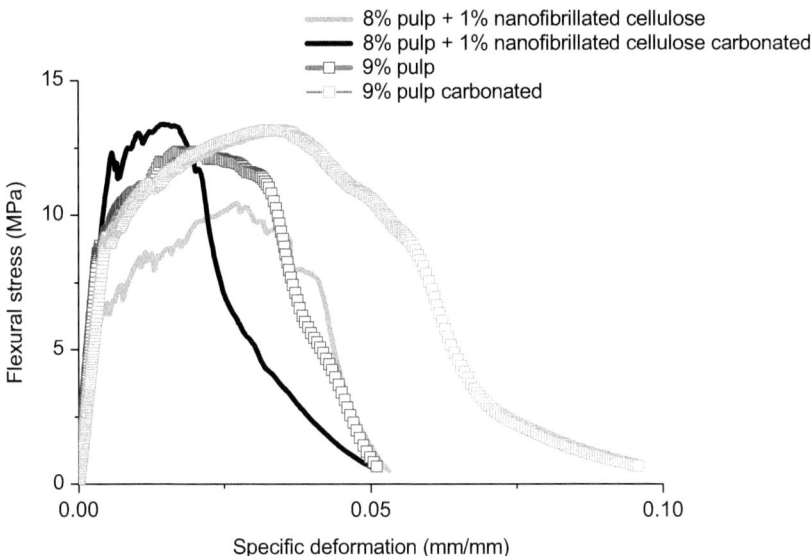

Figure 4.8 Typical stress—strain curves under flexure tests of the composites reinforced with 8% pulp + 1% nanofibrillated cellulose, and 9% pulp in carbonated and noncarbonated conditions.

Reproduced from Correia, V.C., Santos, S.F., Savastano Jr., H., 2015. Effect of the accelerated carbonation in fibercement composites reinforced with eucalyptus pulp and nanofibrillated cellulose. World Academy of Science, Engineering and Technology: International Journal of Civil, Architectural, Structural and Construction Engineering 9, 7—10.

The present work contributes to the widespread use of the vegetable fibers as a source of raw material that may be used to reinforce composites and give them new properties for applications as insulating panels, roofing tiles, siding, etc. The mechanical and physical characteristics of the different types of vegetable fibers available must be determined to better explore their potential as reinforcement. Similarly, their morphological properties (length, diameter, and fibrillation) and chemistry of their surface lead to important changes in the fiber to cement bond in vegetable fiber—reinforced cement materials. Fibrillation and hydrophilicity of the fiber surface improves the fiber to cement adherence. The use of regional agro-industrial residues and local vegetable fibers is a constructive way to minimize transport energy and simplify logistics in the global civil construction market. The possibility of using carbon dioxide for accelerating the hardening and stabilization of products made from Portland cement is also attractive. Accelerated carbonation curing has been studied to mitigate the cellulose fiber degradation in the cementitious composites and for the maintenance of their mechanical performance under weathering. Initial efforts to explore nanotechnology and/or nanoscience in the development of the construction industry have focused on understanding the nanoscale phenomena and improving the performance of commercial materials and products. The use of nanoreinforcements and functional nanostructures in

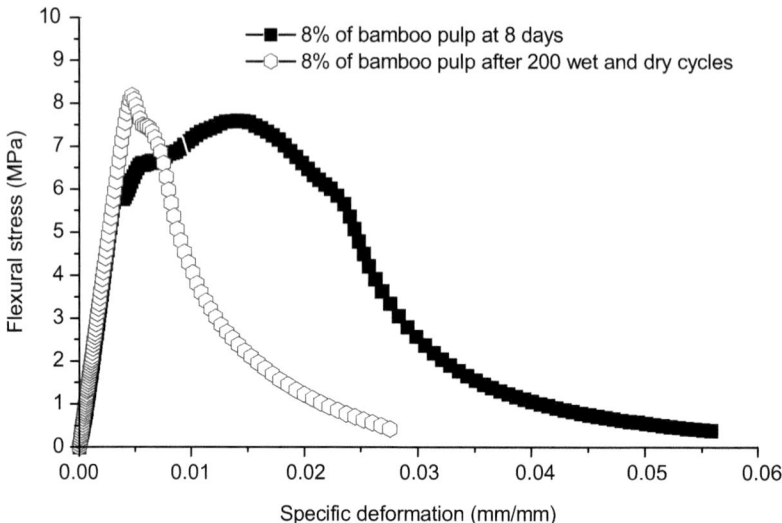

Figure 4.9 Typical stress—strain curves during flexural tests of the composites reinforced with 8% bamboo organosolv pulp at 8 days and after 200 wet and dry cycles.
Adapted from Correia, V.C., Santos, S.F., Mármol, G., Curvelo, A.A.S., Savastano Jr., H., 2014a. Potential of bamboo organosolv pulp as reinforcement element in fiber-cement. Construction and Building Materials 72, 65—71.

cement-based composites has been scarcely explored in the literature, and therefore further investigation is merited. Further development and optimization of the mechanical defibrillation process could lower the production costs of micro/nanofibers for engineering of fiber—cement materials. The development of standard pull-out test procedures and standardization of equipment suitable for testing particular vegetable fibers should be considered and better explored. Additionally, new methods for producing the cementitious composites may also be considered having the goals of low energy consumption, varied geometry of products, lower initial investment requirements for a new plant, the use of simpler machines for continuous production, the possibility of partial alignment of fibers, and new concepts and technologies to improve durability.

Acknowledgments

The authors acknowledge financial support from the São Paulo Research Foundation (FAPESP, Grants n°: 2009/10614-0, 2009/17293-5; 2010/16524-0; 2011/01128-5), the Brazilian National Council for Scientific and Technological Development (CNPq, Grants n°: 472133/2009-8, 305792/2009-1, 303061/2009-0 and 484082/2013-2), and the Minas Gerais Research Foundation (FAPEMIG). We would also like to extend special thanks to the Brazilian companies Infibra Ltda and Imbralit Ltda.

References

Abdul Khalil, H.P.S., Bhat, I.U.H., Jawaid, M., Zaidon, A., Hermawan, D., Hadi, Y.S., 2012. Bamboo fibre reinforced biocomposites: a review. Materials and Design 42, 353−368.

Abe, K., Iwamoto, S., Yano, H., 2007. Obtaining cellulose nanofibers with a uniform width of 15 nm from wood. Biomacromolecules 8 (10), 3276−3278.

Agopyan, V., 1988. Vegetable fibre reinforced building materials developments in Brazil and other Latin American countries. In: Swamy, R.N. (Ed.), Natural Fibre Reinforced Cement and Concrete. Concrete Technology and Design 5. Blackie Academic & Professional, Glasgow, pp. 208−242.

Agopyan, V., Cincotto, M.A., Derolle, A., 1989. Durability of vegetable fibre reinforced materials. In: International Council for Research and Innovation in Building and Construction, 11, 1989. Paris. Proceedings… Paris, pp. 353−361.

Agopyan, V., John, V.M., 1989. Building panels made with natural fibre reinforced alternative cements. In: Swamy, R.N., Barr, B. (Eds.), Fibre Reinforced Cements and Concretes: Recent Developments. Elsevier, London, pp. 296−305.

Agopyan, V., Savastano Jr., H., John, V.M., Cincotto, M.A., 2005. Developments on vegetable fibre−cement based materials in São Paulo, Brazil: an overview. Cement & Concrete Composite 27, 527−536.

Akers, S.A.S., Studinka, J.B., 1989. Ageing behavior of cellulose fiber cement composites in natural weathering and accelerated tests. International Journal of Cement Composites and Lightweight Concrete 11 (2), 93−97.

Alemdar, A., Sain, M., 2008. Isolation and characterization of nanofibers from agricultural residues − wheat straw and soy hulls. Bioresource Technology 99, 1664−1671.

Alila, S., Besbes, I., Vilar, M.R., Mutjé, P., Boufi, S., 2013. Non-woody plants as raw materials for production of microfibrillated cellulose (MFC): a comparative study. Industrial Crops and Products 41, 250−259.

Almeida, A.E.F.S., Tonoli, G.H.D., Santos, S.F., Savastano Jr., H., 2013. Improved durability of vegetable fibre reinforced cement composite subject to accelerated carbonation at early age. Cement & Concrete Composites 42, 49−58.

Ardunay, M., Claramunt, J., Arévalo, R., Parés, F., 2012. Nanofibrillated cellulose (NFC) as a potential reinforcement for high performance cement mortar composites. BioResources 73 (3), 3883−3894.

Atchison, J.E., 1993. Data on non-wood plant fibers. In: Hamilton, F., Leopold, B. (Eds.), Secondary Fibers and Non-wood Pulping Pulp and Paper Manufacture, third ed., Series Pulp and paper manufacture. The Joint Textbook Committee of the Paper Industry, TAPPI, Atlanta, pp. 4−16.

Aziz, S., Sarkanen, K., 1989. Organosolv pulping − a review. Tappi Journal 72, 169−175.

Banthia, N., Nandakumar, N., 2003. Crack growth resistance of hybrid fiber reinforced cemet composites. Cement & Concrete Composites 25, 3−9.

Bentur, A., Akers, S.A.S., 1989. The microstructure and ageing of cellulose fibre reinforced cement composites cured in a normal environment. The International Journal of Cement Composites and Lightweight Concrete 11, 99−109.

Blanco, A., Fuente, E., Alonso, A., Negro, C., 2010. Optimal use of flocculants on the manufacture of fibre-cement materials by the Hatschek process. Construction and Building Materials 24, 158−164.

Bufalino, L., Mendes, L.M., Tonoli, G.H.D., Rodrigues, A., Fonseca, A.S., Cunha, P.I., Marconcini, J.M., 2014. New products made with lignocellulosic nanofibers from Brazilian amazon forest. IOP Conference Series: Materials Science and Engineering 64, 012012.

Chen, Y., Wan, J., Zhang, X., Ma, Y., Wang, Y., 2012. Effect of beating on recycleb properties of unbleached eucalyptus cellulose fiber. Carbohydrate Polymers 87, 730−736.

Chinga-Carrasco, G., 2011. Cellulose fibres, nanofibrils and microfibrils: the morphological sequence of MFC components from a plant physiology and fibre technology point of view. Nanoscale Research Letters 6, 417−424.

Claramunt, J., Ardanuy, M., Garcia-Hortal, J.A., Toledo, R.D., 2011. The hornification of vegetable fibers to improve the durability of cement mortar composites. Cement & Concrete Composites 33, 586−595.

Claramunt, J., Ardanuy, M., García-Hortal, J.A., 2010. Effect of drying and rewetting cycles on the structure and physicochemical characteristics of softwood fibres for reinforcement of cementitious composites. Carbohydrate Polymers 79, 200−205.

Correia, V.C., Santos, S.F., Savastano Jr., H., 2015. Effect of the accelerated carbonation in fibercement composites reinforced with eucalyptus pulp and nanofibrillated cellulose. World Academy of Science, Engineering and Technology: International Journal of Civil, Architectural, Structural and Construction Engineering 9, 7−10.

Correia, V.C., Santos, S.F., Mármol, G., Curvelo, A.A.S., Savastano Jr., H., 2014a. Potential of bamboo organosolv pulp as reinforcement element in fiber−cement. Construction and Building Materials 72, 65−71.

Correia, V.C., Siqueira, F.M., Dias, R.D., Savastano Jr., 2014b. Macro, micro and nano scale bamboo fiber as a potential reinforcement of composites. In: International Conference on Non-Conventional Materials and Technologies, 15, 2014, Pirassununga. Brazil.

Correia, V.C., Sain, M., Leão, A.L., Santos, S.F., Savastano Jr., H., 2014c. Nanofibrillated cellulose by *grinding* method from bamboo organosolv pulp as nanoreinforcement in composites. In: Congresso Luso-Brasileiro de Materiais de Construção Sustentáveis, 1, 2014, Guimarães. Materiais de Construção Sustentáveis. University of Minho, Minho, p. 816 p.

Coutts, R.S.P., 1986. Sticks and stones…!! Forest Products Newsletter. CSIRO Division of Chemical Wood Technology 2, 1−4.

Coutts, R.S.P., 1992. From forest to factory to fabrication. In: Swamy, R.N. (Ed.), Proceedings of the 4th International Symposium Fibre Reinforced Cement and Concrete. E&FN Spon, London, pp. 31−47.

Coutts, R.S.P., Ni, Y., 1995. Autoclaved bamboo pulp fibre reinforced cement. Cement & Concrete Composites 17, 99−106.

Coutts, R.S.P., 2005. A review of Australian research into natural fibre cement composites. Cement & Concrete Composites 27, 518−526.

D'Almeida, A., Toledo Filho, R., Melo Filho, J., 2010. Cement composites reinforced by short curauá fibers. Revista Matéria 15, 151−156.

Dias, C.M.R., Savastano Jr., H., John, V.M., 2010. Exploring the potential of functionally graded materials concept for the development of fiber cement. Construction and Building Materials 24, 140−146.

Duff, S.J.B., Murray, W.D., 1996. Bioconversion of forest products industry waste cellulosic to fuel ethanol: a review. Bioresource Technology 55, 1−33.

Escamilla, E.Z., Habert, G., 2014. Environmental impacts of bamboo-based construction materials representing global production diversity. Journal of Cleaner Production 69, 117−127.

Fernández Bertos, M., Muntean, A., Simons, S.J.R., Hills, C.D., Carey, P.J., 2004. A review of accelerated carbonation technology in the treatment of cement-based materials and sequestration of CO_2. Journal of Hazardous Materials 112, 193–205.

Foelkel, C., 2007. The Eucapyptus fibers and the kraft pulp quality requirements for paper manufacturing. Eucalyptus Online Book & Newsletter. Brazilian Pulp and Paper Technical 42.

Fonseca, A.S., Mori, F.A., Tonoli, G.H.D., Savastano Jr., H., Ferrari, D.L., Miranda, I.P.A., 2013. Properties of an Amazonian vegetable fiber as a potential reinforcing material. Industrial Crops and Products 47, 43–50.

Fordos, Z., 1988. In: Swamy, R.N. (Ed.), Natural Fibre Reinforced Cement and Concrete (Concrete Technology and Design 5). Blackie, Glasgow, UK, pp. 173–207.

Galao, O., Baeza, F.J., Zornoza, E., Garcés, P., 2014. Strain and damage sensing properties on multifunctional cement composites composites with CNF admixture. Cement & Concrete Composites 46, 90–98.

Garcia-Santos, A., 2004. Constructive applications of composite gypsum reinforced with Typha Latifolia fibres. Materiales de Construcción 54, 1–7.

Gardner, D.J., Oporto, G.S., Mills, R., Samir, M.A.S.A., 2008. Adhesion and surface issues in cellulose and nanocellulose. Journal of Adhesion Science and Technology 22, 545–567.

González, M., Tejado, A., Pena, C., Labidi, J., 2008. Organosolv pulping process simulations. Industrial & Engineering Chemistry Research 76, 1903–1909.

Gram, H.E., 1983. Durability of Natural Fibers in Concrete. Swedish Cement and Concrete Research Institute, Stockholm, 255 p.

Guimarães, S.S., 1984. In: Proceedings of the International Conference Development of Low-cost and Energy Saving Construction Materials and Applications; Rio de Janeiro: Envo, Brazil.

Guimarães Jr., M., Botaro, V.R., Novack, K.M., Flauzino Neto, W.P., Mendes, L.M., Tonoli, G.H.D., 2015a. Preparation of cellulose nanofibrils from bamboo pulp by mechanical defibrillation for their applications in biodegradable composites. Journal of Nanoscience and Nanotechnology 15.

Guimarães Jr., M., Botaro, V.R., Novack, K.M., Teixeira, F.G., Tonoli, G.H.D., 2015b. Starch/PVA-based nanocomposites reinforced with bamboo nanofibrils. Industrial Crops and Products 70, 72–83.

Han, G., Lei, Q., Wu, Y., Kojima, S., Suzuki, S., 2008. Bamboo-fiber filled high density polyethylene composites: effect of coupling treatment and nanoclay. Journal of Polymers and the Environment 16, 123–130.

Hassan, M.L., Mathew, A.P., Hassan, E.A., El-Wakil, K.O., 2012. Nanofibers from bagasse and rice straw: process optimization and properties. Wood Science and Technology 46, 193–205.

Heinricks, H., Berkenkamp, R., Lempfer, K., Ferchland, H.J., 2000. In: Moslemi, A.A. (Ed.), Proceedings of the 7th International Inorganic-Bonded Wood and Fiber Composite Materials Conference. Siempelkamp Handling Systems Report; University of Idaho, Moscow, 12 p.

Henriksson, M., Berglund, L.A., Isaksson, P., Lindstrom, T., Nishino, T., 2008. Cellulose nanopaper structures of high toughness. Biomacromolecules 9, 1579–1585.

Higgins, H.G., 1996. Paper Physics in Australia. CSIRO - Division of Forestry and Forest Products, Melbourne, Australia.

Hoyos, C.G., Cristia, E., Vázquez, A., 2013. Effect of cellulose microcrystalline particles on properties of cement based composites. Materials and Design 51, 810–818.

Hurter, A.M., 1991. Nonwood Plant Fiber Pulping. Progress Report No. 19. Tappi Press.

Institute of Research Construction (IRC), 2002. Innovations of construction. National Council of Research from Canada 7.

Iwamoto, S., Nakagaito, A.N., Yano, H., Nogi, M., 2005. Optically transparent composites reinforced with plant fiber-based nanofibers. Applied Physics A: Material Science & Processing 81, 1109–1112.

Iwamoto, S., Nakagaito, A.N., Yano, H., 2007. Nano-fibrillation of pulp fibers for the processing of transparent nanocomposites. Applied Physics A: Materials Science & Processing 89 (2), 461–466.

Jarabo, R., Monte, M.C., Blanco, A., Negro, C., Tijero, J., 2012. Characterisation of agricultural residues used as a source of fibres for fibre-cement production. Industrial Crops and Products 36, 14–21.

Jarabo, R., Monte, M.C., Fuente, E., Santos, S.F., Negro, C., 2013. Corn stalk from agricultural residue used as reinforcement fiber in fiber–cement production. Industrial Crops and Products 43, 832–839.

Joaquim, A.P., Tonoli, G.H.D., Santos, S.F., Savastano Jr., H., 2009. Sisal organosolv pulp as reinforcement for cement based composites. Materials Research 12, 305–314.

Kamel, S., 2007. Nanotechnology and its applications in lignocellulosic composites, a mini review. Express Polymer Letters 1 (9), 546–575.

Khorami, M., Ganjian, E., 2013. The effect of limestone powder, silica fume and fibre content on flexural behavior of cement composite reinforced by waste Kraft pulp. Construction and Building Materials 46, 142–149.

Konsta-Gdoutos, M.S., Metaxa, Z.S., Shah, S.P., 2010a. Highly dispersed carbon nanotube reinforced cement based materials. Cement and Concrete Research 40, 1052–1059.

Konsta-Gdoutos, M.S., Metaxa, Z.S., Shah, S.P., 2010b. Multi-scale mechanical and fracture characteristics and early-age strain capacity of high performance carbon nanotube/cement nanocomposites. Cement & Concrete Composites 32, 110–115.

Lavoine, N., Desloges, I., Dufresne, A., Bras, J., 2012. Microfibrillated cellulose – its barrier properties and applications in cellulosic materials: a review. Carbohydrate Polymers 90, 735–764.

Lesti, M., Tiemeyer, C., Plank, J., 2013. CO_2 stability of Portland cement based well cementing systems for use on carbon capture & storage (CCS) wells. Cement and Concrete Research 45, 45–54.

Li, G.Y., Wang, P.M., Zhao, X., 2005. Mechanical behavior and microstructure of cement composites incorporating surface-treated multi-walled carbon nanotubes. Carbon 43, 1239–1245.

Li, M.-F., Sun, S.-N., Xu, F., Sun, R.-C., 2012. Formic acid based organosolv pulping of bamboo (*Phyllostachys acuta*): comparative characterization of the dissolved lignins with milled wood lignin. Chemical Engineering Journal 179, 80–89.

Lima, H., Willrich, F., Barbosa, N., Rosa, M., Cunha, B., 2008. Durability analysis of bamboo as concrete reinforcement. Materials and Structures 41, 981–989.

Macvicar, R., Matuana, L.M., Balatinecz, J.J., 1999. Aging mechanisms in cellulose fiber reinforced cement composites. Cement & Concrete Composites 21, 189–196.

Makar, J.M., Margeson, J.C., Luh, J., 2005. Carbon nanotube/cement composite—early results and potential applications. In: Proceedings of the 3rd International Conference on Construction Materials: Performance, Innovation and Structural Implications. Vancouver, pp. 1–10.

Mandal, S., Alam, S., Varma, I.K., Maiti, S.N., 2010. Studies on bamboo/glass fibre reinforced USP and VE resin. Journal of Reinforced Plastic and Composites 29, 43–51.

Mármol, G., Santos, S.F., Savastano Jr., H., Borrachero, M.V., Monzó, J.M., Payá, J.J., 2013. Mechanical and physical performance of low alkalinity cementitious composites reinforced with recycled cellulosic fibres pulp from cement kraft bags. Industrial Crops and Products 49, 422—427.

Melo Filho, J.A., Silva, F.A., Toledo Filho, R.D., 2013. Degradation kinetics and aging mechanisms on sisal fiber cement composite systems. Cement & Concrete Composites 40, 30—39.

Millogo, Y., Morel, J.-C., Aubert, J.-E., Ghavami, K., 2014. Experimental analysis of Pressed Adobe Blocks reinforced with *Hibiscus cannabinus* fibers. Construction and Building Materials 52, 71—78.

Mohr, B.J., Nanko, H., Kurtis, K.E., 2005. Durability of kraft pulp fiber—cement composites to wet/dry cycling. Cement & Concrete Composite 27, 435—448.

Mohr, B.J., Biernacki, J.J., Kurtis, K.E., 2007. Supplementary cementitious materials for mitigating degradation of kraft pulp fiber—cement composites. Cement and Concrete Research 37, 1531—1543.

Nakagaito, A.N., Yano, H., 2004. The effect of morphological changes from pulp fiber towards nano-scale fibrillated cellulose on the mechanical properties of high-strength plant fiber based composites. Applied Physics A: Materials Science & Processing 78, 547—552.

Nexant, 2000. United state production of propylene and its major derivatives — energy and environmental profile of the U.S. Chemical Industry. Chapter 3, 22 p.

Nishiyama, Y., 2009. Structure and properties of the cellulose microfibril. Journal of Wood Science 55, 241—249.

Osorio, L., Trujillo, E., Van Vuure, A.W., Verpoest, I., 2011. Morphological aspects and mechanical properties of single bamboo fibers and flexural characterization of bamboo/ epoxy composites. Journal of Reinforced Plastics and Composites 30, 396—408.

Pääkkö, M., Ankerfors, M., Kosonen, H., Nykänen, A., Ahola, S., Österberg, M., Ruokolainen, J., Laine, J., Larsson, P.T., Ikkala, O., Lindström, T., 2007. Enzymatic hydrolysis combined with mechanical shearing and high-pressure homogenization for nanoscale cellulose fibrils and strong gels. Biomacromolecules 8, 1934—1941.

Pacheco-Torgal, F.P., Jalali, S., 2011a. Cementitious building materials reinforced with vege-table fibres: a review. Construction and Building Materials 25, 575—581.

Pacheco-Torgal, F., Jalali, S., 2011b. Nanotechnology: advantages and drawbacks in the field of construction and building materials. Construction and Building Materials 25, 582—590.

Pereira, C.L., Savastano Jr., H., Payá, J.J., Santos, S.F., Borrachero, M.V., Monzó, J.M., Soriano, L., 2013. Use of highly reactive rice husk ash in the production of cement matrix reinforced with green coconut fiber. Industrial Crops and Products 49, 88—96.

Peter, M.A., Muntean, A., Meier, S.A., Böhm, M., 2008. Competition of several carbonation reactions in concrete: a parametric study. Cement and Concrete Research 38, 1385—1393.

Peyvandi, A., Sbiam, L.A., Soroushianm, P., Sobolev, K., 2013. Effect of the cementitious paste density on the performance efficiency of carbon nanofiber in concrete nanocomposite. Construction and Building Materials 48, 265—269.

Pizzol, V.D., Mendes, L.M., Frezzatti, L., Savastano Jr., H., Tonoli, G.H.D., 2014a. Effect of accelerated carbonation on the microstructure and physical properties of hybrid fiber—cement composites. Minerals Engineering 59, 101—106.

Pizzol, V.D., Mendes, L.M., Savastano Jr., H., Frías, M., Davila, F.J., Cincotto, M.A., John, V.M., Tonoli, G.H.D., 2014b. Mineralogical and microstructural changes promoted by accelerated carbonation and ageing cycles of hybrid fiber—cement composites. Construction and Building Materials 68, 750—756.

Rehsi, S.S., 1988. In: Swamy, R.N. (Ed.), Natural Fibre Reinforced Cement and Concrete (Concrete Technology and Design 5). Blackie, Glasgow, UK, pp. 243–255.

Ruiz, H.A., et al., 2011. Development and characterization of an environmentally friendly process sequence (Auto hydrolysis and organosolv) for wheat straw delignification. Applied Biochemistry and Biotechnology 164, 629–641.

Santos, S.F., Schmidt, R., Almeida, A.E.F.S., Tonoli, G.H.D., Savastano Jr., H., 2015. Supercritical carbonation treatment on extruded fibre–cement reinforced with vegetable fibres. Cement & Concrete Composites 56, 84–94.

Satyanarayana, K.G., Guimarães, J.L., Wypych, F., 2007. Studies on lignocellulosic fibers of Brazil. Part I: source, production, morphology, properties and applications. Composites: Part A 38, 1694–1709.

Satyanarayana, K.G., Arizaga, G.G.C., Wypych, F., 2009. Biodegradable composites based on lignocellulosic fibers – an overview. Progress in Polymer Science 34, 982–1021.

Savastano Jr., H., Warden, P.G., Coutts, R.S.P., 2003. Mechanically pulped sisal as reinforcement in cementitious matrices. Cement & Concrete Composites 25, 311–319.

Savastano Jr., H., Warden, P.G., Coutts, R.S.P., 2005. Microstructure and mechanical properties of waste fibre–cement composites. Cement and Concrete Composites 27, 583–592.

Savastano Jr., H., Santos, S.F., Agopyan, V., 2009a. Sustainability of vegetable fibres in construction. In: Khatib, J. (Ed.), (Org.). Sustainability of Construction Materials, first ed., vol. 1. Woodhead Publishing Limited (Cambridge, England); CRC Press LLC (New York; Washington DC, USA), Cambridge, pp. 55–78.

Savastano Jr., H., Santos, S.F., Radonjic, M., Soboyejo, W.O., 2009b. Fracture and fatigue of natural fiber-reinforced cementitious composites. Cement & Concrete Composites 31, 232–243.

Sena Neto, A.R., Araujo, M.A.M., Barboza, R.M.P., Fonseca, A.S., Tonoli, G.H.D., Souza, F.V.D., Mattoso, L.H.C., Marconcini, J.M., 2015. Comparative study of 12 pineapple leaf fiber varieties for use as mechanical reinforcement in polymer composites. Industrial Crops and Products 64, 68–78.

Silva, M.C., Takahashi, J.A., Chaussy, D., Belgacem, M.N., Silva, G.G., 2010. Composites of rigid polyurethane foam and cellulose fiber residue. Journal of Applied Polymer Science 117, 3365–3672.

Siró, I., Plackett, D., 2010. Microfibrillated cellulose and new nanocomposite materials: a review. Cellulose 17, 459–494.

Spence, K.L., Venditti, R.A., Rojas, O.J., Habibi, Y., Pawlak, J.J., 2011. A comparative study of energy consumption and physical properties of microfibrillated celulose produced by different processing methods. Cellulose 18, 1097–1111.

Spinacé, M.A.S., Lambert, C.S., Fermoselli, K.K.G., De Paoli, M.A., 2009. Characterization of lignocellulosic curaua fibres. Carbohydrate Polymers 77, 47–53.

Sridach, W., 2010. The environmentally benign pulping process of non-wood fibers. Suranaree Journal of Science and Technology 17, 105–123.

Stenstad, P., Andresen, M., Tanem, B., 2008. Chemical surface modifications of microfibrillated cellulose. Cellulose 15, 35–45.

Sundar, S.T., Sain, M.M., Oksman, K., 2010. Characterization of microcrystalline cellulose and cellulose long fiber modified by iron salt. Carbohydrate Polymers 80, 35–43.

Svagan, A.J., Samir, M.A.S.A., Berglund, L.A., 2008. Biomimetic foams of high mechanical performance based on nanostructured cell walls reinforced by native cellulose nanofibrils. Advanced Materials 20, 1263–1269.

Tan, T., Santos, S.F., Savastano Jr., H., Soboyejo, W.O., 2012. Fracture and resistance-curve behavior in hybrid natural fiber and polypropylene fiber reinforced composites. Journal of Materials Science 47, 2864−2874.

Technical Association of the Pulp and Paper Industry, 1995. Tappi T 273 cm-95: Wet Zero-span Tensile Strength of Pulp. Atlanta, GA, USA.

Technical Association of the Pulp and Paper Industry, 1996. Tappi T 231 cm-96: Zero-span Breaking Strength of Pulp (Dry Zero-span Tensile). Atlanta, GA, USA.

Teixeira, R.S., Tonoli, G.H.D., Santos, S.F., Savastano Jr., H., Protassio, T.P., Toro, E.F., Maldonado, J., Lahr, F.A.R., Delvasto, S.A., 2013. Different ageing conditions on cementitious roofing tiles reinforced with alternative vegetable and synthetic fibre. Materials and Structures 46, 1−14.

Thwe, M.M., Liao, K., 2002. Effects of environmental aging on the mechanical properties of bamboo-glass fiber reinforced polymer matrix hybrid composites. Composites: Part A 33, 43−52.

Toledo Filho, R.D., Ghavami, K., Sanjuán, M., England, G., 2005. Free restrained and drying shrinkage of cement mortar composites reinforced with vegetable fibres. Cement & Concrete Composites 27, 537−546.

Tomczak, F., Satyanarayana, K.G., Sydenstricker, T.H.D., 2007. Studies on lignocellulosic fibers of Brazil. Part III − morphology and properties of Brazilian curauá fibers. Composites: Part A 38, 2227−2236.

Tonoli, G.H.D., Joaquim, A.P., Arsène, M.-A., Bilba, K., Savastano Jr., H., 2007. Performance and durability of cement based composites reinforced with refined sisal pulp. Materials and Manufacturing Processes 22, 149−156.

Tonoli, G.H.D., Fuente, E., Monte, C., Savastano Jr., H., Rocco Lahr, F.A., Blanco, A., 2009. Effect of fibre morphology on flocculation of fibre−cement suspensions. Cement and Concrete Research 39, 1017−1022.

Tonoli, G.H.D., Santos, S.F., Joaquim, A.P., Savastano Jr., H., 2010. Effect of accelerated carbonation on cementitious roofing tiles reinforced with lignocellulosic fibre. Construction and Building Materials 24, 193−201.

Tonoli, G.H.D., Santos, S.F., Savastano Jr., H., Delvasto, S., Mejía de Gutiérrez, R., Lopez de Murphy, M.M., 2011. Effects of natural weathering on microstructure and mineral composition of cementitious roofing tiles reinforced with fique fibre. Cement & Concrete Composites 33, 225−232.

Tonoli, G.H.D., Santos, S.F., Teixeira, R.S., Pereira-da-Silva, M.A., Lahr, F.A.R., Pescatori Silva, F.H., Savastano Jr., H., 2013. Effect of eucalyptus pulp refining on the performance and durability of fibre-cement composites. Journal of Tropical Forest Science 25, 400−409.

Turbak, A., Snyder, F., Sandberg, K., 1983. Microfibrillated cellulose: a new cellulose product: properties, uses, and commercial potential. Journal of Applied Polymer Science 37, 815−827.

Wan, J., Yang, J., Ma, Y., Wang, Y., 2011. Effects of the pulp preparation and papermaking processes on the properties of OCC fibers. BioResources 6, 1615−1630.

Xu, Y., Lee, S.-Y., Wu, Q., 2011. Creep analysis of bamboo high-density polyethylene composites: effect of interfacial treatment and fiber loading level. Polymer Composites 32, 692−699.

Yakovlev, G., Kerienė, J., Gailius, A., Girniene, I., 2006. Cement based foam concrete reinforced by carbon nanotubes. Materials Science 12 (2), 147−151.

Ying-Chen, Z., Hong-Yan, W., Yi-Ping, Q., 2010. Morphology and properties of hybrid composites based on polypropylene/polylactic acid blend and bamboo fiber. Bioresource Technology 101, 7944−7950.

Yuan, L., Wan, J., Ma, Y., Wang, Y., Huang, M., Chen, Y., 2013. The content of diferente hydrogen bond models and crystal structure of eucalyptus fibers during beating. BioResources 8, 717−734.

Zakikhani, P., Zahari, R., Sultan, M.T.H., Majid, D.L., 2014. Extraction and preparation of bamboo fibre-reinforced composites. Materials and Design 63, 820−828.

Natural fibre-reinforced noncementitious composites (biocomposites)

S.J. Christian
Carnegie Mellon University, Pittsburgh, PA, United States

5.1 Introduction

Biocomposites are natural fibre-reinforced biopolymers. Researchers have been developing these materials as an alternative to conventional materials that may be nonrenewable, recalcitrant, or manufactured by pollution emitting processes. While industrial-scale production of biocomposites is becoming more viable, the durability of these natural materials limits application in many environments. This chapter defines biocomposites and provides examples of particular biocomposites and their material properties. The chapter then focuses on durability challenges and efforts being made to improve resistance to various environmental challenges.

The definition of biocomposites varies greatly throughout the literature with some definitions including composites composed of synthetic fibres or polymers. Compilations of research on biocomposites (Mohanty et al., 2005a; Wool and Sun, 2005) and many review articles (Billington et al., 2014; John and Thomas, 2008; Kozlowski and Wladyka-Przybylak, 2008; Chapple and Anandjiwala, 2010; Dittenber and GangaRao, 2012; Sahari and Sapuan, 2012; Azwa et al., 2013; Koronis et al., 2013; Dicker et al., 2014; Faruk et al., 2014; Mngomezulu et al., 2014; Thakur et al., 2014; Netravali and Chabba, 2003; Mohanty et al., 2000c) provide an excellent overview of composites made from natural fibres and/or biopolymers. This chapter focuses on composites made only of natural materials, that is, natural fibres *and* biopolymers.

5.2 Description of biocomposites

Composites are structural materials made by combining two or more constituent materials with significantly different properties. Fibre-reinforced polymers (FRP) are composites in which fibres are embedded in a polymeric matrix. The polymer binds the fibres together giving shape to the composite, distributes stresses between fibres, resists a small amount of stress and typically protects the fibres from exposure to the environment. Typically, the fibres are stronger and stiffer than the polymeric matrix, which means that the fibre behaviour dominates the mechanical behaviour, while the matrix tends to dominate the durability performance. The materials used

Nonconventional and Vernacular Construction Materials. http://dx.doi.org/10.1016/B978-0-08-100038-0.00005-6

to form conventional composites are often synthetic, for example, glass and carbon fibres and petroleum-based polymers. These materials are used in many industries from aerospace to automotive, sports and construction. Applications within the construction industry include reinforcing bars for concrete, external reinforcement used to repair or strengthen existing components of concrete and steel, and pultruded cross sections. While these materials show significant strength and do not corrode, their constituents and their manufacturing processes, like many construction materials, require the use of nonrenewable materials and fossil fuels.

Biocomposites comprised of natural fibres and bio-based polymers are being developed to provide an environmentally responsible alternative to conventional materials. Although, some researchers extend the definition to include composites made with natural fibres and petroleum-derived polymers and composites made with synthetic fibres and biopolymers (Mohanty et al., 2005a; Wool and Sun, 2005), in this chapter, biocomposites are defined as being entirely constituted of bio-based materials. Benefits of biocomposites include that the materials are derived from renewable resources and in many cases can completely biodegrade when subjected to an appropriate environment. Current uses of biocomposites tend to be nonstructural and located in controlled environments due to their variable mechanical behaviour and unknown long-term durability.

Maximum tensile strength and stiffness of biocomposites range from about 20 to 200 MPa and 1 to 4 GPa, respectively (Mohanty et al., 2005a; Netravali et al., 2007). Most impressively, unidirectionally oriented ramie fibre/soy protein concentrate have exhibited tensile properties comparable to steel (Netravali and Chabba, 2003). For most biocomposites to compete with conventional construction materials, their structural shapes are often modified using, for example, cellular cross sections, sandwich panels and idealised cross sections like rectangular tubes.

5.2.1   Natural forms of fibre/polymer biocomposites

Nature has provided us with examples of biocomposites in plants, wood, bird feathers and bone. Filaments within the stems of plants are composed of cellulose and hemicellulose bonded together by a matrix of lignin or pectin. Wood is composed of long cellulose fibres bonded together by lignin. Bone is comprised of hydroxyapatite, a hard material, and collagen, a flexible material.

Early uses of natural fibre biopolymer composites have been recorded as early as the 6th century BCE. Linothorax is a laminated armour of linen and animal fat used by Greek and Macedonian warriors from the 6th to 3rd centuries BCE. Because the material biodegrades, the only evidence of linothorax is literary and iconic depictions in mosaics and other artwork from the time (Aldrete, 2013). Mongols fabricated bows from wood, bone and 'animal glue' around 1200 (Oestmoen, 2002). The pneumatic tyre made from natural rubber reinforced with cotton cording was invented by R.W. Thomson in 1845 (Staiger and Tucker, 2008). Linoleum, a mixture of linseed oil, rosin, wood flour, cork powder and pigments rolled onto jute fibre backing, was patented in 1860 by Frederick Walton (Staiger and Tucker, 2008).

In 1942, Henry Ford produced a prototype automobile body of soybean, hemp, flax, wheat and ramie (Soybean Car, 2015).

5.2.2 Natural fibres

Many natural fibres have excellent mechanical properties, although, in comparison to synthetic fibres, exhibit greater variation in properties. Natural fibres can be derived from plants, animals and minerals. Animal fibres include hair and silk. Natural mineral fibres, like asbestos, are nonrenewable and in some cases carcinogenic (Riedel and Nickel, 1999); so are not used in biocomposites. Plant fibres have lower density than many synthetic fibres, are renewable and biodegradable, and can exhibit high tensile strength and stiffness (Table 5.1).

One of the biggest obstacles to the use of natural fibres in the composites industry is their uncertain durability affecting both application and processing. Natural fibres have high moisture absorption, can be microbially degraded, and are susceptible to rotting. Additionally, natural fibres tend to degrade under thermal loads at relatively low temperatures, around 200°C. Processing of composites typically involves heating a polymer to melt or cure, and the low thermal degradation temperature for natural fibres limits the potential matrices to those polymers that melt or cure below 200°C. Finally, physical properties of natural fibres are inconsistent, varying by harvest season and region with differences in temperature, moisture, soil, sun, and can additionally be dependent upon the part of the plant from which the fibre is extracted (Koronis et al., 2013).

Natural fibres have high moisture absorption (hydrophilic) because hydroxyl (OH) groups located on the backbone of the cellulose structure readily bond with water (Herrera Franco and Valadez-Gonzalez, 2005), which limits the hydrogen bonding sites available for bonding with a matrix (Dittenber and GangaRao, 2012). The difference in polarity between the highly polar natural fibres and nonpolar polymer matrices

Table 5.1 **Mechanical properties of natural fibres (Mohanty et al., 2000a; Okubo et al., 2004; Dittenber and GangaRao, 2012)**

Fibre	Tensile strength (MPa)	Tensile modulus (GPa)	Elongation at break (%)
Flax	340−2000	28−103	1.2−3.3
Hemp	270−900	24−90	1.0−3.5
Jute	320−800	8−78	1.2−1.8
Kenaf	220−930	14−53	1.5−2.7
Ramie	400−1000	24−128	1.2−4.0
Sisal	360−700	9−38	2.0−7.0
Bamboo	140−800	11−49	1.3−3.7

leads to poor interfacial bonding (Herrera Franco and Valadez-Gonzalez, 2005). Poor adhesion at the interface limits stress transfer at the fibre-matrix interface resulting in low composite strength and stiffness because the fibres are not fully utilised. Compatibilising and coupling agents introduce chemical bonds between fibres and matrices to improve adhesion and consequently mechanical properties (Herrera Franco and Valadez-Gonzalez, 2005; eg, Khan et al., 2001; Lee et al., 2003; Mohanty et al., 2000b). Popular fibre treatments include alkali (mercerisation), acetylation, stearic acid, benzylation, peroxide, anhydride, permanganate, silane, isocyanate and plasma (La Mantia and Morreale, 2011). The properties and processing of composites vary greatly depending upon the type of fibre that is used. Long fibres and fabrics are more often used for structural purposes than short fibres because they are easier to align and rarely pull-out from the polymer, allowing the fibre to reach its full strength. Physical properties of natural fibres vary between harvest seasons and regions due to variable sun, rain and soil conditions (Koronis et al., 2013). To reduce the effects of fibre variability and provide consistent composite behaviour, batches of fibres from different harvests could be mixed (Mussig et al., 2006), or genetic transformation may be employed to provide a more reliable source of fibres (Koronis et al., 2013).

5.2.3 Biopolymers

Biopolymers are polymers that are produced by or derived from living organisms, such as plants and microbes, rather than from petroleum, the traditional source of polymers. The primary sources of biopolymers are renewable. Many, but not all, biopolymers are biodegradable, which means they 'are capable of decomposing into carbon dioxide, methane, water, inorganic compounds or biomass by the enzymatic action of microorganisms' (Kandola, 2012). While biopolymers make up only a small percentage of the polymer market, it has been predicted that they have the ability to replace 30−90% of petroleum-based polymers (Shen and Patel, 2009; USDA, 2008). Table 5.2 shows mechanical properties for select biopolymers often considered for biocomposite matrices.

Choice of biopolymer to be used in construction often depends on mechanical behaviour, moisture absorption, photodegradation stability, availability, compatibility and cost. Polylactic acid (PLA) and Polyhydroxyalkoanate (PHA) are the most common biopolymers in terms of production and use but are mainly used for short-term applications (Reddy, 2012). Starch-based biopolymers are the only biopolymers that can be produced without the use of advanced biotechnology such as advanced fermentation techniques or microbiological methods. Biodegradable polymers include PLA, cellulose esters (cellulose acetate), starch plastic, poly ε-caprolactone and aliphatic polyester-copolyesters which are all hydrophilic. While initial commercialisation has focused on short-term applications for biopolymers, research and development is now directed to increasing the durability of these polymers (Chen and Patel, 2011). Researchers are beginning to use genetically modified crops to produce higher quality polymers (Reddy, 2012). The biggest challenge for biopolymers remains their relatively high price (Reddy, 2012).

PHAs are a group of biopolymers that has shown promise for use in construction. PHA is the only biopolymer that can be completely synthesised by microorganisms.

Table 5.2 Mechanical properties of biopolymers (Mohanty et al., 2005b; Netravali et al., 2007; Mohanty et al., 2004; Dicker et al., 2014)

Biopolymer	Tensile strength (MPa)	Tensile modulus (MPa)	Elongation at break (%)
PLA	21–60	350–3800	2.5–6
PHB	24–40	1700–4000	5–9
PHBV	25	1000	25
SPC	50	2000	12
Cellulose acetate	30–40 (flexural strength)	1000–2000	–
Starch	5–6	130–850	31–44

PLA, polylactic acid; *PHB*, polyhydroxybutyrate; *PHBV*, polyhydroxybutyrate-co-hydroxyvalerate; *SPC*, soy protein concentrate.

Carbon-based feedstocks for PHAs include sugarcane, whey and methane. Researchers have focused significantly on polyhydroxybutyrate (PHB) and polyhydroxybutyrate-co-hydroxyvalerate (PHBV) for construction-based applications. PHAs have been blended with other biopolymers, such as PLA and starches, to improve mechanical, biodegradation and morphological properties (Reddy et al., 2013). PHB is water-insoluble, which should be a benefit as most biopolymers are not.

Biopolymers for composites can also be derived from vegetable oils and proteins (eg, soybean oil or linseed oil). Khot et al. (2001) synthesised acrylated epoxidised soybean oil, maleinised soybean oil monoglyceride and maleinised hydroxylated soybean oil monomers from triglycerides. Netravali et al. (2007) produced biocomposites from soy protein isolate and soy protein concentrate (SPC) matrices that exhibit superior mechanical properties.

5.2.4 Composite forms

Fibres and matrices can be combined into bulk or laminate composites. Bulk composites consist of randomly oriented fibres distributed in three dimensions, intended to result in nearly isotropic behaviour. However, the mixing process is known to chop fibres into smaller pieces, and fibres often clump together, resulting in variable properties throughout composite. Laminate composites are orthotropic, consisting of multiple layers of oriented or randomly oriented fibres bound together by a matrix in which each layer of fibres has a two-dimensional (ie, planar) orientation.

5.2.5 Concerns about behaviour

The incompatibility of natural fibres and polymers, poor moisture resistance of natural fibres, propensity for short-fibre clumping during fabrication and low thermal decomposition temperature limit the potential applications of biocomposites (Saheb and Jog, 1999). Application of biocomposites is typically limited to interior and nonstructural applications due to their poor mechanical properties, fire resistance, fibre/matrix adhesion and durability, susceptibility to moisture-induced changes, variation in fibre quality and manufacturing difficulties (Dittenber and GangaRao, 2012). Engineering biocomposites through fibre treatment, biopolymer blending, additives, coatings and efficient processing are described as methods for producing commercially viable biocomposites (Mohanty et al., 2002; Dittenber and GangaRao, 2012).

Long-term behaviour, including fatigue and creep behaviours, and lifetime prediction for biocomposites is unknown and potentially of major concern (Dittenber and GangaRao, 2012). Most extant durability studies have focused on moisture and weathering (Dittenber and GangaRao, 2012).

Commercial production of biopolymers is still in its infancy and improved processing is a major focus of research. While some biopolymers may be naturally synthesised, many must be entirely synthesised from the renewable resources. Research has focused on increasing scale, improving yields and reducing production costs. As the industry becomes established, the price of biopolymers is expected to decrease. This trend has been shown for PLA (Reddy et al., 2013).

Growth and processing of natural fibres can be relatively low-tech utilising the skills of local farmers, weavers and workers. Suitable natural fibres grow quickly at low cost and can be produced locally. Composite manufacturing varies by material. While some processes require a capital investment and energy consumption, others are relatively simple with little upfront cost. Fabrication by pultrusion, injection moulding and compression moulding, for example, require a capital investment, while vacuum injection, hand lay-up and vacuum pressing are relatively inexpensive and simple.

Heat and pressure are usually applied for manufacturing composites. Pressure is applied to consolidate the composite to achieve a higher fibre volume fraction and to induce flow of the liquid matrix around the fibres (ie, improve and ensure wet-out). Heat performs different functions depending on the matrix type. Thermoset polymers are heated to induce crosslinking of the polymer chains. Heat is added to thermoplastics to melt the plastic and lower its viscosity so that the polymer can flow around the fibres. Upon cooling, the thermoplastic solidifies in the new shape.

5.2.6 Applications

Acceptance of composite materials, and especially biocomposites, in the construction industry has been slow, in part due to the uncertainty in long-term performance of the materials. Nonstructural applications are more likely to precede structural applications, and use in controlled environments will be required until confidence in the long-term durability of the materials is established. Proposed uses of biocomposites in the construction industry include light-weight structures, internal/secondary structural elements, façade panels, door frames and other architectural uses. Biocomposites have been used to replace wood fittings, fixtures, furniture and noise insulating panels (Mohanty et al., 2005a) and as tiles, flower pots and marine piers (Yan et al., 2012). Toyota Motors has used kenaf-reinforced PLA for spare tyre covers and interior door trim (Toyota Boshoku, 2015). Fatima and Mohanty (2011) investigated the noise-reducing properties of jute and natural rubber latex jute felt composites.

5.3   Durability concerns

This section discusses the performance of biocomposites when exposed to conditions that include moisture, elevated temperature, UV, bacteria and fire. Natural weathering conditions, such as temperature, humidity, ultraviolet radiation and rain can negatively affect the behaviour of the composites. Changes in physical properties (eg, colour, surface roughness and weight) and mechanical properties (eg, strength and stiffness) of conditioned specimens in previous research are reviewed.

5.3.1   Hygrothermal effects

Exposure to moisture almost always has a negative effect on the performance of biocomposites. Water can plasticise hydrophilic polymers and natural fibres causing reduction in stiffness. Differential expansion between the fibres and the matrix can introduce permanent damage to the fibre-matrix interface and to hydrophobic matrices. While pure thermal effects have not been a major focus of study for in-service biocomposites, typical variation in polymer behaviour with temperature suggests that stiffness will decrease near the melting temperature and failure will become more brittle near the glass transition temperature.

5.3.1.1   Moisture

Expected humidity and water exposure varies by climate; therefore potential exterior uses of biocomposites are limited. Interior conditions can be controlled and, therefore, provide a better environment for successful use of biocomposites.

Natural fibre composites absorb more moisture and exhibit more severe water-related degradation than composites made with synthetic fibres mainly due to the hydrophilic nature of the natural fibres. Natural fibres swell as they absorb moisture and, once the fibres are saturated, moisture enters the fibre-matrix interfacial region. Indeed, if bonding between fibres and matrix are poor, water may enter this region prior to saturation of fibre. Fibre swelling results in poor dimensional stability, and moisture at the fibre-matrix interface may also cause delamination. Moisture absorption has been shown to increase with an increase in fibre volume fraction (Cinelli, 2003). Upon absorption of moisture, natural fibres plasticise, resulting in lower fibre stiffness.

The effects of moisture absorption vary greatly depending upon the absorption properties of the matrix. Biocomposites with hydrophilic matrices (eg, starch-, cellulose- and soy protein-based) have been shown to absorb substantially more moisture than those with hydrophobic matrices (Christian and Billington, 2012; Alvarez et al., 2004). Moisture acts as a plasticiser, reducing the stiffness of hydrophilic matrices. A 30–50% loss in flexure modulus and a 70% loss in tension modulus were observed for sisal/MaterBi-Y (derived from cellulose and starch) and hemp/cellulose acetate biocomposites, respectively, when exposed to high humidity and/or water immersion (Alvarez et al., 2004; Alvarez and Vasquez, 2004; Christian and Billington, 2012). Yet, hydrophobic matrices (eg, PHB and PHBV) may sustain more permanent damage than their hydrophilic counterparts because fibre swelling induces higher internal stresses in the stiffer materials, resulting in cracking. Christian and Billington (2012) observed more significant cracking and lower recovery of mechanical properties upon drying for hemp/PHB as compared to hemp/cellulose acetate biocomposites.

Improving biocomposite performance when exposed to moisture has been a focus of many researchers. Most research has focused on treatment of natural fibres to reduce their hydrophilicity and to improve the fibre-matrix interfacial bond. Treatment of natural fibres may remove hemicellulose, the most hydrophilic component of natural fibres, reduce the number of hydroxyl groups free to bond with water, coat the fibre with a hydrophobic layer and/or coat the fibre with a compatibiliser to improve the interfacial bond between the fibre and the matrix. When the fibre—matrix bond is improved, the moisture absorption of the resulting biocomposite is lowered. Fibre treatments including alkalisation, pectin enzyme treatment and Duralin steam treatment (Pott, 2001) have been shown to reduce moisture absorption in natural fibres (Dittenber and GangaRao, 2012). Barrier systems, such as waterproof coatings, are typically only effective in the short to medium term (Dittenber and GangaRao, 2012).

5.3.1.2 Temperature

Temperature effects will be most pronounced in extreme climates where ambient temperatures are near the glass transition and melting temperatures and where absorption of solar radiation leads to material temperatures well above ambient temperature. Dimensional stability may impact in-service appearance and mechanical performance, and differential expansion may lead to material damage similar to that due to moisture absorption (Azwa et al., 2013). Studies on thermal effects have been limited to small, laboratory-scale testing via thermogravimetric analysis (TGA), differential scanning calorimetry and dynamic mechanical thermal analysis (Holloway, 2010; Azwa et al., 2013). While TGA often focuses on degradation at high temperatures, which is important for manufacturing and fire exposure, degradation at lower temperatures may be used to compare the thermal stability of biocomposites in the typical service ranges (Azwa et al., 2013).

5.3.2 Ultraviolet (UV) radiation

UV radiation causes photodegradation of exposed materials. Solar radiation cleaves covalent bonds in organic polymers, which may cause discolouration, surface roughening, mechanical property degradation and embrittlement of biocomposites (Azwa, 2013). Photodegradation of the lignin in natural fibres results in chemicals that impart a yellowing discolouration (Beg and Pickering, 2008). Campos et al. (2012) found that UV exposure of thermoplastic starch/sisal and polycaprolactone/sisal biocomposites led to embrittlement, colour change, holes in the surface and reduction in mechanical properties. In the laboratory, materials are rarely exposed to UV alone (Campos et al., 2012), but are often exposed to a combination of UV, moisture and temperature intended to simulate actual conditions at an accelerated rate (eg, Chen et al., 2011).

5.3.3 Accelerated weathering

Actual environmental conditions vary constantly. While understanding the effects of individual conditions is imperative to improving biocomposite durability, in service, materials are exposed to constantly varying conditions. In the laboratory, researchers expose biocomposite materials to a combination of UV, moisture and temperature intended to simulate actual conditions at an accelerated rate. While researchers have not correlated laboratory results to expected in-service performance, they are able to compare relative performance of materials exposed to the same conditions. Nonetheless, relatively few studies have been conducted for fully bio-biocomposite materials.

Michel and Billington (2012) tested PHB/hemp composites in an accelerated test and observed significant damage: increased thickness due to fibre swelling, mass loss due to cyclic hygrothermal fibre expansion/contraction and polymer erosion, cracking in biopolymer matrix becoming more severe as cracks further exposed fibres to moisture and flaking of the PHB polymer due to polymer embrittlement and cracking. Mechanical deterioration was attributed to incompatible deformations between fibres and matrix because the primary failure mode changed from fibre fracture to pull-out following exposure.

5.3.4 Fire

Combustibility and significant loss of strength and stiffness at elevated temperatures pose a technical barrier to the widespread use of FRPs in structural engineering applications (Azwa et al., 2013). Ignited polymer matrices decompose into combustible and noncombustible gases, smoke and liquid, which cause hazards due to inhalation of toxic gases, loss of structural integrity and ignition of other sources due to dripping (Azwa et al., 2013). In addition to adverse effects of the matrix, the natural fibres in biocomposites pose additional fire safety threats both individually and in conjunction with the matrix.

When biopolymers undergo thermal decomposition, they release heat, smoke and volatiles. Typical combustible volatiles that may be released include hydrocarbons and carbon monoxide. Noncombustible volatiles include carbon dioxide and hydrogen halides (Chapple and Anandjiwala, 2010). Quite often, exposure to these volatiles is more damaging to human health than the fire itself. Another risk that presents itself is flammable volatiles coming into contact with oxygen and producing highly reactive hydroxyl radicals which can lead to further decomposition of the polymer and sustain the burning of the fire (Mouritz and Gibson, 2006). Of course, the type and degree of decomposition is dependent on the type of polymer.

Variation in natural fibre flammability is in part due to differences in fibre microstructure and chemical composition (Chapple and Anandjiwala, 2010; Dittenber and GangaRao, 2012). In general, fibres with high crystallinity and lower polymerisation are less flammable (Chapple and Anandjiwala, 2010; Dittenber and GangaRao, 2012; Azwa et al., 2013). The composition of a natural fibre largely influences performance at elevated temperatures because cellulose, lignin and hemicellulose, the major fibre components, behave in markedly different ways.

Most natural fibres begin to degrade at temperatures exceeding 170−200°C (Dicker et al., 2014). The pyrolysis of hemicellulose occurs rapidly from 220 to 315°C, while pyrolysis of cellulose occurs from 315 to 400°C. Lignin, on the other hand, decomposes over a wider range from 160 to 900°C (Dorez et al., 2014). Among the components that make up natural fibres, lignin is the most difficult to decompose, as it has a low decomposition rate and a broad decomposition temperature range (Zhou et al., 2013). Also, lignin is responsible for char formation, which protects the integrity of underlying biocomposites by acting as a layer of insulation (Azwa et al., 2013; Dittenber et al., 2012). A high content of cellulose can result in an increased flammability of the fibre. Its decomposition can emit flammable volatiles, noncombustible gases and tars and allow char formation. When hemicellulose decomposes, it adds significantly to the amount of noncombustible gases produced but produces less tar (Chapple and Anandjiwala, 2010).

Like FRPs in general, biocomposites experience creep, softening and distortion at relatively low temperatures (100−200°C) leading to a significant loss in strength and stiffness and the potential for buckling of load bearing components (Azwa et al., 2013). While fibres help to maintain the integrity of the composite in fire, which otherwise may be compromised by softening and melting of the polymer matrix, flammability of natural fibres result in generally poorer performance than synthetic fibres (Azwa et al., 2013). Likewise, the 'scaffolding effect' that may enhance decomposition and production of hazardous chemicals in composites when the polymer remains in contact with the heat source due to the continued integrity of the fibre reinforcement will not be as pronounced with natural fibres.

Fire resistance of composites can be improved by reducing flammability of the composite components and the composite as a whole (Chapple and Anandjiwala, 2010). Flame retardants are added to natural fibres through soaking and polymers by additive mixing. Upon exposure to flame, these retardants may char improving the insulation at the surface, produce water and react with H and −OH radicals in the flame to stop combustion, create a vapour-rich gas phase to limit oxygen available for burning at the surface, and reduce dripping of the melting polymer. Fibre modification by acetylation and acrylonitrile grafting has been shown to improve thermal stability. Finally, intumescent systems painted on the surface of the biocomposite char and expand upon exposure to flame, which creates a surface layer that limits heat transfer and oxygen access (Chapple and Anandjiwala, 2010).

5.3.5 Bacteria and fungus

Bacteria growth on biocomposite materials is evidence of biodegradation. High exposure to and absorption of water lead to conditions that may promote bacterial growth (Azwa et al., 2013). Laboratory-scale studies have shown that biocomposites made from PLA, PBS and PHB with various fibres biodegrade through enzymatic degradation and/or anaerobic digestion (Morse, 2009; Lee and Wang, 2006).

Soil burial tests for sisal/PLA and flax/PHB have shown that exposure to microorganisms present in soil (eg, bacteria and fungi) causes holes and cracking on the biocomposite surface as well as a reduction in mechanical properties (Alvarez et al., 2006; Christian, 2015). These studies suggest that in-service exposure to microorganisms will likely cause biodegradation of these biocomposites.

5.4 Managing durability

This section briefly discusses how design for specific durability performance criterion, affects material performance with respect to other design or performance considerations including mechanical properties, aesthetic quality, constructability dimensional stability, thermal stability, moisture stability, UV stability, sustainability and anticipated lifespan. For each application, specific requirements are set for these criteria. Unless, the environmental conditions can be controlled (eg, some interior locations), materials must often be designed to meet competing performance requirements.

5.4.1 Performance consequences of design decisions

Given the sensitivity of biocomposite materials to environmental conditions, material design choices to improve performance for specific criteria have repercussions with respect to other performance criteria. Many studies have called attention to the consequences of targeted design decisions or treatments on other important performance requirements. Examples of potential interactions are discussed.

The most fundamental design decision is the choice of fibre volume fraction. Designers must balance the need for high mechanical properties with the need to limit material damage due to moisture absorption. High fibre volume fraction corresponds to improved mechanical performance but also to greater moisture-related damage. When mechanical performance is specified, fibre volume fraction may be chosen only to meet but not exceed those requirements, thereby minimising potential moisture absorption (Azwa et al., 2013).

Treatment of fibres to improve moisture performance often strip the lignin present in the fibres. Because lignin is the component within the fibre responsible for charring, many moisture treatments may reduce the fire performance of the biocomposite.

Alkali treatment used to reduce the moisture absorption of natural fibres can also degrade cellulose fibres resulting in reduced strength of the natural fibre and additionally results in polluted wastewater (Dittenber and GangaRao, 2012).

Positive side effects of the Duralin process used to reduce moisture absorption include improved fibre yield and quality, stability, resistance to fungal attacks and mechanical properties, which are balanced by increased embodied energy due to steaming of the natural fibres (Pott, 2001; Dittenber and GangaRao, 2012).

Fire retardants added to wood fibre−reinforced high-density polyethylene (HDPE) caused reduction in colour retention when exposed to UV, but the addition of light stabiliser reduced the effect (Garcia et al., 2009). The addition of fire retardants to coconut fibre/PLA has been shown to reduce tensile properties (Suardana et al., 2011).

By bleaching natural fibres to remove lignin and hemicellulose, Beg and Pickering (2008) improved thermal stability at lower temperatures (275−400°C). Consequently, the removal of lignin reduced the amount of charring which acts as insulation leading to increased thermal degradation at higher temperatures (400−500°C). Further studies into the optimisation of lignin content are needed to achieve required thermal stability within different temperature ranges (Azwa et al., 2013).

5.4.2   Optimisation for tuned performance

The most appropriate way to manage durability is to intelligently choose the location and function of each material component so that the strengths of the material are taken advantage of and the weaknesses do not pose a risk or fail to meet desired performance. Understanding the interactions between the material components and the durability treatments is a good first step toward successful implementation of biocomposites. Designers must also clearly define the performance criteria and priorities for a specific application. Given the number of variables with respect to performance and biocomposite processing, the development of a single 'standard' biocomposite material is not practical. Rather, fine-tuning of biocomposite material properties for specific applications may allow for more immediate implementation of these materials.

Relatively little work has been done to optimise biocomposite materials to meet competing performance requirements. Systematic studies of the interactions are required to define clear relationships between both individual and multiple variables. Given the complexity of the systems of treatment methods and performance and the wide variation of potential biocomposite materials, optimisation programs for design decision-making are a current research focus (Srubar et al., 2014; Miller et al., 2015).

5.5   Conclusions and future research

While biocomposites have been the subject of research for use in infrastructure, significant challenges with regard to durability have limited the application of these materials. To achieve an impact in the construction industry, we must either modify the materials to meet performance criteria or choose less-controlled applications that do not require high performance. Further study and development is required to improve the performance of biocomposites. Choosing applications with low performance criteria and changing design criteria to allow a material to be used rather than just trying to replace a material/component that is currently used are two approaches to developing acceptance. Most of the knowledge regarding durability of biocomposites is based on laboratory experiments. The behaviour of the material in situ has not been used to validate laboratory results. Thus, long-term performance of the materials is not yet understood.

References

Aldrete, G.S., Scott, B., Alicia, A., 2013. Reconstructing Ancient Linen Body Armor: Unraveling the Linothorax Mystery. Johns Hopkins University Press.

Alvarez, V.A., Fraga, A.N., Vazquez, A., 2004. Effects of the moisture and fiber content on the mechanical properties of biodegradable polymer-sisal fiber biocomposites. Journal of Applied Polymer Science 91, 4007−4016.

Alvarez, V.A., Vazquez, A., 2004. Effects of water sorption on the flexural properties of a fully biodegradable composite. Journal of Composite Materials 38 (13), 1165−1182.

Alvarez, V.A., Ruseckaite, R.A., Vazquez, A., 2006. Degradation of sisal fibre/Mater Bi-Y biocomposites buried in soil. Polymer Degradation and Stability 91, 3156−3162.

Azwa, Z.N., Yousif, B.F., Manalo, A.C., Karunasena, W., 2013. A review on the degradability of polymeric composites based on natural fibres. Materials and Design 47, 424−442.

Beg, M.D.H., Pickering, K.L., 2008. Accelerated weathering of unbleached and bleached kraft wood fibre reinforced polypropylene composites. Polymer Degradation and Stability 93, 1939−1946.

Billington, S.L., Srubar III, W.V., Miller, S.A., Michel, A.T., 2014. Renewable biobased composites for civil engineering applications. In: Netravali, A., Pastore, C. (Eds.), Sustainable Composite and Advanced Materials.

Campos, A., Marconcini, J.M., Martins-Franchetti, S.M., Mattoso, L.H.C., 2012. The influence of UV-C irradiation on the properties of thermoplastic starch and polycaprolactone biocomposite with sisal bleached fibers. Polymer Degradation and Stability 97 (10), 1948−1955.

Chen, G.-Q., Patel, M.K., 2011. Plastics derived from biological sources: present and future: a technical and environmental review. Chemical Reviews 112 (4), 2082−2099.

Cinelli, P., Lawton, J.W., Gordon, S.H., Imam, S.H., Chiellini, E., 2003. Injection molded hybrid composites based on corn fibers and poly(vinyl alcohol). Macromolecular Symposium 197, 115−124.

Chapple, S., Anandjiwala, R., 2010. Flammability of natural fiber-reinforced composites and strategies for fire retardancy: a review. Journal of Thermoplastic Composite Materials 23 (6), 871−893.

Chen, D., Li, Ren, J., 2011. Influence of fiber surface-treatment on interfacial property of poly(l-lactic acid)/ramie fabric biocomposites under UV-irradiation hydrothermal aging. Materials Chemistry and Physics 126, 524−531.

Christian, S.J., Billington, S.L., 2012. Moisture diffusion and its impact on uniaxial tensile response of biobased composites. Composites Part B: Engineering 43 (5), 2303−2312.

Christian, S.J., 2015. Unpublished research.

Dicker, M.P.M., Duckworth, P.F., Baker, A.B., Francois, G., Hazzard, M.K., Weaver, P.M., 2014. Green composites: a review of material attributes and complementary applications. Composites Part A: Applied Science and Manufacturing 56, 280−289.

Dittenber, D.B., GangaRao, H.V.S., 2012. Critical review of recent publications on use of natural composites in infrastructure. Composites Part A: Applied Science and Manufacturing 43 (8), 1419−1429.

Dorez, G., Ferry, L., Sonnier, R., Taguet, A., Lopez-Cuesta, J.M., 2014. Effect of cellulose, hemicellulose and lignin contents on pyrolysis and combustion of natural fibres. Journal of Analytical and Applied Pyrolysis 107, 323–331.

Faruk, O., Bledzki, A.K., Fink, H.P., Sain, M., 2014. Progress report on natural fiber reinforced composites. Macromolecular Materials and Engineering 299 (1), 9–26.

Fatima, S., Mohanty, A.R., 2011. Acoustical and fire-retardant properties of jute composite materials. Applied Acoustics 72 (2–3), 108–114.

Garcia, M., Hidalgo, J., Garmendia, I., Garcia-Jaca, J., 2009. Wood-plastics composites with better fire retardancy and durability performance. Composites Part A: Applied Science and Manufacturing 40 (11), 1772–1776.

Herrera Franco, P.J., Valadez-Gonzalez, A., 2005. Fiber-matrix adhesion in natural fiber composites. In: Mohanty, A.K., Mishra, M., Drzal, L. (Eds.), Natural Fibers, Biopolymers, and Biocomposites. CRC Press, Taylor & Francis Group, Boca Raton, FL, pp. 177–230.

Holloway, L.C., 2010. A review of the present and future utilisation of FRP composites in the civil infrastructure with reference to their important in-service properties. Construction and Building Materials 24, 2419–2445.

John, M.J., Thomas, S., 2008. Biofibres and biocomposites. Carbohydrate Polymers 71 (3), 343–364.

Kandola, B.K., 2012. Flame retardant characteristics of natural fibre composites. Natural Polymers vol 1: Composites M. J. John and T. Sabu: 86–117.

Khan, M.A., Kopp, C., Hinrichsen, G., 2001. Effect of vinyl and silicon monomers on mechanical and degradation properties of bio-degradable jute-biopol composites. Journal of Reinforced Plastics and Composites 20 (16), 1414–1429.

Khot, S., LaScala, J., Can, E., Morye, S., Williams, G., Palmese, G., Kusefoglu, S., Wool, R., 2001. Development and application of triglyceride-based polymers and composites. Journal of Applied Polymer Science 82, 703–723.

Koronis, G., Silva, A., Fontul, M., 2013. Green composites: a review of adequate materials for automotive applications. Composites Part B: Engineering 44 (1), 120–127.

Kozlowski, R., Wladyka-Przybylak, M., 2008. Flammability and fire resistance of composites reinforced by natural fibers. Polymers for Advanced Technologies 19 (6), 446–453.

La Mantia, F.P., Morreale, M., 2011. Green composites: A brief review. Composites: Part A 42, 579–588.

Lee, S.G., Choi, S.S., Park, W.H., Cho, D., 2003. Characterization of surface modified flax fibers and their biocomposites with PHB. Macromolecular Symposia 197, 89–99.

Lee, S.H., Wang, S., 2006. Biodegradable polymers/bamboo fiber biocomposite with bio-based coupling agent. Composites Part A 37, 80–91.

Michel, A.T., Billington, S.L., 2012. Characterization of poly-hydroxybutyrate films and hemp fiber reinforced composites exposed to accelerated weathering. Polymer Degradation and Stability 97 (6), 870–878.

Miller, S.A., Srubar, W.V., Billington, S.L., Lepech, M.D., 2015. Integrating durability-based service-life predictions with environmental impact assessments of natural fiber-reinforced composite materials. Resources, Conservation and Recycling 99, 72–83.

Mngomezulu, M.E., John, M.J., Jacobs, V., Luyt, A.S., 2014. Review on flammability of biofibres and biocomposites. Carbohydrate Polymers 111, 149–182.

Mohanty, A., Misra, M., Hinrichsen, G., 2000a. Biofibres, biodegradable polymers and biocomposites: an overview. Macromolecular Materials and Engineering 276 (3–4), 1–24.

Mohanty, A.K., Khan, M.A., Sahoo, S., Hinrichsen, G., 2000b. Effect of chemical modification on the performance of biodegradable jute-yarn/Biopol composites. Journal of Materials Science 35, 2589−2595.

Mohanty, A.K., Khan, M.A., Hinrichsen, G., 2000c. Surface modification of jute and its influence on performance of biodegradable jute-fabric/Biopol composites. Composites Science and Technology 60 (7), 1115−1124.

Mohanty, A.K., Misra, M., Drzal, L.T., 2002. Sustainable bio-composites from renewable resources: opportunities and challenges in the green materials world. Journal of Polymers and the Environment 10 (1−2), 19−26.

Mohanty, A.K., Misra, M., Drzal, L.T. (Eds.), 2005a. Natural Fibers, Biopolymers, and Biocomposites. CRC Press, Taylor & Francis Group, Boca Raton, FL.

Mohanty, A.K., Misra, M., Drzal, L.T., Selke, S.E., Harte, B.R., Hinrichsen, G., 2005b. Natural fibers, biopolymers, and biocomposites: an introduction. In: Mohanty, A.K., Mishra, M., Drzal, L. (Eds.), Natural Fibers, Biopolymers, and Biocomposites. CRC Press, Taylor & Francis Group, Boca Raton, FL, pp. 1−36.

Mohanty, A.K., Wibowo, A., Misra, M., Drzal, L.T., 2004. Effect of process engineering on the performance of natural fiber reinforced cellulose acetate biobased composites. Composites Part A 35, 363−370.

Morse, M.C., 2009. Anaerobic Biodegradation of Biocomposites for the Building Industry (Ph.D.). Stanford University.

Mouritz, A., Gibson, A., 2006. Fire Properties of Polymer Composite Materials, 11th ed. Springer, Dordrecht.

Mussig, J., Schmehl, M., von Buttlar, H.B., Schonfeld, U., Arndt, K., 2006. Exterior components based on renewable resources produced with SMC technology − considering a bus component as example. Industrial Crops and Products 24 (2), 132−145.

Netravali, A., Chabba, S., April 2003. Composites get greener. Materials Today 24−31.

Netravali, A.N., Huang, X., Mizuta, K., 2007. Advanced 'green' composites. Advanced Composite Materials 16 (4), 269−282.

Oestmoen, P.E., 2002. The Mongol Bow. http://www.coldsiberia.org/monbow.htm.

Okubo, K., Fujii, T., Yamamoto, Y., 2004. Development of bamboo-based polymer composites and their mechanical properties. Composites Part A 35, 377−383.

Pott, G.T., 2001. Reduction of moisture sensitivity in natural fibres. In: MRS Proceedings 702. Materials Research Society.

Reddy, M.M., Misra, M., Mohanty, A.K., 2012. Bio-based materials in the new bio-economy. Chemical Engineering Progress 108 (5), 37−42.

Reddy, M.M., Vivekanandhan, S., Misra, M., Bhatia, S.K., Mohanty, A.K., 2013. Biobased plastics and bionanocomposites: current status and future opportunities. Progress in Polymer Science 38 (10−11), 1653−1689.

Riedel, U., Nickel, J., 1999. Natural fibre-reinforced biopolymers as construction materials. Die Angewandte Makromolekulare Chemie 272, 34−40.

Sahari, J., Sapuan, S.M., 2012. Natural fibre reinforced biodegradable polymer composites. Reviews on Advanced Materials Science 30 (2), 166−174.

Saheb, D.N., Jog, J.P., 1999. Natural fiber polymer composites: A review. Advances in Polymer Technology 18, 351−363.

Shen, L., Haufe, J., Patel, M.K., 2009. Product Overview and Market Projection of Emerging Bio-based Plastics. Utrecht, Netherlands.

Soybean Car, The Henry Ford. Retrieved 2 January 2015, from: https://www.thehenryford.org/research/soybeancar.aspx.

Srubar, W.V., Miller, S.A., Lepech, M.D., Billington, S.L., 2014. Incorporating spatiotemporal effects and moisture diffusivity into a multi-criteria materials selection methodology for wood—polymer composites. Construction and Building Materials 71, 581—601.

Staiger, M.P., Tucker, N., 2008. Natural-fibre composites in structural applications. In: Pickering, K. (Ed.), Properties and Performance of Natural-fibre Composites. England, Woodhead Publishing Limited, Cambridge.

Suardana, N.P.G., Ku, M.S., Lim, J.K., 2011. Effects of diammonium phosphateon the flammability and mechanical properties of bio-composites. Materials & Design 32 (4), 1990—1999.

Thakur, V.K., Thakur, M.K., Gupta, R.K., 2014. Review: raw natural fiber-based polymer composites. International Journal of Polymer Analysis and Characterization 19 (3), 256—271.

Toyota Boshoku, Kenaf. Retrieved 15 September 2015, from: http://www.toyota-boshoku.com/global/about/development/eco/kenaf/index.html.

USDA, 2008. US Biobased Products Market Potential and Projections through 2025.

Wool, R.P., Sun, X.S. (Eds.), 2005. Bio-based Polymers and Composites. Elsevier, Inc., Amsterdam.

Yang, Z.Y., Wang, X.W., Lei, D.P., Fei, B., Xin, J.H., 2012. A durable flame retardant for cellulosic fabrics. Polymer Degradation and Stability 97 (11), 2467—2472.

Zhou, F., Cheng, G., Jiang, B., 2014. Effect of silane treatment on microstructure of sisal fibres. Applied Surface Science 292, 806—812.

Straw bale construction

P. Walker, A. Thomson, D. Maskell
University of Bath, Bath, United Kingdom

6.1 Introduction

Straw, the stem of cereal crops such as wheat, barley and rice, has been used in construction for thousands of years. Historical uses include roofing thatch, linings for internal plasters and reinforcement for traditional earthen building techniques, including adobe and cob. In contrast, the late-19th-century origins of straw bale construction in the USA are much more recent, following the invention of mechanical baling machines (King, 2006). Following World War II, straw was for a time also widely used in compressed straw panels for walls and short-span roofs.

In its simplest form, straw bale wall construction is comprised of individual bales laid in courses, as large lightweight masonry units but generally without mortar, and coated for both protection and additional strength by protective plasters. Though bales were developed for agricultural uses, the traditional bale, measuring approximately $1000 \times 450 \times 350$ mm and weighing around 20 kg, forms an ideal building unit that can be stacked to form load-bearing walls and provide excellent thermal insulation in external walls. Though primarily used for walling, straw bales may also be used for roof insulation.

The features of a typical load-bearing straw bale wall are shown in Fig. 6.1. Bales are laid horizontally to form a wall measuring around 500-mm wide once plastered. Timber stakes, such as hazel or in some cases broom handles, are used to secure the base of the wall to the footing and pin the bales together. Once the bales have been stacked to storey height (around seven or eight courses high), the wall plate is placed on top, and then the wall is compressed (prestressed), typically using external strapping wrapped around the wall, such as fencing wire or packaging tape, which greatly improves robustness of the wall. Thereafter, the internal plaster and external render coats of between 20- and 80-mm thick when complete are applied directly onto the straw in two or three coats. The simplicity of this form of construction is one of its great attractions but also a source of concern for a construction industry unfamiliar with it. Plastered straw bale walling has been likened to structural insulated panel construction (King, 1996), in which two structural skins are separated by lightweight insulation material (Kermani, 2006). Though bales are often trimmed before use, they typically undergo no further treatment or processing prior to use.

The enduring interest and continued development of straw bale construction is based on its beneficial properties in comparison to other forms of construction. First, straw bale walls offer excellent thermal insulation; 450- to 500-mm-thick walls provide U-values of $0.13-0.19$ W/m^2 K (depending on bale density and thermal

Figure 6.1 Load-bearing straw bale wall.

conductivity). The beneficial thermal properties of straw are further enhanced by its hygroscopic properties, which are discussed further in the following text. As a co-product of cereal production, straw is a renewable, inexpensive, and readily available resource for construction. Like other plant-based materials, cereal crops use carbon dioxide through the natural process of photosynthesis. Sodagar et al. (2011) estimated that 1 kg of straw sequesters 1.35 kg of carbon dioxide. As a renewable resource, plant-based materials are an effective means of removing and storing excess carbon dioxide from the atmosphere. Crop-based materials, such as straw, are a co-product of food production, so unlike crops such as *Miscanthus*, grown solely for bio-fuel, they do not place pressure on or compete with food production. Though the price of straw bales fluctuates depending on cereal production, straw bales remain one of the most affordable building materials. Straw has many other uses, including animal bedding, mushroom horticulture, and as low-grade biomass. However, the quantity of straw ploughed back into the land each year in the UK is sufficient to build at least 200,000 new straw bale houses (DEFRA, 2013). The technology of straw bale construction is relatively simple and readily transferred, making it popular among self-builders seeking low-carbon and affordable solutions.

The earliest surviving examples of straw bale construction are found in Nebraska, USA (King, 2006). Settlers seeking an affordable construction solution suited to the harsh winter climate of the Midwest, USA, adopted the technique. Following its initial success, straw bale construction was largely ignored until the 1970−80s, when self-builders exploring alternatives to conventional industrialised and fossil fuel−based materials started to experiment once again with straw bale construction (King, 2006). Over the past 20 years, straw bale construction has slowly gained increasing respectability from a sceptical construction industry. The introduction of innovations such as prefabricated panelised building has, without doubt, driven this wider acceptance. Straw bale buildings can now be found in all major inhabited continents, in countries such as South Africa, Mongolia, Pakistan, Chile, as well as throughout Europe, North America and Australasia. One of the oldest examples of European straw bale construction is the Feuillette House (Lamarche, 1921), built in 1921 in Montargis, France (Fig. 6.2).

Many straw bale buildings are low-rise, single- or two-storey construction (Fig. 6.3). The modest structural properties of straw bales allow load-bearing wall construction suited primarily to domestic use. Using an alternative structural system, often timber framing with straw as in-fill insulation, offers greater freedom of application. Though straw bale construction is presently a niche industry compared with many other forms of construction, with just an estimated few thousand buildings worldwide to date, modern innovations in straw bale construction have seen remarkable developments and wider acceptance in recent years.

This chapter aims to provide an overview of the current state of the art in understanding of materials and straw bale construction technologies. Stories of the three little pigs are rarely very far from news articles, reinforcing common conceptions that straw bales do not provide a durable or structurally resilient solution suited to modern construction. The properties of straw, including structural and hygrothermal properties, are outlined in the following sections. This is followed by a summary of work on characterisation, including resistance to fire and decay as well as assessment of the acoustic properties and life cycle of straw bales. Finally, the chapter outlines current and potential future applications of straw bales in construction.

Figure 6.2 Feuillette house, Montargis, France (built 1921).
Photo courtesy of RFCP.

Figure 6.3 Load-bearing straw bale wall under construction.

In the following sections the terms render and plaster refer to wet-applied coatings, typically cement-, lime- or earth-based, applied to straw bale walls. Render generally relates to external coating, while plaster is the interior coating but may be used more generically in reference to both coatings. North American texts also commonly refer to 'stuccos' when describing straw bale coatings.

6.2   Material properties

6.2.1   Plant growth

Straw bales are made up of the compacted stems of a cereal crop following harvest. The principal cereal crops that produce straw as a co-product are wheat, barley, oats, rye and rice. Across Europe, wheat straw is commonly used for straw bale construction due to its relative abundance, while the use of rice straw is more common in areas of the USA and in Asia. Barley and oat straw contains sufficient amounts of nutrition to be used as a feed supplement for livestock. However, wheat has a low nutritional value and is therefore mainly used for livestock bedding.

During the last century, cereal crops have been selectively bred to provide an increased yield. The result has been that modern plants have also been bred to be shorter than traditional varieties to prevent crops collapsing and spoiling. However, some contemporary hybrid varieties such as triticale (a cross between wheat and rye) are grown especially to provide long straws for use in the traditional roof thatching industry. The modern wheat crops that make up the straw bales used in construction are typically 500-mm tall compared with traditional varieties, which can grow to over a metre in height.

In addition to changes in the physical properties of straw, modern agricultural practice has led to a change in the purity of straw bales (Staniforth, 1979). The widespread use of herbicides means that most straw bales now have minimal weed content, while straw bales that are sourced from an organic production site will typically contain a greater proportion of grasses and broad-leaved weeds such as docks and thistles. These weeds are of higher nutritional value than the cereal straw and may therefore contribute to an increased susceptibility to mould growth or pests.

6.2.2   Structure and chemistry of straw

Cereal straw is made up of two distinct parts: the nodes and the internodes. The nodes are the point at which the plant's leaves grow upwards, and the internodes are the hollow tube sections synonymous with dried cereal straw (Fig. 6.4). As a whole the straw is made up of three main chemical compounds: cellulose, hemicellulose and lignin. These three constituents make up around 90% of the dry mass of straw, with the remainder formed of hot water-soluble elements and ash (Harper and Lynch, 1981). The insoluble ash fraction contains silica, which has several advantages for the use of straw in construction. For instance, the silica content makes combustion of straw difficult in power generation applications as well as reducing its digestibility.

6.2.3   Hygrothermal properties

Straw is a hygroscopic material, which means that it will adsorb water vapour from the air and absorb liquid water when exposed to a suitable source. Adsorption is the process by which water vapour molecules are held on the surface of a material through polar attraction. Materials that store water vapour in this way are termed hydrophilic. In the field of wood science, the point at which a material can store no more water vapour through adsorption is termed the fibre saturation point. At moisture content values greater than this point, liquid water will begin to form within the pores of the material, and this liquid moisture will be stored through capillary suction (Berry and Roderick, 2005). King (2006) notes that this behaviour is also applicable to straw.

An understanding of the hygrothermal properties of straw is important when using the material in a building. Moisture storage relationships between straw and the local microclimate within a wall are used for computational modelling of moisture transfer within building fabrics as well as for interpreting findings from field monitoring studies.

The adsorption behaviour of straw can be described through sorption isotherms, which are plots of the relationship between straw moisture content and ambient relative humidity. Isotherm sorption data have been reported by Lawrence et al. (2009a), Carfrae (2011) and Hedlin (1967) and are summarised in Fig. 6.5. Isotherm relationships are influenced by temperature and whether the straw is absorbing or desorbing water vapour; hysteresis is observed upon drying. The isotherm data presented in Fig. 6.5 are from adsorption tests that were completed between 20 and 23°C. Lawrence et al. (2009a) presented a model that can be used to determine the equilibrium moisture content of wheat straw for a given relative humidity. A plot of the model presented by Lawrence et al. (2009a) is shown in Fig. 6.5. This model can be a useful tool when reviewing relative humidity monitoring data. However, due to the dynamic nature

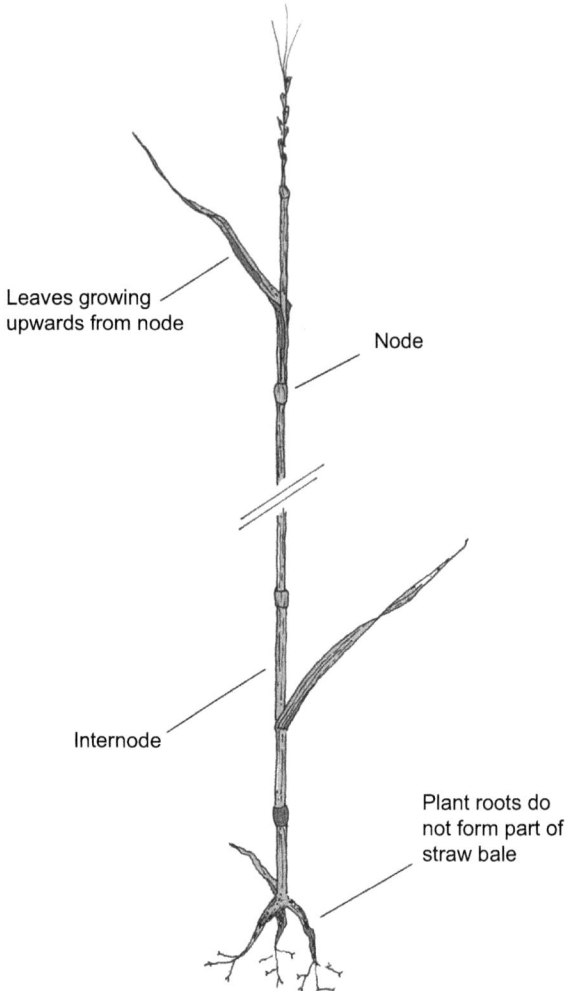

Figure 6.4 Structure of wheat straw.

of moisture transfer between hydrophilic materials and their surroundings and the asymptotic nature of the isotherm, this method is most suitable for assessing only general trends in straw moisture content.

6.2.4   Durability

When used appropriately, straw will not degrade within a building. The oldest surviving straw bale buildings are over 100 years old (King, 2006). However, similar to the use of untreated timber, prolonged exposure to damp or excessively humid conditions

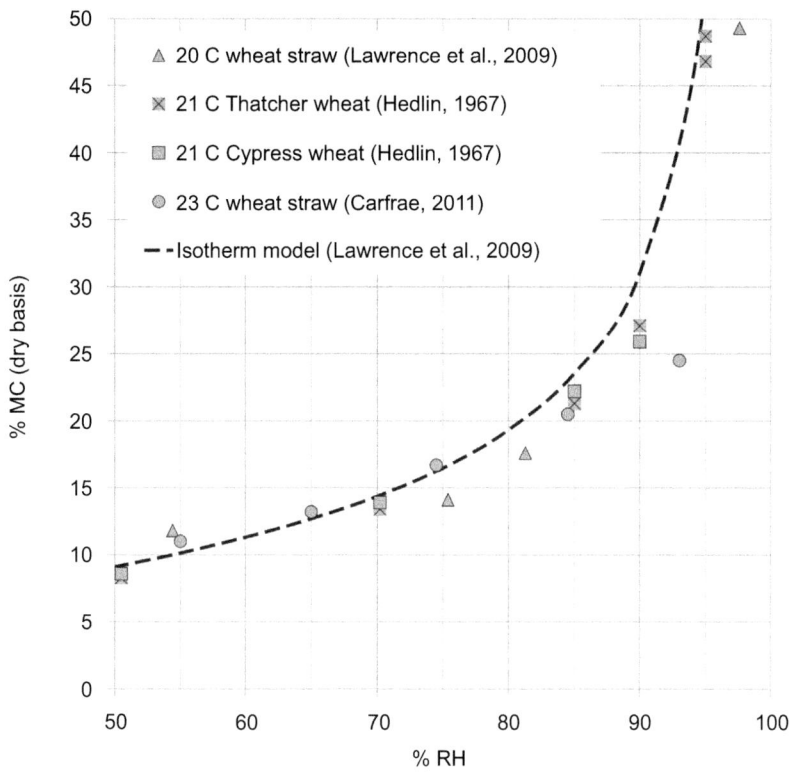

Figure 6.5 Isotherms for wheat straw.

can cause straw to degrade. Degradation risk may be associated with concentrated water ingress during construction or from leaking roofs or defective cladding.

In extreme conditions, straw may be subject to either aerobic or anaerobic degradation. Anaerobic degradation will occur where there is insufficient oxygen for the growth of microflora and is therefore associated with high levels of water and saturation of the straw. This form of degradation can be alarming if it occurs in a straw bale wall, but it is typically limited to a small area and is easily repaired once the water source has been removed.

Aerobic degradation of straw occurs as a result of mould or microflora growth on the straw. The extent and duration of mould growth is predominantly affected by the moisture content of the straw, which in turn is influenced by the prevailing microclimatic conditions around the straw. Experimental testing by the Fraunhofer Institute of Building Physics describes the different boundaries at which varying degrees of mould growth will initiate on wheat straw (Sedlbauer et al., 2011). The findings of the study were presented using a graphical isopleth system, which provides a simple guide to climatic conditions that can support mould growth on wheat straw and is shown in Fig. 6.6.

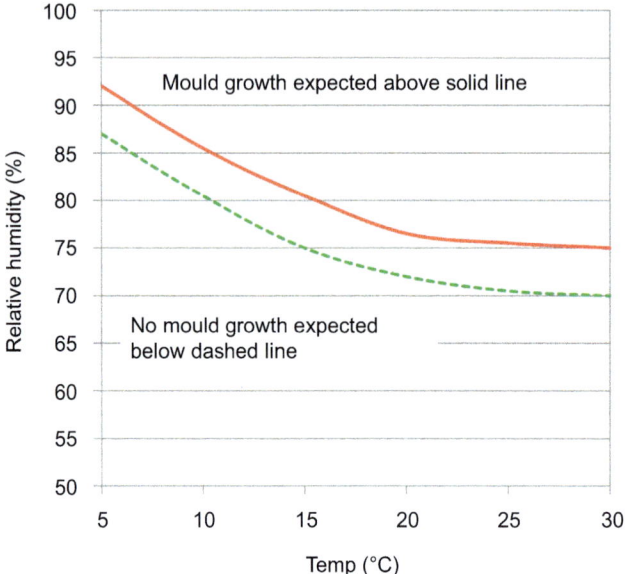

Figure 6.6 Wheat straw isopleth (Sedlbauer et al., 2011).

The Fraunhofer Institute study was initially undertaken to investigate risk of surface mould development on different building substrates, but it has been developed further to assess the degradation risk of straw bale insulation. Within the 100-day test exposure period, mould growth was not observed below 80% RH at 10°C or below 70% RH at 25°C. Above these levels, mould growth is reported to begin and to become progressively more aggressive as relative humidity increases.

The isopleth is a useful tool for assessing the risk of mould growth on straw bales. However, it does not provide an indication of the durability of straw over time. Research completed by Thomson and Walker (2014) used carbon dioxide sensors to monitor the extent of aerobic mould growth on straw stored in high-humidity conditions. The study found that when straw is kept in a high-humidity environment, initial mould growth is observed both visually and as an increase in carbon dioxide production (Fig. 6.7). However, after approximately 6 days, this growth slows to a much lower rate. Furthermore if the straw is dried and then re-exposed to high humidity, the level of microflora growth remains at this low level of growth (Fig. 6.8).

6.2.5 Materials used with straw bales

Straw can be baled into a variety of standard sizes, with the most common size used in construction being the two-string bale that measures approximately $1000 \times 450 \times 350$ mm. The minimum uncompressed density of the bales has been shown to be a key factor in its performance and should be approximately 100 kg/m³. The orientation of the bale has been reported to influence both the structural and thermal properties of the assembly (Vardy and MacDougall, 2006; FASBA 2009).

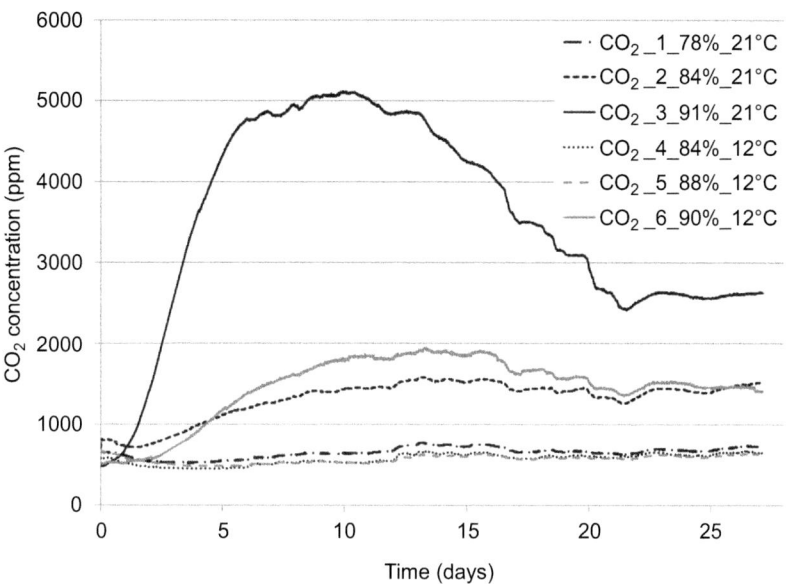

Figure 6.7 CO_2 evolution with time for fresh straw exposed to a first cycle of elevated humidity levels (Thomson and Walker, 2014).

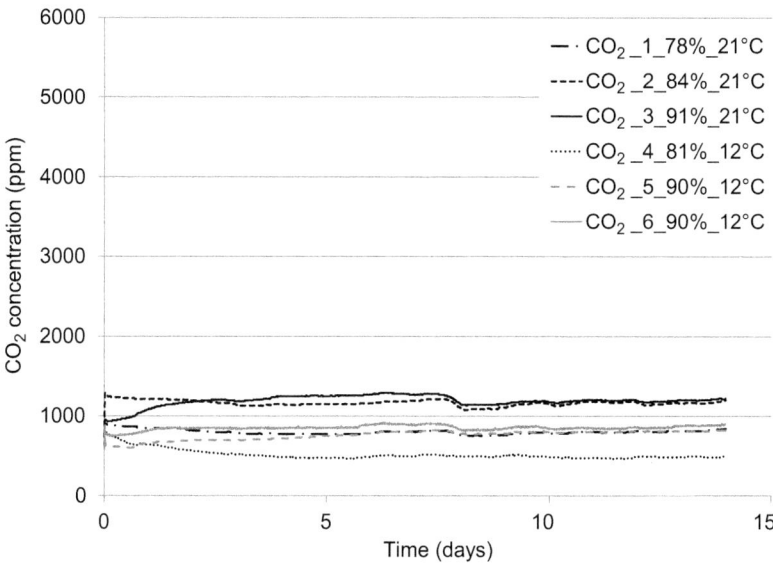

Figure 6.8 CO_2 evolution with time for the same straw exposed to a second cycle of elevated humidity levels (Thomson and Walker, 2014).

The wall is finished with a timber plate, which is used to precompress the bales to the required height. Following compaction during construction, this density can increase to around 130 kg/m^3. The wall is then typically rendered in several coats of a mineral-based render or could be dry lined with a panel product.

While straw bales can be used to provide vertical load-bearing capacity, vertical timber members may be used instead. The presence of a timber-framed structure can enable prefabrication off site, which leads to a panelised construction methodology and architecture. Prefabrication allows for the benefits of straw bale construction while removing some of the barriers to adoption while still utilising many of the standard materials.

The surface of a straw bale wall is typically covered in wet coatings (render and plaster) or dry lined with panel products such as oriented strand board, high-density wood fibre sheathing board or cross-laminated timber. When dry-lined solutions are adopted, it is necessary to use an internal finish that provides adequate fire protection such as fire-rated gypsum or cement board. The choice of surface covering plays a crucial role, not only in the aesthetic and textural finish of the wall but also in ensuring a wall's durability, fire performance, and resistance to pests and accidental damage as well as contributing hygrothermally and structurally.

Renders and plasters are a mineral-based coating that is wet-applied, often to both faces of the straw bale wall. Wet-applied coatings are typically applied directly to the straw or with the inclusion of a mesh or a render carrier with a thickness varying between 20 and 80 mm. The render is a mix of a binding agent, fine aggregate and water but could also include additives to enhance either the workability or the resulting performance.

The render type is defined by the binding agent and typically is lime, clay or cement based. Gypsum plasters have also been occasionally used on internal surfaces. The amount of water used is dependent on the chemical and physical makeup of the constituent dry mix, with the suitable amount of water defined by workability. Cement and hydraulic lime binders derive primary strength from hydraulic reaction when mixed with water, with further strength gain through carbonation. Weaker nonhydraulic lime binders derive strength only from carbonation. While there is no chemical reaction with water in earth (clay)-based coatings, water is essential for suction forces within the material that provides the binding forces.

The properties of the different renders can vary, with each requiring unique considerations. In Europe, lime- and earth-based coatings are most commonly used, while in North America cement (often combined with nonhydraulic lime) coatings are more common. While cement render is typically stronger and more resistant to erosion, it can also be more brittle, and so all except the leanest mixes are generally considered incompatible with the flexible straw substrate beneath. Both lime and earth are more flexible than cement and therefore are less likely to crack on the straw substrate. Due to the carbonation process of lime, if hairline cracks do occur, then lime exposed to the atmosphere can carbonate, effectively healing itself. As earth does not undergo any chemical change, the surface can be wetted and reworked to remove cracking. As the properties of the straw, specifically degradation, are sensitive to temperature and humidity, a breathable (water vapour—permeable) wall construction that allows for vapour diffusion is required. Both lime and clay have generally higher water vapour permeability than cement-based renders. Cement-based coatings are generally considered less vapour permeable, so they restrict the movement of moisture, which can trap moisture and lead to internal degradation of the straw; therefore their use is often not advised. Many hydraulic lime-based plasters are indeed blends of white cement and nonhydraulic lime.

In addition to providing a barrier to moisture for the straw bale wall, the render coating is the stiffest element in the straw bale assembly and as such attracts and transfers a significant proportion of the applied load. The thin skins of render are restrained from buckling through the bonding action with the straw bale, resulting in a composite stressed skin structure. The thickness of the render has been shown to be the greatest contributor to the strength of the unframed wall systems, with the render compressive strength having a smaller effect (Vardy, 2009).

6.3　Characterisation of straw bale construction materials

6.3.1　Physical and mechanical properties

Much of the original research into straw bale construction focused on the physical and mechanical properties intended for structural applications. While there has been a lack of standardisation with respect to both testing methodologies and design of the wall, there is consensus on structural properties.

Unplastered straw bale walls exhibit orthotropic (King, 2006) visco-inelastic behaviour (Walker, 2004) with single bales having a highly variable elastic modulus ranging between 0.05 and 0.9 N/mm^2 (Bou-Ali,1993; Watts et al., 1995; Vardy, 2009). Typical within straw bale construction is the addition of internal pins and application of precompression. Both of these additions help increase ultimate compression capacity and stiffness of the bales.

An applied render provides the most significant contribution to the strength and stiffness of the straw bale wall assembly. The render skins act as stiff elements attracting the applied load. Although these render skins themselves are very slender and susceptible to buckling, when there is good adhesion between the render and the straw, then the render is effectively restrained and buckling prevented (King, 2006; Vardy and MacDougall, 2012).

Experimental studies have shown that a rendered straw bale system is capable of sustaining a substantial load of up to 90 kN/m (Grandsaert, 1999; Vardy and MacDougall, 2006). These higher strengths, however, used a 38-mm-thick, weak cement-based render, with a compressive strength 1.72 N/mm^2. A straw bale wall with a 40-mm average thickness lime render (composed of two lime coats of compressive strengths of 3.2 and 2.6 N/mm^2) achieved an ultimate strength of 42 kN/m (Walker, 2004). The strength and stiffness of earth rendered walls were about 30% for a comparable thickness of lime (Brojan and Clouston, 2014).

Although a render is always recommended for load-bearing straw bale construction, there are potential problems with this design approach. Reliance on a brittle finish for structural resistance may not be acceptable to some approval authorities. Alternatively, a timber frame is often used with both on-site and prefabricated straw bale construction. While the timber frame can be designed to resist the vertical loading, the rendered straw bale still contributes to the overall structural system through improvement of the in-plane horizontal loading of the system.

6.3.2   Fire resistance

The various compounds that make up straw decompose at different temperatures. The initial decomposition of the lignin begins at 160°C, with the hemicellulose decomposition beginning at 180°C, followed by the cellulose proportion of the straw decomposing at approximately 260°C (Kozłowski and Władyka-Przybylak, 2008).

On building sites, there are common concerns regarding the fire resistance of straw bale buildings due to the risk associated with loose straw during construction. While loose straw can be a concern during construction, the risk of fire can be easily managed through safe site practice. Additionally, straw that is densely baled and used within a wall construction is a much lower risk on site as there is insufficient oxygen within the bale to support rapid combustion.

The application of render or dry-lining panels to a straw bale wall assembly provides a fire performance at least equal to that of many conventional construction methods. Additionally, testing has shown that even after a render coat has completely collapsed, the exposed straw bale wall retains its integrity. Straw exposed for 45 min to furnace temperatures of 1000°C did not ignite and was observed only to char (Wall et al., 2012) (Fig. 6.9). In a similar manner as timber, the charring of the outer layer of straw is thought to inhibit the rate of further straw decomposition as the fire progresses. The flameless combustion is due to the lack of available oxygen, and the rate of the smouldering reaction is dependent on the rate of oxygen transferred through the straw bale (Apte et al., 2008).

Render, or an alternative surface finish, is always recommended to provide initial thermal resistance and reduced spread of flame. Experimental results have shown that the temperature behind an approximately 45-mm-thick render is approximately half of that compared to the surface (Apte et al., 2008). The render skin can last up to 90 min before collapsing and exposing the straw beneath (Wall et al., 2012). This performance surpasses the 30-min minimum period of fire resistance required for many low-rise domestic applications in the UK. There are often further criteria for approved fire resistance, including the prevention of sustained flame and maximum temperature rises on the unexposed faces, which straw bale panels have been shown to meet.

Figure 6.9 Straw bale panel after fire test.

6.3.3 Hygrothermal properties and risk of decay

Straw bale walls are constructed with vapour-permeable materials, which allow water vapour to passively diffuse through them. Examples include lime-based renders, clay plasters and high-density wood fibre boards. The use of vapour-permeable coatings and finishes with straw bale insulation helps to mitigate the risk of degradation associated with accumulation of moisture within the straw.

Building fabrics are subjected to a constantly changing set of climatic conditions both internally and externally. This in turn influences the temperature, moisture content and moisture distribution through the thickness of a wall and between its constituent parts. The use of vapour-permeable construction accommodates dynamic hygrothermal conditions within a construction build-up. In a climate with large diurnal and annual variation in temperature and humidity, the distribution and level of moisture within a wall will be in a state of flux much of the time. Equally, in a predominantly warm and humid environment, the bulk moisture content of a vapour-permeable wall will likely be greater than that in a dry, arid region. The result is that consideration must be given to the hygrothermal performance of a straw bale building fabric in the context of the environment in which it is situated, including the likely heating and/or air conditioning demands of the enclosed space. There is, of course, scope for significant temperature, and relative humidity, gradients across the full thickness of a straw bale wall. Internal temperatures of $20-25°C$ could very easily be contrasted with external temperatures below zero. This has given rise to concerns about condensation forming at the external face within the straw, prompting concerns for durability. Evidence from practice suggests this is not a significant problem; however, further research is needed to better understand this complex behaviour.

There are several options available to designers of straw bale walls when considering the hygrothermal performance of a wall. The first is to seek local precedence. In sites exposed to wind-driven rain, it is typically recommended that a rain-screen cladding be used on the exposed face of the straw bale wall to prevent direct wetting of the wall finish (Wihan, 2007). If there are no straw bale buildings local to a proposed site, then a simulation of the wall's hygrothermal performance can be completed. This assessment will give an indication of whether there is a risk of moisture build-up within the straw bale insulation based on the local climate and building use. There are two main methods available. The first approach is to use a steady-state model of vapour flow through the wall following the method set out in ISO 13788:2012. However, this method does not make allowances for rainfall, which can be very significant if a rain-screen cladding is not used. The second approach is to use dynamic modelling software as described in EN 15026:2007, which can include the effects of heat and moisture storage, moisture penetration due to driving rain exposure and liquid transport under realistic initial and boundary conditions.

An increasing precedence of straw bale construction around the world has shown straw bale walls to be a durable form of construction in a wide range of climates. Monitoring studies of the hygrothermal conditions within straw bale walls have provided insight into the levels of moisture that straw within a wall may experience, and experimental data and site investigations have shown straw insulation to retain its integrity over time (Thomson and Walker, 2013). Fig. 6.10 shows a photograph from an opening-up survey completed at a straw bale exposure test facility located in Cornwall, UK. The wall panel shown was finished using a 30- to 40-mm-thick formulated lime render on the internal and external faces. The unheated test facility faced towards the prevailing southwesterly weather direction. The opening-up survey was completed 3 years after construction of the facility was built. It can be seen from inspection of Fig. 6.10 that the straw was found to be in a satisfactory condition.

The wall panel shown in Fig. 6.10 was deliberately positioned in an exposed location to maximise the risk of degradation. Indeed, hygrothermal monitoring of the panel showed that during some months the moisture content of the straw exceeded levels at which mould growth would be expected to occur. However, in line with the observations presented in Figs 6.7 and 6.8, the subsequent inspection has shown that any mould growth that may have occurred did not adversely affected the integrity of the straw bale insulation.

Figure 6.10 External render core removed from straw bale wall panel.

6.3.4 Acoustic properties

Acoustic properties are generally concerned with the transmittance of sound through the wall and with room acoustics considering the reverberation time and level. Sound reduction is not uniform across the whole sound spectrum. Most materials are better at reducing high-frequency noise. Since straw bale walls are a form of composite construction with a stiff surface finish and more flexible internal bales, good acoustic properties are achieved.

The weighted reduction parameter R_w is a common way to classify sound insulation. It takes account of typical spectra of disturbing noises and our sensitivity to different frequencies. Several studies have indicated that straw bale walls can achieve a sound reduction of 43−55 dB (Goodhew et al., 2010). Wall et al. (2012) compared to sound reduction of prefabricated panels in laboratory and in situ conditions in accordance with ISO 140-3 (1995) and ISO 140-5 (1998) series of acoustic testing. The panel tested in laboratory conditions achieved a sound reduction of 48 dB compared to 44 dB in the field. The values of sound reduction provided by straw bale walls equate to reducing loud speech to faint, nondistinguishable sound.

6.3.5 Life-cycle assessment of straw bale construction

Life cycle analysis (LCA) is the only method of considering the environmental impact that has been standardised through ISO 14040:2006 and ISO 14044:2006 and consists of four components: goal and scope definition, inventory analysis, impact assessment and interpretation. LCA is an iterative and bespoke technique. Therefore, any LCA must be considered in the context of the building, requiring project-specific inventory analysis and assumptions made typically useful.

Straw used for construction is typically a co-product from the agricultural industry and as such the allocation of its impact should be considered and transparent; an economic weighting is commonly used when attributing environmental impacts. Approximately 60% of the grains are used as straw, which is equivalent to 40% of the biomass; whereas the economic value of straw is approximately 15% of the grain (Stoddart and Watts, 2012). However, sensitivity analysis has indicated that choice of allocation of straw has limited impact on the LCA outcomes (Seguret, 2009).

The main advantages of straw bale construction of low embodied impact and low heating consumption have been confirmed by LCA. Due to photosynthesis during growth, the straw absorbs approximately 1.35 kg (Sodagar et al., 2011) of carbon dioxide for every kilogram of straw. This allows straw bale construction to be considered carbon-negative as a material and allows the potential to offset the carbon footprint of the building as a whole. However, the extent of carbon off-setting is dependent upon other potential impacts, including those associated with harvesting and transportation. A complete inventory analysis of straw indicates that it has a relatively high eutrophication potential, which can be attributed largely to the fertilisers and pesticides used during the growth (Seguret, 2009). Carbon stored within the straw will eventually be released back into atmosphere when it decomposes or is burnt after use as insulation. The key to carbon storage through use of straw, and other plant-based building materials, is to design and build robust, durable, well-performing and beautiful buildings that we will want to retain for many years before their demolition.

The operational (in-use) phase of a building contributes the most to its overall life-cycle impacts, predominantly due to heating requirements. This gives straw bale construction its main advantage: reduced heating requirements due to the high levels of thermal insulation.

Various end of life scenarios are usually considered during LCA. These include recycling, disposal by incineration and disposal by landfill. As expected, incineration of straw bales has a high global warming potential, while disposal to landfill has a high eutrophication potential. Recycling scenarios need to consider the allocation of impact from one project to the other. As straw bale construction is a composite form of construction, often using renders and a timber frame, these materials also need to be considered in any LCA. In addition to the superstructure, the choice of ground works and installed building services will have a large impact on the embodied and operational phases, respectively. The use of lightweight materials, such as straw bales, has potential further impact benefits through reduced material usage with smaller foundations, though this will be project specific.

Straw bale construction can have significantly lower environmental impact across the spectrum of categories considered by LCA when compared to conventional buildings. Compared to a conventional building, a straw bale building can have a reduction of 40−60% in global warming potential (Seguret, 2009).

6.4   Applications

In this section the uses of straw bales in construction are described, including a summary of key aspects of their use as an insulation material and practical detailing. Straw is unique as a natural insulation material in that it is used to form both load-bearing and non-load-bearing walls; both techniques are outlined below. Recent innovations in straw bale building, including the development of prefabricated panelised systems, are also described.

6.4.1 Straw bales as insulation

Straw bales provide excellent levels of thermal insulation by virtue of their relatively low thermal conductivity and wall thickness. In addition, straw, in common with other moisture responsive hygrothermal materials, such as hemp-lime, has further beneficial and complex dynamic thermal qualities. An accurate model of straw bale insulation requires a dynamic hygrothermal model.

The thermal conductivity of straw is relatively insensitive to bale density (Fig. 6.11). Shea et al. (2013) used a least squares regression analysis to determine apparent straw bale thermal conductivity of 0.064 W/m K at a density of 120 kg/m^3.

The German national organisation of straw bale building (FASBA) has published two different values for thermal conductivity of straw bales based on apparent orientation of the straw (FASBA, 2009). When the individual straws are oriented parallel to the direction of heat flow, the thermal conductivity is taken as 0.067 W/m K, whereas when the straw is aligned perpendicular to the direction of heat flow, the measured apparent thermal conductivity reduces to 0.043 W/m K. These values were derived from tests on small specimens under steady-state conditions in which all the individual straws where oriented in one direction. These findings are based on the misconception that straw in a typical bale is similarly oriented in one direction or another. In fact, it is much more randomly distributed.

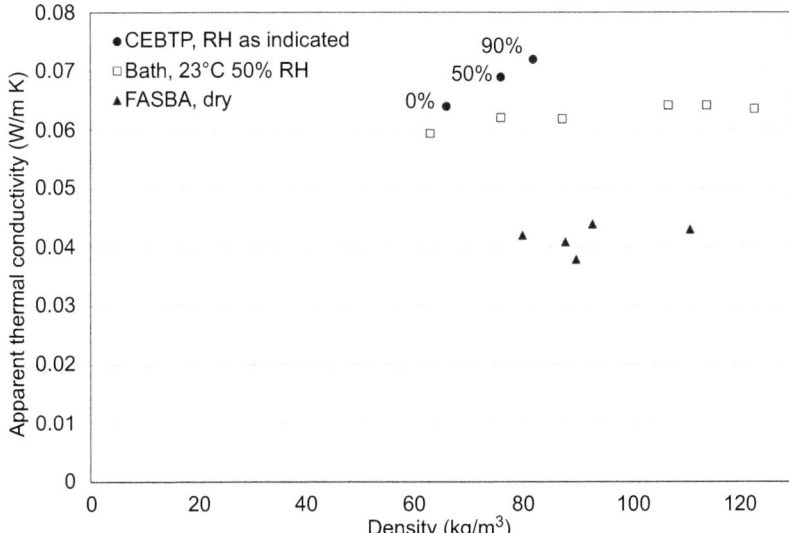

Figure 6.11 Variation of straw bale thermal conductivity with density (Shea et al., 2013).

The thermal conductivity of straw is also dependent on its moisture content, with conductivity increasing as moisture levels rise (Shea et al., 2013). Values are generally quoted for 'dry' material; whereas in practice, bales will be subject to varying relative humidity as weather conditions change. In tests by the Centre d'Expertise du Bâtiment et des Travaux Publics (CEBTP, 2004), when the straw relative humidity was increased from 0 to 90%, the thermal conductivity increased from 0.064 to 0.072 W/m K.

Straw bales are most commonly used for wall insulation; however, they can also be used for roof and under-floor insulation. The same properties as for wall insulation apply, although the same measures are required to protect straw in both applications.

6.4.2 Design of straw bale buildings

When designing the layout of straw bale buildings, it is beneficial to recognise that the bales are a modular unit, albeit of varying size. This will minimise cutting during construction, waste and reduce construction time. Bales are an agricultural co-product, not a standardised construction product, and consequently do vary in size. Bales from one batch using one baling machine should have consistent height and width but may vary in length by up to 100 mm. It is important to recognise straw bales are not supplied to the same tolerances required of many other construction products. Where bales are short, the gaps can be filled by hand with loose straw. Oversized bales can be cut with a hedge trimmer or chain saw, although cutting the twine will cause the bale to fall apart. Arguably more important than size is ensuring compliance with minimum density ($100-110$ kg/m^3) and permitted moisture content (not more than 20% by mass) in supply. Bales can be sourced from local farmers, although larger users will tend to go to straw bale merchants.

Details for straw bale buildings should protect them from build-up of excessive moisture content that could potentially lead to decay. Base and roof details should protect the straw from the ingress of moisture, mostly in the form of wind-driven rain. External wall faces rely on rendered or dry-lined cladding systems. Renders should be vapour permeable and dry-cladding systems vented and breathable to prevent build-up of moisture. In field trials, a vented timber rain-screen provided the most effective protection (Lawrence et al., 2009b). Like other natural materials, such as cob, suggested roof details for straw bale buildings generally specify large eave overhangs to provide further protection (King, 2006). However, more modern details have also been shown to provide effective protection (Fig. 6.12).

Window and door openings need to be detailed to accommodate straw as the insulation. When straw is load-bearing, this includes accommodating potentially large settlement movements. Increasingly designers are specifying full-height openings with straw to simplify the details. External details, such as sills around windows, should shed water away from the wall.

The relatively large thickness of straw bale walls has often been a barrier to adoption, with clients and developers keen to maximise usable floor area on a given site.

Figure 6.12 BaleHaus at the University of Bath.

However, as insulation specifications for modern buildings have increased, driven by legislative requirements, the thickness of straw bale walls is no longer at such odds with many competing solutions.

Plastered straw bale walls will accept mechanical attachments to support domestic features such as picture frames and bookcases, though larger items should normally be supported independently.

The fire resistance of rendered straw walls is normally in excess of 60 min. For dry-lined systems, the fire-resistant properties of the lining materials, including gypsum plasterboard, will typically govern resistance.

Electrical services can be readily incorporated into straw bale walls, concealing conduit, switches and other electrical items. However, due to the potential for leaks and condensation, plumbing is best detailed away from walls or surface mounted.

Infestation by rodents is often cited as a concern of straw bale buildings. Rodents might be attracted by a potential source of food and warm bedding. Straw should provide minimal food but like other forms of construction will provide warm and dry shelter. Robust detailing, including metal mesh and screening, together with regular maintenance should provide sufficient protection. Incidence of book lice (psocids) outbreaks have been noted in some plastered projects, as the book lice are attracted by fungal growth when the plaster dampens the straw. Experience suggests outbreaks of book lice are short-lived and easily dealt with.

6.4.3 In situ straw bale construction

Straw bale construction developed as an in situ method of building; like conventional forms of construction such as masonry, the walls are erected on site in their final location. While this is an efficient form of construction, minimising handling of materials, it poses challenges for wider acceptance of straw bale construction. The most significant of these is the need to keep the bales protected from wet weather while stored on site and during the construction process. In climates where there is a reliable dry season, this can be relatively straightforward if construction can be programmed to fit in with the weather, but in climates where the weather is more variable, consideration for keeping bales dry must be made; these are discussed below. Other specific considerations when using straw bales on site include minimising fire risk of loose cut straw; storage of a bulky material; and handling an agricultural co-product of variable size, shape and quality.

There are two main methods of in situ straw bale construction: load-bearing and non-load-bearing (or in-fill) construction.

6.4.3.1 Load-bearing straw bale construction

Straw bale construction developed as a load-bearing form of construction, and this still represents the most efficient use of the material in low-rise buildings. As previously described, the bales are laid in 'running bond' courses as large masonry units (Fig. 6.13). The vertical (perpend) joints are broken by using cut bales, which can be readily created on site using twine and a bale needle. During construction the straw bale walls are quite flexible and potentially unstable, possibly requiring propping, until the wall plate is in place and the vertical precompression is applied. The irregular and slightly rounded nature of straw bales can further contribute to the temporary instability of walls during construction. However, propping and realignment is

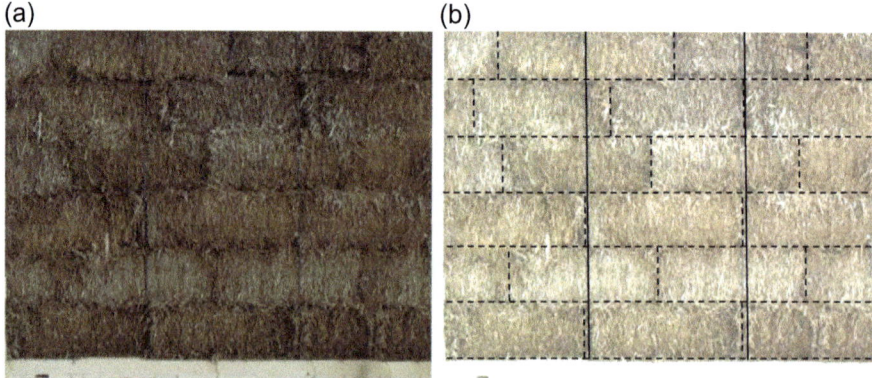

Figure 6.13 Running bond coursing of straw bales. (a) Photo. (b) Enhanced photo highlighting running bond and precompression cables.

relatively easy as the wall is very lightweight and flexible. Precompression is usually applied to the wall by tightening fencing wire or packaging tape wrapped around the base plate and upper wall plate. The amount of precompression needed depends largely on the initial density of the straw bales used, but a single-storey-height wall (2.4–2.8 m) will generally be compressed by at least 20 mm and sometimes by up to 80 mm. This movement needs to be incorporated in the design of any openings and other features. Once the bale walls have been compacted, they can be rendered; however, some guidance suggests leaving the walls to settle for 6–8 weeks before rendering to minimise risk of further movement and possible cracking of the coating (Jones, 2009).

Load-bearing straw bale walls are generally rendered, as the render forms an integral structural skin for the walls (Fig. 6.14). In North America, it is common to reinforce the render coats with a metallic or polymer mesh to minimise cracking, although this is a less common practice in the UK. Unrendered straw bale walls can be load-bearing, though their capacity is significantly reduced by the lack of render coatings. The design service loading of a single-storey (less than 3-m-high), fully rendered, 450-mm-thick straw bale wall is approximately 8–10 kN/m (King, 2006). Vertical loads should be applied uniformly, avoiding large concentrations of load to reduce risk of excessive localised settlement. Vardy and MacDougall (2006) proposed a method for designing strength of cement rendered walls based on compressive

Figure 6.14 Rendering of straw bale wall. (a) Manual finishing off of lime render. (b) Lime render sprayed directly onto surface of the straw bales.

strength of the render skins. Where the render skins are effectively restrained by the straw, there seems to be little risk of the slender render coatings buckling under the vertical load, though the longer term effectiveness of this straw restraint is unclear with weathering cycles. The slenderness (height to thickness) ratio of load-bearing straw bale walls is generally around 5−7, low compared with many other forms of walling. The global buckling of rendered straw bale walls is presently unknown; consequently there are no guidelines for the effects of wall slenderness ratio on vertical compression resistance.

The flexible tensile and in-plane shear (racking) resistance of straw bale walls relies on the render. The out-of-plane flexural strength of straw bale walls can be estimated based on a structural double skin and modulus of rupture of the render. For general low-rise load-bearing walls, however, flexural resistance is unlikely to be a major design concern. Knowledge of the racking resistance of load-bearing straw bale walls is also not well developed, although there have been tests reported on prefabricated timber panels (discussed later). Nonetheless, shake table testing of a simple building has shown resilience of straw bale building to seismic loading (http://www.paksbab.org).

Load-bearing straw bale construction relies on using the insulated walls to support the roof structure. Consequently, alternative approaches are needed during construction to protect the bales from wet weather. For modest-sized projects the most common solution is to use temporary tarpaulin sheet covers to protect the construction. For larger scale projects an independent temporary shelter may be used (Fig. 6.15);

Figure 6.15 Temporary shelter for straw bale construction.

however this can add significant additional expense. Jones (2009) and others have developed an innovative solution to this problem by using temporary supports for the permanent roof structure, and then lowering the roof as one onto the walls when ready. The roof structure also provides uniform compression of the bale walls before rendering.

6.4.3.2 Non-load-bearing applications

Non-load-bearing or in-fill straw bale construction seeks to use the material's excellent thermal insulation qualities while relying on an alternative structural system, typically timber framing. The framing system should be modular to account for the size of the bales used to minimise the need for cutting. There are a number of advantages of this approach. First, it minimises the risk, as the structural resilience of the building is not reliant on the straw. To date, mortgage and insurance companies have been more likely to finance this form of construction. In-fill also enables construction of the protective frame and roof before insertion of the straw on site, minimising the risk of weather damage to the bales during construction (Fig. 6.16). The use of an independent structural frame also provides potentially greater flexibility in architecture and engineering design. While load-bearing straw bales rely on the heavy lime- or cement-based renders to provide structural capacity, non-load-bearing straw bales can receive a much wider variety of dry-cladding systems.

Figure 6.16 Non-load-bearing straw bale building under construction.

6.4.4 *Prefabricated panel construction*

To overcome the limitations of in situ straw bale construction, a variety of panelised construction systems have developed independently around the world over the past 10−15 years. Though solutions vary, including fully off-site techniques and on-site tilt-up construction approaches, they all seek to maximise the benefits of prefabricated construction and open up the wider construction industry to straw bale construction. Panels are typically timber frames with straw bale insulation inserted off site (Fig. 6.17). Some panels are comprised of external frames only, while others use internal studwork as well. Most panels are finished with a dry-cladding system; some approaches have also used hydraulic lime renders.

Panelised construction provides a system that is more acceptable to a wider range of stakeholders and that now has potential to be certified as a construction product. The benefits of prefabricated straw bale construction include reduced risk of programme delays due to inclement weather; improved quality control; faster on-site construction programme; and removal of straw bales from the construction site. These benefits can sometimes be offset, however, by increased costs of construction compared to traditional on-site load-bearing construction.

Panels may be simply cladding attached to a structural frame, as shown in Fig. 6.18, but they may also form the entire structural system (Fig. 6.19). In panel systems the primary structural resistance is generally derived from the timber frames. Non-load-bearing panels need sufficient structural resilience for lifting and handling

Figure 6.17 Fabrication of prefabricated straw bale panels.

Figure 6.18 Straw bale cladding panels at the Inspire Bradford project, Bradford, UK, 2012.

Figure 6.19 Load-bearing straw bale panels at the LILAC project, Leeds, UK, 2013. Courtesy of ModCell.

during construction and to resist wind pressures without panel failure. Load-bearing panels use the timber frame to carry vertical load as well as resist lateral loading (wind, seismic). Vertical loading resistance is normally derived from the timber elements only; however under racking, interaction between the frame, straw and render is evident.

There have been significant developments in prefabricated straw bale construction in recent years. Recently in France, the eight-storey, straw bale panel—clad Résidence Jules Ferry building in St-Dié-des-Vosges has been completed.

6.5 Future of straw bale construction

For many years, straw bale construction was largely dismissed by the mainstream construction industry, fed by concerns and some misconceptions about durability, fire risk, infestations and inability to fit into modern procurement processes. However, advances in scientific understanding of straw bale construction combined with technological advances, especially prefabricated panel construction, have seen significant developments over the past 10 years. Panel development has recently enabled the certification of straw bale products in the UK. Certified products will enable further market development as it will support wider acceptance and enable financial support and warranty of projects, particularly important for growth in a domestic housing market. Wider certification of straw bale products will, however, require a supply chain of construction-certified straw bales, which are currently only available from very limited sources.

Unlike many other construction materials and products, which are supported by great many national and international standards and design codes, no such framework currently exists for straw bale construction. This will require lobbying and coordinated action by straw bale industry stakeholders to move forward with the appropriate authorities worldwide.

Continued research and development of straw bale construction will lead to improved understanding of complex performance aspects such as dynamic hygrothermal performance and durability. Performance monitoring of a growing number of straw bale buildings is supporting this knowledge development.

Growth in straw bale construction may well now also support growth in other straw-related construction products. Compressed straw board is one such product which may now be ready for significant market expansion. Despite previous success in the 1950—80s, the use of compressed straw board in the UK has almost ceased, with only imported products being available. Other straw-based composite building products, using compatible binders and resins, can also be expected.

As a renewable resource and co-product of an essential agricultural process, with excellent building performance qualities, straw is uniquely well placed as a modern construction material and product to deliver sustainable buildings for the 21st century and beyond. Straw bale construction offers a solution that combines both low embodied energy and carbon with low operational energy and carbon impact.

The available supply of straw should support significant market development, such that all future housing in the UK, for example, could, in principle at least, be built using sustainable straw bale technologies.

References

Apte, V., Griffin, G.J., Paroz, B.W., Bicknell, A.D., 2008. The fire behaviour of rendered straw bales. Fire and Materials 32 (5), 259−279.

Berry, S., Roderick, M., 2005. Plant-water relations and the fibre saturation point. New Phytologist 168, 25−37.

Bou-Ali, G., 1993. Straw Bale Wall Systems. University of Arizona (Civil Engineering, Thesis).

Brojan, L., Clouston, P.L., 2014. Effect of plaster type and loading orientation on compression behaviour of straw bales for construction. Journal of Engineering & Applied Sciences 9 (9).

Carfrae, J., 2011. The Moisture Performance of Straw Bale Construction in a Temperate Maritime Climate. University of Plymouth, UK (PhD thesis).

Centre d'Expertise du Bâtiment et des Travaux Publics (CEBTP), 2004. Utilisation de la Paille en Parois de Maison Individuelles à Ossaure Bois (Use of Straw for Walling of Detached Timber Framed Houses).

DEFRA, 2013. https://www.gov.uk/government/uploads/system/uploads/attachment_data/file/141626/defra-stats-foodfarm-landuselivestock-nonfoodcrops-latestrelease-130125.pdf (accessed 18.02.15.).

EN 15026, 2007. Hygrothermal Performance of Building Components and Building Elements − Assessment of Moisture Transfer by Numerical Simulation.

Fachverband Strohballenbau (FASBA), 2009. Thermal Performance: Strawbale Building Research Development 2003−2009. FASBA, Germany.

Goodhew, S., Carfrae, J., De Wilde, P., 2010. Briefing: challenges related to straw bale construction. Proceedings of the ICE-Engineering Sustainability 163 (4), 185−189.

Grandsaert, M., 1999. A Compression Test of Plastered Straw-bale Walls. University of Colorado at Boulder, Colorado (MS thesis).

Harper, S., Lynch, J., 1981. The chemical components and decomposition of wheat straw leaves, internodes and nodes. Journal of the Science of Food and Agriculture 32 (11), 1057−1062.

Hedlin, C., 1967. Sorption isotherms for five different types of wheat straw. Canadian Agricultural Engineering 9 (1), 37−42.

ISO 13788, 2012. Hygrothermal Performance of Building Components and Building Elements. Internal Surface Temperature to Avoid Critical Surface Humidity and Interstitial Condensation. Calculation Methods. International Standards Organization (ISO).

ISO 140-3, 1995. Acoustics − Measurement of Sound Insulation in Buildings and of Building Elements − Part 3: Laboratory Measurement of Airborne Sound Insulation of Building Elements. International Standards Organization (ISO).

ISO 14040, 2006. Environmental Management − Life Cycle Assessment − Principles and Framework. International Standards Organization (ISO).

ISO 14044, 2006. Environmental Management − Life Cycle Assessment − Requirements and Guidelines. International Standards Organization (ISO).

ISO 140-5, 1998. Acoustics — Measurement of Sound Insulation in Buildings and of Building Elements — Part 5: Field Measurements of Airborne Sound Insulation of Façades Elements and Façades. International Standards Organization (ISO).

Jones, B., 2009. Building with Straw Bales: A Practical Guide for the UK and Ireland. Green Books, 160 pp.

Kermani, A., 2006. Performance of structural insulated panels. Proceedings of the ICE — Structures and Buildings 159 (1), 13—19.

King, B., 1996. Buildings of Earth and Straw. Ecological Design Press, Sausalito, California, 169 pp.

King, B., 2006. Design of Straw Bale Buildings. Green Building Press, San Rafael, California, 259 pp.

Kozłowski, R., Władyka-Przybylak, M., 2008. Flammability and fire resistance of composites reinforced by natural fibers. Polymers for Advanced Technologies 19 (6), 446—453.

Lamarche, G., 1921. Fraiches en ete, chaudes en hiver, les maisons de paille sont avant tout economiques. La Science et la Vie 56, 481—486.

Lawrence, M., Heath, A., Walker, P., 2009a. Determining moisture levels in straw bale construction. Construction and Building Materials 23, 2763—2768.

Lawrence, M., Heath, A., Walker, P., 2009b. The impact of external finishes on the weather resistance of straw bale walls. In: 11th International Conference on Non-conventional Materials and Technologies, NOCMAT 2009. University of Bath, UK.

Sedlbauer, K., Hofbauer, W., Krueger, N., Mayer, F., Breuer, K., 2011. Material specific isopleth-systems as valuable tools for the assessment of the durability of building materials against mould infestation — the 'isopleth-traffic light'. In: International Conference on Building Materials and Components, Porto, Portugal.

Seguret, A., 2009. Life Cycle Assessment of BaleHaus at Bath: Environmental Impacts of a Straw-bale Building over its Life-cycle. Imperial College London, London (MSc thesis).

Shea, A., Wall, K., Walker, P., 2013. Evaluation of the thermal performance of an innovative prefabricated natural plant fibre building system. Building Services Engineering Research 34 (4), 369—380.

Sodagar, B., Rai, D., Jones, B., Wihan, J., Fieldson, R., 2011. The carbon-reduction potential of straw-bale housing. Building Research & Information 39 (1), 51—65.

Staniforth, A., 1979. Cereal Straw. Oxford University Press, Oxford, 169 pp.

Stoddart, H., Watts, J., 2012. Energy Potential From UK Arable Agriculture: Straw — What Is It Good for? HGCA, Stoneleigh.

Thomson, A., Walker, P., 2013. Condition Monitoring and Durability Assessment of Straw Bale Construction. Portugal SB13: Contribution of Sustainable Building to Meet EU 20-20-20 Targets, pp. 791—798.

Thomson, A., Walker, P., 2014. Durability characteristics of straw bales in building envelopes. Construction and Building Materials 68, 135—141.

Vardy, S., MacDougall, C., 2006. Compressive testing and analysis of plastered straw bales. Journal of Green Building 1 (1), 63—79.

Vardy, S., MacDougall, C., 2012. Concentric and eccentric compression experiments of plastered straw bale assemblies. Journal of Structural Engineering 139 (3), 448—461.

Vardy, S., 2009. Structural Behaviour of Plastered Straw Bale Assemblies under Concentric and Eccentric Loading. Queen's University, Kingston, ON, Canada (PhD thesis).

Walker, P., 2004. Compression Load Testing of Straw Bale Walls. Department of Architecture and Civil Engineering, University of Bath, Bath, UK.

Wall, K., Walker, P., Gross, C., White, C., Mander, T., 2012. Development and testing of a prototype straw bale house. Proceedings of the ICE-Construction Materials 165 (6), 377–384.

Watts, K.C., Wilkie, K.I., Thompson, K., 1995. Thermal and Mechanical Properties of Straw Bales as They Relate to a Straw House. Canadian Society of Agricultural Engineering.

Wihan, J., 2007. Humidity in Straw Bale Walls and its Effect on the Decomposition of Straw. University of East London, UK (PhD thesis).

Part Three

Concrete and masonry

Utilization of industrial by-products and natural ashes in mortar and concrete: development of sustainable construction materials

7

R. Siddique, Kunal
Thapar University, Patiala, Punjab, India

7.1 Introduction

Mortar and concrete are among the most durable and widely used building materials due to their low cost, general availability, and wide applicability. Advancement in concrete technology in recent years is generally focused on the strength and durability of structures resulting from the improved design of mortar and concrete. However, this advancement carries with it environmental considerations, and it is necessary to move toward consideration of the sustainability of mortar and concrete construction. In addition to the sheer volume of materials required to produce billions of tons concrete worldwide, energy and water consumption and the generation of construction and demolition waste is not compatible with the demands of sustainable construction. Worldwide, the cement industry alone is estimated to generate about 7% of all CO_2 and other greenhouse gases (GHGs). It is estimated that production of one ton of cement releases about one ton of CO_2 and other GHGs into the atmosphere (Kunal et al., 2012). Use of natural sand as fine aggregate also creates environmental burdens in terms of mining and quarrying of river beds (Subramanian and Kannan, 2013). This indicates that the construction industry has become a victim of its own success and ubiquity, and therefore, it is now facing tremendous challenges for sustainable development of building and construction materials.

Solid waste management has also become a principle environment concern with ever-increasing quantities of industrial by-products and waste materials generated globally. Due to the scarcity of landfill space and its ever-increasing cost, recycling of by-products and waste has become an attractive alternative to disposal. Several types of by-products and waste material are considered for use in cement-based materials, and each of these have specific effects on the properties of the materials. The utilization of by-products and waste material in concrete is not only economical, but it also helps in reducing disposal problems by removing the products from the waste stream altogether. In recent developments, the increasing use of nonconventional cementitious materials and industrial by-products, also known as supplementary cementitious materials, such as

Nonconventional and Vernacular Construction Materials. http://dx.doi.org/10.1016/B978-0-08-100038-0.00007-X

cement kiln dust, fly ash, silica fume, waste foundry sand, coal bottom ash, etc., and natural ashes such as palm oil fuel, rice husk ash, etc., have been made. Each impacts the mechanical properties and durability of the resulting mortar and concrete. This chapter provides a review of the current state of the art and of practice for the use of nonconventional cementitious materials and industrial by-products in mortar and concrete.

7.2 Cement kiln dust

In the process of continuous feeding of raw materials and rapid flow of combustion gases typical of the production of Portland cement, large quantities of particulate material trapped in the gases are removed from the kiln by air pollution control devices. Cement kiln dust (CKD) is a fine powdery, micron-sized by-pass dust collected in the control devices such as cyclones, bag houses, or electrostatic precipitators during the production of cement clinker. CKD is similar in appearance to Portland cement. The development of the cement manufacturing process is characterized by wet- or dry-process routes. When the raw materials are wet or exhibit large fluctuations in chemical composition of the individual raw mix components, wet kilns are used for easier handling and homogenization of raw materials. This process accepts feed material in the form of slurry containing 30–40% water. The CKD removed from a wet kiln is termed waste dust. In dry-process kilns, raw materials are in a dry-grounded form, ie, without the addition of water to prepare slurry. The dust collected from dry-process kilns has less embodied energy than the wet process kilns due to the lower fuel consumption of the dry kiln process. In either process, the dust removed from the clinker cooler end of the kiln, termed by-pass dust, may contain heavy metals, alkalis, and chlorides that may adversely affect its subsequent use.

In 2013, over 4 billion tons of cement was manufactured globally (Oss, 2014). India is the second largest cement producer in the world after China (2300 million tons) with the total capacity of 280 million tons. According to the US Environmental Protection Agency (EPA, 1993), the generation of CKD is estimated to be 15–20% of clinker or cement production, which puts worldwide CKD generation at an estimated 700 million tons and Indian production at 42 to 56 million tons in 2013. The most common uses of CKD are soil stabilization, cement replacement, waste treatment, and asphalt pavement.

The chemical composition of CKD is quite varied due to variation in raw feed materials, fuels, and kiln type at each plant (Hawkins et al., 2004). The concentration of alkalis (Na_2O and K_2O) and free lime sulfates in CKD mainly depend upon the particle size. Coarser particles contain high free lime content, while fine particles exhibit higher concentrations of sulfates and alkalis and a lower free lime content, as shown in Table 7.1 (Corish and Coleman, 1995). The alkalis and loss on ignition (LOI) in CKD are higher than in ordinary Portland cement (OPC), while the calcium oxide (CaO) content is lower than that of OPC. The high LOI indicates that less free lime is available for hydration. Analysis of cement kiln dust with X-ray diffraction studies revealed that limestone ($CaCO_3$) is the major component of CKD; whereas, quartz (SiO_4) together with small quantities of gypsum ($CaSO_4$), sodium chloride (NaCl), arcanite (K_2SO_4), spurite $[2(C_2S) \cdot CaCO_3]$, and sulfospurite $[2(C_2S) \cdot CaSO_4]$ are typically present.

Table 7.1 **Effect of particle size on alkali content of CKD (Corish and Coleman, 1995)**

Particle size (microns)	Mass (%)	Na$_2$O (%)	K$_2$O (%)
46−68	0.3	0.3	3.6
34−46	0.4	0.3	3.5
24−34	0.7	0.4	4.5
17−24	1.8	0.4	5.1
12−17	5.1	0.4	5.2
6−12	27.3	0.3	5.4
0−6	64.4	0.4	10.7

Table 7.2 **Typical physical properties of CKD**

Property	Collins and Emery (1983)	Maslehuddin et al. (2008)	Peethamparan (2006)	Marku et al. (2012)
Gradation (75% passing)	0.030 mm (no. 450 sieve)	−	−	−
Maximum particle size	0.300 mm (no. 50 sieve)	−	−	−
Specific surface area (cm^2/g)	4600−14,000	−	−	−
Specific gravity	2.6−2.8	2.4	2.42	2.96

7.2.1 Physical properties

CKD is a fine, powdery material, gray to tan in color, of relatively uniform size, having specific gravity ranging from 2.40 to 2.96, less than that of Portland cement (specific gravity ≈ 3.15) as reported in several published literature (Table 7.2). Particle size distribution is an important physical characteristic of CKD and depends on the processing technology, method of dust collection, chemical composition, and alkali content of CKD (Corish and Coleman, 1995).

7.2.2 Chemical properties

CKD is derived from the same raw materials as ordinary Portland clinker, but still a significant variation in physical and chemical composition of CKDs obtained from different cement plants has been observed. Compounds of lime, iron, silica, and alumina constitute the major chemical composition of CKD. Generally, CKDs are characterized by higher alkali and sulfur content. The pH of CKD−water mixtures

Table 7.3 **Typical chemical composition of CKD**

Composition (%)	Cement kiln dust (CKD)				
	Udoeyo and Hyee (2002)	Al-Harthy et al. (2003)	El-Aleem et al. (2005)	Sreekrishnavilasam et al. (2006)	Maslehuddin et al. (2008)
CaO	52.72	63.80	42.99	43.99	49.3
SiO_2	2.16	15.80	13.37	15.05	17.1
Al_2O_3	1.09	3.60	3.36	6.75	4.24
MgO	0.68	1.90	1.90	1.64	1.14
Na_2O	–	0.30	3.32	0.69	3.84
K_2O	0.11	3.00	3.32	4.00	2.18
Fe_2O_3	0.54	2.80	2.29	2.23	2.89
SO_3	0.05	1.70	5.10	6.02	3.56
LOI	42.39	–	15.96	21.57	15.8

is generally more than 12 and is considered to be caustic. Certain trace metals such as cadmium, lead, selenium, and radionuclides are generally found in concentrations less than 0.05% by weight in CKD. Table 7.3 lists some typical chemical properties of CKD reported by several researchers.

7.2.3 Effect of CKD on the properties of mortar and concrete

CKD's cement-like properties make it a potential replacement for Portland cement, although its inclusion affects the properties of the resulting mortar and concrete. Several studies have been reported on aspects of CKD utilization in cement paste, mortar, and concrete (eg, Marku et al., 2012; Pavia and Regan, 2010; Maslehuddin et al., 2009; Al-Jabri et al., 2006).

7.2.3.1 Setting time

The density of CKD is lower than that of Portland cement (PC) meaning that more CKD particles are required to replace cement, affecting properties of cement−CKD blends. Initial setting time of PC is important and desirable for concrete to harden and develop strength (Marku et al., 2012). Generally, with increase in water demand, setting time increases. However, it was observed that replacement of up to 10% PC with CKD increased water demand and decreased setting time (Kunal et al., 2014a,b; Maslehuddin et al., 2008; El-Aleem et al., 2005; Ramakrishnan, 1986). El-Aleem et al. (2005) concluded that decrease of both initial and final setting times was almost linear as a function of CKD replacement. The initial and final set time in control OPC was approximately 135 and 230 min, respectively, which decreased to 65 and 110 min with 10% CKD replacement. The decrease in setting time was due to the presence of high amounts of free lime and alkalis in CKD which resulted in accelerated hydration and faster setting time.

Bhatty (1984, 1986) published a series of reports on the addition of CKD, and they reported that CKD−cement blends reduced strength, setting time, and workability due to the presence of high alkali content. This was improved by the addition of fly ash with CKD, which lowered the alkali content. Bhatty (1984) stated that if CKD is characterized by high sulfate, high free lime, and low chloride (0.26%) contents, then 20% CKD addition would have shorter initial setting time. On the contrary, Daous (2004) reported that addition of CKD as partial replacement of cement increased setting time of the cement blend. Udoeyo and Hyee (2002) found that the values were, nonetheless, within the relevant BS and ASTM standards.

CKD also affects the rheological properties of slag−cement composites. Heikal et al. (2002) reported that addition of 2.5% CKD by mass to cement mixes containing 30% and 50% granulated slag reduced the initial and final setting time; whereas in 70% granulated slag−30% cement mix, setting time was increased. Konsta-Gdoutos and Shah (2003) observed shorter setting time for blends having high sulfate content (17%) in CKD.

7.2.3.2   Hydration properties

The high alkali and sulfate content in CKD makes it an excellent activator for pozzolanic materials. Taylor (1997) reported that class F fly ash (FA) reacts slowly at low temperature due to low pozzolanic reactivity. Blending alkaline CKD was reported to activate this pozzolanic reaction resulting in a better performing cementitious material. Dyer et al. (1999) also prepared ternary blends containing two types of CKD, pulverized fuel ash, and OPC. It was reported that the addition of CKD accelerated the binder hydration.

Decrease in hydration of CKD−cement pastes was observed by Kunal et al. (2014a) at early ages of curing (7 days); whereas with increasing curing age (28 and 91 days), increased hydration was observed up to 10% CKD replacement levels. Further increase in CKD content, greater than 10%, resulted in a decrease in hydration. The reduction in hydration was due to the leaching of calcium and silica from the kiln dust during mixing with water resulting in less material available for the formation of hydration products such as calcium silicate hydrate (CSH). Konsta-Gdoutos and Shah (2003) reported the use of four different types of CKD ('E,' 'P,' 'A,' and 'X') as activators for ground granulated blast furnace slag (GGBFS) and investigated the effect of CKD on the heat of hydration. CKD 'E' was the coarsest, with a mean diameter of 60 µm; whereas CKD 'A' was the finest material used with a specific surface area of 8270 cm^2/g and a mean equivalent spherical diameter of 9 µm. Approximately 75% of CKD 'P' and 'X' particles were in a range between 1 and 45 µm. X-ray diffraction (XRD) analysis showed that in CKD 'E' activation of slag calcium hydroxide (CH) was not observed after 7 days of hydration, indicating the consumption of all of the available lime. In the CKD 'X' activated slag mixture after 28 days of hydration, unreacted CH and formation of ettringite was detected.

With increasing calcination temperature of CKD, slag activation increases (Amin et al., 1995). Calcination of CKD at 1300°C with 15% anhydrite is suitable for the production of supersulfated cement (El-Didamony et al., 1997). Wang et al. (2004) reported that up to 120 days of curing (at 24, 38, and 50°C) the ettringite formation in a CKD-FA paste was quite stable due to higher content of sulfur and aluminum in CKD. Increasing curing temperature from 38 to 50°C reduced the formation of ettringite. With addition of NaOH (0%, 2%, and 5%), ettringite and CH formation was observed to decrease under similar curing temperature; additionally, a greater reduction was observed at lower temperatures.

7.2.3.3 Strength properties

Although there are some contradictory results reported, the replacement of cement with CKD up to about 5−10% has little impact on compressive strength (Maslehuddin et al., 2008; El-Sayed et al., 1991; Batis et al., 2002); at replacement rates greater than 10%, compressive strength is generally reported to fall (El-Aleem et al., 2005; Shoaib et al., 2000; Wang et al., 2002; Daous, 2004; Udoeyo and Hyee, 2002; Al-Harthy et al., 2003). The addition of FA has been reported to mitigate this effect (Bhatty, 1984, 1986). The exception to this is when CKD is used in combination with slag to replace cement; in this case the CKD, due to its high alkali and sulfate content, serves as an activator for the pozzolanic slag materials, and increases in compressive strength are observed (Konsta-Gdoutos and Shah, 2003; Shi, 2001). Table 7.4 shows compressive strength data of CKD mortar and concrete published in different studies.

In most of the studies, up to 10% replacement of cement by CKD has no significant effect on the compressive strength of hardened mortar and concrete up to 90 days (Kunal et al., 2014a,b; Maslehuddin et al., 2009, 2008; Wang et al., 2002). The increased strength in CKD mortars and concrete was attributed to an appropriate alkalinity that increases the dissolution of silicate species and the formation of CSH (Wang et al., 2002). Above 10% replacement levels the reduction in compressive strength was due to the poor hydraulic property of CKD, reduction in the cement content, an increase in free lime content and the water-to-binder (w/b) ratio, increased porosity, and the formation of chloro- and sulfo-aluminate phases which lead to the softening and expansion of the hydration products (El-Aleem et al., 2005; Daous, 2004; Wang et al., 2002; Shoaib et al., 2000). Taha et al. (2004) observed that mortars containing copper slag (CS) and CKD as partial replacements for PC at all curing ages improved the compressive strength, but when using CS alone or lime as an activator instead of CKD, lower compressive strength was observed.

Al-Harthy et al. (2003) reported that at a w/b ratio of 0.70 the decrease in strength was greater compared to w/b of 0.60 and 0.50 at the age of 28 days. At 0.70 w/b ratio, up to 8% and 13% reduction in compressive strength in mixes having 5 and 10%, respectively, CKD was observed, whereas for 0.60 and 0.50 w/b ratio 12% and 18%; and 1.8% and 4.5% reduction in strength was observed at similar CKD replacement levels compared to respective control treatments. Similarly, at 30% CKD substitution, 31−22% reductions in strength were observed with decreasing w/b ratios from 0.70 to 0.50. Udoeyo and Hyee (2002) and Batis et al. (1996) reported a decrease in

Table 7.4 Compressive strength development in CKD concrete and mortar studied by several researchers

Study	CKD replacement (%)	w/b	Notes	Compressive strength (MPa)					
				3d	7d	14d	28d	56d	90d
Kunal et al. (2014a)	0	0.50	Concrete	—	23.2	—	34.8	—	40.4
	5			—	23.8	—	35.8	—	41.9
	10			—	24.3	—	36.3	—	44.1
	15			—	23.0	—	34.5	—	40.0
Maslehuddin et al. (2009)	0	NM	Concrete (type I cement)	—	—	—	55.0	—	63.0
	5			—	—	—	57.0	—	60.5
	10			—	—	—	48.0	—	56.0
	15			—	—	—	47.0	—	53.5
	0	NM	Concrete (type V cement)	—	—	—	51.0	—	60.0
	5			—	—	—	50.5	—	60.0
	10			—	—	—	50.0	—	57.0
	15			—	—	—	46.8	—	52.0
Al-Jabri et al. (2006)	0	0.50	Concrete with 13.5% copper slag	—	36.7	—	45.8	—	—
	1.5			—	31.5	—	38.8	—	—
	0	0.60		—	35.8	—	42.7	—	—
	1.5			—	28.4	—	32.9	—	—
	0	0.70		—	24.1	—	27.9	—	—
	1.5			—	19.5	—	22.9	—	—

Continued

Table 7.4 Continued

Study	CKD replacement (%)	w/b	Notes	Compressive strength (MPa)					
				3d	7d	14d	28d	56d	90d
Kunal et al. (2014b)	0	0.50	Mortar	–	15.3	–	22.6	–	30.5
	10			–	17.4	–	25.8	–	32.9
	20			–	12.6	–	19.5	–	24.3
	30			–	10.7	–	15.2	–	20.7
Pavia and Regan (2010)	0	0.60	Mortar with calcium lime binder instead of cement	–	–	–	0.89	–	–
	5			–	–	–	0.73	–	–
	10			–	–	–	0.92	–	–
	15			–	–	–	1.03	–	–
	20			–	–	–	1.05	–	–
Maslehuddin et al. (2008)	0	NM	Mortar	15.0	22.9	–	33.2	–	–
	5			21.6	27.6	–	34.8	–	–
	10			22.7	27.7	–	36.9	–	–

OPC, ordinary Portland cement; *NM*, not mentioned.

strength of concrete with increased w/b of 0.65−0.75 and at very high replacement levels of CKD.

Higher alkali and sulfate content in CKD makes it an excellent activator for GGBFS to create nonconventional cementitious binders for concrete (Konsta-Gdoutos and Shah, 2003). Increased compressive strength of all the CKD−slag blends with curing time (7, 28, and 56 days) were observed compared to ordinary PC−slag blends due to the formation, precipitation, and accumulation of hydration products (such as CSH), and the availability of Ca^{2+} ions provided by the free lime content (Shi, 2001).

Wang et al. (2004, 2007) concluded that the addition of 2% and 5% NaOH and elevated curing temperature (38 and 50°C) activate the CKD-FA (1:1) blends and improved the binder strength development significantly at early ages of curing. On the contrary, no improvement in the strength of CKD−FA pastes without NaOH was observed when cured at 38−50°C temperature due to the effect of hydration temperature on the microstructure of binder pastes (Kjellsen et al., 1992; Gebauer, 1981; Verbeck and Helmuth, 1968). Bhatty (1984, 1986) concluded that the addition of FA into a CKD−OPC mix lowered the alkali content and resulted in improved strength.

The flexural strength development follows a similar pattern as compressive strength. Replacement of up to 10% cement by CKD shows no significant change in flexural strength (Al-Harthy et al., 2003); whereas increase in strength (3.7%) was reported in specimens with up to 15% CKD content (Wang et al., 2002). Shoaib et al. (2000) and Udoeyo and Hyee (2002) concluded that as the amount of CKD increased above 10%, tensile strength decreases due to poor bond strength between aggregate and cement. Tables 7.5 and 7.6 show the flexural and splitting tensile strength data of CKD mortar and concrete reported in different literature.

Table 7.5 Flexural strength development in CKD concrete and mortar studied by several researchers

Study	CKD replacement (%)	w/b	Notes	Flexural strength (MPa)		
				7d	28d	90d
Marku et al. (2012)	0	0.50	Mortar	7.3	8.4	9.5
	15			6.5	8.7	8.7
	30			5.5	6.9	7.0
	45			4.0	4.9	5.1
Al-Jabri et al. (2006)	0	0.50	Concrete with 13.5% copper slag	–	5.4	–
	1.5			–	4.3	–
	0	0.60		–	4.8	–
	1.5			–	3.9	–
	0	0.70		–	4.0	–
	1.5			–	3.7	–

Continued

Table 7.5 **Continued**

Study	CKD replacement (%)	w/b	Notes	Flexural strength (MPa)		
				7d	28d	90d
Al-Harthy et al. (2003)	0	0.50		−	4.7	−
	5		Concrete	−	5.1	−
	10			−	5.1	−
	15			−	4.7	−
	20			−	4.3	−
	25			−	4.3	−
	30			−	3.6	−
	0	0.60		−	3.9	−
	5			−	3.7	−
	10			−	3.7	−
	15			−	3.3	−
	20			−	3.3	−
	25			−	2.8	−
	30			−	3.1	−
	0	0.70		−	3.8	−
	5			−	2.9	−
	10			−	2.6	−
	15			−	3.4	−
	20			−	2.4	−
	25			−	2.9	−
	30			−	2.8	−

7.2.3.4   Durability properties

The influence of 15% CKD—cement replacement on the electrical resistivity of concrete mixes was observed by Maslehuddin et al. (2008) and Konsta-Gdoutos et al. (2001). Both studies concluded that incorporation of CKD decreased the electrical resistivity of the resulting concrete due to the increase of free chloride ions. This would affect the corrosion of reinforcing steel. But at lower replacement levels by CKD (5% w/w) in cement mortars, pore size reduced and thus the permeability, thereby enhancing resistance to corrosion (El-Sayed et al., 1991). The OH^- ions generated during the hydration reaction help in maintaining the passive oxide layer that protects the

Table 7.6 Splitting tensile strength development in CKD concrete and mortar studied by several researchers

Study	CKD replacement (%)	w/b	Notes	Splitting tensile strength (MPa)		
				7d	28d	90d
Abdulabbas (2013)	0	0.47	Concrete with ordinary Portland cement	1.3	2.1	—
	10			0.3	0.7	—
	20			0.9	1.4	—
	0		Concrete with sulfate-resisting Portland cement	1.7	2.9	—
	10			0.5	0.8	—
	20			1.4	2.0	—
Al-Jabri et al. (2006)	0	0.50	Concrete with 13.5% copper slag	—	3.5	—
	1.5			—	2.4	—
	0	0.60		—	3.0	—
	1.5			—	2.4	—
	0	0.70		—	2.3	—
	1.5			—	2.3	—

steel. Batis et al. (1996) concluded that the elevated chloride and sulfate ions resulting from the use of CKD accelerated the corrosion rate; whereas the fineness and relatively higher alkalinity of CKD acts as a barrier for corrosion.

Durability can also be assessed by determining the permeability of the mortar and concrete mixtures. Addition of up to 10% CKD decreased the sorptivity of mortar (Al-Harthy et al., 2003), and water absorption and porosity of concrete specimens (Kunal et al., 2014a), therefore, enhanced the durability. Decrease in permeability also provides resistance to sulfate ingress, which in turn protects reinforcing steel from corrosion.

Ramakrishnan and Balaguru (1987) and Batis et al. (1996) observed that 5−6% replacement of cement with CKD did not appreciably affect the freeze−thaw durability of concrete; whereas Wang and Ramakrishnan (1990) reported some additional mass loss compared to plain concrete specimens. Bhatty (1984) investigated the effect of CKD on alkali aggregate reaction in mortars (ASTM C227) and concluded that CKD contains varying amounts of alkalis, sulfates, and chlorides, which may lead to durability problems which may be overcome by blending CKD with pozzolanic materials such as FA (Bhatty, 1986; Ramakrishnan, 1986).

Conclusion

While the replacement of PC with CKD may be beneficial in serving as an activator for concrete and mortar mixes containing pozzolanic materials, in other applications the replacement rate should be limited to 10—15%. At replacement rates above this, concrete strength and durability properties may be adversely affected. Inclusion of CKD is also known to reduce concrete setting time, which must be a consideration in its use.

7.3   Silica fume

Silica fume (SF) is a by-product of the smelting process (reduction of high-purity quartz with coal in electric furnaces) in the production of silicon and ferrosilicon alloys. It is also collected as a by-product in the production of other silicon alloys such as ferromanganese, ferromagnesium, ferrochromium, and calcium silicon (ACI 226-3R-87). It contains extremely fine amorphous particles of silicon dioxide (SiO_2) which usually make up more than 90% of SF constituents. SF is also known as microsilica, volatized silica, and condensed SF or silica dust. SF, because of its extreme fineness and high silica content, has been recognized as a pozzolanic material conforming to specifications of ASTM C1240 for use as supplementary cementitious material in cement mortar and concrete to enhance mechanical and durability properties. According to the Florida Department of Transportation, the quantity of SF should be between 7% and 9% by mass of cement replacement for mortar and concrete production (Panjehpour et al., 2011). The use of SF is well established in concrete industries throughout the world and, perhaps, represents the most deeply entrenched and accepted use of industrial by-products in the construction industry.

7.3.1   Physical properties

SF particles are extremely small and form a grayish black powder with more than 95% of the particles finer than 1 μm and having a specific area around 20,000 cm²/g. The specific gravity of SF is generally in the range of 2.20—2.5 (compared to PC at 3.15). The bulk density of SF varies from 130 to 430 kg/m³ (Silica Fume Association, 2005). Table 7.7 shows the typical physical properties of SF investigated by several researchers.

Table 7.7 Typical physical properties of silica fume samples

Property	Guneyisi et al. (2012)	Mardani-Aghabaglou et al. (2014)	Haruehansapong et al. (2014)	Lilkov et al. (2014)
Particle size (μm)	<1	—	0.1	<1
Specific gravity	2.20	2.10	—	—
Specific surface area (cm²/g)	21,080	18,000	20,000	18,600

Table 7.8 **Chemical composition of silica fume samples**

Composition (%)	Guneyisi et al. (2012)	Mardani-Aghabaglou et al. (2014)	Haruehansapong et al. (2014)	Lilkov et al. (2014)
SiO_2	90.36	87.29	88.3	89.50
Al_2O_3	0.71	0.47	1.17	1.13
Fe_2O_3	1.31	0.63	4.76	2.31
CaO	0.45	0.81	0.48	0.98
MgO	–	4.47	2.14	1.55
K_2O	1.52	1.28	–	0.60
Na_2O	0.45	1.25	–	0.42
SO_3	0.41	0.22	1.05	0.40
LOI	3.11	2.70	2.1	2.40

7.3.2 Chemical properties

SF is composed primarily of pure silica in noncrystalline form. SF has a very high content of amorphous silicon dioxide and consists of very fine spherical particles. Small amounts of iron, magnesium, and alkali oxides are also found. Table 7.8 shows the chemical properties of SF as reported by several researchers.

7.3.3 Effect of SF on properties of mortar and concrete

7.3.3.1 Water demand and workability

The water demand of cement paste increases with the increase in SF content (Rao, 2003). The influence of SF on the water demand of cement pastes and mortars containing from 0% to 30% SF by weight of cement was investigated, and it was concluded that cement pastes containing 20–30% SF require up to 40% additional water to maintain a workable consistency. Addition of nanosilica (NS) to cement paste incorporating SF increased the water demand, making the cement paste thicker compared to silica fume–cement paste (Qing et al., 2007).

Workability is defined as the ease with which cement pastes and concrete can be placed under the action of gravity. Workability is mainly dependent on particle size: small particle size leads to higher water demand to achieve the same workability. Sellevold and Redjy (1983) and Alshamsi et al. (1993) reported that with a high concentration of silica fume in concrete, water demand fell due to reduction in contact points between different SF grains, and hence the desired consistency can be achieved. On the contrary, Rao (2003) reported that the addition of small amounts of SF (up to 10% by weight of cement) does not require additional water or superplasticizers; whereas replacement of higher amounts of cement ($>10\%$) with SF makes the mixture stiff and reduces its workability. Wong and Razak (2005) suggested that dosage of superplasticizer improved the workability characteristics of SF–concrete with the same w/c ratio. Luther (1989) and Tenoutasse and Marion (1987) proposed that concrete containing SF reduces bleeding because of its effect on rheological properties. Additionally, the pozzolanic nature of SF reduced the alkali-sulfate resistance.

7.3.3.2 Setting time

Lohtia and Joshi (1996) and Uchikawa (1986) observed delay in setting time in SF concrete compared to non-SF concrete of equal strength in the absence of a water reducer. Addition of 15% SF with superplasticizer delayed both the initial and final setting time by approximately 60 and 120 min, respectively. Addition of 10% microsilica had very little effect on setting times; whereas with higher percentages (20% microsilica), an increase of setting time of 6–20% was observed. Rao (2003) contradicted the earlier studies and suggested that due to the pozzolanic nature of SF, initial setting time decreased with an increase in SF content.

7.3.3.3 Hydration

SF, due to its large specific surface area and amorphous nature, is highly reactive and helps in accelerating the hydration of C_3S, C_2S, and C_4AF (Uchikawa and Uchida, 1980; Kurdowski and Nocun-Wczelik, 1983). ACI Committee 234 (1987) reported that SF accelerates the hydration of cement during early stages as it provides nucleation sites for cement hydration products. Grutzeck et al. (1983) observed that silica in SF dissolves in the presence of $Ca(OH)_2$ which then acts as a substitute for the formation of CSH hydration product. Partial replacement of cement by silica fume results in a reduction of heat of hydration without affecting the strength (Lohtia and Joshi, 1996; Scott and Singh, 2011), while addition of lingosulfonate along with SF lowered the cumulative heat evolution of cement pastes (Meland, 1983).

7.3.3.4 Strength properties

Partial replacement of cement with SF decreased the compressive strength of mortars at early ages of curing (2 and 7 days); whereas with increasing age (28, 90 and 360 days) the strength increases. Addition of 5%, 7.5%, and 10% SF increased the 360-day compressive strength of OPC by 5%, 6%, and 10%, respectively (Senhadji et al., 2014). Gleize et al. (2003) concluded that SF acts mainly at the paste–aggregate interface in PC mortars and increased the compressive strength by improving bond between the hydrated cement matrix and sand (Huang and Feldman, 1985). Higher concentration of CH and greater porosity in cement paste provides surfaces on which SF can react, thus improving concrete compressive strength.

Incorporation of NS also improves the strength properties. Shaikh et al. (2014) observed that addition of NS (up to 2%) as partial replacement of cement enhanced the compressive strength by 16% and 14% at 7 and 28 days, respectively. The addition of higher NS content (6%) decreased the strength due to the agglomeration of NS particles in the wet mix resulting in a significant reduction of effective specific surface area.

Concrete made with 15% SF along with superplasticizer (Cong et al., 1992) and glass fibers (De Gutierrez et al., 2005) as partial replacement of cement showed increased compressive strength and improved resistance to sulfate attack (Sakr, 2006). Wolseifer (1984) and Siddique (2011) reported that addition of SF as partial replacement of PC improved the strength properties of mortar and concrete and, thus, can be used to produce concrete with very high strength and low permeability. Tables 7.9 and 7.10 show the compressive strength and splitting tensile and flexural strengths of SF mortar and concrete investigated by several researchers.

7.3.3.5 Durability properties

Water absorption is defined as the amount of water absorbed by a material and is calculated as the ratio of the weight of water absorbed to the weight of the dry material. Partial replacement of up to 15% cement by SF in concrete resulted in significant reductions in porosity (Igarashi et al., 2005), and concrete becomes stronger but also more brittle in nature (Ozyildirim, 1986; Plante and Bilodeau, 1989). Poon et al. (2006) reported a decrease in porosity with age with the addition of 5% and 10% SF. Additionally, SF increased the matrix density and reduced the chloride permeability.

Cwirzen and Penttala (2005) investigated the capillary porosity of eight non-air-entrained concrete mixes with water-to-binder (w/b) ratios of 0.3, 0.35, and 0.42 containing different percentages of SF. Concrete having a w/b ratio of 0.3 showed a decrease in capillary porosity with SF; whereas for w/b ratios of 0.35 and 0.42, capillary and total porosities appeared to be quite similar. Khan and Siddique (2011) reviewed the durability properties of concrete containing silica fume and concluded the effectiveness of SF for the development of high-strength concrete.

Conclusion

Inclusion of up to 10% SF does not require additional water or superplasticizers and thus can be used to achieve desired workability. Partial replacement of cement by SF results in a reduction of heat of hydration and increased setting times without affecting the strength properties. SF can be used as partial replacement of cement to produce concrete and mortar mixes with high strength and low permeability.

7.4   Waste foundry sand

Foundry sand is high-quality uniform-sized silica sand used to form molds for ferrous (iron and steel) and non-ferrous (copper, aluminum, brass) metal castings. In the casting process, foundry sand is recycled and reused multiple times. However, eventually the sand can no longer be reused in the casting process and becomes 'waste foundry sand' (WFS). WFS is also referred to as spent foundry sand or used foundry sand. The physical and chemical characteristics of foundry sand depend mainly on the type of casting process and the industry sector from which it originates. A major user of foundry sand is the automotive industry and its parts suppliers.

Table 7.9 Compressive strength development in SF concrete and mortar studied by several researchers

Study	SF replacement to cement (%)	w/b	Notes	Compressive strength (MPa)					
				3d	7d	14d	28d	56d	90d
Gleize et al. (2003)	0	1.85	Mortar (cement + lime)	–	3.3	–	6.6	–	–
	10			–	2.93	–	7.11	–	–
Wong and Razak (2005)	0	0.27	Concrete	68.0	72.5	–	84.0	86.5	87.5
	5			63.0	75.5	–	88.5	93.0	96.5
	10			61.0	79.0	–	95.5	100.0	104.0
	15			59.5	76.5	–	101.0	103.5	106.0
	0	0.30		63.5	72.0	–	83.5	84.5	85.5
	5			62.0	81.0	–	91.0	95.5	95.5
	10			61.5	78.5	–	95.0	97.0	99.0
	15			57.5	74.5	–	98.5	101.5	104.0
	0	0.33		58.0	62.5	–	75.0	78.0	79.0
	5			55.0	69.5	–	83.0	85.0	90.0
	10			53.0	70.5	–	89.5	90.5	92.0
	15			47.5	70.5	–	88.5	93.0	95.5

Reference	W/C	Type	Replacement (%)						
Poon et al. (2006)	0.30	Concrete	0	68.5	81.1	—	96.5	—	102.2
			5	67.0	79.3	—	106.5	—	110.2
			10	63.2	76.9	—	107.9	—	115.6
	0.50		0	28.6	41.2	—	52.1	—	60.4
			5	27.4	47.0	—	54.3	—	67.5
			10	25.8	47.4	—	58.4	—	69.1
Behnood and Ziari (2008)	0.30	High-strength concrete	0	—	55.3	—	67.4	—	—
			6	—	69.1	—	80.3	—	—
			10	—	74.1	—	84.2	—	—
	0.35		0	—	NM	—	NM	—	—
			6	—	61.5	—	73.9	—	—

NM, not mentioned.

Table 7.10 Splitting tensile and flexural strength development in SF concrete and mortar studied by several researchers

Study	SF replacement to cement (%)	w/b	Notes	Splitting tensile strength (MPa)					
				Splitting tensile			Flexural		
				14d	28d	90d	14d	28d	90d
Almusallam et al. (2006)	0	0.35	Concrete with calcareous limestone aggregate	2.3	2.5	2.7	–	–	–
	10			3.2	3.4	3.7	–	–	–
	15			3.3	3.6	3.9	–	–	–
	0		Dolomitic limestone aggregate	2.7	3.2	3.4	–	–	–
	10			2.8	3.4	3.7	–	–	–
	15			2.9	3.4	4.4	–	–	–
	0		Quartzitic limestone aggregate	3.0	3.2	3.9	–	–	–
	10			4.1	4.3	4.5	–	–	–
	15			4.2	4.5	4.6	–	–	–
	0		Steel slag aggregate	3.9	4.0	4.1	–	–	–
	10			4.0	4.3	4.5	–	–	–
	15			4.1	4.5	4.8	–	–	–

Bhanja and Sengupta (2005)	0	0.34	Concrete	—	4.5	—	—	7.4	—
	5			—	5.6	—	—	8.8	—
	10			—	5.7	—	—	8.9	—
	15			—	5.8	—	—	9.9	—
	20			—	6.0	—	—	10.9	—
	25			—	5.8	—	—	8.8	—
Koksal et al. (2008)	0	0.38	Concrete	—	3.5	—	—	5.7	—
	5			—	3.8	—	—	6.1	—
	10			—	5.4	—	—	8.1	—
	15			—	6.5	—	—	9.3	—

(a)   (b)

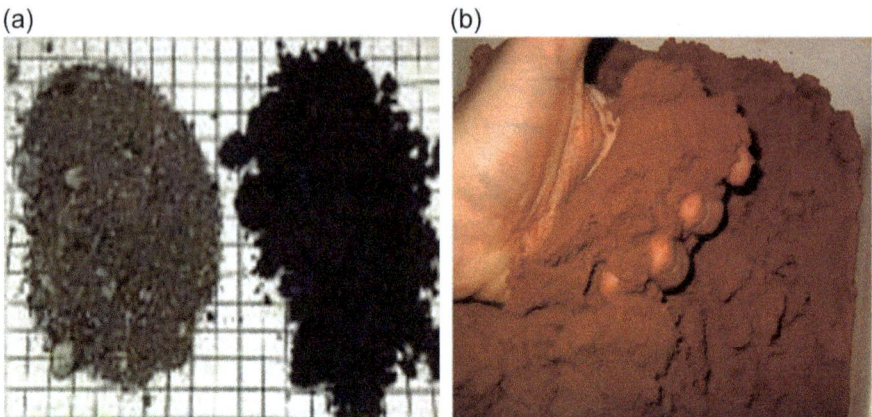

Figure 7.1 (a) Clay-bonded sand (www.alf-cemind.com) and (b) chemically bonded sand (www.mechanicalengineeringblog.com).

WFS, a by-product, has the potential to partially replace natural sand as a fine aggregate in mortar and concrete, providing a recycling opportunity and potentially improving the strength and durability of the resulting concrete. The typical physical and chemical properties of WFS depend upon the type of metal being cast, the casting process and technology used, the type of furnace (induction, electric arc, or cupola) and finishing process (grinding, blast cleaning, or coating). Based on the type of binder systems used in metal casting, WFS is categorized as either clay-bonded sand (green sand) or chemically bonded sand. Clay-bonded sand is composed of naturally occurring high-quality silica sand (85−95%), bentonite clay (4−10%) as a binder, a carbonaceous additive (2−10%) to improve the casting surface finish, and water (2−5%). The high carbon content in WFS gives it a black color (Siddique et al., 2010).

Clay-bonded sand is the most commonly used molding media in foundries. The coating of clay binds the silica sand together (acting as a bulk medium that resists high temperatures), and further addition of water improves plasticity. Addition of carbonaceous material prevents the "burn-on" or fusing of sand onto the casting surface.

Chemically bonded sands are used in mold making where high resistance is necessary to withstand the heat of molten metal. Chemically bonded sand consists of 93−99% silica and 1−3% chemical binder such as phenolic-urethanes, epoxy-resins, furfuryl alcohol, or sodium silicates. Chemically bonded sands are generally lighter in color and in texture than clay-bonded sands (Fig. 7.1). Silica sand is thoroughly mixed with the chemicals; a catalyst initiates the reaction that cures and hardens the mass (Siddique and Singh, 2011).

7.4.1   Physical properties

WFS is generally subangular to round in shape with color varying from medium tan or off-white color (chemically bonded sand) to black or gray (clay-bonded sand). Javed and Lovell (1994) reported that the particle size distribution of WFS is uniform, with

Table 7.11 **Typical physical properties of waste foundry sand**

Properties	Javed and Lovell (1994)	Naik et al. (2001)	Guney et al. (2010)	Siddique et al. (2011)
Specific gravity	2.4–2.5	2.8	2.4	2.6
Fineness modulus	–	2.3	–	1.8
Absorption (%)	0.4	5.0	–	1.3
Moisture content (%)	0.1–10.1	–	3.2	–
Materials finer than 75 μm (%)	–	1.1	24.0	18.0

85–95% of the material between 0.6 and 0.15 mm, and 5–12% smaller than 0.075 mm. The specific gravity of WFS varies between 2.39 and 2.55, and it has low water absorption capacity. Table 7.11 shows the physical characteristics of WFS reported by several researchers.

Carey and Sturtz (1995) and Deng and Tikalsky (2008) suggested that the workability and suitability of WFS in flowable fill depends upon its physical properties such as particle gradation, grain shape, fineness, density, absorption, specific gravity, bleeding, setting time, hydraulic conductivity, and leaching characteristics. Pure clay-based WFS samples have moisture contents of 1–4% and require approximately 10% water content to "activate" bentonite binding; 2–3% water is needed as a solvent or catalyst to activate organic binders in the organic-based chemically bonded sands (Winkler and Bol'shakov, 2000).

Deng and Tikalsky (2008) reported that the variation in the bulk density ($1052–1554 \ kg/m^3$), specific gravity (2.38–2.72), and absorption (0.38–4.15%) largely depends on sand mineralogy, particle gradation, grain shapes, and fine contents. The highest absorption (4.15%) was found in WFS samples obtained from a copper/aluminum foundry.

Naik et al. (2001) found that WFS had a higher content of particles finer than 75 μm relative to the clean foundry sand. Test results showed lower values of soundness than conventionally permitted ASTM C88 limits.

7.4.2 Chemical properties

The chemical composition of WFS depends upon the type of metal molded at the foundry and the types of binder and combustible used. WFS is mainly composed of silica sand, coated with a thin film of burned carbon, residual binder (bentonite, sea coal, resins/chemicals), and dust. Table 7.12 lists the chemical composition of WFS reviewed by Siddique and Singh (2011). Johnson (1981) suggested the pH value of WFS can vary between 4 and 8. Furfuryl alcohol, phenolic urethane, phenolic acid-activated resin, phenolic resole-ester, phosphate, alkyd (oil) urethane, sodium silicate, shell liquid/powered and flake resins are used as chemical binder system in

Table 7.12 **Typical chemical composition of waste foundry sand**

Constituent (%)	American Foundry men's Society (1991)	Guney et al. (2010)	Etxeberria et al. (2010)	Siddique et al. (2011)
SiO_2	87.91	98.00	95.10	78.81
Al_2O_3	4.70	0.80	1.47	6.32
Fe_2O_3	0.94	0.25	0.49	4.83
CaO	0.14	0.03	0.19	1.88
MgO	0.30	0.02	0.19	1.95
SO_3	0.09	0.01	0.03	0.05
Na_2O	0.19	0.04	0.26	0.10
K_2O	0.25	0.04	0.68	–
TiO_2	0.15	–	0.04	–
Mn_2O_3	0.02	–	–	–
SrO	0.03	–	–	–
LOI	5.15	–	1.32	2.15

Adapted from Siddique, R., Singh, G., 2011. Utilization of waste foundry sand (WFS) in concrete manufacturing. Resources, Conservation and Recycling 55, 865–1118.

foundry industries and may be present in WFS. The majority of binders used in the foundry industry are self-setting chemical binders.

Ministry of Natural Resources (MNR, 1992) reported that some WFSs can be corrosive to metals due to the presence of phenols. Therefore, it is necessary to monitor foundry sand sources and stockpiles to assess the need to establish controls for potential phenol discharges. According to the Federal Highway Administration (2004), the maximum phenol content in foundry sand should be less than 2 mg/kg.

7.4.3 Applications of WFS

According to Winkler and Bol'shakov (2000), approximately 9 million metric tons of WFS was disposed of in landfills by metal casting foundries in the United States in 2000. Annual cost of WFS disposal was around US $135–675 million including an average landfill tipping fee of foundry by-products of US $15–75 per ton inclusive of storage, transportation, and labor costs (Winkler et al., 1999). This not only imposes a financial burden on the foundries but also causes potential remediation problems. Alternate applications of WFS offer cost savings for both foundries and user industries, and an environmental benefit at the local and national levels. Utilization of WFS in concrete and concrete-related products like bricks, blocks, asphalt concrete, and paving stones has been reported by Khatib and Ellis (2001), Naik et al. (2004), Bakis et al. (2006), Siddique et al. (2007), Etxeberria et al. (2010), and Guney et al. (2010). Additionally, several studies (Ham et al., 1990; Javed and Lovell, 1994; Mast and Fox, 1998; Kleven et al., 2000; Abichou et al., 2004; Vipulanandan et al., 2000; Santurde et al., 2011) have been reported on reuse of WFS in road construction and embankments, hydraulic barriers and layer construction, flowable fills, controlled low-strength materials (CLSM) and landfill cover, soil reinforcement, hot mix asphalt and smelting, PC manufacturing, concrete and mortars, agricultural/soil amendments, vitrification of hazardous materials, and rock wool and fiberglass manufacturing.

7.4.4 Effect of WFS on properties of mortar and concrete

7.4.4.1 Strength properties

Guney et al. (2010) replaced fine aggregate with 0%, 5%, 10%, and 15% WFS in concrete and observed a decrease in fluidity and the slump value of the fresh concrete. This may be due to the presence of clayey type fine materials in the WFS, which are effective in decreasing the fluidity of the fresh concrete. Similar observations were also observed by Etxeberria et al. (2010) in concrete mixes containing chemically bonded and clay-bonded foundry sand.

Siddique et al. (2007, 2009), Naik et al. (1994), and Khatib and Ellis (2001) concluded that up to 35% WFS additions in concrete as partial fine aggregate replacement increased compressive and splitting tensile strength marginally; whereas beyond this replacement strength decreases (Guney et al., 2010). Siddique et al. (2009) partially replaced fine aggregate (regular sand) with WFS in three percentages (10%, 20%, and 30%) by weight and evaluated the compressive strength, splitting tensile strength, flexural strength, and modulus of elasticity at 28, 56, 91, and 365 days. Marginal increases in the strength properties of concrete containing WFS as partial replacement of fine aggregate were observed, and it was concluded that WFS can be effectively used in making good quality concrete and construction materials. On the other hand, asphalt concrete mixes containing 0−20% WFS replacement of fine aggregate showed decreased compressive and splitting tensile strength values (Bakis et al., 2006).

Etxeberria et al. (2010) observed 28-day compressive strength of 28.4 and 25 MPa and modulus of 27.9 and 27.4 GPa in concrete containing chemically bonded and clay-bonded foundry sand, respectively. It was also observed that both chemically and clay-bonded foundry sand mixes achieved the same tensile strength (2.9 MPa). Singh and Siddique (2012) investigated the abrasion resistance and strength properties of concrete containing WFS replacing sand (fine aggregate) with 0%, 5%, 10%, 15%, and 20% by mass. The w/c ratio was maintained at 0.40, and the workability of mixtures was maintained at a slump of 85 ± 5 mm. Enhancement of 8.3−17% in terms of compressive strength, 3.6−10.4% in terms of splitting tensile strength, and 1.7−6.4% in terms of modulus of elasticity was observed at 28 days in concrete containing different percentages of WFS. It was also observed that inclusion of WFS led to improved abrasion resistance of concrete at all ages.

Siddique et al. (2009) observed an increase in splitting tensile strength of concrete mixtures with the increase in WFS content. The 28-day splitting tensile strength of control mixture (0% WFS) was 2.75 MPa; whereas 10%, 20%, and 30% WFS concrete achieved strengths of 2.85, 2.9, and 3.0 MPa, respectively. Splitting tensile strength was found to increase with age. With the increase in age from 56 to 365 days, increase in splitting tensile strength for control mixture was between 6.5% and 12.7%; while for 10%, 20%, and 30% WFS, it was 8.7% and 13%; 9.3% and 14.5%; and 8% and 15%, respectively. Guney et al. (2010) observed that concrete specimens prepared with 10% WFS as fine aggregate replacement had slightly higher values than a control mix.

7.4.4.2   Durability properties

Naik et al. (2003) investigated the freezing and thawing resistance of bricks, paving stones, and blocks containing WFS according to ASTM C1262. It was observed that brick mixtures reached the critical value (0.2%) of weight loss following 40 and 12 freeze/thaw cycles for mixes having WFS replacement ratios of 25% and 35%, respectively, compared to 92 cycles for 0% WFS control mixtures. Partial substitution of fine aggregate with WFS caused a sharp drop in freeze/thaw durability. In case of paving stones, mixtures with 25% and 35% WFS reached the critical value of weight loss (0.2%) following 95 and 45 freeze/thaw cycles, compared to 192 cycles for 0% WFS control mixtures. Overall, the freeze/thaw resistance of the paving stones was about 2.3 times that of the bricks due to the lower water—cementitious materials ratio of paving stones. With increasing WFS in bricks, the freeze/thaw resistance decreased (Naik et al., 2004); while a significant amount of mass loss due to surface spalling between 60 and 150 cycles was observed in the case of paving stones.

Naik et al. (2001) observed the permeability of CLSM mixtures containing FA and WFS and determined that 30% replacement of FA by foundry sand reduced the permeability; whereas an addition of 80% foundry sand abruptly increased the permeability.

Conclusion

WFS can be suitably used in making good quality concrete and construction materials with improved strength properties. However, limited studies indicate that WFS may be detrimental to durability in some respects, and its use must be considered accordingly.

7.5   Rice husk ash

Pozzolans from agricultural waste are receiving greater attention since their uses generally improve the properties of blended cement concrete and reduce the environmental impact of both the source agriculture industry and the energy-intensive cement industry. Palm oil fuel ash and rice husk ash are two promising pozzolans that are available in many parts of the world (Chindaprasirt et al., 2008). Rice husk ash (RHA) has been used as a highly reactive pozzolanic material to improve the microstructure of the interfacial transition zone between the cement paste and the aggregate in self-compacting concrete (SCC). Mechanical experiments of RHA-blended PC concretes revealed that in addition to the pozzolanic reactivity of RHA (chemical aspect), the particle grading (physical aspect) of cement and RHA mixtures also exerted significant influences on the blending efficiency.

Rice husk is abundant in many parts of the world. RHA is a highly siliceous material that can be used as an admixture in concrete if burned in a specific manner. The characteristics of the ash are dependent on the components, temperature, and duration of burning. When properly burned at temperatures lower than 700°C, reactive amorphous silica is obtained (Chindaprasirt et al., 2008). The silica content in RHA is high at approximately 90%. Silica in amorphous form is suitable for use as a pozzolan. With proper burning and grinding, ground RHA can be produced and used as a pozzolan. Even at higher burning temperature with some crystalline formation of silica, good RHA can still be obtained by fine grinding (Chindaprasirt et al., 2008). The reactive RHA is used to produce good quality concrete with reduced $Ca(OH)_2$ and higher resistance to sulfate attack.

7.5.1   Physical properties

RHA is grayish-black in color due to unburned carbon. At burning temperatures of 550−800°C, amorphous silica is formed, while crystalline silica is produced at higher temperatures. The specific gravity of RHA varies from 2.11 to 2.27; it is highly porous and light weight, with a very high specific surface area. Table 7.13 shows the physical properties of RHA reported by several researchers. Fig. 7.2 shows images of RHA as received and after burning at 700°C for 6 h (Della et al., 2002). Typically, RHA is used in the form of ground RHA, having typical particle sizes generally less than 10 μm; natural RHA (NRHA) has larger sizes of approximately 100 μm.

7.5.2   Chemical properties

RHA is rich in silica content with levels ranging from 85 to 96%. Presence of a high amount of silica makes it a valuable material for use in industrial applications and as pozzolanic material in construction. Apart from silica (SiO_2), K_2O, Al_2O_3, CaO, MgO, Na_2O, and Fe_2O_3 are also present at typically less than 1% each. Table 7.14 shows the chemical composition of RHA studied by several researchers.

Table 7.13 Physical properties of RHA

Property	Chindaprasirt et al. (2008)	de Sensale (2010)	Chatveera and Lertwattanaruk (2011)	Gastaldini et al. (2014)
Particle size (μm)	10	8	−	−
Specific gravity	2.23	2.06	2.02	2.11
Blaine fineness (cm^2/g)	−	−	6200	−
BET-specific surface area (cm^2/g)	−	2,880,000	1,100,000	188,900

(a)

(b)

Figure 7.2 Rice husk ash (a) as received; (b) after burning out at 700 °C for 6 h.
Adapted from Della, V.P., Kuhn, I., Hotza, D., 2002. Rice husk ash as an alternate source for active silica production. Materials Letters 57, 818−821.

7.5.3 Effect of RHA on properties of mortar and concrete

7.5.3.1 Strength properties

The partial replacement of PC up to 20% by RHA resulted in lower compressive strength at early ages (less than 7 days), while with increasing age, compressive strength increases (Gastaldini et al., 2014; Ferraro and Nanni, 2012; Madandoust et al., 2011). Ferraro and Nanni (2012) reported that 28-day compressive strength of

Table 7.14 **Chemical composition of RHA**

Composition (%)	Rice husk ash (RHA)			
	Chindaprasirt et al. (2008)	de Sensale (2010)	Chatveera and Lertwattanaruk (2011)	Gastaldini et al. (2014)
CaO	1.1	0.55	0.08	0.79
SiO_2	93.2	87.2	78.12	93.54
Al_2O_3	0.4	0.15	0.31	0.52
MgO	0.1	0.35	0.34	0.49
Na_2O	0.1	1.12	0.17	0.03
K_2O	1.3	3.60	0.82	1.65
Fe_2O_3	0.1	0.16	0.23	0.20
SO_3	0.9	0.32	0.09	0.05
LOI	3.7	6.55	8.31	2.32

concrete containing off-white RHA increased 13.6% and 15.6% with 7.5% and 15% OWRHA replacement of cement for a control concrete (0% OWRHA) having a strength of 36 MPa. Similarly, increases of 16.9% and 20.3% in 91-day compressive strength were observed. Madandoust et al. (2011) found that 20% RHA replacement of cement is beneficial for long-term strength development. At 3 days of curing, the compressive strength of RHA concrete was 65% of normal concrete, increasing to 96% and 98% at 90 and 180 days of curing. Strength at 270 days was comparable to normal concrete; whereas at 360 days, strength was slightly greater (2%). It was concluded that RHA concrete cannot be used for fast track construction due to its reduced strength at early age. At early ages, lower compressive strength in RHA concrete was due to the reduction in water availability because RHA absorbs some water during mixing, which becomes available at later ages for continued cement hydration. The greater eventual strength is attributed to the filler effect of smaller RHA particles in concrete. Gastaldini et al. (2014) observed higher strength values in 20%, 30%, and 20% RHA concrete with w/c ratio of 0.35, 0.50, and 0.65, respectively.

Particle size of RHA also influences the compressive strength of concrete. Tuan et al. (2011), in a study of ultra-high performance concrete (UHPC), observed that with increasing mean particle size, strength decreases linearly. A 28-day compressive strength of 150 MPa was achieved with the mean particle size of about 8 μm, while strength of 180 MPa was achieved with 3.6 μm particle size. Blends containing 10% grounded RHA (GRHA) and 10% SF showed better compressive strength (~160 MPa) compared to control UHPC at the age of 28 days. NRHA is coarser than GRHA, and its incorporation in concrete results in reduced 28-day strength and stiffness. Replacement of OPC by 15% NRHA achieved 90% of the strength of concrete with 15% GRHA (Zerbino et al., 2012) and 0% RHA (Nor Atan and Awang, 2011).

Comparable strength values were also observed when 30% OPC was replaced with limestone powder (LP)−NRHA and FA−NRHA blends, while lower compressive strength was observed with 30% SF−NRHA blends (Nor Atan and Awang, 2011).

RHA blended with sugar cane bagasse ash (SCBA) (Juma et al., 2012) and meta-kaolin (MK) (Yu et al., 1999) significantly increased the compressive strength of con-cretes due to the microfilling ability of RHA, pozzolanic reaction between CH and silica, and hydration of silica itself. Due to the fine particle size, RHA blended with SCBA and MK filled the microvoids within the cement particles and readily reacted with water and CH to produce additional CSH. With curing time, as hydration pro-ceeds, more hydration products and cementing materials are formed, leading to an in-crease in compressive strength of cement pastes (Shatat, 2013). Above 10% replacement of cement by RHA in concrete, capillary pores increase and accumulate CH on the interface (Chao-Lung et al., 2011). This made the matrix less compact and produced lower early compressive strength than control specimens. After 28 days, the pozzolanic reaction started and decreased the CH content, resulting in densification of the matrix, consequently enhancing the strength at later ages. Table 7.15 shows the compressive strength development in RHA concrete investigated by several researchers.

Ahmadi et al. (2007) studied the compressive strength of SCC mix containing RHA at replacement percentages of 10% and 20% and w/b ratios of 0.40 and 0.35. RHA−SCC mixes showed higher compressive strength (31−41%) than normal concrete. In addition, w/b ratio has more impact on normal concrete than on SCC. Finally, increasing the amount of cement replacement with RHA requires an increase in w/b ratio. It was observed that SCC mixes exhibited flexural strength about 12−20% greater than normal concrete, and the mixes containing 20% RHA had the highest flex-ural strength in all cases; while reduced modulus of elasticity (MOE) was observed in all the RHA−SCC mixes. Memon et al. (2011) calculated the compressive strength of SCC mixes with different w/b ratios (0.4, 0.38, and 0.36) and percentages of superplas-ticizer (3.5, 4.0, and 4.5 by weight of binder content). The control mix with 3.5% superplasticizer (CC3.5) developed the highest compressive strength of 10.5 and 28.4 MPa at 7 and 28 days, respectively. For control concrete and mixes with 5% RHA, the compressive strength decreased with increased dosage of superplasticizer; whereas with 10% RHA, the strength increased with increased dosage due to improved workability and self-compactibility. For an equal dosage of superplasticizer, RHA mixes showed higher compressive strength compared to control mixes. This increase is due to the effectively reduced w/b ratio resulting from RHA absorption, dense par-ticle packing, and pore and grain size refinement.

Khadiry et al. (2014) compared SCC incorporating RHA and shell lime powder (SL), both locally available mineral admixtures, as an additional cementing material for a period of 7, 14, and 28 days. SCC−RHA mix gave a higher strength by 24% for 7 days. At 28 days, the strength of RHA−SCC was essentially the same as that of SL SCC at 28 days. It was concluded that since RHA contains silica and SL contains calcite, the silica reacts better with cement since cement contains lime, which itself consists of calcites.

Table 7.15 Compressive strength development in RHA concrete studied by several researchers

Study	RHA replacement (%)	w/b	Notes	Compressive strength (MPa)					
				3d	7d	14d	28d	56d	90d
Chao-Lung et al. (2011)	0	0.35	Concrete	46.0	50.0	54.0	56.0	60.0	67.0
	10			41.0	47.0	52.0	61.0	62.0	67.0
	20			38.0	47.0	52.0	60.0	61.0	69.0
	30			32.0	43.0	51.0	54.0	60.0	64.0
Ferraro and Nanni (2012)	0	0.44	Concrete	—	27.6	—	36.0	—	37.9
	7.5			—	28.0	—	40.9	—	44.3
	15			—	29.2	—	41.6	—	45.6
Gastaldini et al. (2014)	RHA replacement	w/b	Notes	3 days	28 days air curing including curing in 95% humidity at…		7 days		
	0	0.35	Concrete	54.0			58.0		
	5			56.0			60.3		
	10			59.3			65.0		
	20			61.0			72.0		
	30			65.7			72.7		
	0	0.50		37.3			40.0		

Continued

Table 7.15 Continued

Study	RHA replacement (%)	w/b	Notes	Compressive strength (MPa)						
				3d	7d	14d	28d	56d	90d
	5			43.7			47.7		
	10			40.5			42.0		
	20			40.7			43.0		
	30			47.0			49.2		
	0	0.65		29.0			33.3		
	5			33.8			36.0		
	10			29.0			35.7		
	20			30.0			32.0		
	30			31.0			31.7		

Increasing the RHA content in concrete increased the 28 day MOE ($\sim$9–16%), flexural ($\sim$7–20%), and splitting tensile ($\sim$4–30%) strengths. At 7.5% and 15% replacement levels, unground low-carbon RHA concrete exhibited increases of 7% and 15% in flexural, and 16% and 4% in splitting tensile strengths, while GRHA concrete showed increases of 9 and 20% and 16 and 30%, respectively (Venkatanarayanan and Rangaraju, 2015). Rahman et al. (2014) concluded that 20% RHA replacement was acceptable, as its splitting tensile strength was similar to that of the control mix.

7.5.3.2 Durability properties

Ramasamy (2012) calculated porosity test values of control concrete and those with different percentages of RHA after 60 days. It was observed that the porosity decreases with increase in RHA content. The porosity of concrete decreased from 4.7% to 3.9% when the RHA replacement level increased from 5% to 20%. The small RHA particles improved the particle packing density of the concrete mixture leading to a reduced volume of larger pores.

Increasing the RHA content in concrete as cement replacement decreased the charge passed. Ramasamy (2012) reported that with incorporation of the RHA in concrete, the chloride ion permeability values fell from "low" (1000–2000 coulombs) in the control mixes to the "very low" (100–1000 coulombs) category. This was attributed to the finer pore structure in the hydrated cement paste especially at the aggregate–paste interface. A similar trend was also observed by Nehdi et al. (2003) and Zhang et al. (1996).

Immersion of 15% RHA concrete specimens (w/b 0.50) in chloride solution (5%) for 21 days showed a decrease in chloride penetration, while specimens containing NRHA and 15% combusted RHA (CRHA) with w/b 0.40 exhibited a reduction in chloride ion penetration (de Sensale, 2010). Substitution of PC by 15% CRHA showed the greatest resistance to chloride penetration (w/b 0.50), while in concrete with w/b 0.40 and 0.32, NRHA showed resistance to chloride ion penetration due to the reduction in pore size.

Incorporation of RHA increases the drying shrinkage and depth of carbonation of concrete mixtures compared to OPC concrete. Chatveera and Lertwattanaruk (2011) measured the change in length of concrete specimens periodically for 150 days. Higher dry shrinkage and depth of carbonation was observed in RHA concrete than in OPC concrete. This was due to lower cement content and increased porosity through which CO_2 infiltrates easily to react with water present in capillary pores to form carbonic acid. On the contrary, Rukzon et al. (2009) observed a decrease in drying shrinkage with increasing RHA content due to refinement of pore structure and increased formation of hydration products.

Conclusion

RHA can effectively be used as a constructional material for long-term strength development in mortar and concrete. Replacement of cement up to 20% by RHA reduces the early age strength, but its low pozzolanic activity and water absorbability may be beneficial for long-term strength development and reduced chloride penetrability. At replacement levels of 10–20%, RHA–SCC mixes may prove beneficial for achieving higher compressive and flexural strength with improved workability and self-compactibility.

7.6   Palm oil fuel ash

Palm oil is one of the most important agricultural industries in Southeast Asia, with Malaysia being the largest exporter (47%) of palm oil to the world. The palm oil industry generates large amounts of solid waste and residues such as palm shells, fibers, and husk. These wastes and residues are burned in a boiler under high temperature (900−1000°C); the ash generated is known as palm oil fuel ash (POFA). Due to minimal utilization of POFA, its quantity increases annually and is disposed of on open landfills, causing environmental concerns. The utilization of POFA in cement and concrete has been studied by several researchers (Tangchirapat et al., 2009; Kroehong et al., 2011; Jaturapitakkul et al., 2011; Noorvand et al., 2013).

Similar to RHA, a high content of silica in amorphous form makes POFA a pozzolanic material. According to ASTM C618 (2001), pozzolanic material contains siliceous and aluminous material by composition and has little or no cementitious property, but when it reacts with CH in the presence of water, it acts as a cementitious material. Tangchirapat et al. (2003) reported that POFA contains a large amount of silica and acts as pozzolanic material for cement replacement.

7.6.1   Physical properties

POFA is generated as the by-product of burning oil palm husk and palm kernel shell in the palm oil mill boiler. The ash is grayish-black in color due to unburned carbon. The collected ash is dried in an oven and sieved to get a wide range of sizes. Jaturapitakkul et al. (2007) found that the large size (183 µm) particles of POFA (LP) were spherical and porous. The size of the particles can be reduced by grinding to medium (MP) and small (SP) with median particle sizes of 15.9 and 7.4 µm, respectively. After grinding, the MP and SP exhibited irregular particle geometry with a crushed shape (Fig. 7.3). Table 7.16 shows the physical properties of POFA reported by several researchers.

7.6.2   Chemical properties

X-ray fluorometry analysis revealed that POFA contains a high composition of silica (SiO_2), in the range 46−64%, followed by calcium oxide (6.5−14%) and aluminum oxide (0.9−4.5%). It was observed that finer particle sizes have greater silica content and lower the LOI (Yusuf et al., 2014). Table 7.17 shows the chemical composition of the POFA studied by several researchers. XRD of ultrafine palm oil fuel ash (UPOFA; 1 µm size) showed that UPOFA has both amorphous and crystalline phases. Fig. 7.4 shows the XRD pattern of UPOFA, and the peaks include quartz (SiO_2), calcite ($CaCO_3$), cristobalite (SiO_2), and potassium aluminate phosphate ($KAl_2(PO_4)_3$).

7.6.3   Effect of POFA on properties of mortar and concrete

7.6.3.1   Strength properties

As shown in Table 7.18, several studies report that addition of POFA in mortar and concrete improved strength properties. Altwair et al. (2012) observed improved

(a)

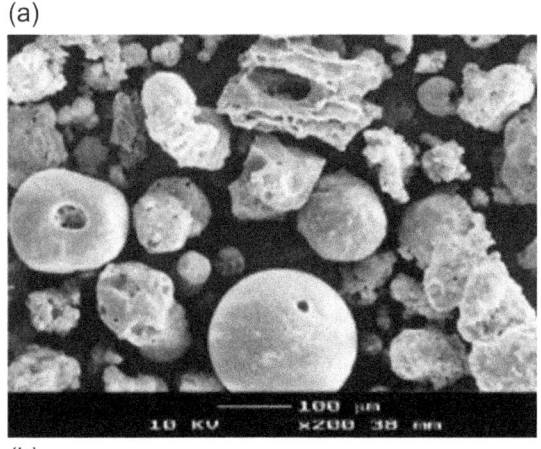

Figure 7.3 Scanning electron microscopy images of (a) large size (LP) palm oil fuel ash (POFA) (median particle size 183 μm); (b) medium size (MP) POFA (median particle size 16 μm); (c) small size (SP) POFA (median particle size 7 μm).
Adapted from Jaturapitakkul, C., Kiattikomol, K., Tangchirpat, W., Saeting, T., 2007. Evaluation of the sulfate resistance of concrete containing palm oil fuel ash. Construction and Building Materials 21, 1399–1405.

(b)

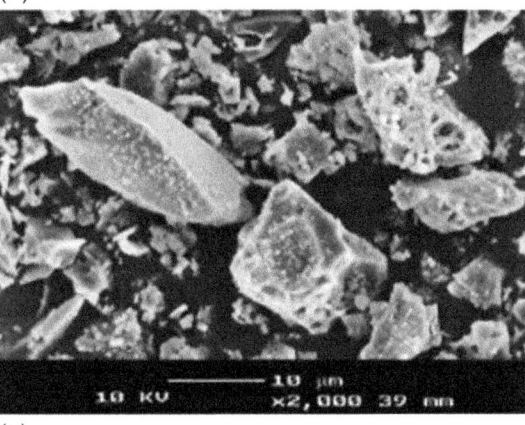

(c)

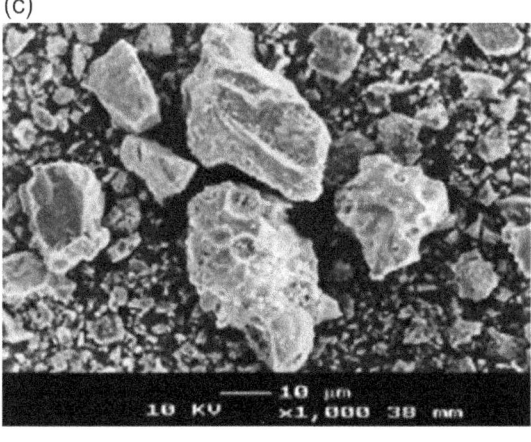

Table 7.16 **Physical properties of POFA**

Property	Kroehong et al. (2011)	Jaturapitakkul et al. (2011)	Noorvand et al. (2013)	Yusuf et al. (2014)
Specific gravity	2.36–2.48	2.05–2.39	1.97	2.6
Median particle size, d_{50} (μm)	2.1–15.6	12.3–30.8	82	1.07
Blaine fineness (cm^2/g)	6,700–14,900	6,605–12,285	–	–
Specific surface area (cm^2/g)	–	–	255,000	134,000

Table 7.17 **Chemical composition of POFA**

Composition (%)	Palm oil fuel ash (POFA)			
	Kroehong et al. (2011)	Jaturapitakkul et al. (2011)	Noorvand et al. (2013)	Yusuf et al. (2014)
CaO	12.5	6.4	13.9	8.5–11.0
SiO_2	54.0–55.7	65.3	48.9	46.0–60.4
Al_2O_3	0.9	2.6	2.7	3.1–4.3
MgO	5.1	3.1	2.7	4.4–5.3
Na_2O	1.0	0.4	0.1	0.1–0.2
K_2O	11.9	5.7	7.1	4.1–5.0
Fe_2O_3	2.0	1.9	6.5	2.4–3.3
SO_3	2.9	0.5	1.5	0.3–0.4
P_2O_5	–	–	3.7	3.9–4.5
TiO_2	–	–	–	0.13–0.19
LOI	4.7	10.0	11.3	2.5–21.6

28-day compressive strength in mixtures containing POFA–cement ratio of 0.4 compared to control mixtures at similar w/c ratios of 0.33–0.38. Lower first cracking strength (flexural) was also observed when increasing POFA–cement ratio from 0 to 1.2 at w/c ratio 0.33, due to decreased available w/c ratio with increased POFA content. Kroehong et al. (2011) and Jaturapitakkul et al. (2011) partially replaced cement with POFA (up to 40%) having smaller and larger particle size. It was observed that mixes with small POFA particles produced higher strength than those with larger particles, and this increased with age (7–90 days). Some specimens produced higher strength than control specimens (Jaturapitakkul et al., 2011); whereas Kroehong et al. (2011) reported lower strength values of POFA-blended cements. The increased

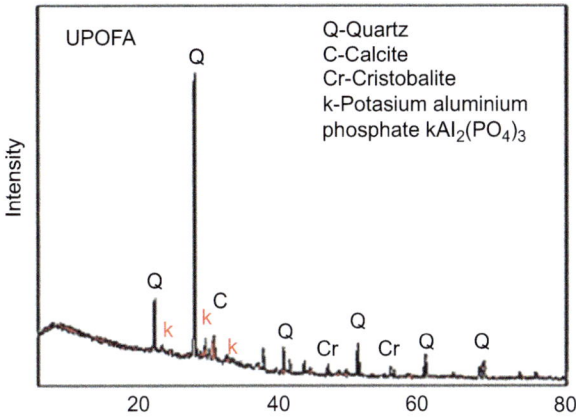

Figure 7.4 X-ray diffraction of ultrafine palm oil fuel ash.
Adapted from Yusuf, M.O., Johari, M.A.M., Ahmad, Z.A., Maslehuddin, M., 2014. Influence of curing methods and concentration of NaOH on strength of the synthesized alkaline activated ground slag-ultrafine palm oil fuel ash mortar/concrete. Construction and Building Materials 66, 541−548.

compressive strength in mixes containing smaller POFA particles was due to improved hydration reaction, nucleation effect, packing effect, and pozzolanic reaction.

Yusuf et al. (2014) achieved 32.8%, 28.9%, and 28.3% increases of compressive strength in oven-cured samples than GSS-free mortar with Na_2SiO_3/10 M NaOH mass ratio of 1.0 at 3, 14, and 28 days, respectively. Strength development in mortars prepared with low NaOH concentration (4 M) was not significant due to low dissolution of active precursors, namely Si−O, Al−O, and CaO. The highest 3-day strength (54.5 MPa) was achieved in mortars prepared with 10 M NaOH, and it was concluded that use of NaOH concentration above 10 M in mortars became corrosive and expensive, and there was no increase in strength beyond 10 M NaOH concentration.

Incorporation of NS in mortars containing UPOFA also increased the compressive strength (Noorvand et al., 2013). Improvement in strength for the blended pozzolan mixtures (POFA + FA and RHA + FA) was observed compared to individual pozzolanic materials up to 40% (Chindaprasirt et al., 2008) and 20% (Chindaprasirt et al., 2007; Tangchirapat et al., 2009) replacement levels; this is attributed to the filler effect of POFA and the synergistic effect of the blended pozzolans that formed the additional hydration products.

Exposing the concrete to sulfate decreased the compressive strength, and this effect can be minimized using POFA as cement replacement (Jaturapitakkul et al., 2007). Incorporation of small-sized POFA particles (SP) showed higher 28-day compressive strength and lower expansion values (after immersing in 5% $MgSO_4$ solution for 24 months) than medium (MP) and large (LP) POFA particles.

7.6.3.2 Durability properties

Kroehong et al. (2011) observed reduced porosity in 20% and 40% finely ground POFA (GPOFA)−OPC pastes which further reduced with age (28−91 days). Pastes containing

Table 7.18 Compressive strength development in POFA mortar and concrete studied by several researchers

Study	POFA replacement (%)	w/b	Notes	Compressive strength (MPa)		
				7d	28d	90d
Chindaprasirt et al. (2008)	0	0.50	Mortar	43.5	57.0	60.0
	20			43.5	57.5	62.0
	40			32.5	53.5	61.5
Noorvand et al. (2013)	0	0.35	Mortar	54.4	67.3	88.5
	10			50.8	62.0	84.2
	20			43.3	52.5	70.3
	30			38.7	45.0	58.0
Kroehong et al. (2011)	0	0.35	OPC control	53.0	75.0	99.1
	20		Cement paste with POFA particle size 15.6 µm	48.3	72.0	102.0
	40			41.0	61.5	88.1
	20		POFA particle size 2.1 µm	51.9	77.3	109.6
	40			44.0	66.5	94.1
Tangchirapat et al. (2009)	0	0.32	Concrete	54.9	58.5	64.7
	10			55.6	59.5	67.5
	20			54.6	60.9	69.4
	30			53.2	58.8	66.1
Altwair et al. (2012)	0	0.33	Concrete with different POFA-cement ratios	–	39.5	–
	0.4			–	40.6	–
	0.8			–	39.2	–
	1.2			–	37.3	–
	0	0.36		–	37.4	–
	0.4			–	39.8	–
	0.8			–	36.5	–
	1.2			–	32.8	–
	0	0.38		–	36.7	–
	0.4			–	35.6	–
	0.8			–	31.9	–
	1.2			–	30.0	–

20% GPOFA showed lower porosity values than 40% GPOFA, and this was due to the higher fineness of POFA which was ground to a particle size smaller than the OPC, having a good filler effect in reducing the pore size of the blended OPC pastes. The permeability of OPC concrete falls with the curing age and can be reduced by incorporating POFA particles. Chindaprasirt et al. (2007) reported lower permeability in 20% and 40% GPOFA concretes after 28 days of curing, while similar observations were also reported by Tangchirapat et al. (2009) in concrete with up to 30% GPOFA content after the age of 90 days compared to control concrete. Chindaprasirt et al. (2008) found that incorporation of pozzolans (POFA, RHA, and FA) reduced the charge passed (200–3050°C) in mortars compared to OPC mortar (7450°C). Blends of 20% and 40% POFA + FA further reduced the charge passed and showed better resistance to chloride permeability than RHA + FA blends at 28 days.

Replacement of cement up to 10% UPOFA content in mortars showed lower values for water absorption and permeable void ratio (Noorvand et al., 2013), and this can further be reduced by the addition of NS (0.5–1.5%). The reduction in water absorption and permeable void ratio in NS modified mortars was due to the filling effect of nanomaterials, nucleation effect of NS (hydration products envelop the nanoparticles) and more formation of CSH that densify the matrix, hence formed the homogenous microstructure.

Conclusion
High silica content in POFA makes it a suitable constructional material for producing high-strength mortar and concrete. Partial replacement of cement with POFA (up to 20%) significantly increases the strength properties of concrete due to fine particle size and pozzolanic activity of POFA; in another application, addition of 10 M NaOH, NS (0.5%), and other pozzolanic materials (FA, RHA, etc.) may have beneficially enhanced the strength properties and produced mortars and concretes with reduced water absorption, porosity, and water and chloride permeability.

7.7   Conclusions

Mortar and concrete are the most widely used construction materials because of their versatility, economy, availability of raw materials, strength, and durability. Advances in concrete technology in recent years have generally focused on the strength and durability of structures resulting from the improved design of mortar and concrete. Researchers are continuously working to improve concrete strength and durability with the help of innovative chemical admixtures and supplementary cementitious materials (SCMs). The use of SCMs conserves energy and reduces global carbon dioxide emission associated with the manufacture of PC. SCMs used in cement mortar and concrete mixes are often industrial by-products or natural ashes, which not only prevents them from being landfilled, but also enhances the fresh and hardened state properties of the resulting concrete.

A substantial amount of research has been dedicated to improving the understanding of the behavior of a broadened range of industrial by-products and natural ashes in mortar and concrete products. Utilization of industrial by-products (CKD, SF, WFS) and natural ashes (RHA, POFA) has beneficial effects on the strength and durability properties of mortar and concrete. CKD is effective as an activator for concrete and mortar mixes containing pozzolanic materials at replacement levels of 10−15%. Partial replacement of cement by SF up to 10% can effectively be used to achieve desired workability, reduced heat of hydration, and increased setting times without affecting the strength and durability properties. Studies report that inclusion of WFS up to 35% by weight of cement can be used in making good quality mortar and concrete with improved strength. High silica content in natural ashes (RHA and POFA) makes them suitable constructional material for long-term strength development in mortar and concrete. Partial replacement of up to 20% cement with RHA and POFA significantly increases the strength properties of concrete at later ages due to fine particle size and pozzolanic activity. Incorporation of RHA in SCC and blending of POFA with other pozzolanic materials (FA, SF, etc.) may prove beneficial in improving the strength properties with reduced water absorption, porosity, and water and chloride permeability.

While some SCMs are widely used (SF), others are in their infancy and require additional study to better establish long-term performance characteristics. Effects and leaching of trace heavy metals (CKD) and other chemicals (WFS) detrimental to the environment or human health need to be considered in future studies; indeed, it may well be determined that sequestering these in concrete is preferred to other disposal options, although their presence may limit the use of these waste-containing concrete mixes.

References

American Foundry men's Society, AFS, 1991. Alternative Utilization of Foundry Waste Sand. Final Report (Phase I) prepared by American Foundry men's Society Inc. for Illinois Department of Commerce and Community Affairs, Des Plaines.

Abdulabbas, Z.H., 2013. Utilization of cement kiln dust in concrete manufacturing. Jordan Journal of Civil Engineering 7, 111−125.

Abichou, T., Edil, T.B., Benson, C.H., Bahia, H., 2004. Beneficial use of foundry by-products in highway construction. In: Geotechnical Engineering for Transportation Projects, vol. 126. Geotechnical Special Publications, ASCE, Reston, pp. 715−722.

ACI Committee 226 3R-87, 1987. Fly ash in concrete. ACI Materials Journal 11, 381−409.

ACI Committee 234, 1987. Guide for the use of silica fume in concrete (ACI 234R). ACI Materials Journal 92, 437−440.

Ahmadi, M.A., Alidoust, O., Sadrinejad, I., Nayeri, M., 2007. Development of mechanical properties of self compacting concrete contain rice husk ash. International Journal of Computational Systems Engineering 1, 259−262.

Al-Harthy, A.S., Taha, R., Al-Maamary, F., 2003. Effect of cement kiln dust (CKD) on mortar and concrete mixtures. Construction and Building Materials 17, 353−360.

Al-Jabri, K.S., Taha, R.A., Al-Hashmi, A., Al-Harthy, A.S., 2006. Effect of copper slag and cement by-pass dust addition on mechanical properties of concrete. Construction and Building Materials 20, 322−331.

Almusallam, A.A., Beshr, H., Maslehuddin, M., Al-Amoudi, O.S.B., 2006. Effect of silica fume on the mechanical properties of low quality coarse aggregate concrete. Cement and Concrete Composites 26, 891−900.

Alshamsi, A.M., Sabouni, A.R., Bushlaibi, A.H., 1993. Influence of set retarding super-plasticizers and microsilica on setting time of pastes at various temperatures. Cement and Concrete Research 23, 592−598.

Altwair, N.M., Johari, M.A.M., Hashim, S.F.S., 2012. Flexural performance of green engineered cementitious composites containing high volume of palm oil fuel ash. Construction and Building Materials 37, 518−525.

Amin, A.M., Ebied, E., El-Didamony, H., 1995. Activation of granulated slag with calcined cement kiln dust. Silicates Industriels 60, 109−115.

ASTM C1240, 2014. Standard Specification for Silica Fume Used in Cementitious Mixtures. Annual Book of ASTM Standards [04.02].

ASTM C1262, 2010. Standard Test Method for Evaluating the Freeze-Thaw Durability of Dry-Cast Segmental Retaining Wall Units and Related Concrete Units. Annual Book of ASTM Standards [04.05].

ASTM C227, 2010. Standard Test Method for Potential Alkali Reactivity of Cement-Aggregate Combinations (Mortar-Bar Method). Annual Book of ASTM Standards [04.02].

ASTM C618, 2001. Standard Specification for Coal Fly Ash and Raw or Calcined Natural Pozzolan for Use Mineral Admixture in Portland Cement Concrete. Annual Book of ASTM Standards, pp. 310−313 [04.02].

ASTM C88, 2013. Standard Test Method for Soundness of Aggregates by Use of Sodium Sulfate or Magnesium Sulfate. Annual Book of ASTM Standards [04.02].

www.alf-cemind.com Technology guide Alf-Cemind − Supporitng the use of alternative fuels in cement industry, http://www.alf-cemind.com/cd/AF_and_ARM_foundry_sand.htm.

Bakis, R., Koyuncu, H., Demirbas, A., 2006. An investigation of waste foundry sand in asphalt concrete mixtures. Waste Management and Research 24, 269−274.

Batis, G., Katsiamboulas, A., Meletiou, C.A., Chaniotakis, E., 1996. Durability of reinforced concrete made with composite cement containing cement kiln dust. In: Dhir, R.K., Dyer, T.D. (Eds.), Concrete for Environment Enhancement and Protection: Proceedings of the International Conference, Concrete in the Service of Mankind. University of Dundee, Dundee, United Kingdom, pp. 67−72.

Batis, G., Rakanta, E., Sideri, E., Chaniotakis, E., Papageorgiou, A., 2002. Advantages of simultaneous use of cement kiln dust and blast furnace slag. In: Proceedings of the International Conference on Challenges of Concrete Construction. University of Dundee, Dundee, United Kingdom, pp. 205−212.

Behnood, A., Ziari, H., 2008. Effects of silica fume addition and water to cement ratio on the properties of high-strength concrete after exposure to high temperatures. Cement and Concrete Composites 30, 106−112.

Bhanja, S., Sengupta, B., 2005. Influence of silica fume on the tensile strength of concrete. Cement and Concrete Research 35, 743−747.

Bhatty, M.S.Y., 1984. Use of cement Kiln dust in blended cements: alkali- aggregate reaction expansion. World Cement Technology 16, 386−392.

Bhatty, M.S.Y., 1986. Properties of blended cements made with Portland cement, cement kiln dust, fly ash, and slag. In: Proceedings of the International Congress on the Chemistry of Cement. Communications Theme-3, vol. 1, pp. 118−127.

Carey, P.R., Sturtz, G., 1995. Sand binder systems part IV urethane binders. Foundry Management & Technology 123, 25−29.

Chao-Lung, H., Anh-Tuan, B.L., Chun-Tsun, C., 2011. Effect of rice husk ash on the strength and durability characteristics of concrete. Construction and Building Materials 25, 3768−3772.

Chatveera, B., Lertwattanaruk, P., 2011. Durability of conventional concretes containing black rice husk ash. Journal of Environmental Management 92, 59−66.

Chindaprasirt, P., Homwuttiwong, S., Jaturapitakkul, C., 2007. Strength and water permeability of concrete containing palm oil fuel ash and rice husk-bark ash. Construction and Building Materials 21, 1492–1499.

Chindaprasirt, P., Rukzon, S., Sirivivatnanon, V., 2008. Resistance to chloride penetration of blended Portland cement mortar containing palm oil fuel ash, rice husk ash and fly ash. Construction and Building Materials 22, 932–938.

Collins, R.J., Emery, J.J., 1983. Kiln Dust-Fly Ash Systems for Highways Bases and Sub-bases. Federal Highway Administration, Report No. FHWA/RD-82/167, Washington, DC.

Cong, X., Gong, S., Darwin, D., McCabe, S.L., 1992. Role of silica fume in compressive strength of cement paste, mortar and concrete. ACI Materials Journal 89, 375–387.

Corish, A., Coleman, T., 1995. Cement kiln dust. Concrete 29, 40–42.

Cwirzen, A., Penttala, V., 2005. Aggregate-cement paste transition zone properties affecting the salt-frost damage of high performance concretes. Cement and Concrete Research 35, 671–679.

Daous, M.A., 2004. Utilization of cement kiln dust and fly ash in cement blends in Saudi Arabia. Journal of King Abdulaziz University - Engineering Sciences 15, 33–45.

Deng, A., Tikalsky, P.J., 2008. Geotechnical leaching properties of flowable fill incorporating waste foundry sand. Waste Management 28, 2161–2170.

Dyer, T.D., Halliday, J.E., Dhir, R.K., 1999. An investigation of the hydration chemistry of ternary blends containing cement kiln dust. Journal of Materials Science 34, 4975–4983.

Della, V.P., Kuhn, I., Hotza, D., 2002. Rice husk ash as an alternate source for active silica production. Materials Letters 57, 818–821.

El-Aleem, S.A., Abd-El-Aziz, M.A., Heikal, M., El-Didamony, H., 2005. Effect of cement kiln dust substitution on chemical and physical properties and compressive strength of Portland and slag cements. Arabian Journal of Science and Engineering 30, 263–273.

El-Didamony, H., Aly, A.H., Sharara, A.M., Amin, A.M., 1997. Assessment of cement dust with anhydrite as an activator for granulated slag. Silicates Industriels 62, 31–35.

El-Sayed, H.A., Gabr, N.A., Hanafi, S., Mohran, M.A., 1991. Reutilization of by-pass kiln dust in cement manufacture. In: Proceedings of the International Conference on Blended Cement in Construction.

EPA, 1993. U.S. Environmental Protection Agency. Report to congress on cement kiln dust, EPA-530-R-94-001.

Etxeberria, M., Pacheco, C., Meneses, J.M., Beerridi, I., 2010. Properties of concrete using metallurgical industrial by-product as aggregate. Construction and Building Materials 24, 1594–1600.

Federal Highway Administration, 2004. United States Department of Transportation, Foundry Sand Facts for Civil Engineers. Report No. FHWA-IF-04-004, USA, p. 80.

Ferraro, R.M., Nanni, A., 2012. Effect of off-white rice husk ash on strength, porosity, conductivity and corrosion resistance of white concrete. Construction and Building Materials 31, 220–225.

Gastaldini, A.L.G., da Silva, M.P., Zamberlan, F.B., Neto, C.Z.M., 2014. Total shrinkage, chloride penetration, and compressive strength of concretes that contain clear-colored rice husk ash. Construction and Building Materials 54, 369–377.

Gebauer, J., 1981. Alkali in clinker: influence on cement and concrete properties. In: Proceedings of the Fifth International Conference on Alkali-Aggregate Reaction in Concrete. Cape Town, South Africa, pp. 1–9.

Gleize, P.J.P., Muller, A., Roman, H.R., 2003. Microstructural investigation of a silica fume–cement–lime mortar. Cement and Concrete Composites 25, 171–175.

Grutzeck, M., Atkinson, S., Roy, D.M., 1983. Mechanism of hydration of condensed silica fume in calcium hydroxide solutions. ACI Special Publications 79, 643–664.

Guney, Y., Sari, Y.D., Yalcin, M., Tuncan, A., Donmez, S., 2010. Re-usage of waste foundry sand in high-strength concrete. Waste Management 30, 1705−1713.

Guneyisi, E., Gesoglu, M., Karaoglu, S., Mermerdas, K., 2012. Strength, permeability and shrinkage cracking of silica fume and metakaolin concretes. Construction and Building Materials 34, 120−130.

De Gutierrez, R.M., Diaz, L.N., Delvasto, S., 2005. Effect of pozzolans on the performance of fibre-reinforced mortars. Cement and Concrete Composites 27, 593−598.

Ham, R.K., Boyle, W.C., Blaha, F.J., 1990. Comparison of leachate quality in foundry waste landfills to leach test abstracts. Journal of Hazardous and Industrial Solid Waste Testing and Disposal 6, 29−44.

Haruehansapong, S., Pulngern, T., Chucheepsakul, S., 2014. Effect of the particle size of nanosilica on the compressive strength and the optimum replacement content of cement mortar containing nano-SiO_2. Construction and Building Materials 50, 471−477.

Hawkins, G.J., Bhatty, J.I., O'Hare, A.T., 2004. Cement kiln dust generation and management. In: Bhatty, J.I., Miller, F.M., Kosmatka, S.H. (Eds.), Innovations in Portland Cement Manufacturing. Portland Cement Association, 5420 Old Orchard Road, Skokie, IL 60077, pp. 735−779.

Heikal, M., Aiad, I., Helmy, M.I., 2002. Portland cement clinker, granulated slag and by-pass cement dust composites. Cement and Concrete Research 32, 1805−1812.

Huang, C.Y., Feldman, R.F., 1985. Influence of silica fume on the micro-structural development in cement mortars. Cement and Concrete Research 15, 285−294.

Igarashi, S.I., Kawamura, A., Watanabe, M., 2005. Evaluation of capillary pore size characteristics in high-strength concrete at early ages. Cement and Concrete Research 35, 513−519.

Jaturapitakkul, C., Kiattikomol, K., Tangchirpat, W., Saeting, T., 2007. Evaluation of the sulfate resistance of concrete containing palm oil fuel ash. Construction and Building Materials 21, 1399−1405.

Jaturapitakkul, C., Tangpagasit, J., Songmue, S., Kiattikomol, K., 2011. Filler effect and pozzolanic reaction of ground palm oil fuel ash. Construction and Building Materials 25, 4287−4293.

Javed, S., Lovell, C.W., 1994. Use of Foundry Sand in Highway Construction. Joint Highway Report No. C-36−50N. Department of Civil Engineering, Purdue University, Indiana, USA.

Johnson, C.K., 1981. Phenols in Foundry Waste Sand Modern Casting. American Foundry men's Society.

Juma, A., Sai, R., Prakash, D.V.A.K., Haider, S., Rao, S.K., 2012. An experimental study on synergic effect of sugar cane bagasse ash with Rice husk ash on self compaction concrete. International Journal of Science and Advanced Technology 2, 75−80.

Khadiry, S.M., Nayak, G.P., Aziz, T., Saurav, S., Pai, B.H.V., 2014. Evaluation of properties of self-compacting concrete specimens having rice husk ash and shell lime powder as fillers. American Journal of Engineering Research 3, 207−211.

Khan, M.I., Siddique, R., 2011. Utilization of silica fume in concrete: review of durability properties. Resources, Conservation and Recycling 57, 30−35.

Khatib, J.M., Ellis, D.J., 2001. Mechanical properties of concrete containing foundry sand. ACI Special Publications 200, 733−748.

Kjellsen, K.O., Detwiler, R.J., Gjorv, O.E., 1992. Pore structure of plain cement pastes hydrated at different temperatures. Cement and Concrete Research 22, 112−120.

Kleven, J.R., Edil, T.B., Benson, C.H., 2000. Evaluation of excess foundry system sands for use as subbase material. Transportation Research Record 1714, 40−48.

Koksal, F., Altun, F., Yigit, I., Sahin, Y., 2008. Combined effect of silica fume and steel fiber on the mechanical properties of high strength concretes. Construction and Building Materials 22, 1874–1880.

Konsta-Gdoutos, M.S., Shah, S.P., 2003. Hydration and properties of novel blended cements based on cement kiln dust and blast furnace slag. Cement and Concrete Research 33, 1269–1276.

Konsta-Gdoutos, M.S., Wang, K., Babaian, P.M., Shah, S.P., 2001. Effect of cement kiln dust (CKD) on the corrosion of reinforcement in concrete. In: Banthia, N., Saloi, K., Gjorv, O.E. (Eds.), Third International Conference on Concrete under Service Conditions of Environment and Loading (CONSEC '01), Vancouver, British Columbia, Canada, pp. 277–284.

Kroehong, W., Sinsiri, T., Jaturapitakkul, C., Chindaprasirt, P., 2011. Effect of palm oil fuel ash fineness on the microstructure of blended cement paste. Construction and Building Materials 25, 4095–4104.

Kunal, Siddique, R., Rajor, A., 2012. Use of cement kiln dust in cement concrete and its leachate characteristics. Resources, Conservation and Recycling 61, 59–68.

Kunal, Siddique, R., Rajor, A., 2014a. Strength and microstructure analysis of bacterial treated cement kiln dust mortar. Construction and Building Materials 63, 49–55.

Kunal, Siddique, R., Rajor, A., 2014b. Influence of bacterial treated cement kiln dust on the properties of concrete. Construction and Building Materials 52, 42–51.

Kurdowski, W., Nocun-Wczelik, W., 1983. The tricalcium silicate hydration in the presence of active silica. Cement and Concrete Research 13, 341–348.

Lilkov, V., Rostovsky, I., Petrov, O., Tzvetanova, Y., Savov, P., 2014. Long term study of hardened cement pastes containing silica fume and fly ash. Construction and Building Materials 60, 48–56.

Lohtia, R.P., Joshi, R.C., 1996. Mineral admixtures. In: Ramachandran, V.S. (Ed.), Concrete Admixture Handbook. Noyes Publications, USA, p. 1153.

Luther, M.D., 1989. Silica fume (microsilica): production, materials and action in concrete. In: Advancements in Concrete Materials Seminar, vol. 18. Bradley University, Peoria, pp. 1–21.

Madandoust, R., Ranjbar, M.M., Moghadam, H.A., Mousavi, S.Y., 2011. Mechanical properties and durability assessment of rice husk ash concrete. Biosystems Engineering 110, 144–152.

Mardani-Aghabaglou, A., Sezer, G.I., Ramyar, K., 2014. Comparison of fly ash, silica fume and metakaolin from mechanical properties and durability performance of mortar mixtures view point. Construction and Building Materials 70, 17–25.

Marku, J., Dumi, I., Lico, E., Dilo, T., Cakaj, O., 2012. The characterization and the utilization of cement kiln dust (CKD) as partial replacement in mortar and concrete production. Zastita Materijala 53, 334–344.

Maslehuddin, M., Al-Amoudi, O.S.B., Rehman, M.K., Ali, M.R., Barry, M.S., 2009. Properties of cement kiln dust concrete. Construction and Building Materials 23, 2357–2361.

Maslehuddin, M., Al-Amoudi, O.S.B., Shameem, M., Ibrahim, M., Rehman, M.K., 2008. Usage of cement kiln dust in concrete products — research review and preliminary investigations. Construction and Building Materials 22, 2369–2375.

Mast, D.G., Fox, P.J., 1998. Geotechnical performance of a highway embankment constructed using waste foundry sand. In: Recycled Materials in Geotechnical Applications, vol. 79. Geotechnical Special Publications, ASCE, Reston, pp. 66–85.

Meland, I., 1983. Influence of condensed silica fume and fly ash on the heat evolution in cement pastes. ACI Special Publications 79, 665–676.

Memon, S.A., Shaikh, M.A., Akbar, H., 2011. Utilization of rice husk ash as viscosity modifying agent in self compacting concrete. Construction and Building Materials 25, 1044−1048.

MNR, 1992. Mineral aggregate Conservation Reuse and Recycling. Report prepared by John Emery Geotechnical Engineering Limited for Aggregate and Petroleum Resources Section. Ontario Ministry of Natural Resources, Queen's Printer for Ontario.

www.mechanicalengineeringblog.com Moulding Sand, Silica Sand, Foundry Sand, Mechanical Engineering − A complete online guide for mechanical engineer, http://www. mechanicalengineeringblog.com/2950-moulding-sand-silica-sand-foundry-sand.

Naik, T.R., Kraus, R.N., Chun, Y.M., Ramme, W.B., Siddique, R., 2004. Precast concrete products using industrial by-products. ACI Materials Journal 101, 199−206.

Naik, T.R., Kraus, R.N., Chun, Y.M., Ramme, W.B., Singh, S.S., 2003. Properties of field manufactured cast-concrete products utilizing recycled materials. Journal of Materials in Civil Engineering 15, 400−407.

Naik, T.R., Patel, V.M., Parikh, D.M., Tharaniyii, M.P., 1994. Utilization of used foundry sand in concrete. Journal of Materials in Civil Engineering 6, 254−263.

Naik, T.R., Singh, S.S., Ramme, W.B., 2001. Performance and leaching assessment of flowable slurry. Journal of Environmental Engineering 127, 359−368.

Nehdi, M., Duquette, J., El Damatty, A., 2003. Performance of rice husk ash produced using a new technology as a mineral admixture in concrete. Cement and Concrete Research 33, 1203−1210.

Noorvand, H., Ali, A.A.A., Demirboga, R., Noorvand, H., Farzadnia, N., 2013. Physical and chemical characteristics of unground palm oil fuel ash cement mortars with nanosilica. Construction and Building Materials 48, 1104−1113.

Nor Atan, M.D., Awang, H., 2011. The compressive and flexural strengths of self-compacting concrete using raw rice husk ash. Journal of Engineering Science and Technology 6, 720−732.

Oss, H.G., February 2014. Minerals Information − Cement from United States Geological Survey, online at: http://minerals.usgs.gov/minerals/pubs/commodity/cement/mcs-2014-cemen.pdf.

Ozyildirim, C., 1986. Investigation of Concrete Containing Condensed Silica Fume. Final report, 86−R25 (January). Virginia Highway & Transportation Research Council, Charlottesville (Peckman J, Paul J, Thiel).

Panjehpour, M., Ali, A.A.A., Demirboga, R., 2011. A review for characterization of silica fume and its effect on concrete properties. International Journal of Sustainable Construction Engineering Technology 2, 1−7.

Pavia, S., Regan, D., 2010. Influence of cement kiln dust on the physical properties of calcium lime mortars. Materials and Structures 43, 381−391.

Peethamparan, S., 2006. Fundamental Study of Clay-cement Kiln Dust (CKD) Interactionto Determine the Effectiveness of CKD as a Potential Clay Soil Stabilizer (Ph.D. thesis). Purdue University.

Plante, P., Bilodeau, A., 1989. Rapid chloride ion permeability test: data on concrete incorporating supplementary cementing materials. ACI Special Publications 114, 625−644.

Poon, C.S., Kou, S.C., Lam, L., 2006. Compressive strength, chloride diffusivity and pore structure of high performance metakaolin and silica fume concrete. Construction and Building Materials 20, 858−865.

Qing, Y., Zenan, Z., Deyu, K., Rongshen, C., 2007. Influence of nano-SiO_2 addition on properties of hardened cement paste as compared with silica fume. Construction and Building Materials 21, 539−545.

Rahman, M.E., Muntohar, A.S., Pakrashi, V., Nagaratnam, B.H., Sujan, D., 2014. Self compacting concrete from uncontrolled burning of rice husk and blended fine aggregate. Materials & Design 55, 410−415.

Ramakrishnan, V., 1986. Evaluation of kiln dust in concrete. ACI Special Publications 91, 821—840.

Ramakrishnan, V., Balaguru, P., 1987. Durability of concrete containing cement kiln dust. ACI Special Publication (SP-100) 100, 305—322.

Ramasamy, V., 2012. Compressive strength and durability properties of rice husk ash concrete. KSCE Journal of Civil Engineering 16, 93—102.

Rao, G.A., 2003. Investigations on the performance of silica fume-incorporated cement pastes and mortars. Cement and Concrete Research 33, 1765—1770.

Rukzon, S., Chindaprasirt, P., Mahachai, R., 2009. Effect of grinding on chemical and physical properties of rice husk ash. International Journal of Minerals, Metallurgy and Materials 16, 242—247.

Sakr, K., 2006. Effects of silica fume and rice husk ash on properties of heavy weight concrete. Journal of Materials in Civil Engineering 18, 367—376.

Santurde, R.S., Andres, A., Viguri, J.R., Raimondo, M., Guarini, G., Zanelli, C., Dond, M., 2011. Technological behavior and recycling potential of spent foundry sands in clay bricks. Environmental Management 92, 994—1002.

Scott, R., Singh, S.P., 2011. High performance silica fume concrete and some applications in India. In: Proceedings of the International UKIERI Concrete Congress: New Developments in Concrete Construction, IIT Delhi, pp. 217—238.

Sellevold, E.J., Redjy, F.F., 1983. Condensed silica fume (microsilica) in concrete: water demand and strength development. ACI Special Publications 79, 677—694.

Senhadji, Y., Escadeillas, G., Mouli, M., Khelafi, H., Benosman, 2014. Influence of natural pozzolan, silica fume and limestone fine on strength, acid resistance and microstructure of mortar. Powder Technology 254, 314—323.

de Sensale, G.R., 2010. Effect of rice husk ash on durability of cementitious materials. Cement and Concrete Composites 32, 718—725.

Shaikh, F.U.A., Supit, S.W.M., Sarker, P.K., 2014. A study on the effect of nanosilica on compressive strength of high volume fly ash mortars and concretes. Materials & Design 60, 433—442.

Shatat, M.R., 2013. Hydration behaviour and mechanical properties of blended cement containing various amounts of rice husk ash in presence of metakaolin. Arabian Journal of Chemistry 12, 1—6.

Shi, C., 2001. Studies on several factors affecting hydration and properties of lime pozzolan cements. Journal of Materials in Civil Engineering 13, 441—445.

Shoaib, M.M., Balaha, M.M., Abdel-Rahman, A.G., 2000. Influence of cement kiln dust substitution on the mechanical properties of concrete. Cement and Concrete Research 30, 371—377.

Siddique, R., 2011. Utilization of silica fume in concrete: review of hardened properties. Resources, Conservation and Recycling 55, 923—932.

Siddique, R., Singh, G., 2011. Utilization of waste foundry sand (WFS) in concrete manufacturing. Resources, Conservation and Recycling 55, 865—1118.

Siddique, R., de Schutter, G., Noumowe, A., 2009. Effect of used-foundry sand on the mechanical properties of concrete. Construction and Building Materials 23, 976—980.

Siddique, R., Gupta, R., Kaur, I., 2007. Effect of spent foundry sand as partial replacement of fine aggregate on the properties of concrete. In: 22nd International Conference on Solid Waste Technology and Management. Widener University, Philadelphia, USA.

Siddique, R., Kaur, G., Rajor, A., 2010. Waste foundry sand and its leachate characteristics. Resources, Conservation and Recycling 54, 1027—1036.

Siddique, R., Aggarwal, Y., Aggarwal, P., Kadri, E.H., Bennacer, B., 2011. Strength, durability, and micro-structural properties of concrete made with used-foundry sand (UFS). Construction and Building Materials 25, 1916–1925.

Silica Fume Association, 2005. Silica Fume Manual, 38860 Sierra Lane, Lovettsville, VA 20180, USA.

Singh, G., Siddique, R., 2012. Abrasion resistance and strength properties of concrete containing waste foundry sand (WFS). Construction and Building Materials 28, 421–426.

Sreekrishnavilasam, A., King, S., Santagata, M., 2006. Characterization of fresh and landfilled cement kiln dust for reuse in construction applications. Engineering Geology 85, 165–173.

Subramanian, K., Kannan, A., 2013. An experimental study on usage of quarry dust as partial replacement for sand in concrete and mortar. Australian Journal of Basic and Applied Sciences 7, 955–967.

Taha, R., Al-Rawas, A., Al-Jabri, K., Al-Harthy, A.S., Hassan, H., Al-Oraimi, S., 2004. An overview of waste materials recycling in the Sultanate of Oman. Resources, Conservation and Recycling 41, 293–306.

Tangchirapat, W., Jaturapitakkul, C., Chindaprasirt, P., 2009. Use of palm oil fuel ash as a supplementary cementitious material for producing high-strength concrete. Construction and Building Materials 23, 2641–2646.

Tangchirapat, W., Tangpagasit, J., Waew-kum, S., Jaturapitakkul, C., 2003. A new pozzolanic material from palm oil fuel ash. KMUTT Research and Development Journal 26, 459–473.

Taylor, H.F.W., 1997. Cement Chemistry, second ed. Thomas Telford, London.

Tenoutasse, N., Marion, A.M., 1987. The influence of silica fume in alkali-aggregate reactions. In: Bellew PEG, Concrete Alkali-Aggregate Reactions. Noyes Publications, Park Ridge, NJ, pp. 711–775.

Tuan, N.V., Ye, G., van Breugel, K., Fraaij, A.L.A., Dai, B.D., 2011. The study of using rice husk ash to produce ultra high performance concrete. Construction and Building Materials 25, 2030–2035.

Uchikawa, H., 1986. Effect of blending components on hydration and structure formation. In: 8th International Congress on the Chemistry of Cement, Rio de Janeiro, Brazil, pp. 249–280.

Uchikawa, H., Uchida, S., 1980. Influence of pozzolans on the hydration of C_3A. In: 7th International Congress on the Chemistry of Cement, Paris, vol. 4, pp. 23–29.

Udoeyo, F.F., Hyee, A., 2002. Strengths of cement kiln dust concrete. Journal of Materials in Civil Engineering 14, 524–526.

Venkatanarayanan, H.K., Rangaraju, P.R., 2015. Effect of grinding of low-carbon rice husk ash on the microstructure and performance properties of blended cement concrete. Cement and Concrete Composites 55, 348–363.

Verbeck, G.J., Helmuth, R.H., 1968. Structures and physical properties of cement paste. In: Proceedings of the Fifth International Symposium on the Chemistry of Cements, Tokyo, pp. 1–32.

Vipulanandan, C., Weng, Y., Zhang, C., 2000. Designing flowable grout mixes using foundry sand, clay and fly ash. In: Advances in Grouting Modification, vol. 104. Geotechnical Special Publications, ASCE, Reston, pp. 215–233.

Wang, K., Konsta-Gdoutos, M.S., Shah, S.P., 2002. Hydration, rheology, and strength of ordinary portland (OPC)-cement kiln dust (CKD)-slag binders. ACI Materials Journal 99, 173–179.

Wang, K., Mishulovich, A., Shah, S.P., 2007. Activations and properties of cementitious materials made with cement-kiln dust and class F fly ash. Journal of Materials in Civil Engineering 19, 112–119.

Wang, K., Shah, S.P., Mishulovich, A., 2004. Effects of curing temperature and NaOH addition on hydration and strength development of clinker-free fly ash binders. Cement and Concrete Research 34, 299–309.

Wang, M.L., Ramakrishnan, V., 1990. Evaluation of blended cement, mortar and concrete made from type III cement and kiln dust. Construction and Building Materials 4, 78–85.

Winkler, E.S., Bol'shakov, A.A., 2000. Characterization of Foundry Sand Waste. Chelsea Centre for Recycling and Economic Development, University of Massachusetts, MA.

Winkler, E.S., Kosanovic, B., Genovese, T., Roth, I., 1999. A Survey of Foundry Participation in the Massachusetts Beneficial Use Determination Process. Chelsea Centre for Recycling and Economic Development, University of Massachusetts, MA.

Wolseifer, J., 1984. Ultra high-strength field placeable concrete with silica fume admixture. Concrete International: Design & Construction 6, 25–31.

Wong, H.S., Razak, H.A., 2005. Efficiency of calcined kaolin and silica fume as cement replacement material for strength performance. Cement and Concrete Research 35, 696–702.

Yu, Q., Sawayama, K., Sugita, S., Shoya, M., Isojima, Y., 1999. The reaction between rice husk ash and $Ca(OH)_2$ solution and the nature of its product. Cement and Concrete Research 29, 37–43.

Yusuf, M.O., Johari, M.A.M., Ahmad, Z.A., Maslehuddin, M., 2014. Influence of curing methods and concentration of NaOH on strength of the synthesized alkaline activated ground slag-ultrafine palm oil fuel ash mortar/concrete. Construction and Building Materials 66, 541–548.

Zerbino, R., Giaccio, G., Batic, O.R., Isaia, G.C., 2012. Alkali–silica reaction in mortars and concretes incorporating natural rice husk ash. Construction and Building Materials 36, 796–806.

Zhang, M.H., Lastra, R., Malhotra, V.M., 1996. Rice-husk ash paste and concrete: some aspects of hydration and the microstructure of the interfacial zone between the aggregate and paste. Cement and Concrete Research 26, 963–977.

Dry-stack and compressed stabilised earth-block construction

H.C. Uzoegbo
University of the Witwatersrand, Johannesburg, South Africa

8.1 Introduction

8.1.1 Economic and environmental issues

One of the serious problems faced by the ever-growing population in developing countries is lack of adequate habitat. Central to the cause of inadequate housing in developing countries is the high cost of conventional building materials and an unfair income distribution (Junior et al., 2003). Unskilled labour is highly available for general construction in developing countries; the most significant problem in housing delivery is the relatively high cost of building materials. Since aggregate is generally locally available, the main focus in material cost for concrete production is the cost of the binder (cement). The first observation in this regard is the small number of cement plants established in Africa and other developing countries. Cement is largely imported and thus becomes very expensive for local use. Less than 5% of world cement production is attributed to Africa, thus Africa is a net importer of cement. Fig. 8.1 shows cement consumption by major countries in subtropical Africa compared with Germany. There are more cement plants in Germany than the whole of sub-Sahara Africa. The locations of cement plants are also indicated. Infrastructure development has been largely neglected in Africa due to its history of civil unrest, political instability and poor investment climate. Although the per capita consumption of cement in Africa is very low, it is one of the fastest growing continents in terms of rate of growth of cement consumption per capita. The average per capita cement consumption in Africa is 90 kg per person per annum. This is compared with a global average of over 550 kg per capita.

The slow development in establishment of cement production plants has resulted in the escalation of cement prices in certain regions of Africa. The cost of 50-kg bags of cement in various countries are presented in Table 8.1 for comparison. By contrast, in 2008, 50 kg of cement in the United States or Europe was less than $6.

The real issue is the affordability of cement. For an average worker, the number of working days required to earn the equivalent cost of a ton of cement makes the material not easily affordable in most African countries.

The circumstances that the typical African consumer faces are a combination of low income and high prices of cement. High cement prices should not be an obstacle for the

Nonconventional and Vernacular Construction Materials. http://dx.doi.org/10.1016/B978-0-08-100038-0.00008-1

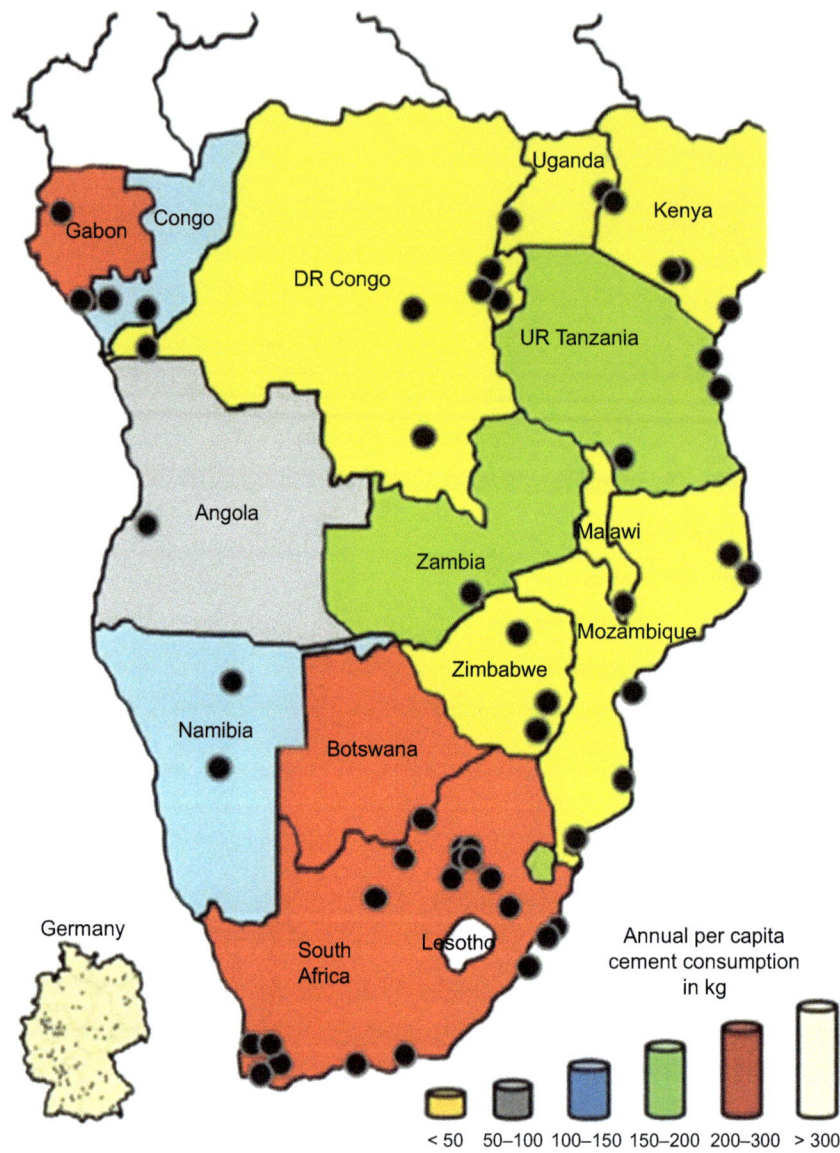

Figure 8.1 Locations of cement plants and annual per capita cement consumption in sub-equatorial Africa (Schmidt et al., 2012).

construction industry in developing countries but should instead motivate the development of concrete technology that does not require large amounts of cement. It is the duty of the local standards organisations to ensure that quality is not compromised in any alternative building solution.

Table 8.1 Cost of cement in selected African countries in 2008

Countries	Retail price (50-kg cement in USD)
Togo	8.40
Burkina Faso	13.10
Cote d'Ivoire	9.40
Nigeria	14.90
Cameroon	13.50
Gabon	13.50
DR Congo	15.60
Central African Republic	39.50
Zambia	11.50
Ethiopia	20.00
Tanzania	15.00
Kenya	10.81
South Africa	8.50

World Bank Report, 2009. Cement Sector Program in Sub-Saharan Africa: Barriers Analysis to CDM and Solutions.

As the people in less-developed countries struggle with the affordability of cement, the environmental concerns relating to the world production of cement becomes a secondary concern. Cement manufacture causes environmental impacts at all stages of the process. These include emissions of airborne pollution in the form of dust, greenhouse gases, noise and vibration when operating machinery and during blasting in quarries and damage to countryside from quarrying. Due to the large quantities of fuel used during manufacture and the release of carbon dioxide from the raw materials, cement production also generates more carbon emissions than any other industrial process. The production of 1 metric ton of Portland cement clinker results in the release of about 850 kg of CO_2 into the atmosphere. About 5–7% of global CO_2 emission is caused by the cement industry (Roskovic and Bjegovic, 2005).

The general goal of sustainable development is to meet the essential needs of people. It is thus geared towards meeting the needs of the present generation without compromising the ability of future generations to meet their own needs (Adedeji, 2010). Sustainable building materials are environmentally responsible because their impacts are considered over the complete life time of the products. Sustainable building materials should pose no or very minimal environmental and human health risks (Calkins, 2009). They should also satisfy the following criteria: rational use of natural resources, energy efficiency, elimination or reduction of generated waste, low toxicity, water conservation and affordability. Soil-cement blocks (or bricks), also known as compressed stabilised earth blocks (CSEBs), largely meet these criteria. The degree of sustainability of CSEB is further increased by the use of interlocking system.

Many block-producing companies are currently developing and marketing dry-stack masonry. Each dry-stack system is unique and therefore will be treated as such. The behaviours of interlocking masonry walls have been studied and reported by several researchers, including Hendry (1981), Monk (1967), Morsy (1968) and Uzoegbo (2011). Various interlocking systems have been developed. Some are made of hollow-block systems such as the Sparlock, Meccano, Sparfil and Haener systems. Others are made of solid blocks such as the Hydraform interlocking blocks. In solid interlocking blocks, substantial cost savings can be achieved due to elimination of bedding mortar in the superstructure. This accelerates the construction duration, thereby reducing both the need for skilled labour and cost.

8.1.2   Summary of the advantages of CSEB construction

Some of the advantages of CSEB construction include:

- Energy saving. Clay bricks require three times more energy (406 kg of CO_2 emission) per square metre than soil-cement blocks (119 kg of CO_2 emission).
- Less cement is used compared with normal concrete products.
- Soil is available in large quantities in every region; it is inexpensive and is easily accessible to low-income groups. In some locations, it is the only material available.
- Requires less transportation of materials.
- Ease of use—usually little specialised equipment is required.
- Fire resistance. Soil-cement products are noncombustible with excellent fire-resistance properties.
- Soil-cement construction has beneficial climatic performance due to its high thermal capacity, low thermal conductivity and porosity; thus, it can moderate extreme outdoor temperatures and maintain a comfortable interior temperature.
- Soil-cement blocks are usually 2.5 times larger than normal burnt clay bricks; the construction is therefore faster as the number of joints is consequently reduced.
- It is the ideal local material as the production is made on site itself or in a nearby area.

8.1.3   Cost comparison

Economy is achieved in soil-cement construction through the use or reuse of in-place or nearby borrow materials. No costly hauling of expensive, granular materials is required; thus, both energy and materials are conserved. Unit production costs for soil-cement blocks will differ in relation to local conditions. The main reasons for cost variations include: availability of soil, whether it is available on site or transported to the site and suitability of the soil for stabilisation (it may be necessary to add sand if the soil has excessively high linear shrinkage); current prices of materials, especially stabilisers; size and type of equipment used, whether manually operated or mechanised; current wage rates; and productivity of the labour force. A cost comparison of stabilised soil-cement blocks and conventional blocks which takes into account the speed of construction, cost of binding mortar, and transportation of materials shows that on average CSEBs are about half the cost of convention concrete masonry of similar strength.

8.1.4 Limitations of CSEB construction

Some of the limitations in the application of soil-cement products in construction are as follows:

- Not all engineers are generally familiar with CSEBs.
- Soil differs from region to region; therefore, soil testing is required to identify suitable soil for block production.
- Dry stacking (with interlocking features) is not common; sometimes, there is resistance from end users.
- International or local standards regulating the production of soil-cement blocks are not commonly available.
- There is reduced durability if not regularly maintained and properly protected, particularly in areas with high humidity.
- They have low tensile strength and poor resistance to bending moments and thus may only be used in compression.
- They have low resistance to abrasion and impact if not sufficiently reinforced or protected.
- There is an unawareness of the need to manage resources for environmental sustainability.
- There is ignorance of the basics for production and use.

8.1.5 Terminology, 'blocks' versus 'bricks'

'Bricks' are rectangular blocks of a maximum standard size. Products that are larger than the maximum size for bricks are called 'blocks'. There is a slight variation in brick standard size from country to country as shown next:

- United States: $203 \times 102 \times 57$ mm (length $\times$ width $\times$ height)
- South Africa: $222 \times 106 \times 73$ mm
- Australia: $230 \times 110 \times 76$ mm

Typically, the number of brick units to produce 1 m^2 of single leaf wall is 60 bricks. Normally, it takes only about 10 blocks to produce 1 m^2 of single leaf block wall.

8.2 Materials selection and block production

Material selection naturally begins with the identification of suitable soil for CSEB production. This is followed by the selection of the stabilising agent and the production processes. The process of manufacture of soil-cement blocks involves the following steps: analysis and selection of the soil; sieving of the soil; preparation of the mix; compaction of the blocks (manual or mechanised); and curing of the blocks.

8.2.1 Materials for CSEB production

Soil on its own can be used for building construction; however, the resulting product will need to be protected from moisture to avoid deterioration. The use of a bonding agent to stabilise the soil results in considerable increase of the wet strength and reduced permeability and improved erosion resistance of the final block product. Materials for the production of stabilised and compressed soil-cement blocks consist of local soil and cement. Soil for block production is typically sourced from the site of production. The soil should be free from organic material. In most cases it is normal to remove the top 0.5 m of topsoil, which generally contains organic matters, and to use the soil below that level. Soil properties vary from one location to another. It is therefore necessary to standardise soil properties suitable for block production. Soils contain particles of different sizes such as gravel, sand, silt and clay. Soil for use in soil-cement block production is usually sieved with a 5-mm sieve to eliminate the gravel particles. Sand and silt are inert materials, but clay is affected by moisture. Sandy soils containing mostly nonexpansive clay minerals are the most suitable for stabilised soil block production.

8.2.2 Soil identification and classification

A typical soil profile is shown in Fig. 8.2. There are broadly four different layers or horizons, which consist of visually and texturally distinct layers. The topsoil layer contains organic matter and is usually dark in colour. This layer is not suitable for stabilised soil-cement block production due to its high content of organic matter. The subsoil layer is an accumulation of iron, clay, aluminium and organic compounds formed through a process referred to as illuviation. It is very sticky if it has high

Figure 8.2 Typical soil profile.

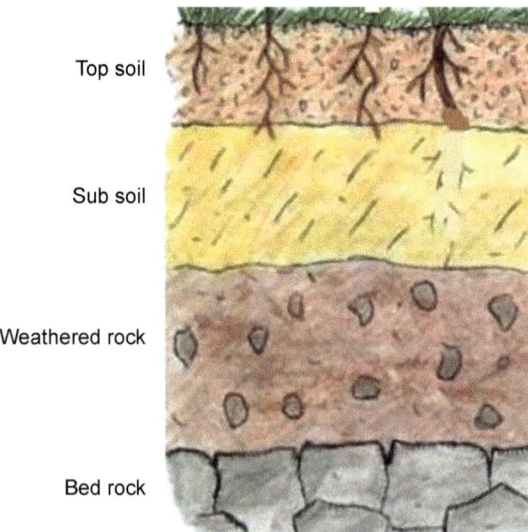

Top soil

Sub soil

Weathered rock

Bed rock

Table 8.2 **Soil particle grading**

Material	Ranges	Particle size
Gravel fraction	Coarse gravel	60–20 mm
	Medium gravel	20–6 mm
	Fine gravel	6–2 mm
Sand fraction	Coarse sand	2–0.6 mm
	Medium sand	0.6–0.2 mm
	Fine sand	0.2–0.06 mm
Silt fraction	Coarse silt	0.06–0.02 mm
	Medium silt	0.02–0.006 mm
	Fine silt	0.006–0.002 mm
Clay fraction	Passes the	0.002 sieve

clay content. This layer is the most suitable layer for soil-stabilised block production. The weathered rock horizon is the third layer, which consists of mostly large broken rocks. This layer usually contains sandy soil that is easier to excavate. It is not suitable for block production due to the size of the gravels and inadequate clay content. The bedrock or parent rock material in bedrock landscapes is a layer of partially weathered bedrock at the base of the soil profile. Unlike the layers above, bedrocks comprise continuous masses of hard rock that cannot be easily excavated.

Soil particles obtained from the subsoil layer are broadly divided into three size classes: clay, silt and sand, as shown in Table 8.2. Soil materials intended for CSEB construction designates the basic material made up of carefully controlled proportions of sand, clay and silt before any mixing with additive or with water. Basic soil particle grading is shown in Table 8.2. Sand components with particle sizes larger than 2 mm are classified as gravel and are generally omitted in the grading of the soil. Gravel is not normally used in soil-cement block production as the large particle size results in poor finishing.

Particles size analysis is used to determine the fraction of each particle size that falls within each of the size ranges shown in Table 8.2. A well-graded soil will produce a better packing of the particles resulting in a denser and less permeable product. The distribution of particle sizes that provides the optimum packing of particles is known as the Fuller curve. It is based on the assumption that the smaller particles fill the void between the larger particles to give the highest density. A soil may be considered well graded if the distribution of particles from fine silt to coarse sand is reasonably uniform.

8.2.3 Soil identification and selection

Soil texture is defined as the relative proportions of each class (clay, silt and sand). Sands give the material strength while clays bind it together and silt fulfils a less clear intermediate function. It is important to get the right texture for soil-cement block production. Good soil-cement blocks can be produced with a sandy soil with clay content between 5% and 20% and silt content of 5% to 25%. Blocks can be produced with higher clay and silt content, but it may be necessary to determine the plasticity index to see if the soil is suitable for block production. Generally, soil with clay and silt portions below 10% will be difficult to handle when coming out the block-making machine. Soil with clay and silt content above 40% will need to be blended with a sandy soil since stabilisation of the material with cement is less effective. Commonly used methods to determine soil texture include hand texture method, separation by sieving and separation by sedimentation.

Soil texture classification by sedimentation is carried out by adding a soil sample to a dispersing solution of sodium hexametaphosphate $(NaPO_3)_6$ in deionised water. The sedimentation tests are based on Stoke's law, which predicts the free fall of any diameter spherical particle of known specific gravity in a fluid of known viscosity at low concentration. It is assumed that the soil particles have approximately the same specific gravity and that the rate of fall is dependent only on the particle size. The larger-diameter particles (sand) fall more quickly and settle at the bottom of the jar; then silt will settle out and finally a clay layer will form on top. Once settled (Fig. 8.3), the relative percentages of sand, silt and clay may then be measured. This test is easily carried out on work site.

The particle size analysis may be used to classify the soil sample into a specific textural class, such as a sand, silt, clay, loam, etc. Soil texture depends on its composition and the relative portions of clay, sand and silt. The soil textural triangle shown in Fig. 8.4 enables one to visually assess textural class based on the three percentage values obtained through particle size analysis. The goal of the particle size analysis may be to classify a soil sample into a specific textural class, such as a sandy clay, sandy silt, clay loam, etc. based on the zone of the material in the textural triangle.

Figure 8.3 Settlement of soil materials.

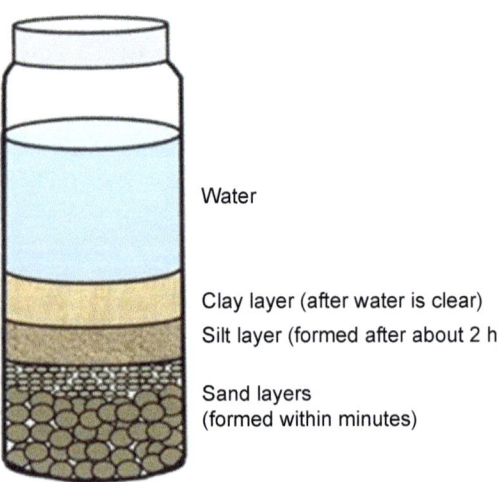

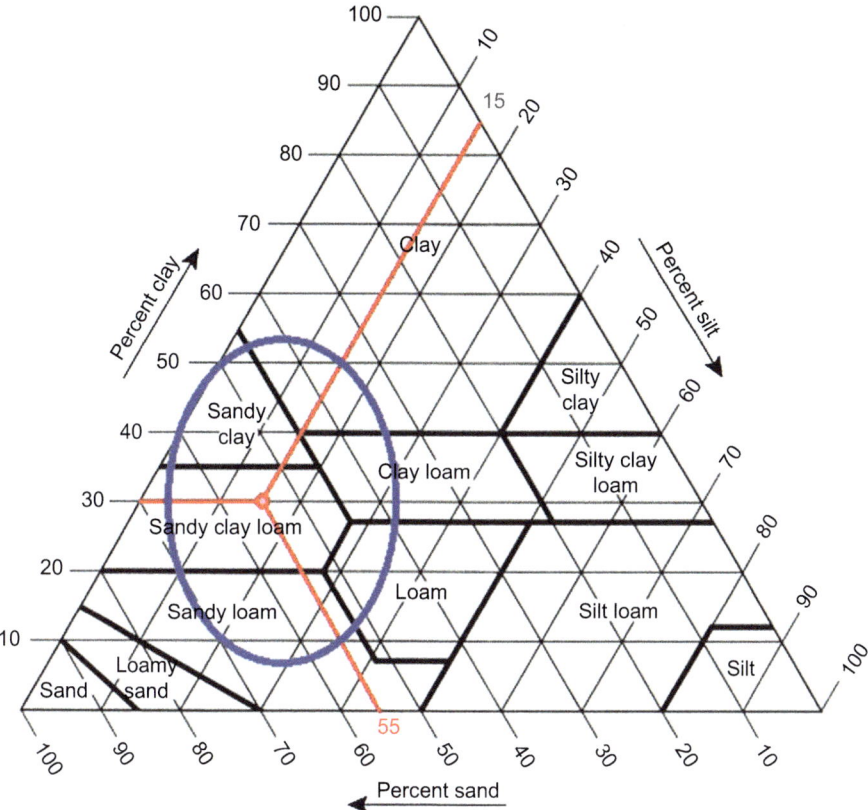

Figure 8.4 Soil textural triangle.

The major textural classes are shown in Fig. 8.4. The area within and about the circle is cantered at around the sandy clay and sandy clay loam region in the textural triangle as shown in Fig. 8.4 and indicates the zone most suitable for stabilisation. Soil properties within the circle in the chart are almost universally available. They are found after removing about 100–150 mm of the topsoil in order to exclude organic matter. Soils are rarely found in the state required for block production. In most cases, they need to be ground and screened through a 5-mm wire mesh.

Clay and cement make different contributions to the material properties of soil-cement blocks; in fact, the two materials will work against each other if the quantities are not carefully selected. Too much clay in the mix will result in the cement not adequately coating all the mix particles and subsequent wetting will cause expansion of the material and cracking. As a rule of thumb, the most suitable soil for stabilised soil block production should contain approximately 30% to 40% clay plus silt and 60% to 70% sand. Spence and Cook (1983) recommended the following ranges for soil-cement block production: sand: 60–90%, silt: 0–25% and clay: 10–25%.

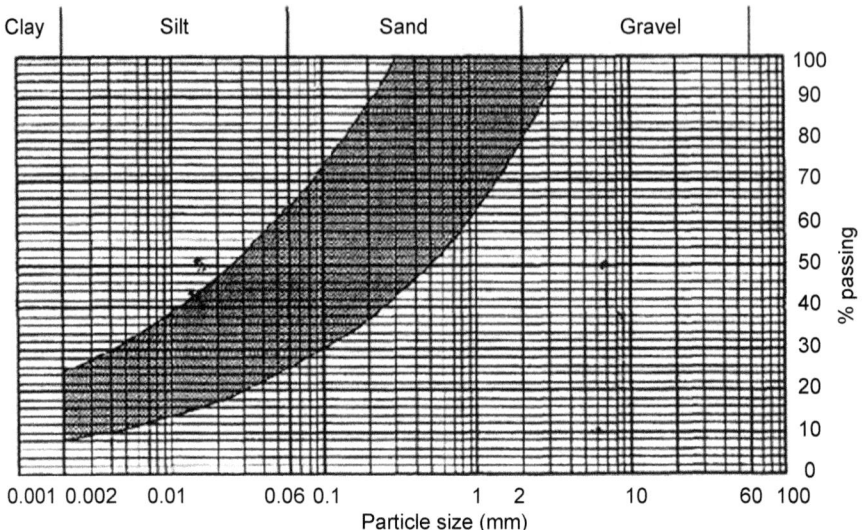

Figure 8.5 Diagram of texture for soil for CSEB production.
African Standards Organisation, 1996. Standards on CEB Construction: Compressed Earth
Block Masonry ARS 670: Part 4: Code of Practice for Production and Construction.

Not all soils are suitable for construction. Soil containing organic matter, highly
expansive soils, and soils containing excessive soluble salts such as gypsum and chalk
should be avoided in the selection of materials for soil-cement block production.

The African Standards Organisation standard ARS 670 of 2014 recommends that
the granular composition of the soil for CSEB production should preferably fall within
the limits of the shaded area on the diagram of texture in Fig. 8.5 and should be similar
in shape. The limits of the recommended shaded area are approximate. Soils with granular composition that fall outside the shaded area may still give acceptable results, but
it is recommended that they be subjected to a series of tests enabling their suitability to
be assessed.

The plasticity of the soil suitable for CSEB should preferably fall within the limits
of the shaded area of the diagram of plasticity in Fig. 8.6 for best results (ARS 670-4:
2014). Types of earth the plasticity of which fall outside the shaded area may still give
acceptable results, but it is recommended that they be subjected to a series of tests
enabling their suitability to be assessed.

8.2.4 Stabilisation of soil for CSEB production

Both stabilised (CSEBs) and unstabilised compressed earth blocks (CEBs) are appropriate for buildings and meet the requirements of international building code standards.
The durability of a properly designed CEB building will allow it to last for centuries.
Ancient earthen structures still stand today in many parts of the world, while in comparison the expected lifespan of wood frame buildings is about 70 years. CEBs are,

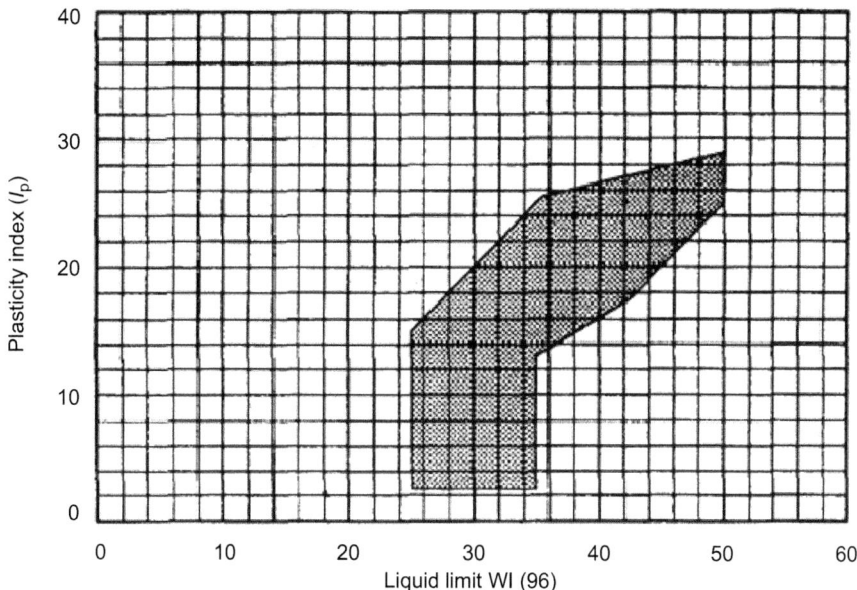

Figure 8.6 Diagram showing zone of suitable plasticity.
African Standards Organisation, 1996. Standards on CEB Construction: Compressed Earth
Block Masonry ARS 670: Part 4: Code of Practice for Production and Construction.

however, vulnerable to moisture and the erosive effects of weather. Walls constructed
out of soil, if well-compacted, give environmental-friendly low-cost housing technology with adequate compressive strength under dry conditions. However, they will lose
a large percentage of their dry strength under sustained moist condition. Alternating
wetting and drying will erode and deteriorate unstabilised soil materials. The African
Standards Organisation (ARSO, 1996) has published several guidelines and standards
for CEB construction. It is known that suction caused by shrinkage of water menisci
between particles in unstabilised clayey soil is the primary source of the strength (Hall
and Djerbib, 2004). Reintroduction of water tends to destabilise the bond created by
suction. Compressed earth construction without a stabilising agent results in structures
that are not durable and require regular maintenance. Fig. 8.7 shows earth construction
without a stabilising agent. The surface erosion of the wall is clearly evident. Residents
need to carry out regular maintenance by replastering the wall with clay material.

In addition to providing proper soil grading, durability and strength can be
improved by adding stabilisers or chemicals, resulting in a stabilised soil. A properly
stabilised, consolidated and well-graded soil that is adequately moisturised, mixed and
cured will provide a strong, stable, waterproof and long-lasting building blocks or
bricks. Stabilisers when introduced in soil will perform the following functions:

- Bind the soil particles together, making the product stronger.
- Reduce the amount of voids and therefore limit the water that can be absorbed by the stabilised soil.

Figure 8.7 Typical nonstabilised soil construction. (a) Weathering of nonstabilised earth construction. (b) Fishing village in Tanzania.
African Standards Organisation, 1996. Standards on CEB Construction: Compressed Earth Block Masonry ARS 670: Part 4: Code of Practice for Production and Construction.

• Reduce the shrinkage and swelling properties of the soil.
• Increase the tensile strength of the soil.

When a stabilising material is added to soil, it improves the engineering properties of the soil as well as those of the resulting bricks and, ultimately, the lifespan of the resulting structure. Different types of soil may require types of stabilisers. The most common stabilising materials include cement, lime, a combination of lime and cement and a combination of lime and pozzolans. When using cement as stabilisers, it is important to cure the cement by keeping the block moist. This may mean watering the bricks twice a day for a couple of weeks and covering the wall with plastic or leaves to keep the sun and wind off in order to limit evaporation. Cement hydration requires water, so letting bricks dry too quickly, especially during the first week or two of curing, may lead to inadequate hydration and strength development. Blocks without stabilisers should also be dried in the shade to avoid excessive drying around the edges which may cause cracking.

When mixing the stabilisers with soil, it is important to make sure they are mixed completely, especially when using small amounts of stabiliser. Using a 3% cement:volume ratio takes a lot of mixing to get the cement distributed thoroughly. Soils are usually stabilised with cement contents ranging from 4% to 15% by dry weight of the soil. Most CSEB building projects were carried out with cement content of about 6% to 8%.

8.2.5 Types of stabilisers

Various materials are available for use in the stabilisation of soil-cement in construction. The most popular used stabilising agents are cement, lime, pozzolanic materials or a combination. It is necessary to apply appropriate level of stabilisation as understabilisation results in low-quality products. Overstabilisation through fear or ignorance may result in wastage. The following are the most commonly used stabilising agents for CSEB construction.

8.2.5.1 Cement

When cement is added to soil, there will be reduction in the following properties with increasing cement content: the liquid limit, the plasticity index and the potential for volume change. Cement enhances the compressive strength of CSEB by replacing suction with chemical bonding. However, the compressive and shear strengths are increased with increased cement content. A variety of compounds and gels are formed by hydration reactions of the cement in the presence of moisture. The products of cement hydration are hydrated calcium silicates, hydrated calcium aluminates and hydrated lime. Calcium silicates and calcium aluminates are the main cementitious products, and the hydrated lime is deposited as a separate phase. The hydrous silica and alumina slowly react with calcium ions liberated from hydrolysis of cement to form insoluble compounds that harden on curing to stabilise the soil. The chemical reaction equations are described by Umesha (2009). Soil stabilisation is mostly done with the use of cement as the stabilising agent. Ordinary Portland cement (OPC) is mostly used for stabilisation purposes and it works best with sandy soils. Stabilisation may be difficult if the clay content is very high. Generally the combined percentage of silt and clay should not be less than 10% and ideally should not be more than 40%. The cement content varies depending on the desired strength of bricks and type of soil, although 5% to 10% by weight of dry soil is most commonly used.

8.2.5.2 Lime

Hydrated lime (calcium hydroxide) is also used as a stabiliser. There are two basic types of lime: high-calcium and high-magnesium lime. Their soil-stabilising efficiency is about the same. Lime will react readily with most plastic soils containing clay but lime does not improve sands or other noncohesive granular materials. Lime makes a good stabiliser for soil with clay content greater than 40%. It reacts with the clay to form strong bonds between soil particles. The recommended amount of lime for stabilisation ranges from 4% to 8% of the dry weight of soil. Soils ranging in plasticity index from 10 to 50 or higher are suitable for lime stabilisation. Lime stabilisation decreases the plastic index and volume change and increases the compressive strength of the soil material.

8.2.5.3   Combination of lime and cement

When a soil has clay content in excess of 40%, cement becomes less efficient as a stabilising agent. A combination of lime and cement stabilisation is recommended. The lime will make the soil easier to work with, and cement will help to improve the strength and water resistance of the product.

8.2.5.4   Combination of lime and pozzolans

Pozzolans are materials that contain a significant amount of silica. Volcanic ash, pulverised blast furnace slag, pumice and silica flour are examples of pozzolans. Lime and pozzolan will react and make cement that may be almost as good as Portland cement. The lime becomes the activator for the pozzolanic action. This can be used for both clayey (>40%) and sandy (<10% clay) soils. This combination of stabilisers results in a relatively low-strength blocks.

8.2.6   Material properties of CSEBs

The main material property requirements for CSEB are identified by the density, compressive strength of individual blocks, absorption, thermal conductivity and compressive strength of masonry. Some of the characteristic properties are described here.

8.2.6.1   Compressive strength

The compressive strength of compressed stablised earth building depends on the soil type, type and amount of stabiliser and the compaction pressure used to form the block. Maximum strengths are obtained by proper mixing of suitable materials and proper compacting and curing. In practice, typical wet compressive strengths for CSEBs may be less than 4 MPa and still meet the minimum requirements particularly for single-storey buildings. The building block is designed to be about 20 times stronger than the average stress under service conditions, which can be as low as 0.2 MPa in single-storey structures (see Fig. 8.8). The South African standards (SANS 10164, 1996) recommend a minimum strength of 3 MPa for block units. British Standards (BS 5628, 1978) recommend a minimum strength of 2.8 MPa.

Most standards prescribe a much higher minimum strength requirement for bricks than for blocks. The reasons for permitting lower strength for blocks include greater

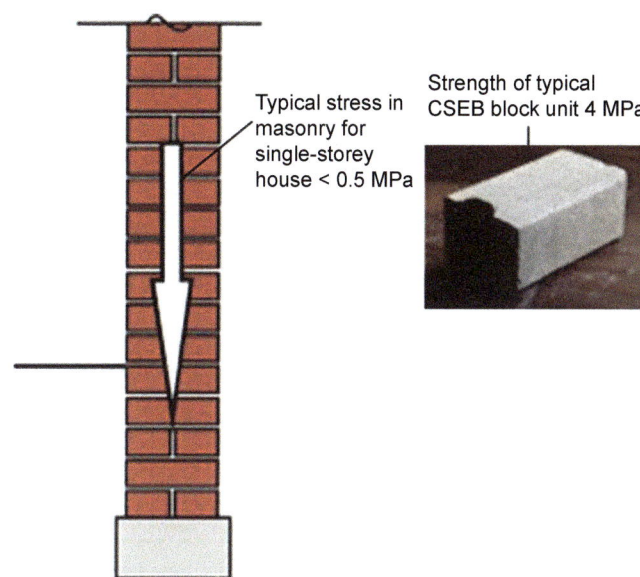

Figure 8.8 Average loading on a typical single-storey wall.

robustness and increased dimensional tolerance in blocks. The average minimum strength prescribed for brick products in most international standards is 7 MPa, while the average minimum strength requirement for blocks is usually less than half of the value for bricks. Since the average strength of CEBs and CSEBs are in the low-strength range, it makes much sense to produce blocks rather than bricks. Where building loads are small (eg, in the case of single storey constructions), a compressive strength of 2–4 MPa may be sufficient for a building block. Many building authorities around the world prescribe minimum values within this range for blocks.

8.2.6.2 Moisture volumetric strain

Most soils contain a fraction of clay as a part of their overall composition. Clay is the finest of the soil particles and can bond other particles together if sufficient clay and moisture is present. Clay has large volumetric expansion when water is added. As the moisture in unstabilised soil increases, swelling occurs. Conversely, drying causes shrinkage and therefore the danger of cracking. This process leads to the breakdown of the soil and internal strength is lost making the material useless for building construction. The balance of clay with respect to the other soil components is important. On one hand clay helps bond particles together and improves the fresh state workability. Yet if a stabilising agent is not introduced, swelling of the clay can be instrumental in driving the soil particles apart should the material get wet.

Proper block manufacture and construction methods, however, will reduce such moisture-induced strain. It is worth mentioning that moisture-induced strain becomes especially important when two materials with different swelling or shrinkage properties are used in a building. Differential movement results in stress, which may break the bond between the materials or cause other damage. For example, cement renderings often peel off earth walls or poorly compressed stabilised earth blocks because of their different expansion properties. With experience many people can easily estimate the correct moisture content for a mix on site. By moulding a lump of moist soil material by hand and dropping it from 1 m above ground, the mode of scatter on the ground determines the suitability of the material. If the drop test results in the ball of soil shattering or breaking into seven or more pieces, then the soil is too dry. If the ball breaks into four to six pieces, water content is suitable, while breaking into three or fewer pieces indicates that the soil is too wet.

When a block machine compresses a block, it reduces the volume of the soil by up to 30%. It does this by mechanical packing the moist clay particles in the spaces between the larger sand particles, removing the air pockets and sticking the clay to the sand. If too much water is introduced in the mix, the particles will be spaced further leaving voids when the brick dries. If there is more clay than is needed to fill the spaces between the sand particles, the block becomes weaker because clay compresses more than sand, especially when wet. However, weaker earth walls can be acceptable, as we know from the use of cob and adobe. Sand particles provide strength and stiffness while the clay provides the bonding between particles as well as the early-age strength required for handling the freshly compressed blocks. Since clay will shrink or swell depending on the relative moisture content, it is necessary to ensure that shrinkage is within allowable limits. A simple test for shrinkage measurement is the Alcock's linear shrinkage test (Houben and Guillaud, 1994) shown in Fig. 8.9.

The test is performed using a wooden box, 600 mm long, 40 mm wide and 40 mm deep. The inside surfaces are oiled before filling the box with moist soil having optimum moisture content. The filled box is then exposed to the sun for a period of 3 days or stored in the shade for 7 days. After this period, the shrinkage is measured as the change in length of the original sample.

A suitable soil should give shrinkage less than the required limit. Minimal shrinkage indicates that the soil material is too sandy and not suitable. Excessive shrinkage is not suitable either. Table 8.3 gives the required shrinkage for dimensional stability. The mode of shrinkage movement is different for sandy and clayey soils. Sandy soils crack at intervals as they shrink (see Fig. 8.9). The mode for clayey soil is indicated by an upward camber of the beam (Fig. 8.9). Typical shrinkage ranges for CSEBs based on the standard sample dimensions are shown in Table 8.3.

8.2.6.3   Density and thermal properties of CSEB

Normally compressed stabilised earth blocks are denser than some concrete masonry products such as aerated and lightweight concrete blocks but are slightly less dense than conventional concrete products. The relative high density of CSEBs may be considered to be a disadvantage when the blocks have to be transported over long distances; however, it is of little consequence when they are produced at or near the construction site.

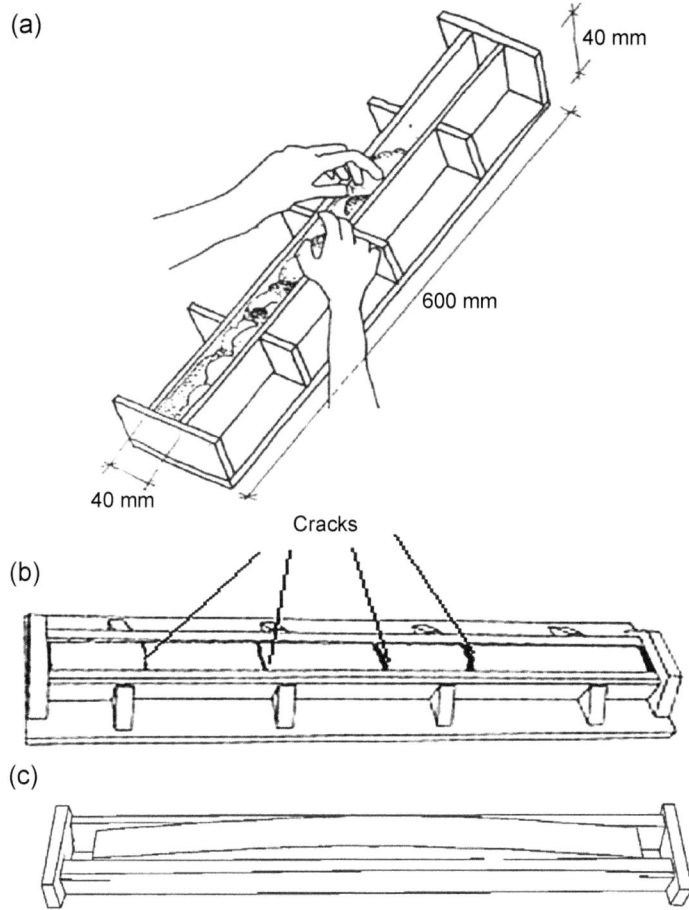

Figure 8.9 Linear shrinkage tests. (a) Linear shrinkage test. (b) Crack pattern of linear shrinkage test of sandy material. (c) Shrinkage of high clay content material (camber effect).

Table 8.3 Shrinkage limits to achieve dimensional stability

Shrinkage	Strain	Comment
<10 mm	<1.7%	Too sandy
20−40 mm	3.3−6.6%	Recommended
>60 mm	>10%	Too clayey

Due to their density, a generally CSEB has good thermal and acoustic properties. This is particularly advantageous in hot dry climates where extreme temperatures can be moderated inside buildings made of CSEBs.

8.2.6.4 Dimensional tolerance of CSEBs

As a guideline, a dimensional tolerance or limit on dimensional deviation should be observed when fabricating soil-cement blocks. The following tolerances are recommended by Boubekeur et al. (1998) for normal blocks. A stricter limit is recommended for interlocking blocks since they are laid without mortar.

- $+1$ to -3 mm for length
- $+1$ to -2 mm for width
- $+2$ and -2 mm for height

Additionally, the difference between any two CEBs should not exceed:

- 3 mm in length
- 2 mm in width
- 3 mm in height

Recommended general limits on properties of CSEB materials are shown in Table 8.4, and comparative properties of CSEB in relation with other materials are shown in Table 8.5. It can be seen that CSEBs are normally produced at a lower strength range compared with fired clay bricks and concrete blocks.

Table 8.4 Recommended limits for properties of soil suitable for CSEBs (Addis and Davis, 1986)

Property	Limits (%)
Clay content	5–20
Silt and clay content	20–50
Sand content	50–80
Liquid limit	25–50
Plasticity limit	10–25
Plasticity index	5–30
Linear shrinkage	<4
Cement content	5–15
Total salts	<0.5

Table 8.5 Comparative properties of CSEBs

Material property	CSEBs	Fired clay bricks	Concrete blocks
Wet compressive strength (MPa)	1−20	5−60	5−50
Density (kg/m³)	1700−2200	1400−2400	1700−2400
Thermal conductivity (W/m °C)	0.81−1.04	0.70−1.30	1.0−1.70
Resistance to weather	Good to very poor	Excellent to poor	Good to poor

8.3 Block production

CSEBs can be produced by hand or using machines. Machines can be manually operated or automated. Automated machines are usually used for bigger projects to manufacture blocks for use in low cost housing to upmarket estates, schools and commercial structures or to start a block yard business in soil-cement blocks. The key input factors that can influence the quality of produced CSEB are soil (type and proportions of main fractions), stabiliser (type and content), mix-water (quantity), compaction pressure and curing conditions.

8.3.1 Raw materials

Soil is the principal raw material for CSEB production. OPC and water are the other two constituents required for the manufacture of soil-cement blocks. Sand and crushed stone dust may also be added to the soil if the clay content is too high. Lime and pozzolan cement are alternative soil-stabilising materials and may also be used. The selection of a stabiliser will depend upon the soil quality and the project requirements. Cement will be preferable for sandy soils (silt + clay < 40%) and to achieve a higher early strength. Lime will be used for very clayey soil (silt + clay > 40%), but will take a longer time to harden and give strong blocks. The most practical soils are those that contain no more than 35% silt and clay and soil that are easily pulverised. Fine-textured soils with 50% or more silt and clay are harder to pulverise and require more cement to harden; a typical material selection with stabilising agent is shown in Table 8.6. It is also possible to use a combination of cement and lime as soil stabiliser.

Table 8.6 Typical material mix proportion

Soil texture			Stabiliser
Sand (%)	Silt (%)	Clay (%)	(as % of soil)
65	15	20	Cement (3−20)
45	20	35	Lime (2−8)

8.3.2 Soil preparation

The main source of soil will be to dig it out of the ground. It will therefore be removed in dense lumps, which will have to be broken up into loose particles. Soil particles will generally be rounded due to the natural illuviation processes. Secondary crushing techniques are generally not practiced in developing countries because of the high cost of the heavy machinery and energy required to crush large aggregate into smaller angular particles.

The soil lumps as dug up from the borrow pit contain various grain sizes, from very fine dust up to pieces that are still too large for use in block production. For the purposes of soil-cement building materials, particles over 5 mm are considered too large and should be removed. By eliminating the gravel, the soil is left with rounded particles in the size range for clay, silt and sand. The oversized material should be removed by sieving. The sieving operation is intended to eliminate all undesirable components (roots, leaves, etc.) together with any components with a diameter greater than required. This operation also enables soil lumps to be loosened in a uniform manner.

The usual sieving device is a screen made from a 5-mm wire mesh, nailed to a supporting wooden frame and inclined at approximately 45 degree to the ground as shown in Fig. 8.10. As the material is thrown against the screen, fine materials pass through and the coarse, oversized materials run down the front and are collected and removed. A horizontally suspended screen could also be used but this method is less effective as the larger particles accumulate on the screen and require frequent removal.

For uniformity of the CSEBs produced, the weight or volume of each material used in the block-making process should be measured at the same physical state for subsequent batches of blocks. The volume of soil or stabiliser should ideally be measured in

Figure 8.10 Sieving process of soil.
African Standards Organisation, 1996. Standards on CEB Construction: Compressed Earth Block Masonry ARS 670: Part 4: Code of Practice for Production and Construction.

dry or slightly damp conditions. It is advisable to mix sufficient materials to allow for 1-h production output of the block-making equipment. The time between mixing of the cement and water and the final finishing is an important factor in the compaction of the mixture. Remixing the materials after about 2 h or more breaks the cementitious bonds through the compacting process and therefore the compacting of the mixture should normally take place before the soil-cement mix passes this critical period.

8.3.3 Batch quantities for CSEB production

The soil-cement mix quantities are determined on the basis of dry weight of soil. Assuming the soil meets all texture requirements for soil-cement mix, only about 5% cement is required to produce blocks of 3- to 4-MPa strength. Typical mix requirement as may be communicated to site workers operating the equipment is described next. Unlike in concrete mix design, it is difficult to prescribe water content. Water is generally added to the soil-cement mix until it reaches the desired consistency, which varies according to the soil particle size distribution (Hall and Djerbib, 2004). The water content could be based on the optimum water requirement for compaction. Usually, stiff mixes with low workability are produced. Water is usually added until a humidity level is obtained that will allow the operator to make a ball of the material in the hand. However, the hand should not be wet after making the ball. The drop ball assessment could also be done on site. Typical mix ratios of soil and cement for the production of 4- to 8-MPa blocks are as shown in Table 8.7.

8.3.4 Mixing of materials

Mixing can be done manually using shovels or a mechanised mixer. It is important to begin with dry mixing for about 5 min and then to wet the mix evenly by sprinkling (with a watering can) or using a fine spray. Wet mixing should take about 3 min. The mixer can be filled using either buckets or measuring boxes which have to be emptied by hand, or using a sloping ramp for wheelbarrow access, or, for large production units, through measuring hoppers as shown in Fig. 8.11.

Formation of lumps of soil during mixing (generally during wet mixing) must be avoided as these are difficult to compact. Lumps can form if the moisture content is too high and/or if the mixing time is too long or if an inappropriate style of mixer is

Table 8.7 Typical components for soil-cement production

Target brick strength	Recommended cement content	Cements (parts)	Soil (parts)
3−4 MPa	5% cement	×1	×20
5−6 MPa	6% cement	×1	×16
7 MPa	8% cement	×1	×12

Figure 8.11 Soil-cement mix being loaded into the hopper.
African Standards Organisation, 1996. Standards on CEB Construction: Compressed Earth
Block Masonry ARS 670: Part 4: Code of Practice for Production and Construction.

used. It is advisable to use the optimum moisture content as determined by proctor test.
Experienced operators can estimate the required water content with reasonable accuracy. However for soils with high clay content, the moisture content should be slightly
higher than the optimum and for sandy soils, it will be slightly lower than the optimum
(Kawamura and Kasai, 2011).

The freshly produced blocks are usually strong enough to be handled immediately
as shown in Fig. 8.12.

Figure 8.12 Handling and temporary stacking of freshly produced block.
African Standards Organisation, 1996. Standards on CEB Construction: Compressed Earth
Block Masonry ARS 670: Part 4: Code of Practice for Production and Construction.

8.3.5 Manual and mechanised production

The soil-cement mix is put into the press chamber for block production. When a force is applied to the soil-cement mix, the material is compressed and the proportion of voids decreases and the density of the material increases. The more the density of the soil can be increased the lower its porosity will be. Moisture content must not be too high. Optimum moisture content may be determined using the proctor test. The Proctor compaction test (ASTM D698-12, 2012) is a laboratory geotechnical testing method used to determine soil compaction properties. The Proctor test shows the relationship between optimum moisture content and optimum dry density for a given amount of compaction energy. The optimum moisture content decreases with increasing compaction force. A typical proctor curve is shown in Fig. 8.13.

The optimum water content is the water content that results in the greatest density for a specified compaction force. Compacting at water contents higher than the optimum water content may result in a relatively dispersed soil structure (parallel particle orientations) that is weaker, more ductile, less pervious, softer and more susceptible to shrinkage.

The production equipment can either be manually operated or mechanised. For manual operation, the Ram consists of a box or mould which is filled with damp soil-cement, and a lever-actuated piston that compresses the earth-binder mix. Once the mould has been loaded with the proper amount of material, the machine's operator forces its long handle down with a pressure of 30−45 kg force; this translates to a pressure of about 4−6 MPa of pressure on the soil that is being compressed. The force classification of the pressure potentially available to compact the soil is given in Table 8.8. Most mechanised presses are programmed to produce blocks with 10 MPa pressure, in the high pressure range.

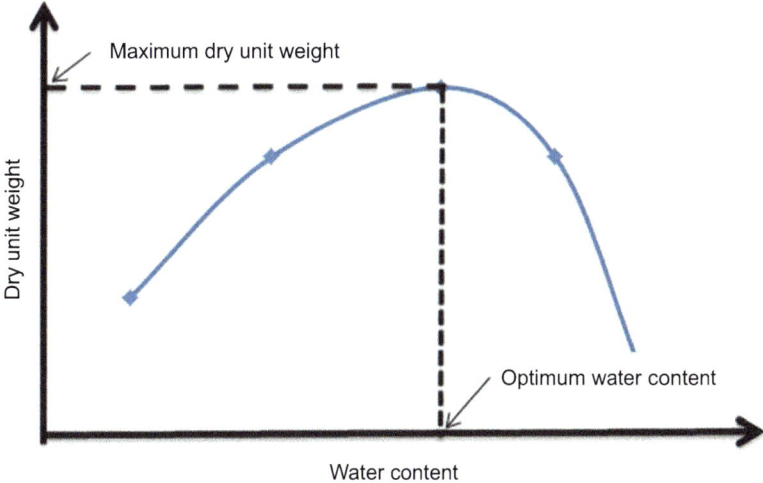

Figure 8.13 Sketch of proctor test result.

Table 8.8 Classification of usable compaction pressure CSEB production

Very low pressure	$1-2$ N/mm^2
Low pressure	$2-4$ N/mm^2
Medium pressure	$4-6$ N/mm^2
High pressure	$6-10$ N/mm^2
Hyperpressure	$10-20$ N/mm^2
Megapressure	≥ 20 N/mm^2

The brick formed by this procedure is then ejected, set in a cool place and left to cure for up to 3 weeks. The first modern version of the manual press is known as the CINVA Ram press and was developed in 1956 in Columbia by Paul Ramirez (Wheeler, 2005). Fig. 8.14 shows a sketch of the CINVA Ram press. Many modifications have been made to the original design by various manufacturers depending on the type, size and geometrical variations of block. Automation of the system has also been introduced for industrial line production. However, for smaller and rural locations, manually operated presses are most commonly used due to their mobility, low initial cost and ability to be used without a source of electricity or fuel.

Though not technically difficult, making construction blocks with a manual press or ram is a labour intensive operation and the amount of pressure exerted on the handle is not usually uniform as it is dependent on the operator at a given time. Four adults working in an organised assembly line with the aid of a manual cement mixer can produce only about 50 blocks an hour. With a mechanised press (see Fig. 8.14), about five times that amount can be produced with the same number of workers.

(a)   (b)

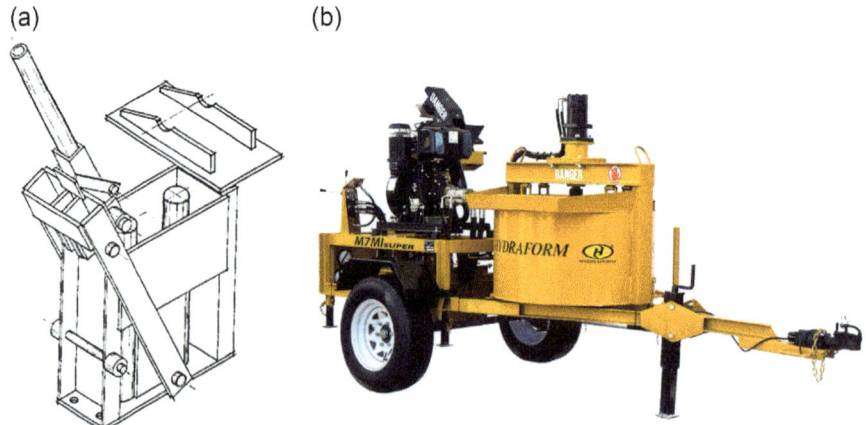

Figure 8.14 Typical presses for compressed soil-cement production. (a) Typical manual press. (b) Portable mechanised press.

Several companies currently manufacture their own version of compressed earth block press with or without interlocking system. Some of the manufacturers are as follows and images of their products are shown in Fig. 8.15:

1. Azar dry-stack blocks (Fig. 8.14(a)): The Azar block was originally developed in Canada and can be interlocked both horizontally and vertically.
2. Stumblebloc Mortarless System (Fig. 8.14(b)): Stumblebloc is hollow core block that reduces heat transmission and has connections at the top and bottom to vertically interlock and form corners (Stumblebloc, 1990).
3. The CVBT block system (Fig. 8.14(c)): The design of the Soeng Thai BP6 press is based on the original CINVA Ram press (Wheeler, 2005). The BP6 is a manually operated vertical block press manufactured by the CVBT in Thailand. The interlocking dowels are intended to provide ease of block alignment during construction and resistance to lateral in and out-of-plane forces. The round reinforcement holes are provided to allow use of grouted vertical reinforcement. The rectangular holes or grout channels are provided to help ensure wall stability and good load distribution.
4. The Hydraform dry-stack block (Fig. 8.14(d–f)): This system consists of a tongue-and-groove system produced by Hydraform Africa. The Hydraform system will be used as a case study in the presentation of the behaviour of dry-stack interlocking blocks. A typical machine for block production is the Hydraform block machine (see Fig. 8.15). The Hydraform block-making machines hydraulically compress soil (earth), mixed with cement into solid blocks. Hydraform machines are ideal for remote sites where transport, cement and sand costs are high. They are equipped with diesel generators for areas with no electricity as well as wheels for attachment as trailers to vehicles. The Hydraform interlocking block presses are manufactured by Hydraform Africa in Johannesburg, South Africa (Agrément South Africa, 1996; Hydraform, 1988). Different press models are available for different budgets and types. The structural performance of masonry buildings using Hydraform interlocking CEBs has been investigated and reported by Uzoegbo et al. (Ngowi and Uzoegbo, 2005; Uzoegbo and Ngowi, 2003, 2004; Uzoegbo et al., 2007; Uzoegbo and Senthivel, 2009). The Hydraform block system is designed to be mostly dry stacked. Hydraform dry stacking blocks are commonly used as infill for confined masonry structures but can also be as full load-bearing walling system. Different interlocking block designs including the Hydraform block are also shown in Fig. 8.15.

8.3.6   Curing of CSEBs

Stabilised soil blocks must be kept in a humid environment for at least 7 days after production. The surface of the blocks should not be allowed to dry out too quickly, as this causes shrinkage cracks. The blocks must be sheltered from direct sun and wind and kept in conditions of RH exceeding 80% by covering them with waterproof plastic sheets as shown in Fig. 8.16, ideally for up to 21 days. After 28 days, there will be no further significant increase in the strength. High temperatures combined with high humidity gives the best results. Covering with black plastic sheeting helps retain the temperature inside the curing chamber. To illustrate the effect of curing temperature, it has been shown that blocks cured for 7 days at 40°C will be about 1.5 times stronger that blocks cured for 7 days at 20°C (Rigassi, 1985).

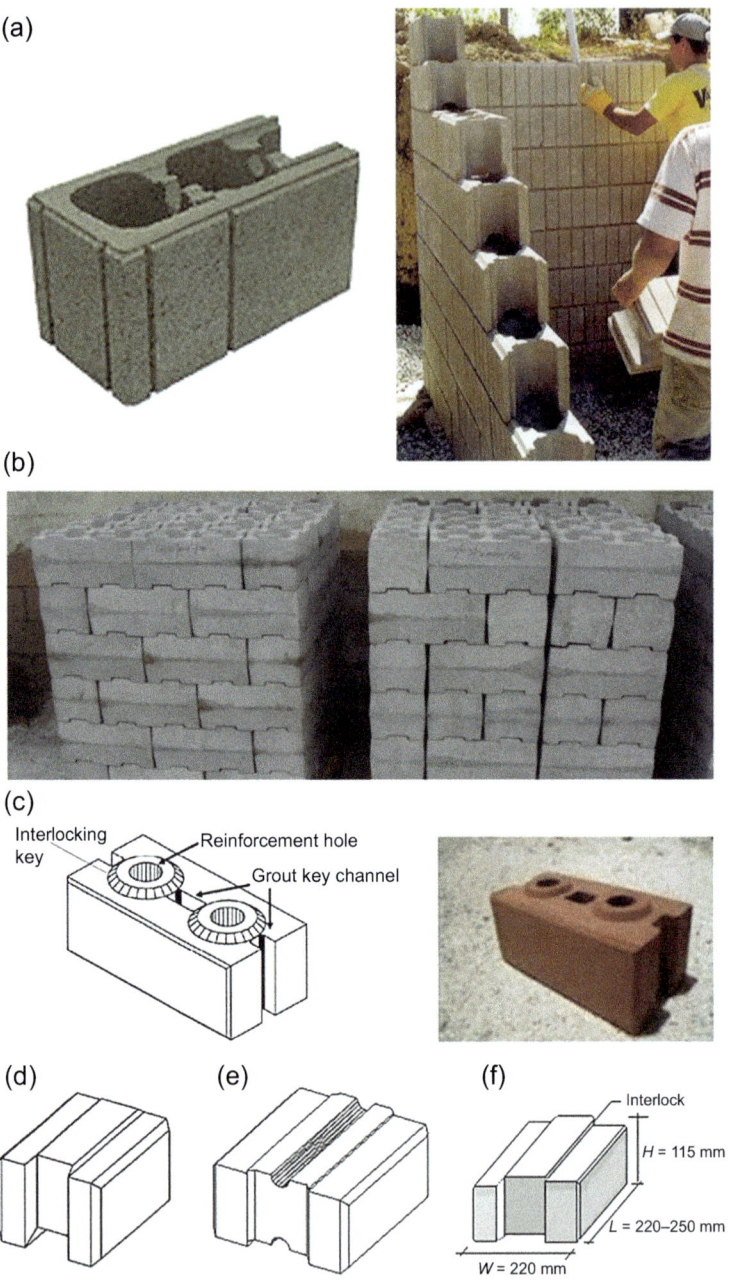

Figure 8.15 Different types of interlocking blocks. (a) The Azar interlocking block system. (b) The Stumblebloc dry-stack system *(http://www.sapropertynews.com/interlocking-concrete-block-system-makes-building-easy/)*. (c) Soeng Thai BP6 interlocking block. (d) Hydraform interlocking block. (e) Hydraform block with reinforcement groove. (f) Hydraform CSEB dimensions.

Figure 8.16 Curing of CSEB in a block yard. (a) Newly produced blocks are sprayed with water in preparation for curing. (b) Blocks are covered with black plastic sheeting.
Hydraform, 1988. Building Breakthrough — Product Information. Hydraform Africa, Johannesburg.

Freshly moulded blocks are fragile and cannot be stacked. A specific area close to the machine is normally provided for wet curing requiring minimal handling or transportation to avoid damage to the blocks. The blocks can be handled after 2 days' curing. The curing areas therefore need to be able to hold the equivalent of 2 days' production.

8.4 Strength evaluation of block units and masonry

8.4.1 Key features of interlocking units

Interlocking blocks are like two adjoining pieces of in a jigsaw puzzle. The main design feature is that each block has a projection at one end and a depression at the other. The projection of one block fits in to the depression of the next such that the blocks align. The blocks can be manufactured with or without holes in them. Hollow blocks usually have vertical holes, which reduce the amount of material required to make the block without compromising on strength. Steel rods can be inserted or mortar poured into the holes to increase the building's strength and stability. Horizontal groves may also be provided for horizontal reinforcement, especially in earthquake-prone areas.

The advantage of hollow interlocking CSEBs, compared with hollow concrete blocks, is that the interlocking keys provide greater resistance to both in- and out-of-plane shear and buildings are therefore stronger without the need for reinforcing materials. Interlocking block walls can resist earthquakes without major damages where normal masonry may experience major damage. Each dry joint allows both rocking and some sliding movement, allowing the structure to absorb shocks due to low-level earthquakes.

8.4.2 Construction using CSEBs

The main construction principle is to confine every panel of dry-stack (unbonded) masonry within boundaries of bonded masonry strips or columns and beams. A typical section through a wall is illustrated in Fig. 8.17. The base course is bonded with mortar up to one course above the floor level. The middle courses are dry-stack up to lintel level. The top three courses are again bonded with mortar to form a ring beam at the top of structure as shown in Fig. 8.17. Alternatively, a reinforced concrete ring beam could be cast at the top. A typical example of the confinement principle in dry-stack masonry construction is shown in Fig. 8.18. Some houses in South Africa built with the Hydraform dry-stack interlocking CSEB system are shown in Fig. 8.19.

CSEB will lose strength under sustained wet conditions. It is therefore recommended to use normal concrete in foundations as the materials below ground are expected

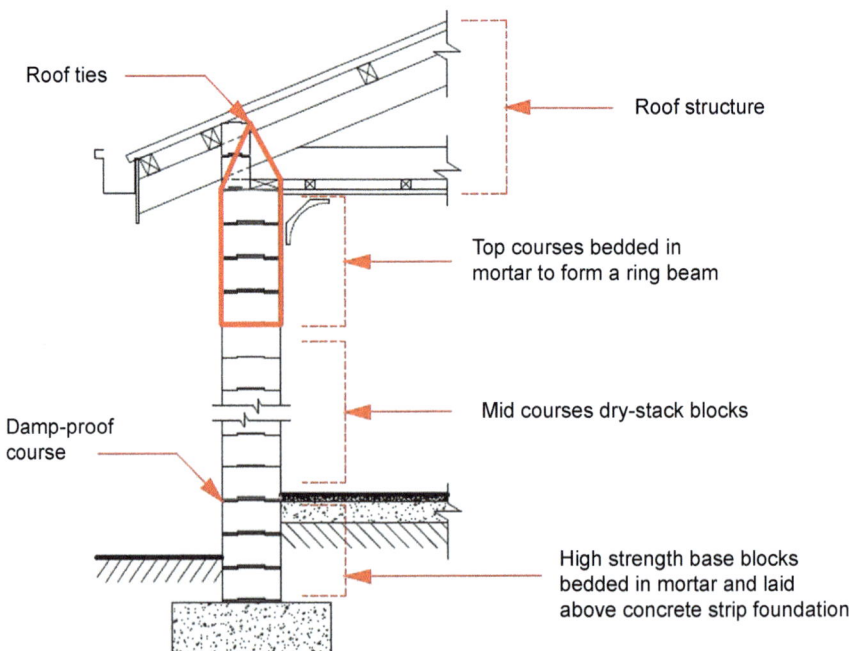

Figure 8.17 Cross section of construction using the dry-stack masonry.

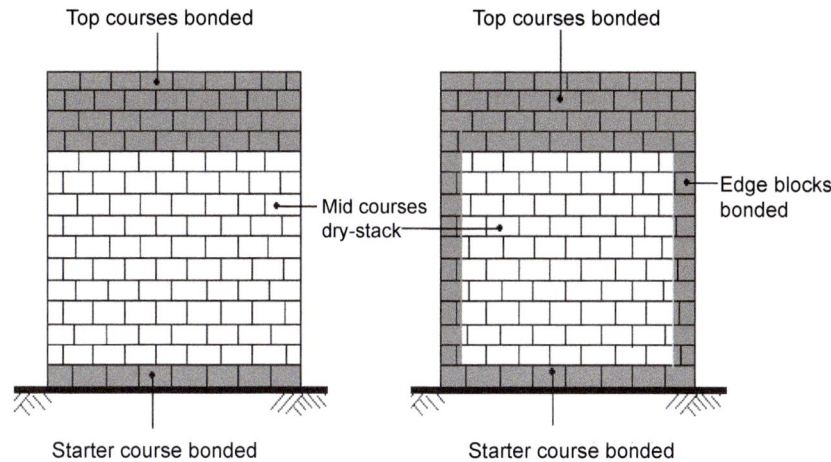

Figure 8.18 Typical elevation details of interlocking wall systems.

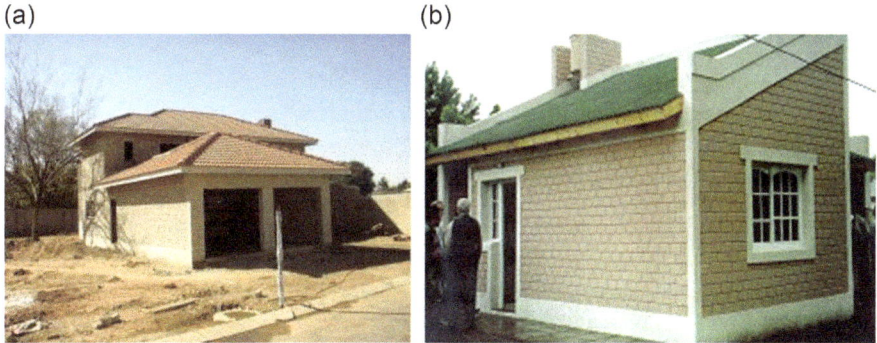

Figure 8.19 Examples of Hydraform dry-stack interlocking CSEB system. (a) Double-story building in Johannesburg using CSEB interlocking system. (b) House construction with confinement of the dry-stack masonry units.

to be wet especially during the wet seasons. Alternatively, higher-strength masonry units could be used at the base to account for possible loss in strength due to moisture in the soil. Examples of CSEB houses on concrete foundations are shown in Fig. 8.20. The CSEB material starts from about 0.5 m above ground level. The best practice for CSEB is to keep it permanently dry.

8.4.3 Dry-stacking of masonry

Dry-stacking is an ancient form of stone masonry construction that has existed in Africa and other parts of the world for centuries. The Egyptian pyramids constructed over 5000 years ago were based on dry-stacking of large stone masonry. Dry-stacking of masonry is a construction method in which most of the masonry units are laid

(a)   (b)

(c)   (d)

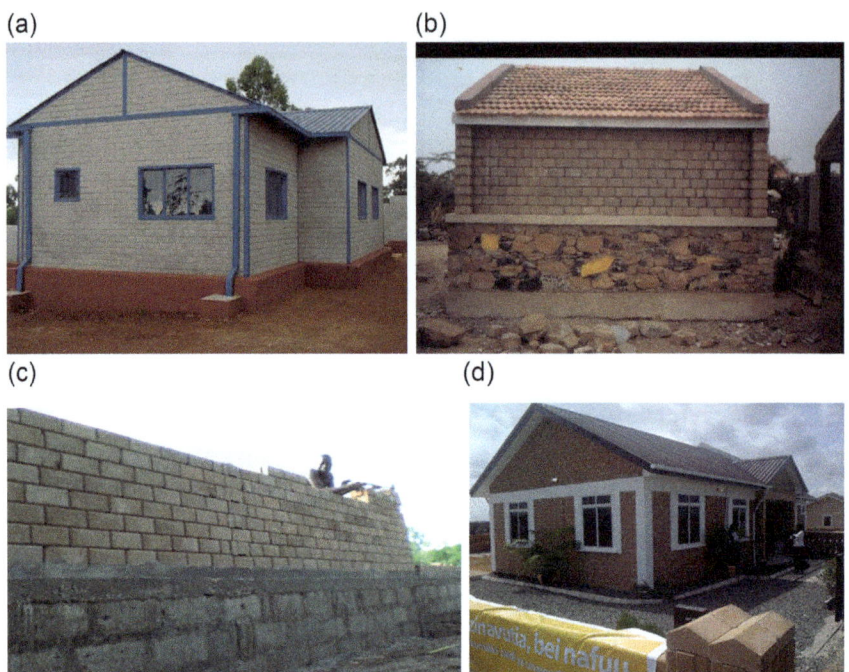

Figure 8.20 CSEB wall and foundation construction. (a) Concrete foundation. (b) Stone masonry foundation. (c) Conventional masonry base with interlocking CSEB superstructure. (d) Foundation and superstructure in CSEBs with higher-strength blocks at base.

without mortar. The construction usually depends on interlocking mechanism of the blocks for stability. The absence of a bonding agent in the joints between masonry units makes the wall structure flexible and assists in alignment during construction. The construction of curved walls, as shown in Fig. 8.21, is easily accomplished with the use of dry-stack technology. The interlocking blocks accommodates more movement than conventional masonry and enables the construction of curved shapes.

The external walls of the great Zimbabwe ruins (Fig. 8.22), built around 400 AD as the capital of the ancient Shona Kingdom (Walker et al., 1991), were made of dry-stack granite stone masonry. The complex consists of 12 groups of buildings spread over about 5 km^2. Existing structures from the ruins are a great tourist attraction in Zimbabwe.

Lessons learnt from the behaviour of the ancient dry-stack structures were applied in designing modern dry-stack masonry construction. Some of the problems experienced in ancient dry-stack masonry structures are described next:

- Toppling of masonry units: The masonry units depend on friction and the interlocking mechanism between the rough surfaces for proper bonding. The top units do not have the critical mass to develop adequate friction and will therefore topple at a small disturbance.

Figure 8.21 Dry-stack interlocking block masonry enables the construction of curved walls. Photograph: Head office building of Hydraform South Africa.

Figure 8.22 Dry-stack masonry construction of the Great Zimbabwe ruins.

- Bulging of wall panels: Sections of the stone walls which retained earth on one side have bulged outwards due to active earth pressure. The bulging generally occurs between the ground level and mid-height, due to a reduction in friction between the units caused by the penetration of clay/silt materials at the block interfaces. A monitor of movement of the Zimbabwe ruins conducted by Walker et al. (1991) revealed a peak movement of 0.3 mm per week.
- Lack of arching action: Arching action is present in masonry walls when there is an opening and sufficient wall width on either side of the opening to resist the arch resulting thrust. A lintel is required to support the weight of the wall material above the opening. When arching action is present, the weight that must be supported is only the weight of material within the triangle formed at a 45 degree angle as shown in Fig. 8.23(a). Dry-stack systems (Fig. 8.23(b)) do not have the benefit of arching action and therefore must always be designed to support the entire load above the opening.

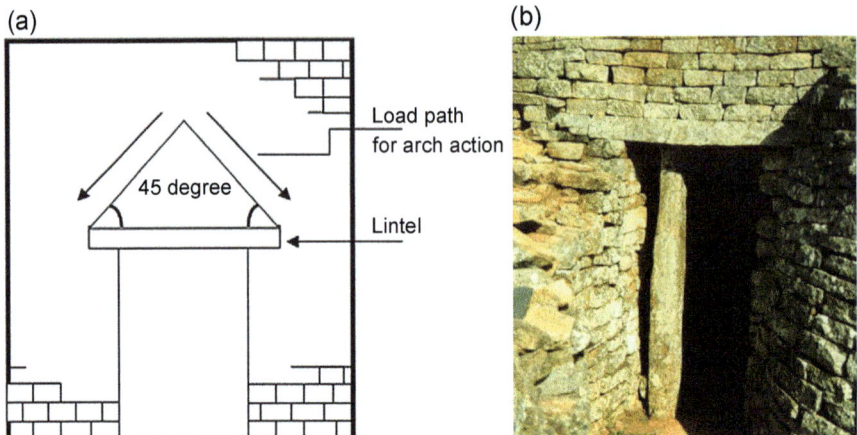

Figure 8.23 Effect of openings in dry-stack masonry. (a) Effect of arching action: lintel to support load within 45 degree angle. (b) Dry-stack stone masonry with no arching action: lintel must support all load above it.

8.5   Strength evaluation of block units and masonry walls

Compressive strength evaluation is the primary method of assessment of masonry for compliance with material and construction requirements. It is generally accepted as a key value for design of structures. Masonry units generally have rectangular sides and are designed with 'match box' configuration. This enables tests to be carried out on masonry units using flat platens and strength results are easily obtained. Interlocking blocks are usually designed differently and therefore require special attention to the method of strength assessment.

8.5.1   General features of the Hydraform interlocking blocks

The interlocking blocks used as a case study in this chapter are developed by Hydraform Africa Pty and are produced by mixing soil and cement in predetermined ratios extruding them vertically under a pressure of about 10 MPa using a hydraulic powered machine (Fig. 8.13(b)). The blocks are shown in Fig. 8.13(d–f). The resulting wall system uses 37 blocks/m^2 of wall requiring 0.5 m^3 soil and approximately one-third of a bag of cement. Views of a stacked wall are shown in Fig. 8.24.

8.5.2   Strength evaluation of Hydraform blocks: a case study

Compression strength evaluation of block units described in this section is based on Hydraform interlocking block system. Each interlocking system is unique and requires specific development of strength assessment methods.

Figure 8.24 Front and side views of stacked blocks.

8.5.2.1 Methods of compression strength evaluation

Compressive strength is accepted as the universal measure of quality of masonry units. Other properties of the material such as durability tend to follow the trends of strength. Due to the irregularity of the surfaces of interlocking block units, each type of block needs to be independently studied to determine the most suitable method of testing. For the Hydraform blocks, compression tests were carried out using four different testing arrangements as shown in Fig. 8.25: (1) capped arrangement, (2) loading on block shoulders, (3) application of central loading on block and (4) testing of cubes cut from the units. The tests were conducted for samples cured under three different humidity conditions: air-dry, oven-dry and wet conditions. The typical failure mode for each loading arrangement is also shown in Fig. 8.25.

8.5.2.2 Capping arrangement

Capping is the preparation of the ends of the specimens in order to ensure that the specimens have smooth, parallel, uniform bearing surfaces that are perpendicular to the applied axial load during compressive strength testing. This method of testing is prescribed by most standards. The main purpose of the compressive strength test is to simulate the actual strength that masonry can withstand in the field. However, capping the masonry will simulate the behaviour of the CSEB bonded with mortar. The specified capping material to be used is high strength gypsum cement. In order to obtain flat bedding for compressive strength tests, the block unit is capped as shown in Fig. 8.26 (1). The test was then conducted as a conventional concrete block.

8.5.2.3 Loading on block shoulders

In this arrangement, a loading platen was designed such that the load path under normal service condition is simulated. The load was applied directly on the shoulders of each block as shown in Fig. 8.25 (2).

Method of load application	Loading arrangement	Mode of failure
1	Load applied to full cross-section	
2	Load applied on shoulders of block	
3	Load applied to grove or tongue of block	
4	Cube test	

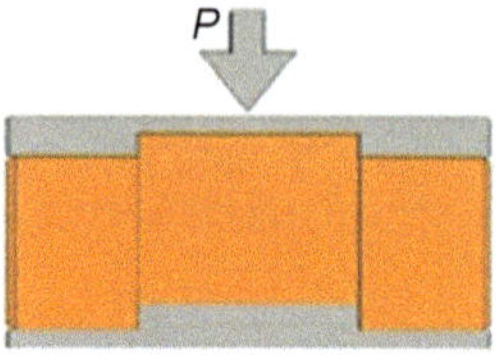

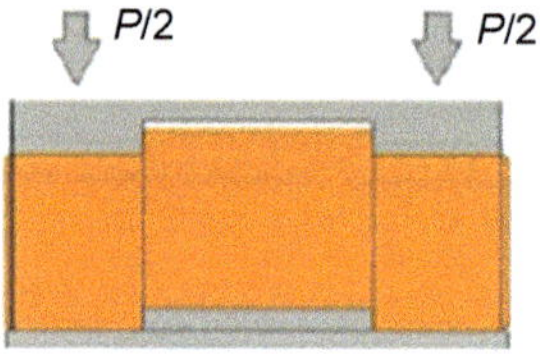

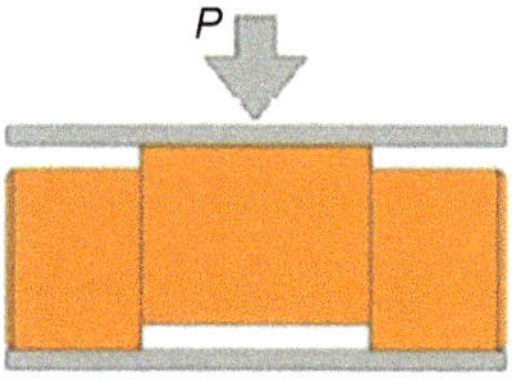

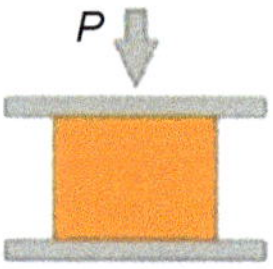

Figure 8.25 Different loading arrangements for compression test.

Figure 8.26 Hydraform block with cubes cut from a block.

8.5.2.4   Direct flat loading on block

Another method of testing involves the use of a flat steel plate as the loading platen both at the top and bottom of the block as shown in Fig. 8.25 (3). This method at best simulates the loading of only the bottom blocks of a wall system. A hard board was used to uniformly distribute the load on the 'tongue' as shown in the figure. The failure mode for this method of testing is described by two vertical shear cracks that divide the block in three separate units at failure (Fig. 8.25 (3)).

8.5.2.5   Testing of cubes cut from blocks

To investigate the uniformity of the strength distribution within the block and to obtain control strength for the verification of other testing methods, 100-mm cubes were cut from full-scale blocks and tested. Up to four cubes can be obtained from one block. The block is shown in Fig. 8.26 as extruded from the production machine with four cubes cut from a block. The compressive cube tests indicate a failure mechanism characterised by vertical and x-cracks similar to those from the concrete cube compressive test.

A relative failure load for each loading arrangement with the cube strength as standard is shown in Fig. 8.26. By testing samples from the upper and lower part of the CSEB as extruded from the press the uniformity of strength was investigated. It was observed that when compaction was applied only in one direction (upwards), the bottom part of the block was about 10% stronger than the upper part. This nonuniformity can be eliminated by pressing from both top and bottom. Some machines are already designed to do that.

8.5.3   Factors affecting the strength in compression of CSEBs

Tests have shown that it is mainly the compression applied during production that is responsible for the improved compressive strength of a CSEB in relation to handmade 'natural' adobe. The key factors affecting the strength of CSEB are described next.

8.5.3.1   Effect of clay content

The 28-day compressive strengths of blocks stabilised with 8% cement content at different clay contents is illustrated in Fig. 8.27. It shows that as the clay content increases the compression strength is increasingly compromised. This is due to interference of clay in the hydration processes of the cement in the mix (Walker, 1995). It is also observed that with an increase in the clay proportion the ratio of the wet to dry strength decreases. Higher clay content increases the block's affinity for water. Clay also expands with water, loosening the existing cement−gel−sand matrix, effectively reducing the strength of a block.

8.5.3.2   Effect of cement content

Tests to investigate the influence of cement stabiliser on the strength of CSEBs were carried out. The clay contents of the mixes were maintained at 25% while varying the

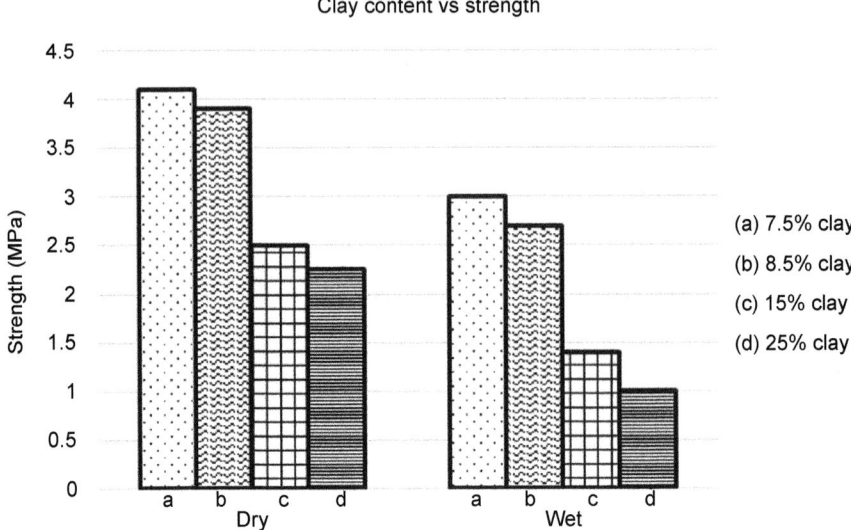

Figure 8.27 The 28-day strength at different clay contents.

cement content. Fig. 8.28 shows that the strength of CSEB increases markedly with cement content and that, for this mix, 12% cement is sufficient to meet the minimum required strength (Uzoegbo et al., 2004).

The wet/dry strength ratio increases with increasing cement content. The more cement in a mix, the more soil-cement paste is produced, covering more clay particles,

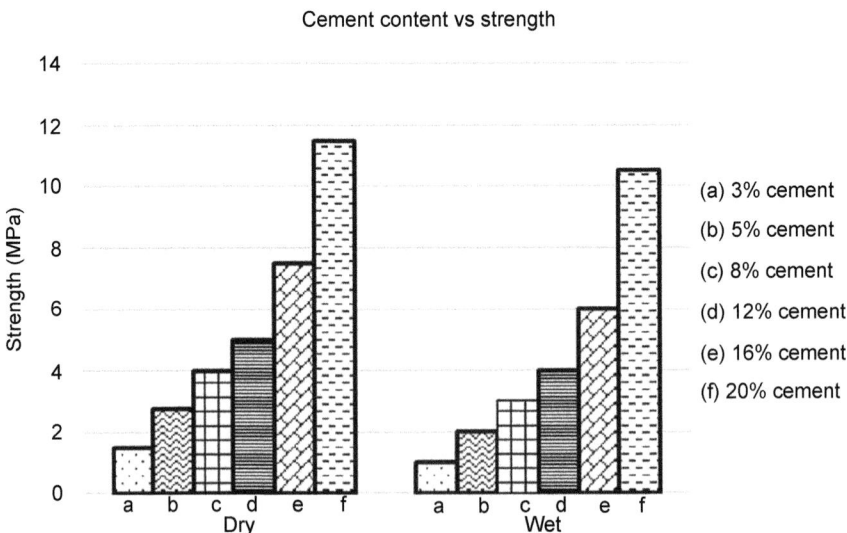

Figure 8.28 28 Day strength at different cement contents.

and reducing the proportion of uncemented clay particles. This in turn reduces the porosity and water absorption properties of the block (Walker, 1995). Effectively with increases in cement content, the availability of water for clay is reduced and the clay has insufficient moisture to expand and weaken the cement-soil paste.

8.5.4 In-plane and out-of-plane tests on interlocking dry-stack walls

Applied loading on a wall structure are usually classified into in-plane loading and out-of-plane loading for the design of masonry structures and each type of loading analysed separately. In-plane loading is mainly due to gravity loads while out-of-plane loading is due to horizontal loads such as wind and earthquakes.

8.5.4.1 Compression tests on dry-stack masonry wall (in-plane loading)

The wall panel compression test can be carried out on full-scale wall or on scale model wall consisting of four or more courses. At the ultimate limit state, cracks appeared on the faces and edges of the specimen (Fig. 8.29). The line of cracks running vertically along the edge of the interlocking groove is similar to that experienced by the block unit when loaded at centre (Fig. 8.24 (3)). This failure mode suggests that the block units have failed by shear before they can develop their full compressive capacity. Failure in low-strength unit walls was characterised by a local crushing of the top courses as shown in Fig. 8.29. The weakest sample (made with 3-MPa blocks) failed in this manner, by the crushing of the top courses which are in direct contact with the loading platen.

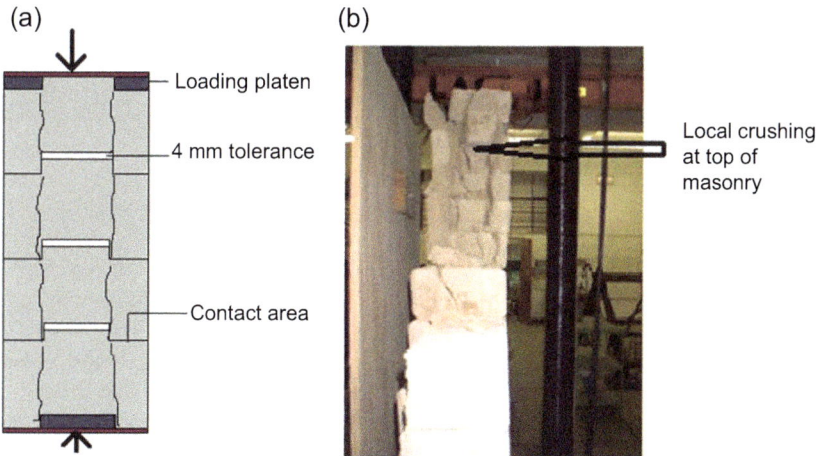

Figure 8.29 Failure modes for different strength of material. (a) Wall plane view: failure line. (b) Top crushing failure in low-strength material.

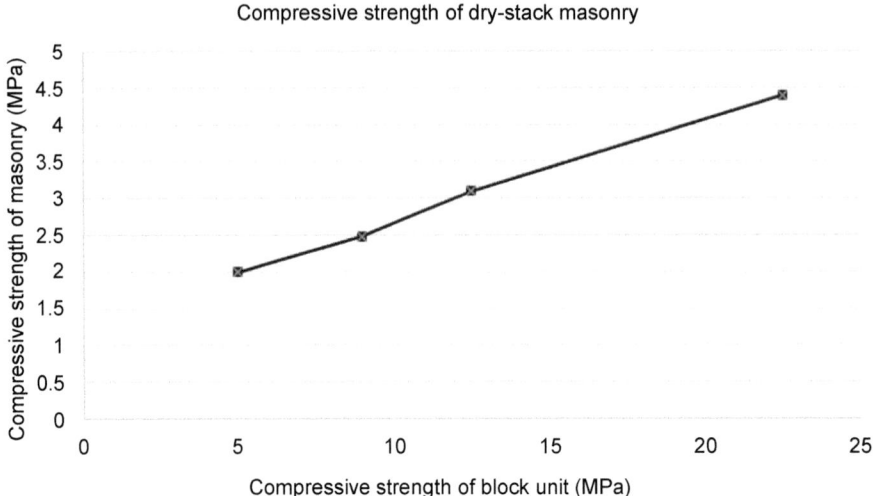

Figure 8.30 Wall panel results in compression.

Standard tests conducted on the Hydraform dry-stack masonry showed that there is an approximately linear relationship between the strength of the block unit and the strength of the masonry (wall strength) in compression. A typical result is shown in Fig. 8.30. The tests were performed a full-scale wall panel 3 m wide and 2.5 m high.

8.5.4.2   Behaviour of conventional masonry wall under lateral load

A standard conventional concrete masonry wall with three sides fixed and the top free to deform out-of-plane when subject to lateral (out-of-plane) uniform increasing load on one side will display a typical inverted Y-crack pattern as shown in Fig. 8.31. At the ultimate limit state, such walls are characterised by sudden failure. Fig. 8.31 shows a load-deflection plot of a typical conventional masonry wall subject to uniform horizontal loading. The load-deflection curve shown in Fig. 8.31 describes a brittle mode of failure.

8.5.4.3   Behaviour of dry-stack masonry wall under lateral load

Due to the tolerance between the interlocked units, it follows that when no precompression forces are applied, there are unavoidable free translations and/or rotations until the interlocking mechanisms are developed. In addition, there is no tensile resistance (or bonding). Thus, the deflection under a small load may be fairly large compared with that observed in conventional masonry. When the deflections due to such free movements exceed a serviceability limit state, the resulting interlocking mechanism may have little remaining structural integrity. Fig. 8.32 shows the typical mode of failure of dry-stack masonry under lateral loading. The failure is characterised by gradual deflection with increased lateral load. This was accompanied by the gradual opening of

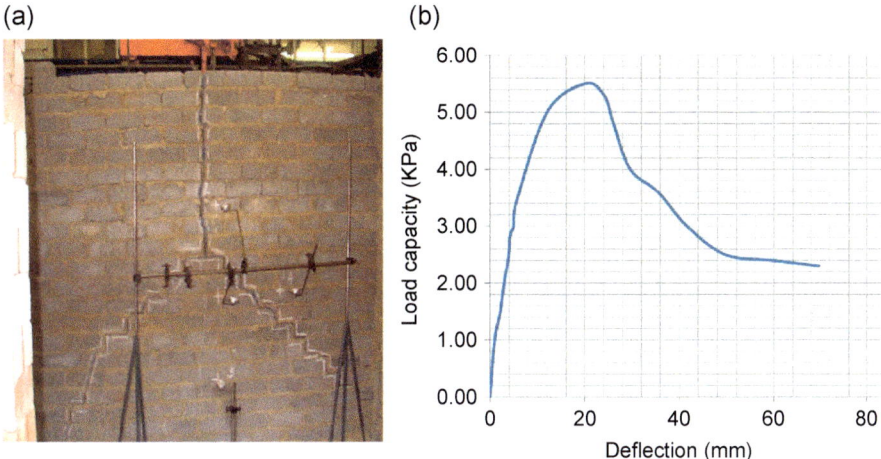

Figure 8.31 Behaviour of conventional masonry construction. (a) Typical failure mode for conventional masonry. (b) Typical load-deflection curve for a conventional masonry subject to lateral loading.

the bed joint above the mid-section (between the loading points) stretching across the entire length of the specimen. At the failure load, the maximum opening of the bed joints due to the rotation of the units along vertical axis resulted in the sliding of the units out of the interlocking positions. The specimen continued to deflect with decrease of lateral pressure allowing further rotation of the wall until failure.

The crack pattern in the wall is shown in Fig. 8.33. If the dry-stack interlocking block masonry is plastered, the failure pattern is more clearly defined as it is similar to that of conventional wall as shown in Fig. 8.32.

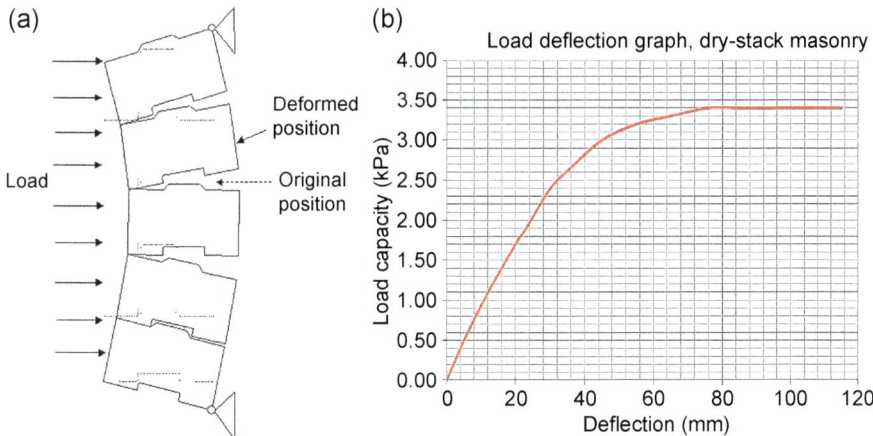

Figure 8.32 Behaviour of dry-stack masonry under lateral load. (a) Out-of-plane load response of dry-stack walls. (b) Typical load-deflection curve for dry-stack masonry subject to lateral loading.

(a)   (b)

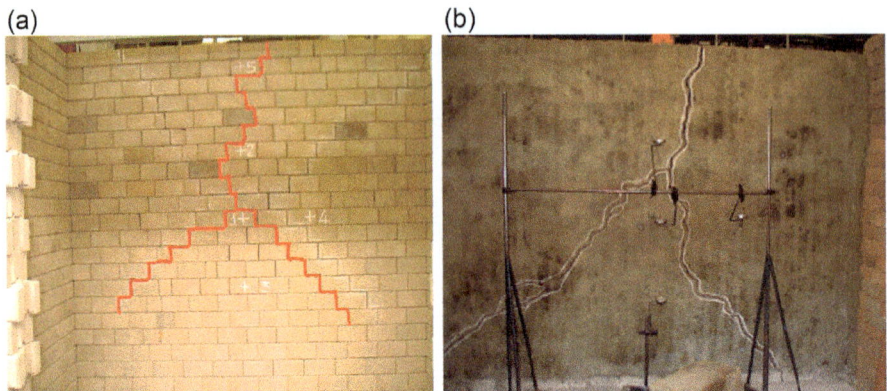

Figure 8.33 Typical crack patterns in dry-stack masonry construction. (a) Crack pattern of unplastered dry-stack masonry under lateral loading. (b) Plastered dry-stack wall crack pattern.

The differences in load-deflection behaviour of conventional and dry-stack systems are shown in Fig. 8.34. It can be seen that for block units of the same strength, conventional masonry will give higher strength but lower ductility. In fact the lateral load capacity for the dry-stack system for the Hydraform interlocking system is about 60% of the lateral strength of conventional masonry of the same block unit strength. This result should not be considered a universal conclusion as every interlocking system has its own unique characteristics.

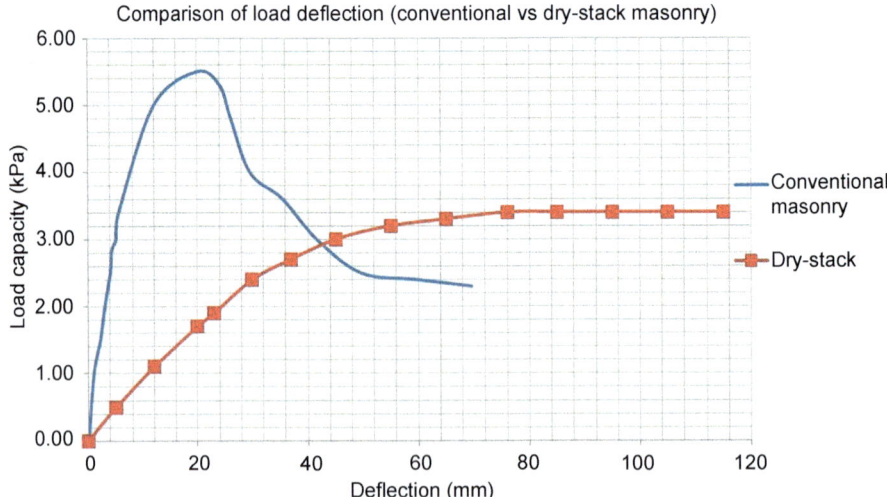

Figure 8.34 Comparison of behaviour of the two masonry types under lateral load.

8.6 Durability of CSEBs

8.6.1 Durability of stabilised soil blocks

Durability is the ability to last a long time without significant deterioration and requiring minimum maintenance. It is an important factor in assessing the sustainability of a material. A durable material helps the environment by conserving resources and reducing waste and the environmental impacts of repair and replacement. Due to the unstable nature of clay as a building material, it is important to study the durability of CSEBs. Different structures require different degrees of durability depending on their exposure environment and the properties desired. In the Indian standards (IS 1725−1982), two methods of durability testing are recommended:

- Water absorption: After immersion in cold water for 24 h, water absorption shall not be more than 15%.
- Weathering: The maximum loss in weight following a standard weathering test shall not exceed 5%.

There are many methods of durability assessment for CSEB. Two of the test methods are described below.

8.6.2 Abrasion test (wet/dry durability)

The wet/dry durability test is a form of abrasion testing that simulates the abrasive effects of water and wind driven erosion on a wall structure. The test is carried out with a stiff wire brush (Fig. 8.35); which is used to brush the blocks cyclically after intermittent drying and wetting periods. The average results from a minimum of three tests are accepted. Blocks are initially oven dried at $\pm 70°C$ (for 19 h) or until a constant dry mass is achieved. The initial oven dried mass is recorded before any brushing commences. Brushing of blocks is to take place with an applied normal force of 13.5 N. The

Figure 8.35 Oven-dried block and wire bristle brush.

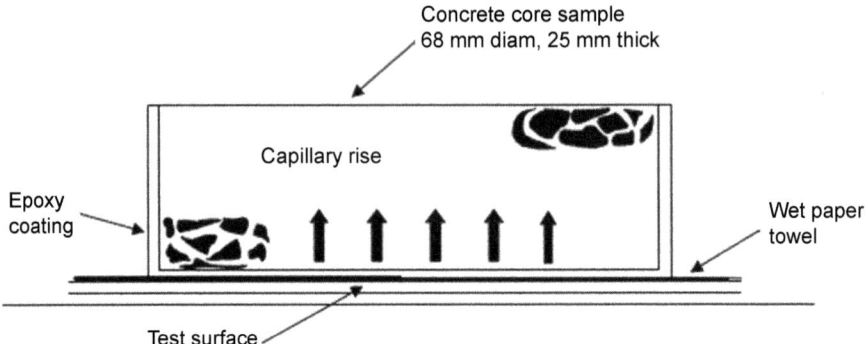

Figure 8.36 Sorptivity test.

blocks are to be brushed with two firm strokes of the wire brush over all surfaces. The percentage mass loss after 12 cycles of wetting, drying and brushing is a measure of the abrasion resistance. The abrasion coefficient (C_a) expresses the ratio of the brushed surface S (in cm^2) to the mass of the material detached by the brushing (in grams).

8.6.3   Water absorption by capillarity rise (sorptivity)

Sorptivity, or capillary suction, is the transport of liquids in porous solids due to surface tension acting in capillaries. It is a function of the viscosity, density and surface tension of the liquid and the pore structure (radius, tortuosity and continuity of capillaries) of the solid. A CSEB is a porous material that interacts with the surrounding environment. The durability of a construction material such as concrete or CSEBs is largely dependent on the transport of fluid through it. Permeability is a measure of a material's ability to transport fluid. Sorptivity is a material's ability to absorb and transmit water through it via capillary suction and provides an engineering measure of microstructure and properties important for durability. Sorptivity is increasingly being used as a measure of concrete resistance to exposure to aggressive environments. Uptake of water by unsaturated, hardened CSEB may be characterised by sorptivity. The water sorptivity test is a unidirectional absorption test. A 68-mm-diameter core is coated around its circumference with epoxy (Fig. 8.36) and placed in a tray with solution of water and Ca(OH)$_2$ in order to measure the rate of water uptake into the cylinder by capillary rise. The quality of the CSEB is assessed based on the values in Table 8.9.

Table 8.9 Interpretation of sorptivity results

Durability class	Sorptivity (mm/$\sqrt{h}$)
Excellent	<6
Good	6–16
Poor	12–15
Very poor	>15

8.7 Conclusions

The processes of production and use in construction of interlocking blocks made with CSEBs have been described and presented in this section. A summary of notable aspects of the selection of materials, production of CSEB construction and general structural behaviour is presented.

The production of CSEBs utilises locally available soil, thus helping to develop the local economy rather than investing on imported materials; there are savings in fuel and replacement parts, thereby reducing transportation costs. The soil used is generally subsoil, leaving the topsoil for agriculture. Building with local materials provides a more sustainable employment for local people. CSEB production does not require kiln-firing; therefore, production of CSEBs reduces carbon emission and the consumption of nonrenewable resources. Soil-cement blocks are usually two to three times larger in size than normal burnt clay bricks; consequently, the number of joints in a wall panel is reduced and construction is faster. CSEBs are the ideal construction material for low-cost housing projects.

A careful selection of soil is needed for CSEB production. Natural soil properties such as texture, plasticity and shrinkage have been shown to be related to the degree of suitability of soil for stabilisation. Optimum composition of soil for CSEB is made up of approximately 70% sand and 30% silt and clay. The clay content should not be less than 10% or more than 40% of the soil. High clay content results in lower strength, less stable product and high cement demand. Low clay content presents handling problem at early ages as sand does not have bonding property when dry; the addition of clay binds the sand to maintain the shape of the block. To get clay to stick together well, it must first be saturated with water.

Most soils that are free from vegetable matter can be satisfactorily stabilised with cement, lime, or cement and lime. Compliance with the acceptable range for linear shrinkage and plasticity index are the best indicators for the suitability of soil for CSEB production. The textural classification of the soil is also an important factor in soil characterisation. Soils with linear shrinkage less than 6% and plastic index less than 30 are considered good for CSEB. Minimum cement content for soil stabilisation is 5% in order to produce CSEB in the strength range of 3–5 MPa.

There is no standard or prescribed water content for CSEB production as the water demand is highly influenced by the type and ratio of clay and silt in the mix. However, low workability range is recommended for the mix as lower moisture content produces higher-strength CSEBs. Ideally, the optimum water content may be determined by the standard Proctor compaction test, according to BS 1377 (1990). This laboratory test may be used to determine the maximum dry density and the optimum moisture content of the soils. On site, operators rely on experience and simple drop test to determine adequate moisture content.

Interlocking block system does not require mortar bonding and as a result construction is faster. Once the base course has been accurately laid, the blocks can be laid by unskilled workers with minimal training. To accommodate geometric variation specific strength test method needs to be developed for each interlocking block system. Cutting standard shaped specimens from solid blocks, such as cubes, might be an alternative solution to this problem. For the dry-stack interlocked block system to be most effective, it has to be confined within a system of beams, columns and bonded base. As with other masonry types, compressive strength is a basic measure of quality for CEBs.

The structural behaviour of dry-stack masonry system under compressive load is similar to conventional masonry. Under lateral loading, however, the dry-stack system has lower ultimate load capacity and higher ductility.

Clay is an unstable material under sustained moisture condition and due to the relatively high clay content of CSEBs; it is estimated to lose between 30% and 50% of its dry strength under sustained wet condition. To account for this, it is recommended that where CSEBs are used below ground level in foundations, it should be assumed to be under sustained ground moisture and a much higher strength material used. Alternatively, normal concrete may be for foundations up to one or two courses above ground level and continued with CSEBs.

References

Addis, B.J., Davis, D.E., 1986. Fulton's Concrete Technology. Portland Cement Institute. 1986-Concrete — 956 pages.

Adedeji, Y., 2010. Technology and standardised composite cement fibres for housing in Nigeria. Journal of the Nigerian Institutes of Architects 1, 19—24.

African Standards Organisation, 1996. Standards on CEB Construction: Compressed Earth Block Masonry ARS 670: Part 4: Code of Practice for Production and Construction.

Agrément South Africa, 1996. Hydraform Building System. Pretoria.

ASTM D698-12, 2012. Standard Test Methods for Laboratory Compaction Characteristics of Soil Using Standard Effort.

Boubekeur, S., Houben, H., comps, 1998. Compressed Earth Blocks Standards. Center for Development of Industry, Brussels.

British Standards Institution BS 1377—4, 1990. Methods of Test for Soils for Civil Engineering Purposes. Compaction-related Tests.

British Standards Institution BS 5628, 1978. Code of Practice for Use of Masonry, Part 1: Unreinforced Masonry. BSI, London.

Calkins, M., 2009. Materials for Sustainable Sites: A Complete Guide to the Evaluation, Selection, and Use of Sustainable Construction Materials. John Wiley & Sons, Hoboken, NJ, USA.

Hall, M., Djerbib, Y., 2004. Rammed earth sample production: context, recommendations and consistency. Construction and Building Materials 18, 281—286.

Hendry, A.W., 1981. Structural Brickwork. The Macmillan Press Ltd, London.

Houben, H., Guillaud, H., 1994. Earth Construction — A Comprehensive Guide, CRATerre-EAG. Intermediate Technology Publications, London.

Hydraform, 1988. Building Breakthrough — Product Information. Hydraform Africa, Johannesburg.

Indian Bureau of Standards, IS: 1725, 1982. Specifications for Soil Based Blockd Used in General Building Construction.

Júnior, H.C., Fábio, L., Willrich, F.L., Barbosa, N.P., 2003. Structural behavior of load bearing brick walls of soil-cement with the addition of ground ceramic waste. Revista Brasileira de Engenharia Agrícola e Ambiental 7 (3), 552−558.

Kawamura, M., Kasai, Y., 2011. Mix design and strength of soil−cement concrete based on the effective water concept. Materials and Structures 44, 529−540.

Monk, C., 1967. A Historical Survey and Analysis of the Compressive Strength of Brick Masonry (Research report No12-Structural Clay Products Research Foundation), Geneva, III.

Morsy, E., 1968. An Investigation of Mortar Properties Influencing Brickwork Strength. University of Edinburgh (Ph.D. thesis).

Ngowi, J.V., Uzoegbo, H.C., 2005. Empirical studies of flexural strength for dry-stack interlocking masonry. Botswana Journal of Technology 14 (2), 26−31.

Rigassi, V., 1985. Compressed Earth Blocks: Manual of Production, CRATerre-EAG. In: Manual of Production, vol. I.

Roskovic, R., Bjegovic, D., 2005. Role of mineral additions in reducing CO_2 emission. Cement and Concrete Research 35, 974−978.

SANS 10164 1, 1996. Structural Use of Masonry: Unreinforced Masonry Walling. South African Bureau of Standards.

Schmidt, W., Hirya NM,Bjegovic, D., Uzoegbo, H.C., Kumaran, S., 2012. Cement technology in sub-Saharan Africa: practical and scientific experiences. American Ceramic Society Bulletin 91 (5).

Spence, R.J.S., Cook, D.J., 1983. Building Materials in Developing Countries. Wiley, Chichester, England.

Stumblebloc, 1990. http://www.sapropertynews.com/interlocking-concrete-block-system-makes-building-easy/.

Umesha, T.S., Dinesh, S.V., Siv, P.V., 2009. Control of dispersivity of soil using lime and cement. International Journal of Geology 3 (1).

Uzoegbo, H.C., Ngowi, J.V., May 2003. Structural behaviour of dry-stack interlocking block walling systems subject to in-plane loading. Concrete Beton 103, 9−13 (Publisher: Concrete Society of Southern Africa).

Uzoegbo, H.C., Ngowi, J.V., December 2004. Lateral strength of a dry-stack wall system. Masonry International, Journal of the British Masonry Society 17 (3), 122−128.

Uzoegbo, H.C., Senthivel, R., December 2009. An effective retrofitting system for dry-stack masonry subject to seismic loading. Masonry International, Journal of the International Masonry Society 22 (3), 71−84. ISSN: 0950-2289.

Uzoegbo, H.C., Senthivel, R., Ngowi, J.V., September 2007. Load capacity of dry-stack masonry walls. The Professional Journal of the Masonry Society of North America 25 (1), 41−52. ISSN: 0741-1294.

Uzoegbo, H.C., 2011. Seismic behaviour of single-story URM buildings with light weight steel roof. International Journal of Advanced Technology in Civil Engineering (IJATCE-2011) 1 (1). Publisher: Interscience Open Access Journals. ISSN: 2231-5721.

Walker, P.J., Maunsel, M.G., Dickens, J.G., 1991. Structural Instability at the Great Zimbabwe Monument, Research Report. University of Zimbabwe.

Walker, P.J., 1995. Strength, durability and shrinkage characteristics of cement stabilised bricks. Cement and Concrete Composites 17, 301−310.

Wheeler, G., 2005. Interlocking Compressed Earth Blocks. In: Manual of Construction, vol. II. Center for Vocational Building Technology, Thailand.

Unfired clay materials and construction

J.M. Kinuthia
University of South Wales, Cardiff, United Kingdom

9.1   Introduction to clay-based materials

The use of clay-based materials has a long and prehistoric background. The association of clay building materials with mud huts is correct but not sufficient. There has been serious rethinking and modernizing, and clay-based materials can now produce aesthetically pleasant construction, such as sometimes witnessed in the careful use of rammed earth as shown in Fig. 9.1. It will be seen later that good workmanship can be achieved for both individual and large-scale housing projects using unfired clay materials, in both developed and developing countries.

Over the last few decades, clay-based materials have started to gain recognition and the respect they once had in the past. This is credit to clay-based material enthusiasts, who have brought to the fore the well-recognized and indisputable benefits of these materials. These benefits include low impact on the environment, in particular if the materials are not fired, and their ability to passively control interior humidity. The downside to the materials include the bad reputation these materials have, especially when compared with the apparent advantages of energy-intensive materials such as Portland cement and, to some extent, lime. Another notable drawback is the often slow speed of construction and the sensitivity clay materials sometimes have when one has to balance cost and immediate- and long-term performance. While some approaches may require little or no prior experience, others such as the modern and beautiful appearance shown in Fig. 9.1 require extra care during construction in order to maintain material mixing regimes and the resultant uniformity in density, strength and colour.

This chapter points out, in brief, the disadvantages of fired and energy-intensive building and construction material systems but, more importantly, highlights the properties and benefits of unfired clay−based systems. The author of this chapter has dealt in detail on the subject of durability of compressed earth materials in a recent book (Pacheco et al., 2015). For this reason, this chapter will only highlight the key durability aspects of these materials, without fully detailing the various tests for the determination of durability of clay-based materials.

Clay-based materials are in either fired or unfired forms. In the fired category, the material is mixed with a significant amount of water and either extruded or moulded in a timber, plastic or steel mould into a regular shape. There are reports of sand moulds also being used (Sutton et al., 2011). Before firing, the cast or extruded bricks or blocks are allowed to condition or dry slightly to reduce the water content, so as to reduce the shrinkage on firing. Firing of clays drives out not only the free water used in the mixing

Nonconventional and Vernacular Construction Materials. http://dx.doi.org/10.1016/B978-0-08-100038-0.00009-3

Figure 9.1 Modern look created by clay-based materials, as in the careful use of rammed earth in construction in Wales, United Kingdom.

process but also any water that is chemically combined or adsorbed in the inter-layers of the clay microstructure. The firing also goes further to dehydroxylate (removal of −OH group) and/or decarboxylate (removal of CO_2) any carbonates present and sinters the materials (if temperature is high enough, typically $>1000°C$) to a very robust fired product. The loss of free and chemically bound water reduces the swelling potential associated with clay soils. Considering that some soils can exhibit excessive swelling potential as high as 2000%, the firing process is very effective in imparting volume stability to a target clay soil material. The strong bonding reduces the weakening effects of porosity and imparts chemical resistance. Firing is, however, expensive due to its high energy intensity and represents about 85% of the energy involved in the manufacture of fired clay materials (Heath et al., 2009). This is a major problem in both developed and developing countries alike. Fig. 9.2 shows the activities of a community project near Mumias in western Kenya that makes hand-made fired bricks. A visit to the area by the author revealed serious problems of tree cutting to provide firewood for firing bricks, causing significant community tensions due to domestic energy requirements and environmental issues. This chapter, however, does not aim to address this category of fired clay−based materials. The aim of the chapter is to address the re-emergence of unfired clay−based materials. These unfired systems, compared to their fired counterparts, have a relatively lower energy consumption due to the absence of the firing process. It is however best to start by addressing the target raw material itself − clay soil.

9.2   Structure and properties of clay soils

The term 'clay' is generally used in two contexts. In the more general context, the term 'clay' is used to refer to generally fine-grained loosely bound natural soil material, irrespective of its composition or origin. In the more strict and scientific context, the term

Figure 9.2 Small holder fired brick community project at Mumias in western Kenya. The absence of trees may be noted, most having been cut to provide energy for firing bricks. Courtesy of the author.

'clay' is used to refer not only to the fine-grained nature of a soil material, but also to the presence of certain properties whose origin is traceable to the presence of certain 'clay' minerals. In this stricter context, clays are distinguished from other fine-grained soils by both the fine particle size and by their mineralogy. For this reason, other fine-grained soils such as silts that do not have a significant proportion of clay minerals are, strictly speaking, not viewed as clays.

The special and easily observable properties that may suggest presence of clay minerals in a soil are primarily evident when the clay soil is in contact with water. These properties include significant cohesion, shrinkage and expansive or swelling behaviour. The genesis of these properties is in the microstructure of the clay, and the composition of trace metal cations. The variable amounts of water molecules trapped in the clay mineral structure and the type of interlayer metal ions combine to result in the familiar clay soil properties mentioned, as well as its colour. For this reason, to both soil experts and nonexperts, it is well known and established that clays are plastic in the presence of water; become hard, brittle and nonplastic on drying or firing and shrink on drying to show significant cracking. Clay soils can appear in various colours depending on their mineral composition, ranging from deep orange-red to brown to grey to white.

Strength, volume stability and overall durability of clay-based construction materials are all very important issues. The solution to swelling of materials, especially clays and stabilized soils, is a major breakthrough in clay-material systems. In order to understand why clay soils swell both when water is added and when stabilized with a stabilizer such as Portland cement or lime, it is necessary to understand further structural details about clay soils.

Two structural forms are involved in the atomic lattices of clay minerals. There is the tetrahedral unit (T) (see Fig. 9.3(a)) where a silicon atom in the middle of the tetrahedron (triangular-base pyramid) is equidistant from four oxygen atoms. There is

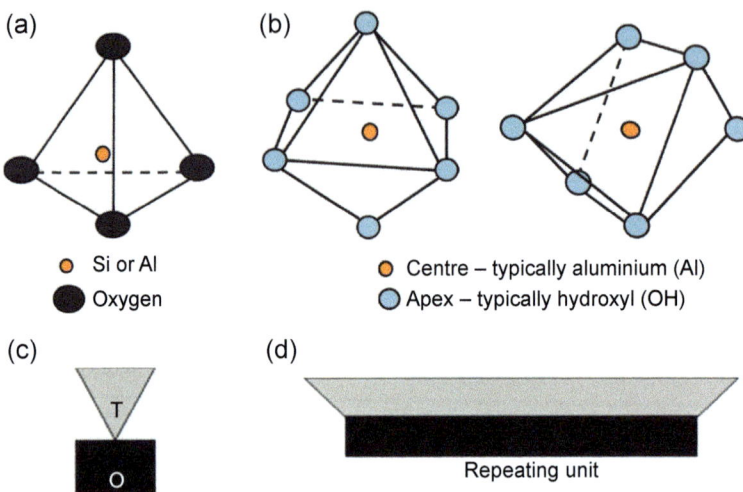

Figure 9.3 Basic structural forms in a clay soil (a) tetrahedral unit (T) − internal silicon atom, associated with four external oxygen atoms, and an (b) octahedral unit (O) − internal aluminium atom, associated with six hydroxyl ($-OH^-$) ions (Grim, 1968), combined in nature into repeating units as schematically demonstrated in (c) or in (d).

evidence that the silicon atoms may be substituted by aluminium to form tetrahedral alumina. These tetrahedral groups are normally arranged to form a hexagonal network in a full-fledged tetrahedral layer.

The second structural form is an octahedral unit (O) (see Fig. 9.3(b)), which consists of sheets or layers of hydroxyl molecules in which aluminium (or sometimes iron or magnesium) atoms are embedded in octahedral coordination, so that they are equidistant from six hydroxyls ($-OH^-$) such as in gibbsite ($[Al_2(OH)_6]$ or brucite $[Mg_3(OH)_6]$). Other cations such as Li, Ca, Mg, Mn, Ni, Cu and Zn occupy these octahedral sites in some species.

The tetrahedral and octahedral units combine to form repeating units (see Fig. 9.3(c) and (d)), which in turn combine in numerous ways to form the various classes of clay soils − kaolinites, illites and montmorillonites − as shown in Table 9.1. Hydrated cations (such as Li, Na, K and Ca) can find their way into the interlayer spaces via various mechanisms (such as adsorption, diffusion and osmosis) (Kinuthia et al., 1999; Kinuthia and Wild, 2001). These cations are the genesis of the differences in the swelling potential of different clay soils (Table 9.1). Thus, when soils are used in an unstabilized form, serious volume instability may be encountered through excessive swelling.

In practice, swelling in soils can be evidenced or suggested by simple tests such as the Atterberg limits. Table 9.2, for example, shows two samples analysed by the author, from Gorom−Gorom in Burkina Faso, West Africa. One sample consisted of a brown soil with a well-distributed particle size range and with significant grass and fibrous/ organic material. This particular soil sample was designated the test specimen code BF − brown and fibrous. The soil was clearly sampled near the ground surface.

Table 9.1 Common clay soil types and different typical free swell values

Clay type	Structure	% Swell
Kaolinite	K K K K K K K K K K K K 1 : 1 clays	5—70
Illite	K K K K K K K K K K K K K K 2 : 1 clays	15—120
Ca-Montmorillonite	As for illite, but with Ca^{2+} cations	45—145
Na-Montmorillonite	As for illite, but with Na^+ cations	1400—2000

The clay structure shows the layering of the tetrahedral and octahedral units, and the interlayer cations (K, potassium cations, or K^+; Na, sodium cations, or Na^+) that contribute to the swelling potential.

The other soil sample was deep red and had no visible organic matter. It was designated RNF — red in colour and nonfibrous. This sample was from a deeper soil horizon than sample BF. The two soils are from two different locations, though not very far from each other. Table 9.2 shows the consistency (Atterberg limits) of the two soil samples. Based on these limits (LL, PL, PI), soil type RNF was observed to be more plastic than sample BF. This fact appears to be corroborated by the more significant linear shrinkage showed by sample RNF relative to that shown by sample BF. This difference is further illustrated by visual observation of the shrinkage magnitudes (note the gaps between the brass mould and the clay test specimens) and by the numerical values for linear shrinkage — 4.5% and 7.9% for BF and RNF, respectively. Interestingly, despite the higher linear shrinkage, sample RNF does not appear to have cracked on shrinking but buckled instead. This is most likely due to the higher clay content and therefore stringer cohesion in sample RNF as suggested by the plasticity indices (PI): 11 versus 15 for samples BF and RNF, respectively.

9.3   Unfired clay material systems

Table 9.3 summarizes the most commonly encountered types of unfired clay-material systems. The table shows that unfired clay materials can be formulated using clay without any additional materials but basic clay—water compaction (mud, cob, sod, adobe, rammed earth, etc.). It also shows that cementitious binders can be used,

Table **9.2** **Consistency and swelling properties of two clay soils from Gorom−Gorom, Burkina Faso, West Africa**

Specimen code	Fibrous	Non fibrous
Liquid limit (LL) (%)	26	34
Liquid limit (LL) (%)	26	34
Plastic limit (PL) (%)	15	19
Plasticity index (PI) (%)	11	15
Linear shrinkage (LS) (%)	4.5	7.9

with or without the addition of one or more other ingredients. The systems that use only water and natural ingredients obviously have the lowest environmental impact but unfortunately also offer a narrower range of 'utility' as almost all have no significant load-bearing capacity. There is a wide range of possible additives, having varying levels of environmental impact.

9.4 Clay materials without additives

The most basic historic clay−based materials capitalize on inherent clay properties, in particular, its cohesive nature. Simple techniques have merely involved the selection of clay deposits. It is possible to find significant deposits of clay without having to remove any coarse materials. However, clay is a versatile material, and this makes

Table 9.3 **Some categories of unfired clay−based material systems**

Category		Additive/ stabilizer	Examples
Unstabilized	Mud/cob/sod	None	
	Adobe	None	
	Rammed earth	None	
Stabilized	Natural materials	Natural ash	Volcanic ash or ground pumice
		Certain soils	Laterites; calcretes; anthill materials
	Industrial/industrial waste or byproduct	Calcined clay	Metakaolin
		Coal ash[a]	Pulverized fuel ash (PFA)
		Ground slag[a]	Ground granulated blastfurnace slag (GGBS)
		Silica[a]	Silica fume
		Bitumen	
		Paper ash[b]	Wastepaper sludge ash (WSA)
		Cement[b]	Ordinary Portland cement
		Lime[b]	Quicklime; Slaked/hydrated lime
	Agricultural/ agricultural waste or byproduct	Agricultural waste	Rice husk ash[b]
			Sugarcane waste (bagasse) ash[b]
			Straw; coir, hemp, sisal, jute, palm

[a]These materials will not ordinarily be used without a hydraulic binder. On their own, they would not add significant strength to clay materials.
[b]These materials possess or have been observed to show hydraulic properties when combined with clay.

it an attractive construction material. It is able to include a significant amount of coarse material while remaining plastic. Traditional houses have been built with clayey silt, clayey sand and boulder clay. In its wet plastic form, clay can be moulded into a regular shape, usually but not limited to a cuboid form. The regular shape is especially necessary if the wet material is to be sun-dried for later use. Alternatively, the wet material can be placed in either panel formwork, as in rammed earth, or in a cage-like open formwork as in traditional huts. The wet cohesive materials sticks together and form a solid irregular mass. The structure can be used with or without further finishing such as plastering. If well plastered, it is hard to believe that a structure is made of irregular mud bulbs.

Clay has been used in this and similar unfired styles in the production of a wide range of earth materials with or without 'reinforcing' strategies such as using straw. If formed in a regular sun-dried (not fired) manner, unfired clay systems comprise mud brick, cob, sod and adobe. However, if used directly without forming into a particular shape, examples include rammed earth.

9.5 Industrial additives

Soils can be strengthened by other unfired processes, such as soil-stabilization techniques. When soils are stabilized with lime and/or Portland cement, a colloidal product predominantly composed of a calcium silicate hydrate (C—S—H) gel is formed, although aluminium phases and traces of iron may also be present. The complex gel gradually changes with time by partially crystallizing, resulting in strength gain in a mechanism very similar to that in Portland cement hydration. The composition of this colloidal product is dependent on material ingredients used: the compounds of calcium, silicon, aluminium and traces of iron coming from the lime or Portland cement used as stabilizer; aluminium and silicon from the soil, and finally the water added for the stabilization process. This colloidal $CaO-Al_2O_3-SiO_2-H_2O$ system is beneficial for strength development, although it is prone to ingress of water and other elements. The quantity and long-term development of this colloidal product influence the total porosity and affect strength and volume stability. The minimization of pores is obviously beneficial to strength in construction materials in general (Benavente et al., 2004; Molina et al., 2011). When well protected, all goes well, and the materials develop significant strength to enable applications in roads, foundations, bricks, blocks and other products of soil-stabilization. Compared with the behaviour of the raw soils, the stabilized materials show little of no expansion potential on stabilization, unless there are other deleterious mechanisms at play, such as the presence of sulphates in the system.

Depending on the prevailing environment, for both soil- and cement-based systems, the hydration products from lime and/or Portland cement (ie, complex C—A—S—H gels) are prone to attack in aggressive solutions (O'Farrell et al., 1999, 2000; Beaudoin et al., 2001; Kinuthia et al., 2003; Snelson and Kinuthia, 2010; Wild et al., 1996, 1998, 1999; Miqueleiz et al., 2012; McCarthy et al., 2014). Such deleterious mechanisms may exist when the target clays for stabilization contain certain compounds, such as sulphates (Kinuthia et al., 1999; Kinuthia and Wild, 2001; Higgins et al., 1998, 2002). The sulphates may also emanate from other sources such as deicing salts or underground water movements caused by either natural flows and/or artificial occurrences. Examples include situations such as broken effluent pipes, especially from industrial developments with resultant liquid or soluble solid wastes containing sulphates or other chemicals and compounds. The deleterious effects of these reactions can be mitigated or eliminated altogether by using some industrial waste and byproduct materials, as will be seen in the next section.

Due to various forces such as pursuance of environmental care, low cost, technological advances and/or other drivers, there is no longer what may be considered conventional or classic materials for clay masonry. Changes have been encountered with either the materials used and/or their use in nonclassic applications. For this reason, marginal natural materials that have hitherto not been considered in building and construction have become viable. The use of marginal naturally occurring materials does not, however, attract much attention compared with the use of industrial and agricultural waste streams, primarily because of the negative environmental impact of these waste and byproduct materials. A few examples will now be discussed.

9.5.1   *Ground granulated blastfurnace slag*

This is an industrial byproduct material that results from the manufacture of steel from iron ore in a blastfurnace. The material has successfully been applied in the concrete industry, where it results in reduced use of Portland cement, an energy-intensive material with a significant negative environmental impact. Use of ground granulated blastfurnace slag (GGBS) also results in improved durability in concrete. The material has had very little impact in masonry until in recent times. Fig. 9.4 shows higher compressive strength values being obtained using formulations containing GGBS, as long as there is significant lime to activate the slag. Kimmeridge clay is a sulphate-bearing stiff clay from Oxfordshire, UK, which has been very difficult to stabilize using lime alone. Fig. 9.4(a) and (b) suggest best stabilization of this difficult soil using a total stabilizer content of at least 6% and a $0.5 \leq$ slag:lime ratio ≤ 2. Considering that slag is a byproduct material that is in most cases less expensive relative to lime or Portland cement, there are benefits to using a blended binder that results in the use of reduced amounts of traditional binders in masonry. The optimal slag:lime ratio changes to unity with higher levels of slag, for clay soils with little or no sulphate, thus increasing profitability. This is demonstrated in Fig. 9.4(c), where the non—sulphate-bearing kaolinite is stabilized with lime—GGBS blends, with and without artificially dosing the kaolinite with small amounts of sulphate (gypsum).

The successful stabilization of both sulphate- and non—sulphate-bearing clay soils using lime-GGBS blends can be exploited for clay-based building materials. This has been demonstrated by various researchers in partnership with the author (Kinuthia et al., 1999; Wild et al., 1996, 1998, 1999; Kinuthia and Oti, 2012; Oti and Kinuthia, 2012; Oti et al., 2008a,b; 2009, 2010a,b,c,d), who have also demonstrated the significant reduction in swelling potential in lime-stabilized clay soils by gradually replacing the lime used in the stabilization process with GGBS. The significance of this outcome is further demonstrated in Fig. 9.5, which shows unfired building bricks made with a sulphate-bearing clay soil (Lower Oxford Clay from Oxfordshire in the United Kingdom), stabilized using blended binder composed of lime and GGBS.

Using industrial byproducts in a sustainable manner so as to achieve robust clay-material systems obviates the brick firing process and also reduces the use of traditional binders of lime or PC. The bricks shown in Fig. 9.5 were made during a pilot industrial trial at Hanson Brick Company Ltd at their fired clay brick plant at Stewartby, in Bedfordshire. Hanson Brick Company Ltd is one of the largest manufacturers of the well-known fired clay 'London' brick in the United Kingdom. The company's mould was used in the trials for the bricks shown in Fig. 9.5. The unfired bricks

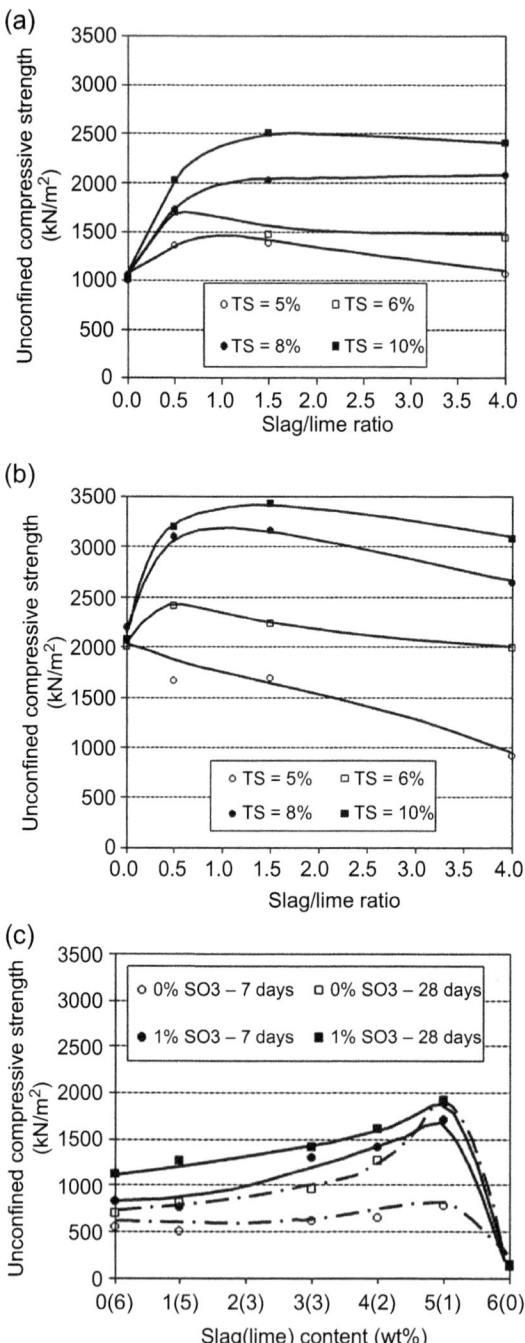

Figure 9.4 Unconfined compressive strength against slag:lime ratio for two types of clay soils stabilized using various lime−GGBS blends, (a) and (b) Kimmeridge Clay moist cured 7 days and 28 days, respectively, and (b) kaolinite stabilized moist cured for 7 days and 28 days (in both cases, moist curing was at 30°C and 100% relative humidity). *TS*, total stabilizer.

Figure 9.5 Unfired building bricks made with stabilized sulphate-bearing clay (Lower Oxford Clay, UK), using lime−GGBS blended binder (Industrial pilot-scale trials were carried out using a steel mould to make a well-known fired clay brick in the United Kingdom: 'London' brick, manufactured by Hanson Brick Company, Ltd.).

have proved to be very robust and have shown phenomenal resistance to repeated freezing and thawing (Oti et al., 2010c), which is one of the most severe tests for durability for both soil- and cement-based materials.

9.5.2 Pulverized fuel ash

For many years, coal has been a dominant source of energy worldwide. The waste from this industry ranges from unusable mining debris, collectively referred to in various terms such as coal waste, colliery spoil, colliery waste, coal mining waste, coal mine tailings and possibly other terms, to the usable pulverized fuel ash (PFA) that results from the burning of coal as a fuel. The burnt waste is predominantly the fine particulate material collected from the flue gasses, mainly by electrostatic precipitation. It is commonly referred to as fly ash (FA) in America and other places or as pulverized fuel ash (PFA) in the United Kingdom and some parts of Europe and beyond. There is also the relatively coarser waste referred to as bottom ash (BA). As the name suggests, BA is collected from the bottom of the coal burning boilers. Although PFA's classic application is in concrete where the benefits include enhanced workability, reduction in the amount of Portland cement used, improved later strength and enhanced durability such as increased resistance to sulphate and chloride forms of attack, its use in clay masonry is not very widespread. As with GGBS, it has been demonstrated that PFA can also be used in the manufacture of durable clay-based masonry (Rahmat et al., 2011; McCarthy et al., 2014). The compressed clay-PFA blocks shown in Fig. 9.6 have not been commercially tested, but research work has suggested that durable compressed clay-based blocks can be made using this industrial waste or byproduct material that is readily available worldwide (Rahmat et al., 2011).

Figure 9.6 Clay—PFA 'Eco-brick' made sustainably by using the already proved sustainable lime—GGBS blended binder.

9.5.3   Wastepaper sludge ash

Most paper is coated with clay and limestone to create smooth surfaces to write on. When the paper is then recycled, the clay and the limestone remains in the sludge from which the recyclable cellulose fibrous material is removed during recycling. When this sludge is combusted to reduce volume of waste going to landfill, the clay and limestone are heated in the process, undergoing more or less a similar heating process as Portland cement, albeit at lower temperatures. Thus, wastepaper sludge ash (WSA) possesses appreciable and useful cementitious potential.

Successful development of a novel cement using wastepaper ash and blastfurnace slag has been reported by Nidzam and Kinuthia (2010, 2011a,b). The cement was developed by combining an industrial waste (WSA) with a byproduct material (GGBS) for replacement of Portland cement. This has enabled the development of 'green' cement for masonry. This cement has performed very well in terms of strength and durability, and sometimes better in terms of appearance compared with the traditional Portland cement. Besides strength development (Nidzam and Kinuthia, 2010, 2011a,b), Fig. 9.7 shows that it was also possible to successfully suppress swelling potential by using WSA-GGBS blended binder on the expansive Lower Oxford clay used in demonstration projects earlier.

Similar approaches are possible elsewhere, using waste and/or byproduct materials that are available in significant quantities, including agricultural waste. Other examples of commonly encountered industrial waste and byproduct materials that have applicability in unfired systems in both soil- and cement-based building and construction materials and components include waste glass (Chidiac and Mihaljevic, 2011); waste types (Snelson et al., 2009); shale, slate, colliery waste and other forms of recycled aggregates (Baojin et al., 2013; Bryson et al., 2012; Corinaldesi, 2009; Debieb et al., 2010; Kinuthia et al., 2009; Oti et al., 2010e,f) and brick dust (O'Farrell et al., 1999, 2000; Kinuthia and Nidzam, 2011). The research work involving the utilization of waste materials is advanced and involves observation of different parameters and informed balancing to optimize performance. In this regard, after observing enhanced strength on stabilizing clay with a waste-based binder such as WSA, the next step was to establish the volume stability as shown in Fig. 9.7.

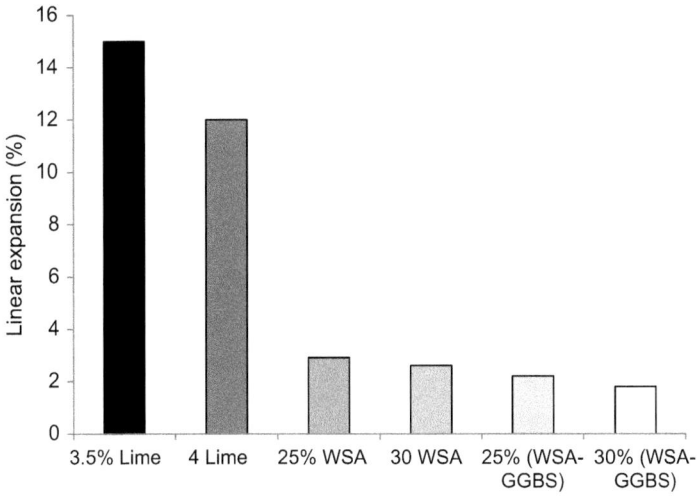

Figure 9.7 Significant reduction in linear expansion by the combined use of WSA and GGBS in the stabilization of an expansive sulphate-bearing clay soil.

Once satisfied with all the parameters, pilot and full-scale trials demonstrated the full realization of success. The whole journey from 'cradle to grave' requires a multi-disciplinary approach, with key stakeholders and players from research, laboratory, field and the entire procurement chain including government. Significant patience is also required, so as to overcome the many hurdles of research and product development and its many frustrations. The reported pilot trials, and as will be seen later in the remaining sections of this chapter some clay-based materials, have gone through such a process, and full-scale factory production of clay-based materials is on the increase. Full-scale construction of local housing schemes using unfired clay products has also started and is on the increase.

9.6 Agricultural additives

For most countries, many major industrial activities are directly or indirectly related to agriculture. For these countries, any major breakthrough in the development of sustainable infrastructure cannot afford to ignore waste from the agricultural sector. The author has been brainstorming on this agenda, in liaison with like-minded co-workers for some time. Initial focus has been in activities that produce large agricultural waste streams, including the growing of palm oil and sugarcane (Mofor et al., 2009), rice (Billong et al., 2011), and fibrous materials for fibre-based soil-cement blocks — bagasse, straw, kenaf, bamboo, jute, coir, durian (Khedari et al., 2005). In Cameroon, inspired researchers at the Mission de Promotion des Materiaux Local (MIPROMALO) — a local materials promotion authority — based in Yaoundé have provided valuable collaborative support. In Kenya, initial collaboration has been with local higher education institutions as well as with environmentally conscious local industries and environmental lobby groups, particularly those in the agricultural sector. In both Kenya and Cameroon, the initial focus has been in sectors that produce large agricultural waste streams, including the palm oil industry in Cameroon, and the sugarcane industry in Kenya.

Enthused by the success of UK-based research into the utilization of waste materials in the development of sustainable civil engineering infrastructure, scoping studies have been carried out to establish the key waste streams in Kenya and Cameroon. The findings in these two African countries are very typical of many other developing countries, in terms of the socioeconomic stature and practices, level of technology, need for basic infrastructure and commonality in future aspirations.

There is an acute shortage of cement in many developing countries. For this reason, the use of fine particulate wastes in cementitious systems by way of replacement of Portland cement (PC) has been a common option for many researchers. This has led to interest in the use of agricultural waste in civil engineering construction. In the more advanced cases, after using the agricultural waste in electric power generation, the secondary waste produced is fine particulate or agglomerated ash. With determination, persistence and combined synergies with equally committed peers of varied backgrounds and expertise, these ashes can be exploited for building and construction materials.

Traditionally, the high-volume changes observed in building materials made with clay systems have been mitigated by enhancing tensile strength by incorporating fibrous constituents (Moropoulou et al., 2005; Khedari et al., 2005). This is an old technology, and unfired bricks have been strengthened with straw in practices that are centuries old. However, the research and full-scale commercial application of masonry products made incorporating agricultural materials and agricultural waste and byproducts is in its formative years. Most research groups have only managed to demonstrate potential. Preliminary work on bagasse ash from the growing of sugarcane by Mofor et al. (2009) has demonstrated possible potential savings in Portland cement of about 50%, and there is also potential for clay-based materials using this waste material.

9.7   Construction using clay-based systems

The increased use of clay-based materials in building and construction has challenging practical implications, relative to cement-based materials, at all stages of their use. Undertaking research in clay-based materials is also challenging, due to low strength and variability in other performance characteristics. There are also equal if not more difficult challenges to execute research findings into practice, at various stages of infrastructure development – materials manufacture or formation, storage, construction and eventually on their durability.

9.7.1   Formation

Unfired clay systems may be formulated in the same manner as fired systems, the only difference being the firing process. Heath et al. (2009), for example, undertook

(a) (b)

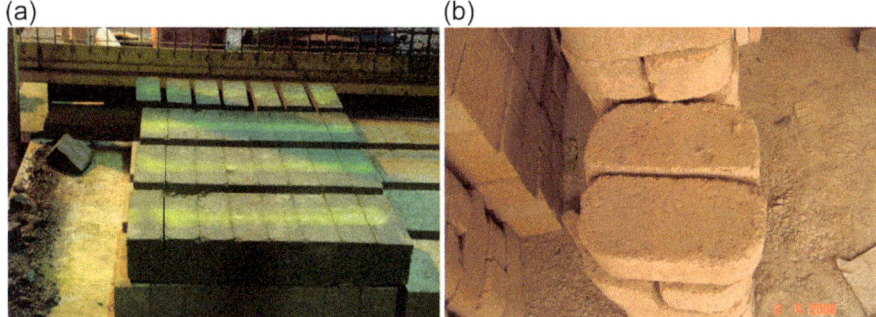

Figure 9.8 Unfired brick illustrating possibility of automated production, with likelihood of edge breakage (bricks produced by author during trials at P D Edenhall concrete brick plant at Bridgend, South Wales, UK) and (b) Rounded or chamfered edges ready for use in clay-based construction.

research work on unfired clay systems by starting with wet extruded bricks that were originally meant for firing. An alternative process to extrusion is the use of purpose-made moulds. For fired bricks, the extruded or wet moulded material is allowed to condition to reduce the excess moisture prior to firing. The moisture conditioning has been observed to minimize volume changes upon firing so as to maintain shape, volume and dimensional stability. After this conditioning, systems meant to be used in unfired applications may be allowed to dry further (Heath et al., 2009). This approach is sometimes unsuitable for unfired systems due to the excessive water content used, and most unfired systems are designed to contain less water from the start, by compacting in a semidry state. This results in denser and stronger material. In the absence of additives such as stabilizers (lime, Portland cement or any other emergent stabilizers), the control of moisture is less restrictive. However, with hydraulic stabilizers, especially Portland cement, careful control of the amount of water used is essential.

Ability to handle the freshly made unfired clay materials is critical, for both manual and automated production (see Fig. 9.8(a)). The automated handling is the more critical one, as massive losses can be incurred if larger production batches are damaged during early movement. In order to minimize breakage, most unfired clay-based materials have rounded or chamfered edges, unlike their fired or cement-based counterparts (see Fig. 9.8(b)).

9.7.2 Storage

Storage of unfired clay products is also critical, as heavy losses can be incurred if the material is left unprotected during storage. For this reason, unfired clay materials, without exception, are routinely protected in some manner during storage, no matter how ineffective the protection might be, as shown in Fig. 9.9. This extra care continues to be critical during service life, depending on the type of use. As most unfired systems are used for either internal walling or inner leafs in outer walls (Birch, 2005; McGregor et al., 2014), the more critical care is during the storage prior to, and during, construction.

Figure 9.9 Protection of cement-stabilized (unfired) clay bricks at a community brick project (cooperative) in Kenya (note grass cover on pile, to reduce the effect of rain).

9.7.3   Construction

Experience with large-scale factory-type production of unfired clay-based materials is rare and recent. During automated production, the control of compaction moisture content, the handleability of the freshly compacted product, the care needed during early transport to storage/curing yard, and the subsequent and continued care during loading on trucks all pose problems. The significant product losses during transport and during construction have all discouraged large-scale investment.

In recent times, clay materials and environmental enthusiasts have faced these problems head on, so as to address the emerging environmental, aesthetic and inherent passive advantages of clay-based materials. This has started to stimulate growth and market in unfired clay-based materials and their acceptance in construction. Heavy losses continue to be incurred during construction, but better understanding of the handing, care and use of these materials will ensure that losses remain minimal may be noted in Fig. 9.10.

Production of unfired clay materials in most developing countries has predominantly remained manual. In both developed and developing countries, there is a slow shift from individual or single applications to larger materials consumption as in local housing projects. The next few decades are likely to witness more research and application of clay-based materials in large housing and construction projects. Achievement of good workmanship is not limited to developed countries, and this can be achieved irrespective of country as shown earlier in Fig. 9.1 for rammed earth in Wales, United Kingdom and on a large-scale housing project in Cameroon as shown in Fig. 9.10.

Innovation in the use of unfired clay materials has already been demonstrated in uses as varied as rammed earth, clay bricks and clay blocks, stabilized in various ways ranging from use of traditional stabilizers such as lime and/or Portland cement, and/or

(a) (b)

Figure 9.10 Low-cost housing using locally available materials (unfired bricks) and labour in Cameroon illustrating (a) impressive workmanship and (b) minimal material losses during construction.

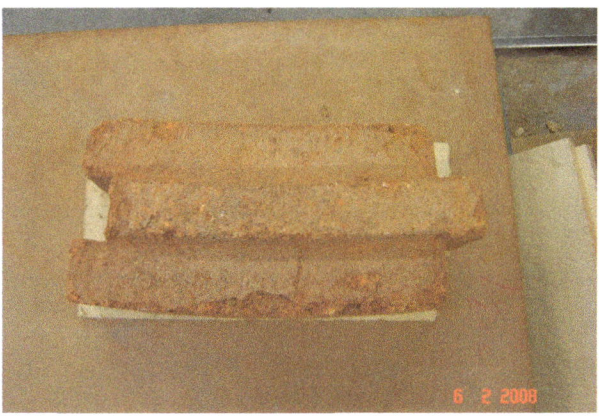

Figure 9.11 Innovative manufacture and applications of unfired clay materials. The curved blocks have been used for the construction of water tanks.

in combination with other natural, industrial or agricultural waste streams. Innovation can also take the form of block shape and type of application of the materials, such as that demonstrated by an interlocking and also curved block shown in Fig. 9.11. When curved, the blocks can be used for the construction of circular houses and/or of water tanks. The interlocking block can also be used in the normal construction of buildings (usually without curvature). The blocks are typically made using cement-stabilized clayey soil, and their use in water-retaining structures demonstrates the durability and robustness of the resulting material if well constructed and protected.

9.7.4   Durability of unfired clay materials

The analysis of the influence of the material ingredients that constitute unfired clay systems, together with the prevailing environmental condition, has suggested that it is very important that the limits of applicability and care during service are critical considerations. Unfired clay systems stand little chance in flooding situations. Therefore, clay systems exposed to the elements demand far higher expectations in terms of robustness compared to those used for internal walls or other protected environments (Dias et al., 2014; Janssen et al., 2012; Obuzor et al., 2011a,b; 2012). On the other hand, the level of care during service is also critical. Careful maintenance regimens are recommended for any unfired clay system. Biological actions including microorganism and plant growth thrive in damp conditions, particularly moulds and fungi. Other effects such as torn or inelastic sealants, cracked pointing mortar, blocked weep holes, standing water, missing damp-proofing and worn coatings, among other critical factors, are key concerns.

Mitigating or remedial steps together with a strict adherence to manufacturer recommendations (Kelman and Spence, 2004) should be taken as early as possible. To prevent the biological actions mentioned, thorough disinfecting of wet zones is necessary to prevent mould growth. In case of problems, a thorough problem analysis is necessary before any remedial action is taken. Covering problems for example, cracks or loose material with render or other cover only serves to delay and exacerbate the problem, leading to more expensive remedial measures later. In addition, expert help may sometimes be necessary, as inappropriate remedy or workmanship can appear an attractive repair and maintenance option but may have consequences at a later date, including serious ones such as loss of life.

For a long time, compressive strength has been used by many users as an indicator of durability. While to some degree this does reflect possible outcome during service life, it has also been established in both soil- and cement-based materials that strength alone is only one of the many factors determining durability. In concrete and unfired clay systems, for example, tests involving ingress of water and chemicals, especially for applications of concrete in either marine or buried environments, form among the most severe test environments under which durability of concrete may be assessed. For unfired clay systems such as compressed earth, absorption of water has a significant influence on the durability, with and without the presence of any dissolved chemicals. For this reason, applications of buried unfired clay materials are extremely rare. Water absorption capacity tests are therefore very strong indicators of durability. Indeed, all the common tests for durability of compressed earth units are water based or related. A detailed analysis of the durability of compressed earth units is covered by the author in Chapter 17 of Pacheco et al. (2015).

9.8 Future trends

There is an observable increase in the researched use of compressed earth masonry units. These units are increasingly being seen as industrialized materials, and no longer considered as only appropriate or applicable to traditional approaches for self-construction (Cid-Falceto et al., 2012). Many examples exist of the increasing use of industrial and agricultural waste, and byproduct materials in compressed earth masonry, and it is not possible to cover or quote them all here. However, a brief survey of the materials, techniques and applications of today's masonry suggests a thriving ongoing research and innovation-led development of compressed earth masonry technology. There is also a gradual shift from the traditional fired clay masonry categories to unfired categories. While the fired categories adopted a blanket approach of firing clay irrespective of clay composition, there is a far wider variety of material selection and technology adopted for the unfired masonry category. This is because the durability imparted by the firing process is harder to mimic or achieve in unfired systems, resulting in the pursuance of alternative approaches to materials usage to include industrial and agricultural waste and byproduct materials. There is also increasing use of synthetic (industrial) and natural fibres for reinforcement, chemical cementation, attention to particle matrix configurations, and variable and careful selection of applicability scenarios. Fortunately, as mentioned earlier, there is also a gradual shift in emphasis from a blanket stipulation of standard properties and material expectations, relying rather on recommendations by the manufacturer for the appropriate use and for the possible consequences of lack of adherence, by the end user, to these recommendations. The future will therefore most likely witness enhanced end-user sensitivity to global changes in materials, technology and material failure mechanisms, and better or closer liaison with material manufacturers, coupled with a keen(er) awareness of relevant government policies, guidance and legislation. This development or trend is likely to enhance reliance on compressed earth masonry, and it is hoped that mortgage lenders and insurance industries will be quick to respond to the fast changing landscape, resulting in an ever increasing adaptation of compressed earth masonry in future social and commercial housing. This scenario suggests an urgent need for a near-universal harmonization, consensus building and characterization and/or classification of compressed earth-based building and construction materials, taking into account the wide range of materials and variable manufacturing methodologies likely to be encountered. In terms of further research work needed, tests on durability need to be well delineated, with those for fired systems adopting considerations that suit fired systems separate from those best suited for unfired systems. Those for unfired systems need in turn to differentiate between stabilized and nonstabilized compressed earth systems. Both test categories should embrace and embed the indicative tests of compressive strength and water absorption. The more severe cyclic freeze—thaw and wet—dry tests should be more in reference to stabilized systems, while the water spray and drip test methodologies being in specific reference to unsterilized systems. This task is beyond the remit intended for this chapter, but the chapter has managed to raise most of the factors that determine the durability of compressed earth units.

References

Baojian, Z., Chisun, O., Caijun, S., 2013. CO_2 curing for improving the properties of concrete blocks containing recycled aggregates. Cement and Concrete Composites 42, 1—8.

Beaudoin, J.J., Catinaud, S., Marchand, J., 2001. Volume stability of calcium hydroxide in aggressive solutions. Cement and Concrete Research 31, 149—151.

Benavente, D., Garcıá, M.A., Fort, R., Ordoñez, S., 2004. Durability estimation of porous building stones from pore structure and strength. Engineering Geology 74, 113−127.

Billong, N.A.B., Melo, U.C., Kamseu, A.B.E., Kinuthia, J.M., Njopwouo, B., 2011. Improving hydraulic properties of lime−rice husk ash (RHA) binders with metakaolin (MK). Construction and Building Materials 25, 2157−2161.

Birch, A., August 5, 2005. Bricks come down to earth: mud goes modern with unfired clay eco-bricks. Building Design (Building.co.uk; http://www.bdonline.co.uk/bricks-come-down-to-earth/3054771.article, last accessed on 12th January 2016).

Bryson, L.S., Gutierrez, I.C.G., Hopkins, T.C., 2012. Development of a new durability index for compacted shale. Engineering Geology 139−140, 66−75.

Chidiac, S.E., Milhaljevic, S.N., 2011. Performance of dry cast concrete blocks containing glass powder or polyethylene aggregates. Cement and Concrete Composites 33, 855−863.

Cid-Falceto, J.C., Mazarron, F.R., Cañas, I., 2012. Assessment of compressed earth blocks made in Spain: international durability tests. Construction and Building Materials 37, 738−745.

Corinaldesi, V., 2009. Mechanical behavior of masonry assemblages manufactured with recycled-aggregate mortars. Cement and Concrete Composites 31, 505−510.

Debieb, F., Courard, L., Kenai, S., Degeimbre, R., 2010. Mechanical and durability properties of concrete using contaminated aggregates. Cement and Concrete Research 32, 421−426.

Dias, J.L., Silva, A., Chai, C., Gaspar, P.L., de Brito, J., 2014. Neural networks applied to service life prediction of exterior painted surfaces. Building Research & Information 42 (3), 371−380.

Grim, R.E., 1968. Clay Mineralogy, second ed. McGraw Hill Book Company.

Heath, A., Walker, P., Fourie, C., Lawrence, M., 2009. Compressive strength of extruded unfired clay masonry units. Ice Proceedings, Construction Materials 162 (CM3), 105−112.

Higgins, D.D., Kinuthia, J.M., Wild, S., 1998. Soil stabilisation using lime-activated ground granulated blastfurnace slag (GGBS). In: Malhotra, V.M. (Ed.), Proceedings of the 6th CANMET/ACI Int. Conf. on Fly Ash, Silica Fume, Slag and Natural Pozzolans in Concrete, 1998, 2, SP 178−55, pp. 1057−1074. Bangkok, Thailand, May 31st−June 5th, 1998, Libr. Of Congr. Catal. Card No. 98-85145. pp. 1057−1074.

Higgins, D.D., Thomas, B., Kinuthia, J., 11−12 April, 2002. Pyrite oxidation, expansion of stabilised clay and the effect of ggbs. In: Zoorob, S.E., Collop, A.C., Brown, S.F. (Eds.), Proceedings of the 4th European Symposium, Bitmat4, on Performance of Bituminous and Hydraulic Materials in Pavements, pp. 161−168. Nottingham, UK, ISBN:90 5809 375 1.

Janssen, H., Derluyn, H., Carmeliet, J., 2012. Moisture transfer through mortar joints: a sharp-front analysis. Cement and Concrete Research 42, 1105−1112.

Kelman, I., Spence, R., 2004. An overview of flood actions on buildings. Engineering Geology 73, 297−309.

Khedari, J., Watsanasathaporn, P., Hirunlabh, J., 2005. Development of fibre-based soil-cement block with low thermal conductivity. Cement and Concrete Composites 27, 111−116.

Kinuthia, J.M., Nidzam, R.M., 2011. Towards zero industrial waste: utilization of brick dust waste in sustainable construction. Waste Management 31, 1867−1878.

Kinuthia, J.M., Oti, J.E., 2012. Designed non-fired clay mixes for sustainable and low carbon use. Applied Clay Science 59/60, 131−139.

Kinuthia, J.M., Wild, S., 2001. Effects of some metal sulfates on the strength and swelling properties of lime-stabilised kaolinite. International Journal of Pavement Engineering (IJPE) 2, 103−120.

Kinuthia, J.M., Wild, S., Jones, G.I., 1999. Effects of monovalent and divalent metal sulphates on consistency and compaction of lime-stabilised kaolinite. Applied Clay Science 14, 27−45.

Kinuthia, J.M., Nidzam, R.M., Wild, S., Robinson, R.B., 2003. Strength and swelling properties of sulphate-bearing soil stabilised utilising wastepaper sludge ash (WSA). In: Dawson, A.R. (Ed.), Proceedings of UNBAR 6-6th International Symposium on Pavements Unbound! Nottingham, England, 6th−8th July, 2004. Balkema Publishers. ISBN:90 5809 699 8.

Kinuthia, J.M., Snelson, D.G., Gailius, A., 2009. Sustainable medium-strength concrete (CS-Concrete) from colliery spoil in South Wales UK. Civil Engineering and Management 15, 149−157.

McCarthy, M.J., Csetenyi, L.J., Sachdeva, A., Dhir, R.K., 2014. Engineering and durability properties of fly ash treated lime-stabilized sulfate-bearing soils. Engineering Geology, Accepted Manuscript. http://dx.doi.org/10.1016/j.engeo.2014.03.001.

McGregor, F., Heath, A., Shea, A., Lawrence, M., 2014. The moisture buffering capacity of unfired clay masonry. Building and Environment 599−607.

Miqueleiz, L., Ramirez, F., Seco, A., Nidzam, R.M., Kinuthia, J.M., Abu Tair, A., Garcia, A., 2012. The use of stabilised Spanish clay soil for sustainable construction material. Engineering Geology 133−134, 9−15.

Mofor, L., Kinuthia, J.M., Melo, U.C., Djialli, D., 2009. Energy in Cameroon: the potential for transporting waste to riches. Materials Solutions 1, 18−21.

Molina, E., Cultrone, G., Sebastián, E., Alonso, F.J., Carrizo, L., Gisbert, J., Buj, O., 2011. The pore system of sedimentary rocks as a key factor in the durability of building materials. Engineering Geology 118, 110−121.

Moropoulou, A., Bakolas, S., Anagnostopulou, S., 2005. Composite materials in ancient structures. Cement and Concrete Composites 27, 295−300.

Nidzam, R.M., Kinuthia, J.M., 2010. Sustainable soil stabilisation with blastfurnace slag (GGBS) − a review. Construction Materials 163, 157−165.

Nidzam, R.M., Kinuthia, J.M., 2011a. Compaction of fills involving stabilisation of expansive soils. Geotechnical Engineering 164, 113−126.

Nidzam, R.M., Kinuthia, J.M., 2011b. Effects of mellowing sulfate-bearing clay soil stabilized with wastepaper sludge ash for road construction. Engineering Geology 117, 170−179.

O'Farrell, M., Wild, S., Sabir, B.B., 1999. Resistance to chemical attack of ground brick-PC mortar part I. Sodium sulphate solution. Cement and Concrete Research 29, 1781−1790.

O'Farrell, M., Wild, S., Sabir, B.B., 2000. Resistance to chemical attack of ground brick-PC mortar part II. Synthetic seawater. Cement and Concrete Research 30, 757−765.

Obuzor, G.N., Kinuthia, J.M., Robinson, R.B., 2011a. Enhancing the durability of flooded low-capacity soils by utilizing lime-activated ground granulated blastfurnace slag (GGBS). Engineering Geology 123, 179−186.

Obuzor, G.N., Kinuthia, J.M., Robinson, R.B., 2011b. Utilisation of lime activated GGBS to reduce the deleterious effect of flooding on stabilised road structural materials: a laboratory simulation. Engineering Geology 122, 334−338.

Obuzor, G.N., Kinuthia, J.M., Robinson, R.B., 2012. Soil stabilisation with lime-activated-GGBS—a mitigation to flooding effects on road structural layers/embankments constructed on floodplains. Engineering Geology 151, 112−119.

Oti, J.E., Kinuthia, J.M., 2012. Stabilised unfired clay bricks for environmental and sustainable use. Applied Clay Science 58, 52−59.

Oti, J.E., Kinuthia, J.M., Bai, J., 2008a. Developing unfired stabilized building materials in the UK. Engineering Sustainability 161, 211−218.

Oti, J.E., Kinuthia, J.M., Bai, J., 2008b. Using slag for Unfired-Clay masonry-bricks. Construction Materials 161, 147−155.

Oti, J.E., Kinuthia, J.M., Bai, J., 2009. Design Thermal values for unfired Clay bricks. Materials Design 31, 104−112.

Oti, J.E., Kinuthia, J.M., Bai, J., 2010a. Design thermal values for unfired clay bricks. Journal of Materials and Design 31, 104−112. http://dx.doi.org/10.1016/j-matdes.2009.07.011.

Oti, J.E., Kinuthia, J.M., Bai, J., 2010b. Engineering properties of concrete made with slate. Proceedings of the Institution of Civil Engineers (ICE). Construction Materials 163, 131−142.

Oti, J.E., Kinuthia, J.M., Bai, J., 2010c. Freeze-thaw of stabilised clay brick. Proceedings of the Institution of Civil Engineers (ICE). Waste and Resource Management 163, 129−135.

Oti, J.E., Kinuthia, J.M., Bai, J., 2010d. Sustainable masonry mortar for brick joint and plaster in the UK. Proceedings of the Institution of Civil Engineers (ICE). Construction Materials 163, 87−96.

Oti, J.E., Kinuthia, J.M., Bai, J., 2010e. Unfired clay masonry incorporating slate waste. Proceedings of the Institution of Civil Engineers (ICE). Waste and Resource Management 163, 17−27.

Oti, J.E., Kinuthia, J.M., Snelson, D.G., Bai, J., 2010f. Applications of slate waste material in the UK. Proceedings of the Institution of Civil Engineers (ICE). Waste and Resource Management 163, 9−15.

Pacheco-Torgal, F., Lourenço, P.B., Labrincha, J.A., Kumar, S., Chidaprasirt, P., 2015. Chapter 17-The Durability of Compressed Earth-based Masonry Bricks, in Eco-efficient Masonry Bricks and Blocks: Design, Properties and Durability. Woodhead Publishing Series in Civil an Structural Engineering (WPCSE 55).

Rahmat, M.N., Ismail, N., Kinuthia, J.M., 2011. The potential of utilising industrial waste as lightweight building components− a preliminary investigation. Modern Applied Science 4, 35−46.

Snelson, D.G., Kinuthia, J.M., 2010. Resistance of mortar containing unprocessed pulverised fuel ash (PFA) to sulphate attack. Cement and Concrete Composites 32, 523−531.

Snelson, D.G., Kinuthia, J.M., Davies, P.A., Chang, S., 2009. Sustainable construction: Composite use of tyres and ash in concrete. Waste Management 29, 360−367.

Sutton, A., Walker, P., Black, D., November 2011. Unfired Clay Masonry: An Introduction to Low-impact Building Materials. Information paper IP 16/11. BRE-University of Bath. ISBN:978-1-84806.

Wild, S., Kinuthia, J.M., Robinson, R.B., Humphries, I., 1996. Effects of ground granulated blastfurnace slag (ggbs) on the strength and swelling properties of lime-stabilised kaolinite in the presence of sulphates. Clay Minerals 31, 423−433.

Wild, S., Kinuthia, J.M., Jones, G.I., Higgins, D.D., 1998. Effects of partial substitution of lime with ground granulated blastfurnace slag (GGBS) on the strength properties of lime−stabilised sulphate bearing clay soils. Engineering Geology 51, 37−53.

Wild, S., Kinuthia, J.M., Jones, G.I., Higgins, D.D., 1999. Suppression of swelling associated with ettringite formation in lime-stabilised sulphate-bearing clay soils by partial substitution of lime with ground granulated blastfurnace slag (GGBS). Engineering Geology 51, 257−277.

Earthen materials and constructions

10

A. Fabbri
University of Lyon, Vaulx-en-Velin, France

J.C. Morel
Coventry University, United Kingdom

10.1 Earth as a building material

The stabilization of the atmospheric greenhouse gas concentrations requires our emissions to drop well below current levels, and thus it requires a drastic reduction in our energy consumption. In this context, the building sector plays a key role, as it accounts for about 40% of global greenhouse gas generation (Dixit et al., 2010). Consequently, increasing research has been performed on the thermal insulation of buildings to optimize and reduce energy consumption. This approach is one way to reduce energy losses from buildings made with industrially manufactured materials. However, most of these building materials, either for insulation (such as rock wool) or for wall manufacturing (such as concrete), are major energy consumers during both their production and implementation (embodied energy), and their recycling is not always practical or feasible (Harris, 1999). Consequently, the development of earth-based buildings appears to be a sustainable alternative to conventional constructions.

The major asset of earth lies in the fact that it is a local material that can be taken and used immediately on the construction site or nearby and does not require industrial processing (Morel et al., 2001). It is not a renewable but is a reusable material; it requires no treatment to be reused and therefore has a very low impact in terms of energy use (Habert et al., 2012). Moreover, the sustainability of this material was recognized long ago, as highlighted by the significant heritage of earth construction all around the world. An iconic example of this is the city of Shibam in Yemen, where buildings with more than eight stories reaching heights of 30 m were built with earth blocks. Shibam is a UNESCO world heritage site and is known as the most ancient skyscraper city in the world (Houben and Guillaud, 1994).

For millennia, human beings have used earth in various forms for construction. The three most common techniques are adobe blocks, cob, and rammed earth (Hall et al., 2012). The technique used mostly depends on local know-how and on the nature of the earth.

Adobe blocks are one of the most widely used building materials, and its use dates back to 8000 BC. (Houben and Guillaud, 1994). They are made of earth mixed with water and, most of the time, with an organic material such as straw or dung. The wet earth-based mixture is molded into a frame, which is removed after initial setting.

Nonconventional and Vernacular Construction Materials. http://dx.doi.org/10.1016/B978-0-08-100038-0.00010-X

After a first stage of drying that lasts a few hours, the bricks are turned on edge to finish drying. Adobe blocks can encompass most any shape or size. Their thickness is chosen to provide uniform drying and good mechanical characteristics.

Adobe blocks are implemented with earth mortar to build masonry. There is less scientific data available on earth mortar than for blocks because the strength of masonry is mainly due to the blocks. However the mortar, depending on the height of its layer, influences the strength of the masonry (see EC6 for conventional masonries). Moreover, it is better to be able to ensure that the fresh mortar will have a suitable consistency according to what the mason needs to implement the masonry, and ensure a suitable behavior after drying (mitigation of cracks and mechanical strength). The adobes walls are most of the time coated with earth renders or plasters. These last materials are close to earth mortar in their composition and behavior. The following references help to understand the specificities of fresh and dried earth mortar mechanical behavior (Walker and Stace, 1997; Azeredo et al., 2008; Pkla et al., 2003; Ciblac and Morel, 2014). And the last is specific to earth plasters (Hamard and Morel, 2013).

Cob is also an ancient building technique that may have been used for building since prehistoric times. It consists of monolithic construction made of unbaked earthen mixtures ladled in courses onto a foundation. During placement, the mixture, composed of earth, water, and usually fibers, is in a plastic state (having water content between 20% and 30% by dry weight of earth). The construction progresses according to the time required for the prior course to dry. Vertical surfaces are eventually drawn up by cutting after a short drying time, while the material is not too hard.

Finally, rammed earth is formed by compacting raw earth with a rammer. The earth is placed in a framework and laid down in successive beds (courses). This technique is relatively new, compared to the others, as it seems to appear for the first time in Carthage (Tunisia) in 814 BC. It spread around the Mediterranean and North Africa. It was exported to Europe with the expansion of Islam since the 8th century.

Whatever the construction technique, strong similarities exist. At first, earth walls are quite thick (around 50 cm) and are separated from the ground by a basement which is designed to prevent capillary rises. The earthen material is composed of aggregates (sand, gravel, fibers, etc.) bonded by a continuous clayey matrix which provides cohesion and strength, but which is also responsible for complex mechanical behavior, such as swelling and shrinkage when subject to hydric changes (Van Damme, 2002; Lei et al., 2014). The clay content proportion should thus be sufficient to ensure a good material stiffness and strength, but the proportion of expansive clay must remain limited (lower than 50% of the total clay amount) to avoid cracking.

Several construction recommendations exist. For example, BS1377 standards (British Standards Institute, 1990) define a range of admissible values for the particle size distribution of earth. Along the same line, a proportion of 15—30% of clays, 10—30% of silts, and 50—75% of sand is commonly recommended for the construction of adobe blocks (Garrison and Ruffner, 1983). Therefore, the study of existing construction indicates that the picture is not so simple, and that the particle size distribution cannot be used as a discriminating parameter (Aubert et al., 2014). Nowadays, no clear objective rule exists to discriminate the ability of an earth to be used as a building material solely on the analysis of its granular and mineralogical properties. It is then still necessary to realize performance-oriented tests that should

Figure 10.1 Picture of the rammed earth house studied in this chapter (a), during its construction, in 2011, (b) in 2013.

depend on the use that will be made of the material. A clear definition of these tests requires a good understanding of the specificities of earth, as well as the identification of the key parameters which drive its behavior.

Following this introduction, this chapter is divided in three main parts. The first deals with the particularities of earthen material and especially on its interaction with water. The consequences of this interaction on its hygrothermal capabilities, on its mechanical behavior, and on design recommendations and constraints are then discussed. In that context, a small focus is made on stabilized earth, as this technique allows reducing the impact of water on the earth behavior. The second part is dedicated to the assessment of the hygrothermal and mechanical performances of the material. Since no clear consensus exists on this point, some propositions and lines of thinking are presented. The last part concludes on the sustainability of earth buildings, and in particular of unstabilized (or lightly stabilized) buildings.

Throughout this chapter, the case study of a recent rammed earth house constructed in 2011 in southeast France by Nicolas Meunier is used for illustration. This house, pictured in Fig. 10.1, is made with a local earth with 16% clay content and mixed with 2.5% (by dry weight) NHL5 lime.

The indoor and outdoor temperatures and hygrometry were measured after the enclosure but before the occupancy of the house. In addition, temperature and water content sensors are placed in the core of the south and west parts of an earthen wall (Fig. 10.2). A complete description of the probes and of the instrumentation plan is reported in (Chabriac, 2014), and the method used to calibrate the water content sensors is explained in (Chabriac et al., 2014).

10.2 Particularities of earthen materials and constructions

10.2.1 Affinity with water

Earth materials are composed of clays, silts, sands, and possibly gravels and fibers. Since the connections between these constituents are not perfect, some small voids, called pores, are embedded within the solid material. The network formed by the

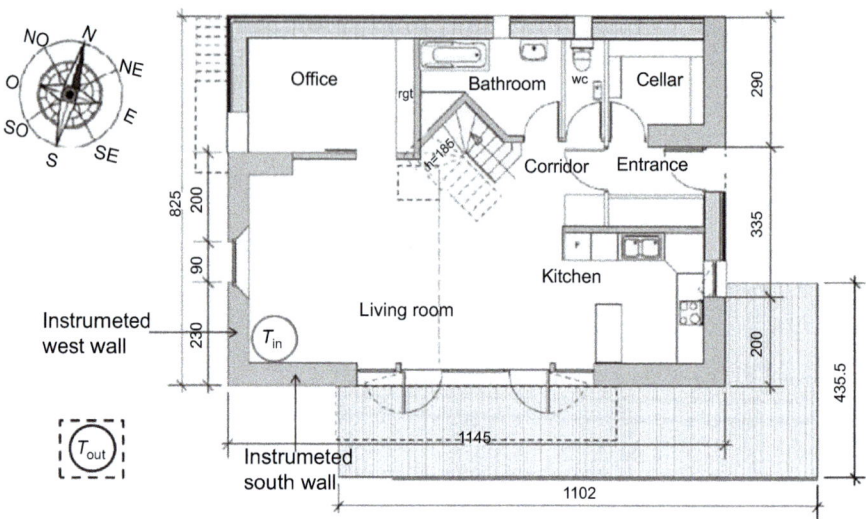

Figure 10.2 Location of the temperature sensors and of the instrumented rammed earth walls.

connection of these voids, called the porous network, enables fluids (either liquid or gas) to flow through the material. This property makes earthen material a porous medium. In addition, some constituents of the solid matrix like clays and fibers are by themselves porous media. Consequently, earth appears to be a double porosity medium, whose morphological description could be summarized by the illustration shown in Fig. 10.3. The size distribution of the pores within an earthen material is particularly broad. Indeed, clay aggregates are commonly formed of interacting particles which are themselves made of 3–10 coupled layers each 1-nm thick. This results an interlayer porosity on the order of magnitude of 1 nm (Pusch and Yong, 2006), while depending on the type of earth, the larger pores can extend 1 mm.

When all the porous space is filled by air or by liquid water, the material is said to be dried or saturated, respectively. However, most of the time, there is coexistence between liquid water and air within the porous network. The air is a mixture of dry

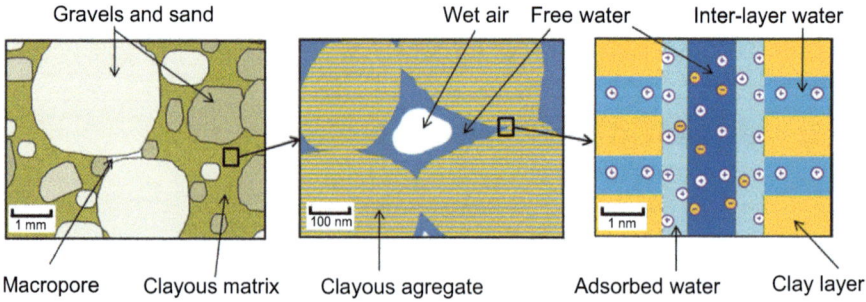

Figure 10.3 Schematic representation of the morphological description of earth.

air and water vapor. The water vapor content in the air is given by the hygrometry or relative humidity (noted φ in the following). The material is then in an unsaturated state. An obvious way to quantify relative proportion of these two phases is to use the saturation ratio, noted S_r, and defined as the volumetric proportion of pore space filled by liquid water. However, since mass measurements are easier than volume measurements, the use of the water content, noted w, defined as ratio between the masses of in-pore water and dried material, is often preferred. The relation between both is as follows:

$$S_r = \frac{\rho_s}{\rho_w} \frac{w}{\phi} \tag{10.1}$$

where ρ_s is the dry density of the material; ρ_w is water density; and ϕ is the porosity (ie, ratio between the volume of the pores and the overall volume of the material).

At the macroscopic scale, the water content depends on ambient conditions (hygrometry and potentially temperature and material strain). This dependency, which is commonly depicted by the isothermal sorption–desorption curves, varies as a function of the morphology of the porous network and of its affinity with water molecules. The sorption isotherm characterizes the water uptake with increasing ambient humidity at a constant temperature, while the desorption isotherms characterize water expulsion with decreasing ambient humidity at a constant temperature. As an illustration, the sorption–desorption curves at 20°C of the earth used for the previously described rammed earth construction (Fig. 10.1) are reported in Fig. 10.4.

The relation between the water content and the relative humidity appears to be strongly nonlinear. Furthermore, a hysteresis can be observed between the sorption and desorption curves. This phenomenon is quite common and has been widely

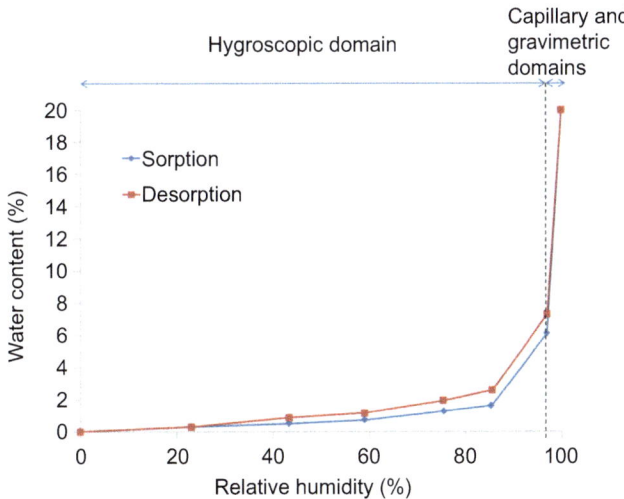

Figure 10.4 Sorption and desorption isotherm of the earth used is the case study.

studied by many authors for a large variety of materials like wood and hemp concrete. Nonetheless, in the case of earthen material, study remains limited, and it will be neglected in the following discussion. In any event, a good understanding of the shape of these curves requires studying the physical processes which occur at the microscopic scale, and these are expected to vary considerably for different earthen materials. Indeed, the impact of the ambient hygrometry and temperature on the earth water content is due to hydric exchanges between the water vapor contained in the gaseous phase that flows within the porous network and liquid water contained in the pores through condensation/evaporation and sorption/desorption phenomena (eg, Barrett et al., 1951).

The in-pore condensation/evaporation processes are governed by equilibrium between the liquid water and the water vapor. Assuming that the air pressure within the porous network remains equal to the atmospheric pressure, it is expressed through Kelvin's Law (eg, Soudani et al., 2015):

$$s = p_{atm} - p_w = -\frac{\rho_w RT}{M_w} \ln R_H \qquad [10.2]$$

where p_{atm} and p_w are the atmospheric and liquid water pressures, respectively; ρ_w the density of liquid water; M_w its molar mass; R the perfect gas constant; T the temperature; and R_H the relative humidity expressed in decimal (and not in percentage), defined as the ratio between the vapor partial pressure and the saturated vapor pressure. This last value is an increasing function of temperature whose expression can be found in (Soudani et al., 2015). The resulting pressure difference at equilibrium between liquid water and air is noted s and is commonly called the matric suction. Since this latter is always positive, Eq. [10.2] underlines that the liquid water must be under strongly negative pressures to be stable (ie, to not evaporate) when the hygrometry decreases. This depression can exceed several tens of MegaPascal when the ambient hygrometry reaches values approaching 90%. The physical meaning of such negative liquid pressures has been investigated by many authors, and it is now well known that it is due to the preponderant affinity between the liquid water and the pore wall of the material. The resulting adhesion forces which tend to maximize the surface contact between both are in competition with the attraction forces between the water molecules (cohesion forces). These two contributions lead to a tensioning of the water molecules and to a curved interface between the in-pore liquid water and air. The link between both is given by the Young–Laplace Law (for a spherical interface):

$$s = p_{atm} - p_w = \frac{2 \cdot \gamma}{r} \qquad [10.3]$$

where γ is the surface tension, and r the radius of the spherical interface.

As it is illustrated in Fig. 10.5, when earthen material is drying, the radius of the liquid–air interface is close to the radius of the pore where the interface is located. As drying progresses, the interface radius becomes smaller and smaller. This property is well depicted by the combination of relations [10.2] and [10.3]. If no

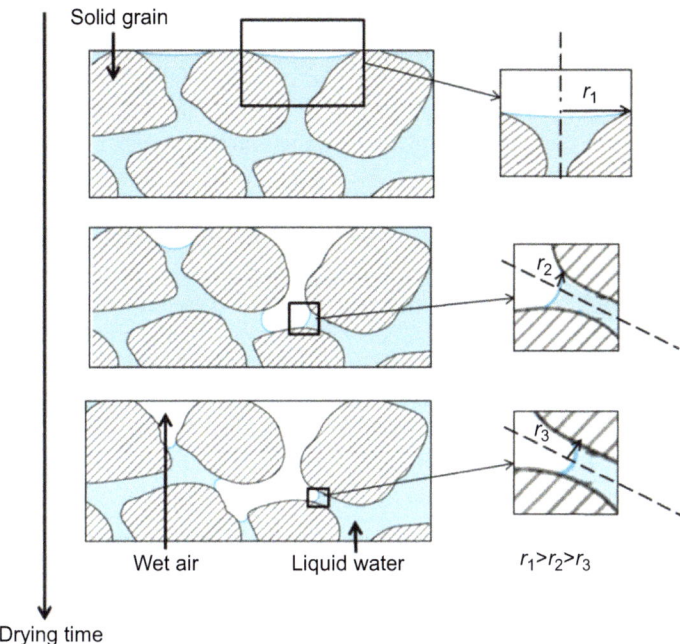

Figure 10.5 Schematic representation of the evolution of the interface during a drying process. Adapted from Delage, P., Cui, Y-J., 2000. L'eau dans les sols non saturés (Water in unsaturated soils). Techniques de l'ingénieur.

sufficiently small pores are present to ensure the interface curvature, all the water evaporates and the porous medium becomes totally dried. In consequence, the progressive variation of water content with hygrometry reported in Fig. 10.4 can be explained by the broad pore size distribution of earthen materials. The sole evaporation/condensation processes are however not sufficient to explain the quite large quantity of water that remains within the material at very low hygrometry. To do so, one must consider the water sorption/desorption processes which occur in interlamellar space of the reactive clay minerals (termed as interlamellar absorbed water), in the smallest pores of the fibers (if any, called intrafiber adsorbed water), and within the first few water layers of the pores walls (called external absorbed water). The adsorbed water molecules are more viscous and have physical properties that deviate significantly from those of free water (Pusch and Yong, 2006). Similarly to condensed water, the amount of adsorbed water tends to increase with hygrometry (eg, Hagymassy et al., 1968). However, the relation between both is not trivial and strongly depends on the nature of the material (mainly, but not entirely, on the specific surface area and on the electrical charging present at the surface of the porous network).

In light of the above, the analysis of the sorption/desorption curves allows three main saturation domains to be defined (eg, Denoth, 1982). When the ambient hygrometry is lower than a value close to 95%, a progressive increase of water content with hygrometry is observed. Within this saturation domain, commonly called the "hygroscopic domain," the liquid phase is mainly adsorbed at the surface of the solid grain and condensed within the smallest pores (radii smaller than 20 nm according to Eqs. [10.2] and [10.3]). As a consequence, most of the porous network remains filled by the gas phase, and the liquid water migration within the porous network is limited.

When the ambient hygrometry goes beyond this threshold value, a significant increase of water content is observed. The material enters the "capillary domain," and a large amount of capillary condensation of water arises within the porous network. Both liquid water and wet air are continuous, and they can freely flow within the porous network. Within this domain, the relation between the hygrometry and the water content is too steep to be accurately measured. The water content is then linked to the matric suction through the retention curve. Closed-form equations exist to reproduce with accuracy this curve for different materials (eg, Brooks and Corey, 1964; Van Genuchten, 1980).

When an earth wall remains within hygroscopic or capillary domains, even small water content gradients lead to quite significant matric suction gradients. This effect tends therefore to homogenize the water content over the height of the wall. When the material becomes close to saturation, the matric suction is no longer sufficient to counter the effect of gravity. In this case, as observed by (Chabriac et al., 2014) in walls of dimensions $1.5 \times 1 \times 0.5$ (m^3), the gravity forces must be taken into account and the resulting flow leads to higher water content in the lower portions of the wall. This domain, which is commonly called the "gravity domain," is typically reached just after wall production or when it is submitted to capillary rises and is very common in the case of vernacular architecture.

It follows from the previous that during their lifetime, earthen buildings materials are subjected to important variations of their water content. Although these variations would be strongly enhanced by pathologies like capillary rises, they remain substantial even in normal operating conditions due to daily and seasonal modifications of both indoor and outdoor ambient air. The next step is then to study the impact of these variations on thermal and mechanical behaviors of earthen materials.

10.2.2   Hygrothermal behavior

It is generally acknowledged that interior temperature control and air quality is significantly enhanced in earthen buildings (eg, Liuzzi et al., 2013). To illustrate this point and to understand its origin, the hygrothermal measurements reported in Fig. 10.6 will be used. They were collected in winter, between January 21, 2013, at 2 pm and January 23, 2013, at 2 pm, on the rammed earth house described previously. The house was vacant and unheated when this data was collected.

The comparison between indoor and outdoor measurements underlines that both relative humidity and temperature variations within the house are significantly reduced, despite the house being vacant and unheated. Moreover, a temperature time-shift of about 3 h, noted δ_1 in Fig. 10.6(a), is observed, and the indoor average

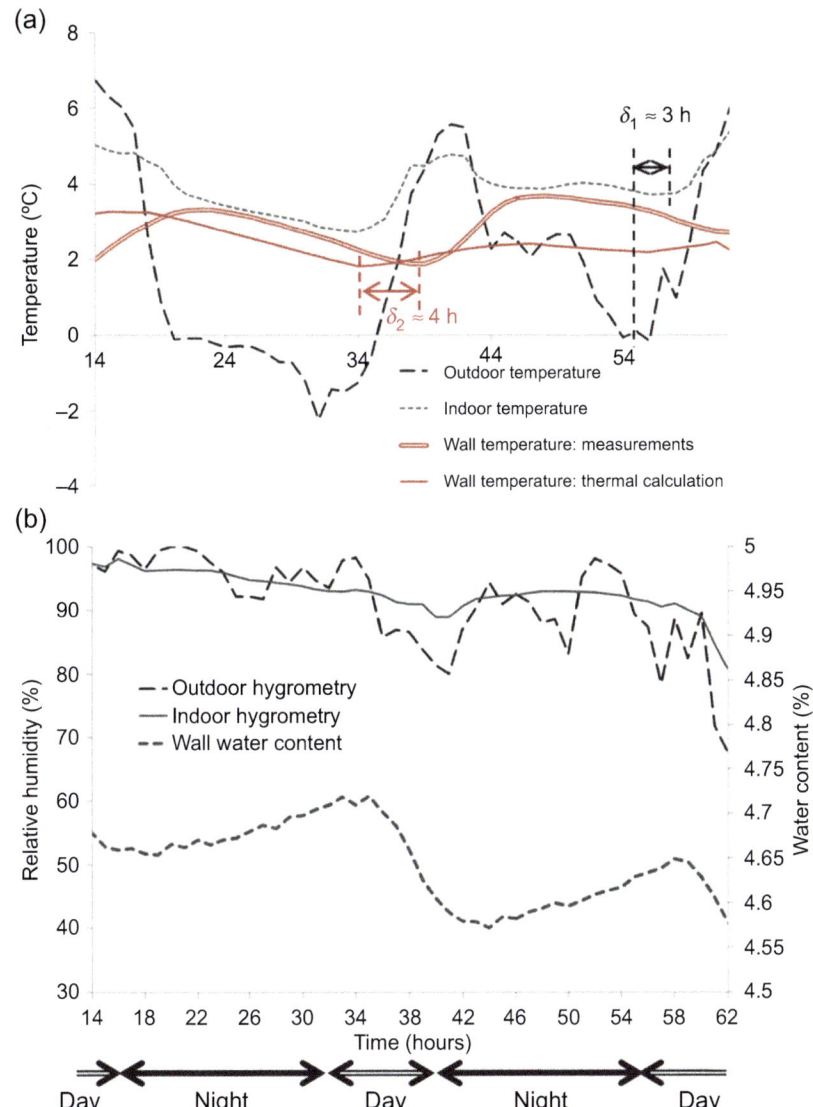

Figure 10.6 Evidences of thermal buffering effect of rammed earth walls. $t = 14$ correspond to January 21, 2013, at 2:00 pm.

temperature is 2.1°C higher than the outdoor temperature. These results demonstrate the passive control capabilities of this building, but more information is needed to conclude on the beneficial role of the earth walls.

Let us then consider the processes which occur within the instrumented rammed earth wall. To ease the analysis, the temperature measured at the middle of the wall is compared to a predicted value through a purely one-dimensional thermal calculation:

$$\rho C \frac{\partial T}{\partial t} = \lambda \frac{\partial^2 T}{\partial x^2}$$
[10.4]

where T is the absolute temperature (in Kelvin), ρC is the average heat capacity, and λ the thermal conductivity. Eq. [10.4] is solved with COMSOL Multiphysics® using the PDE module. The simulations are made for a 1D geometry of length $L = 0.5$ m (x direction) representing a lateral cross section of the tested wall. ρC and λ are taken, respectively, equal to 1100 kJ/m³/K and 1.01 W/m/K, which are the measured values for the constituent earth at 20°C and $w = 4.65\%$ (by dry weight). Boundary conditions of the simulations are set according to indoor and outdoor measurements and are calculated assuming a heat transfer coefficient of 15 W/m²/K, which is a common value for a condition of natural convection (Doumenc, 2009). Finally, the initial condition is extracted from the result of a pre-simulation, to take into account the initial temperature gradient within the wall.

Significant differences can be observed between the calculated and the measured temperatures. Calculation logically leads to a slight variation of temperature at the middle of the wall whose average value is equal to 2.4°C, between the indoor and outdoor average temperatures. This temperature smoothing is due to the density and the thickness of the wall which lead to a strong thermal inertia. In comparison, the temperature measured during the night periods tends to be higher than those calculated, while the opposite occurs during the day periods. A time-shift (noted δ_2 in Fig. 10.6(a)) of about 4 h can also be observed between the measured and the calculated values.

To fully understand the thermal behavior of rammed earth walls, we need to consider the role of water. As mentioned earlier, in-pore water will be subjected to evaporation and condensation processes. The resulting modification in water content will change the material thermal conductivity and thermal capacity (Hall and Allinson, 2009a,b). However, as shown in Fig. 10.6(b), the water content variation is quite limited in this case study, and its impact on the thermal properties of the material would not be sufficient to explain the differences between the calculated and the measured temperatures.

In fact, these phase changes also have implications for the energy balance of the wall. Indeed, to evaporate, a heat quantity must be provided to the liquid water (endothermic reaction). This heat consumption at constant temperature is called the latent heat of evaporation. On the other hand, an amount of heat is released during the condensation of water (exothermic reaction). And, as seen in Fig. 10.6, during the day, temperatures rise and external hygrometry decreases. As a consequence, the equilibrium between the wall and the surrounding air leads to the evaporation of capillary water, which is consistent with the global reduction of the measured water content within the wall. This evaporation induces absorption of heat, and thus limits the increase of the wall temperature. At the building scale, this effect should help to limit the indoor temperature increase caused by heat transfer between the outside and inside of the house through the earthen wall. Conversely, when the temperature decreases during the night,

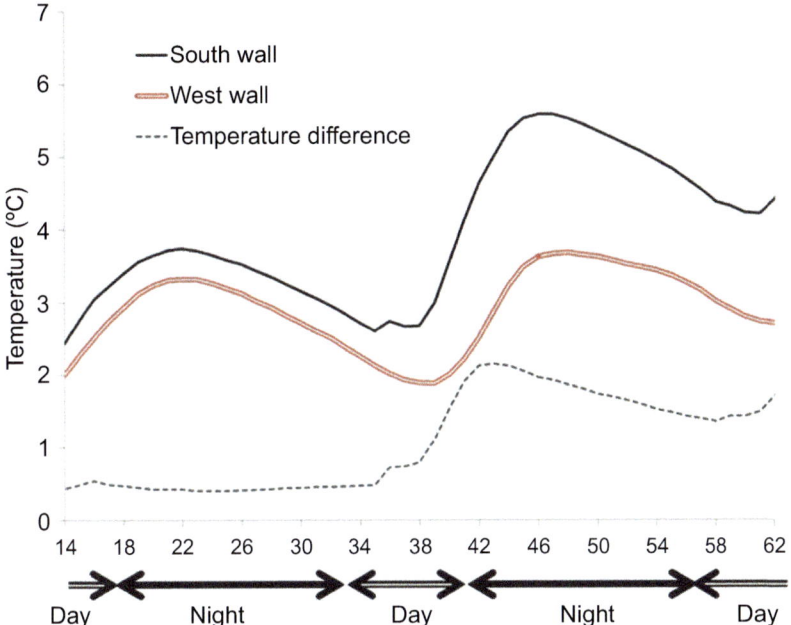

Figure 10.7 Difference between the temperatures measured at the middle of the south wall and at the middle of the west wall. The *x*-axis of the graph represents the time in hours. $t = 14$ correspond to January 21, 2013, at 2:00 pm.

condensation of water vapor occurs within the porous network, which leads to a global increase of the water content and to a release of heat. Thus, again the decrease of wall and indoor temperatures will be limited.

These phase change phenomena can therefore explain the time-shift between the measured and the calculated values as well as their difference in amplitude. But, they cannot explain why the average temperature measured within the wall is higher than the one estimated from the thermal calculation. This last point can be explained by the absorption of solar radiations by the wall. An illustration of this process is seen in Fig. 10.7, which shows a significant difference between the temperature measured within the south and the west walls (up to 2 degree). It is notable that this difference significantly increases during the day periods but slowly decreases during the night. Consequently, even in winter where the sunlight periods are limited, their effect on the wall warming is felt for a quite long duration (more than one night in the example reported in Fig. 10.7). In conclusion, the thermal behavior of earthen buildings is quite complex and is summarized by Fig. 10.8.

10.2.3 Mechanical behavior

The mechanical performance of earthen materials has been widely studied (eg, Bui and Morel, 2009; Kouakou and Morel, 2009; Miccoli et al., 2014; Piattoni et al., 2011;

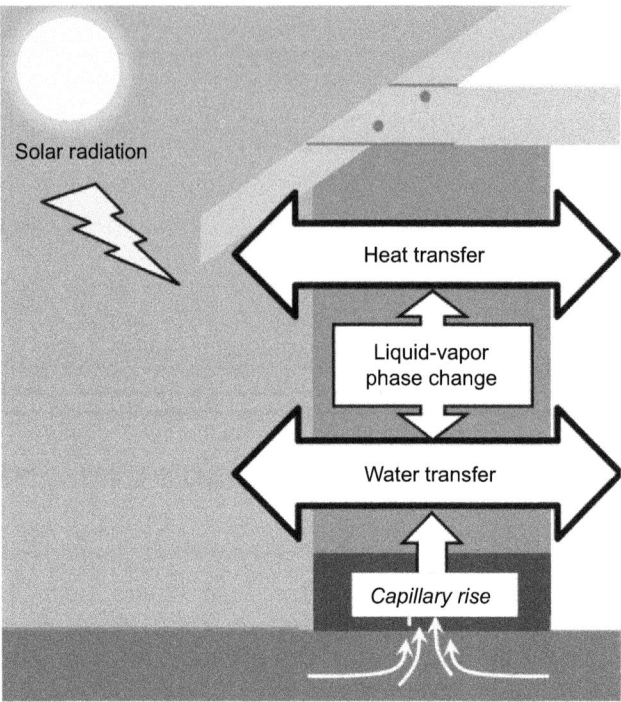

Figure 10.8 Schematic representation of phenomena impacting the thermal behavior of an earthen building.

Silveira et al., 2012; Champiré et al., 2014). Most of these studies are based on laboratory tests and lead to a complex rheological behavior which combines damage and hardening elasto-plasticity. In addition, the decrease in resistance and stiffness with water content, which is well known in soil mechanics has been recently demonstrated for earth materials (Bui et al., 2009a; Jaquin et al., 2009). As an example, consider the study of (Mollion, 2009), whose results are reported in Table 10.1. In this study, 16-cm-diameter × 32-cm-long cylindrical earth samples were tested in compression at several water contents, from their manufactured water content of 10% to the oven-dried state. Five samples at each water content were tested and all samples were compacted at normal proctor density.

These results clearly illustrate the strong effect of water, with a decrease of 81% for compressive strength and of 61% for Young's modulus when the water content increases from 0 to 10%. This illustrates why phenomenon like capillary rises should be taken into account since an abnormal increase of water content could induce strength values too low to ensure the wall integrity.

Of particular note is the fall in compressive strength and Young's modulus between the samples at 3% and 6% water content, equal to 35% and 20%, respectively. According to the sorption—desorption curves reported in Fig. 10.4, these two water content values correspond to ambient relative humidity of approximately 75% and 95%,

Table 10.1 Evolution of the compressive strength in function of the sample water content for compacted earth samples

Water content (%)	10%	9%	8%	7%	6%	5%	4%	3%	dry
Compressive strength (MPa)	0.38	0.59	0.79	1.2	1.3	1.7	1.8	1.9	2.0
Young's modulus (GPa)	0.82	0.97	1.1	1.5	1.6	2.0	2.0	2.1	2.1

Data from Mollion, V., 2009. Etude du comportement mecanique du pisé: étude de l'influence de l'eau sur les propirétés mécanique du matériau (Study of the mechanical behavior of rammed earth: a study of the influence of water on the mechanical properties of the material). ENTPE, Vaulx-en-Velin.

respectively, values commonly encountered by most earthen structures during their lifetime. Thus, the mechanical characteristics of earth walls additionally depend on hygrometry of the surrounding indoor and outdoor environments. This information should always be considered for the design of an earth building. However, provided the material stays in the hygroscopic saturation domain ($w < 6\%$ for this earth material), its strength remains sufficiently high (1.3 MPa for this earth) to ensure good mechanical integrity considering 0.5-m-thick walls used.

A classic way to explain the effect of water content on the mechanical behavior is schematized in Fig. 10.9. This model represents a granular soil where the remaining liquid water surrounds the grain contact point. Since the liquid water pressure is lower than the air (cf. Eq. [10.3]), the overall equilibrium of this system leads to an attraction force between the adjacent grains. The first consequence of this attraction is to increase the frictional force between the grains. As a consequence, the shear stress required to initiate a shift between the grains will be higher, which results in an increase of the uniaxial compressive strength. The second effect is a stiffening of the material which leads to an increase of its mechanical moduli as it is reported for unsaturated clays by (Barden et al., 1969).

However, this schematic representation is not sufficient to explain the entire complexity of earthen materials' behavior. Indeed, in a recent study (Champiré et al., 2014), unconfined compressive tests were conducted on two earth materials (named STR and CRA) sampled from existing rammed earth houses located in the same geographic region (southeast France). Both samples have almost the same particle size distribution (15% of clays, 20% of silts, 65% of sands, almost no gravels) and the same dry density of 1.97, but they exhibited significantly different clay reactivity (methylene blue values and plasticity indexes are respectively equal to

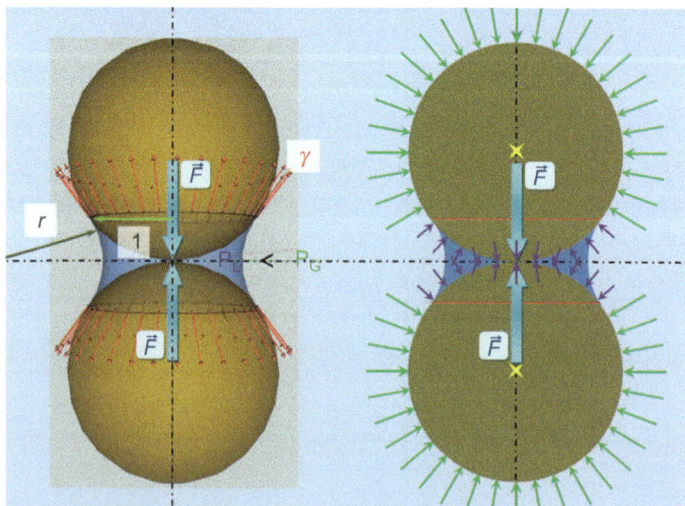

Figure 10.9 Schematic representation of the capillary pressure effect between two spherical grains with a water meniscus at their contact and induced forces.

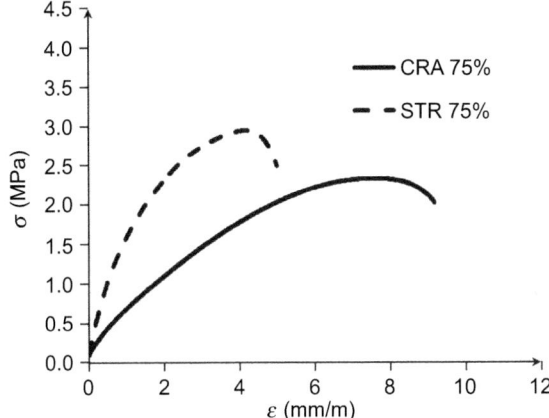

Figure 10.10 Comparison of the stress—strain behavior of two compacted earth with different clay activities (methylene blue value of CRA = 1.95, and methylene blue value of SRT = 1.16).

7% and 1.16 for STR and 14% and 1.95 for CRA). Results are reported in Fig. 10.10 and show that, even if the samples are cured at the same relative humidity of 75%, the sample with the higher clay activity (CRA) appears to be less rigid and resistant than the other one.

This study suggests that the mechanical behavior (both strength and deformability) is strongly influenced by the nature of clay. Moreover, because they are directly in contact with the outdoor conditions, structural earth walls are subject to natural cycles of drying—wetting. These cycles may induce significant and irreversible changes in hydromechanical behavior by changing the porous structure and the solid matrix (reorientation and/or reorganization of grains and sheets, etc.) (Basma et al., 1996; Nowamooz and Masrouri, 2008). This gradually leads to changes of the apparent soil properties (shear strength, etc.) but also results in erosion of the surface layer. The importance of these cycles has been studied experimentally by (Bui et al., 2009b) on different types of unprotected rammed earth walls exposed over 20 years to external climatic conditions in a wet continental environment (Grenoble, France). This work eventually demonstrated an extrapolated lifetime greater than 60 years (mean erosion of 6.4 mm, or 1.6% of the thickness of the 400-mm wall after 20 years of exposure) without any mineral stabilization of the earth.

Finally, a significant number of earth constructions are located in areas of moderate seismic risk. Some of these constructions have resisted historical earthquakes, but to date, little research has been conducted in this area. Nevertheless, the seminal works of (Minke, 2001) and (Hamilton et al., 2006) on the dynamic behavior of unstabilized rammed earth buildings recommend lime stabilization of earth in seismic regions. However, further experimental and theoretical studies are needed before definitive conclusion can be drawn. More recently, the work conducted by (Bui et al., 2011) is a first step toward the confrontation between in situ measurements and empirical formulas for determining damping factors and Eigen frequencies of earthen structures. In this study, a simplified shear beam model demonstrated its effectiveness for rammed earth structures.

10.2.4 Impact on the building design

The observations of the important role of water and solar radiation in earthen building behavior can provide some basis for responding to the recurring question of their thermal insulation. At first, it is quite clear that, even if it provides a significant increase of the thermal resistance of the wall, the addition of an impermeable insulation layer would stop the hygrothermal coupling. The apparent thermal inertia of the wall will thus be reduced as well as its moisture buffering properties. Second, insulation from the outside will prevent the wall from storing the energy obtained from solar radiations. The induced energy shortfall may be significant when south-facing walls are considered.

Beyond limiting the passive control capabilities of earthen walls, adding an additional nonpermeable layer may lead to considerable damage. Indeed, the exchange of vapor with the environment allows a regulation of the water content of the wall. If this process is efficient, the wall will stay within the hygroscopic saturation domain (Fig. 10.4), and thus, as mentioned in the previous section, its strength should remain sufficient to insure the integrity of the construction. If the process of vapor migration is stopped by waterproof insulation layers, two kinds of pathology will occur. The first will be a condensation of water at the interface between the wall and the colder layer (usually the outdoor one). The prolonged stagnation of water at this interface will degrade the connection of the two materials and eventually will lead to the loss of parts of the external coating of the wall.

The second concern, which is even more pronounced in the case of capillary rises, is a global increase of the water content of the wall. Indeed, whatever the basement material and the coating layers, it is nearly impossible to completely prevent water infiltration, especially in the case of vernacular architecture. For a properly constructed building, the kinetics of these infiltrations should be significantly lower than that of natural evaporation in the earthen wall. In this case the increase in wall water content will remain limited. In contrast, the absence of hydric exchange with the environment will prevent water discharged by evaporation. The water content will then increase until the material reaches the water content of its gravimetric saturation domain (Fig. 10.4). And, as is observed in the vast majority of unsaturated soils, when earthen material reaches this greater of water content, it collapses (see, eg, Lai et al., accepted in 2015).

An illustration of this phenomenon is seen in Fig. 10.11. In this construction, the basement was made of slag clinker. The hydric exchange with the outdoor environment was prevented by a cement-based plaster, while the indoor face was insulated by a glass-wood and plasterboard layer. Within hours of the collapse, blocks of the rammed earth wall were sampled. Their weights indicated water contents between 11% and 14%. In comparison, according to (Chabriac, 2014), the water content of the earth from this house at equilibrium in ambient conditions of 95% of relative humidity is about 4%. The wall was thus clearly out of the hygroscopic domain, which should have not occurred if the hydric exchanges with the outside had been prevented.

Figure 10.11 Picture of the collapse of a rammed earth construction in the city of Lyon (France), left: basement after removing the collapse rammed earth, right: the rammed earth wall after its collapse.

It must be emphasized, that totally sealing earthen walls must be prohibited; the use of insulation systems that allow hydric exchange is required. However, more research must be conducted on such systems to draw definitive conclusions.

10.2.5 Earth stabilization: the challenges and limitations

The strong affinity of earthen materials with water and its moderate strength may be prejudicial in some seismic and/or wet climate areas. A way to reduce these drawbacks is using mineral stabilization, which consists in adding a binder, cement, or lime, to the "water-earth" mixture (Cianco and Gibbings, 2012; Venkatarama Reddy and Prasanna Kumar, 2010; Maskell et al., 2014a). This process can also be justified for industrial production of earthen building materials (Cianco and Gibbings, 2012) or for maintenance purposes (Millogo and Morel, 2012).

Stabilization rigidifies the material, increasing both its compressive and tensile strength and limiting the impact of water content on material behavior. It also induces a significant change in its global behavior. Indeed, the elasto-plastic behavior of earth, as depicted in the previous section, tends to change to a more brittle behavior, similar to that of concrete and stone. The consequence of this loss of plasticity on the global behavior of a building is quite difficult to assess, and it should be calculated on a case-by-case basis. It is important to note that, while cement stabilization is recognized to be quite efficient when the clayey proportion is mainly composed by kaolinite, some interaction issues can be seen when the amount of montmorillonite increases (Temimi et al., 1998). Accordingly, it is quite difficult to draw general conclusions regarding mechanical properties resulting from stabilization and a case-by-case basis assessment is recommended.

In addition, since stabilization modifies the interaction between earthen material and water, it also modifies its hygrothermal properties. It is well known (Olivier and Mesbah, 1986) that, for materials having the same soils and manufactured at their optimum water content, stabilization increases the porosity of the material, and in particular the volume fraction of nano- and micropores. As a result, for a given hygrometry, the water content, and thus the thermal conductivity, of the cement-stabilized samples are higher than comparable unstabilized samples. On the other hand, studies on the vapor transfer properties of stabilized earth tend to observe a significant reduction of their permeability (Hall and Djerbib, 2006). The kinetics of the water exchange processes is therefore significantly reduced. The two direct consequences are a reduction of the moisture buffering ability of the material and of the rate of heat supply/consumption due to the condensation/evaporation processes. However, from our knowledge, no integrated study yet exists to evaluate the real impact of stabilization on the thermal performance and on the indoor air quality of an earthen construction.

Finally, the environmental side-effects of stabilized earth are sometimes controversial (Maskell et al., 2014b). Indeed, using cement or lime stabilization increases the embodied energy of the material. Moreover, stabilized earth cannot return to its initial state (as a soil) without additional energy input by simply wetting. This limitation is particularly true in the case of lime stabilization, which leads to an irreversible alteration of the clay mineralogy of the material (Venkatarama Reddy and Latha, 2014). Stabilized earth can be recycled (like most of construction materials), but with no additional benefit in terms of sustainability. In fact, unstabilized earth is one of the few construction materials (with dry stone) which may be reused with the same embodied energy to build again. At that level, an alternative and ecological option, which may be a promising research area in the near future, is organic stabilization with, for example, molasses, cow dung, or sawdust (Vilane, 2010).

The development of mechanically, hygrothermically, and ecologically efficient stabilization methods (either organic and inorganic) is an important issue for the development of earthen construction. However, the heritage of unstabilized earth buildings remains particularly important (Jaquin, 2008), and the existence of these constructions, which have remained intact for decades, is by itself evidence of their sustainability. It is therefore important to continue research efforts on both stabilized and unstabilized materials, and to consider separately these materials in guidelines and standards, since their behaviors are significantly different.

10.3   Assessing the performance of earthen material for construction

10.3.1   Hygrothermal performance

Several researchers have focused on the assessment of the hydrothermal transfers within earthen walls and their modeling (Allinson and Hall, 2010). However, this task is not simple since the load-bearing walls act as barriers between drastically different environments (ie, indoor, outdoor, foundation, and roof). They are subjected to steep thermal and hydric gradients leading to migration of liquid water and water vapor at different temperatures. Proper modeling of the behavior of the interstitial fluid is essential to identify the key parameter which should be measured to quantify the hygrothermal properties of the material. If all these processes and couplings are

theoretically well known by the scientific community, their integration into a complete
and consistent model and the implementation of the latter into a standard of hygrothermal calculation remains a major scientific and technical challenge. Moreover the determination of all material parameters is very difficult as the experimental procedures are
not well established.

To reach this goal, several phenomenological approaches have been developed giving rise to commercial software like WUFI, which can provide reliable results for a
wide range of materials and climatic loads (Fraunhofer, n.d., accessed 08.02.2013),
but which can exhibit some difficulties reproducing with accuracy the hygrothermal
behavior of unconventional materials like earth when they are submitted to significant
hygrometry and temperature variations. Recently, a coupled model, based on heat and
mass balances inside the earthen walls, was proposed (Soudani et al., 2015). This
model considers separately the kinematics of each phase (liquid water, vapor, dry
air, and solid matrix), in interaction with each other. It also takes into account the
impact of pore water confinement on the liquid-to-vapor phase change, and on the
resulting latent heat released or absorbed. The model has been successfully compared
to experimental results on instrumented full-scale rammed earth walls subjected to
natural and forced conditions. The model leads to the following set of equations, based
on the liquid water and vapor mass conservations and on the heat balance of the
system, where the unknowns are the relative humidity within the porous network of
the material, R_H (in decimal), and its absolute temperature, T (in Kelvin):

$$\left(\rho_d \frac{\partial w}{\partial R_H} + \phi_G \frac{M_w p_v^{sat}(T)}{R\,T}\right) \frac{\partial R_H}{\partial t} + \left(\phi_G \rho_v^\alpha + \rho_d \frac{\partial w}{\partial T}\right) \frac{\partial T}{\partial t}$$
$$= \nabla \cdot \left(\mathscr{K}^T \nabla T + \mathscr{K}^R \nabla R_H\right) \tag{10.5}$$

$$\left(\rho C_p - h_v\,\rho_d \frac{\partial w}{\partial T}\right) \frac{\partial T}{\partial t} - \left(h_v\,\rho_d \frac{\partial w}{\partial R_H}\right) \frac{\partial R_H}{\partial t}$$
$$= \nabla \cdot (\lambda \nabla T) + h_v\,\nabla \cdot \left(\mathscr{L}^T \nabla T + \mathscr{L}^R \nabla R_H\right) \tag{10.6}$$

where

$$\rho_v^\alpha = R_H \frac{p_v^{sat} M_w}{RT} \left(\frac{1}{p_v^{sat}} \frac{dp_v^{sat}}{dT} - \frac{1}{T}\right)$$

$$\mathscr{K}^T = \frac{R_H \ln R_H}{T} D_L + \delta_p R_H \frac{dp_v^{sat}}{dT}; \quad \mathscr{K}^R = D_L + \delta_p p_v^{sat}$$

$$\mathscr{L}^T = -\frac{R_H \ln R_H}{T} D_L; \quad \mathscr{L}^R = -D_L$$

with p_v^{sat} the saturated vapor pressure, which is a function of temperature; ρ_L the water
density, which is assumed to be constant; M_w the molar mass of water, and R the

perfect gas constant. In addition to these well-known and tabulated physical parameters, the solution of this system of equation involves six material parameters that must be experimentally determined. The terms $\partial w/\partial \varphi$ and $\partial w/\partial T$, which correspond to the variation of water content with equilibrium relative humidity at constant temperature and variation of water content with temperature at constant hygrometry, can be estimated from sorption isotherms at several temperatures. The variation of $\partial w/\partial \varphi$ with temperature and $\partial w/\partial T$ can often be neglected. In this case, only one sorption isotherm test remains necessary. D_L and δ_p are the hydraulic transport characteristic defined by (Künzel, 1995). They can be determined from the following relations:

$$\delta_p = \frac{\delta_a}{\mu}; \quad D_L = p_V^{sat}\delta_a\left(\frac{1}{\mu^*} - \frac{1}{\mu}\right)$$

where δ_a is the vapor permeability of stagnant air (equal to 2.0×10^{-10} kg/m.s.Pa at 20°C); μ is the vapor resistance factor determined with the dry cup method of NF EN ISO 12572:2001; and μ^* the fictitious vapor resistance factor determined using the same protocol but at higher humidity. The final two parameters, the thermal conductivity, λ, and the heat capacity, ρC_p, are both functions of water content and can be measured by several routine tests, such as the hot wire (Fig. 10.12(a)), hot plate, and kept warm box.

All these parameters are, however, still not sufficient. Indeed, to fix the boundary conditions of the system, the solar absorption must be known in addition to the convective heat transfer and the water vapor transfer coefficients at the internal and external surfaces of the wall. The solar absorption coefficient can be easily measured with a spectrometer using the apparatus seen in Fig. 10.12(b). However, the exact determination of vapor and heat transfer coefficients can be particularly difficult since they strongly depend on the wind direction and velocity. One simple option is to consider these values equal to their average conventional values of 8 W/m^2.K for the heat

Figure 10.12 (a) Measurement of the thermal conductivity of compacted earth block using the hot wire technique. (b) Estimation of the solar radiation absorption factor from the measurement of the reflection factor on a compacted earth block.
After Chabriac, P., 2014. Mesure du comportement hygrothermique du pisé (Measurment of the hygrothermal behavior of rammed earth). (Ph.D. thesis, s.l.: Ecole Nationale des Travaux Publics de l'Etat).

transfer at the internal surface; 15 W/m^2.K for the heat transfer at the external surface; and of 25×10^{-9} kg/m^2.s.Pa for the vapor transfer coefficient (Künzel, 1995).

In conclusion, at least four characterization tests at several temperatures and/or hygrometries are required to properly characterize the heat and mass transport within the material (sorption test, dry cup, DIN 52 615, and thermal conductivity and capacity), and one additional test is necessary to fix the boundary conditions. It is obvious that such an experimental program is time-consuming and expensive. Imposing it may be detrimental for the development of earthen materials. Consequently, an important challenge for the scientific community is to identify the physical processes that drive these parameters at the microstructural scale. In this way, their value at a given temperature and hygrometry could be estimated from routine tests of microstructure and mineral characterization (eg, clay activity, particle size distribution, density, and porosity) though multiscale calculations.

10.3.2 Mechanical stability and durability

The mechanical performance of earthen material is generally evaluated through the compressive strength measured from unconfined compression tests (Morel et al., 2007) and through tensile strength estimated from splitting tests or three-point bending tests (Morel and Pkla, 2002; Morel et al., 2003). However, for earthen materials, the set-up of these routine tests is not trivial. Indeed, as already discussed in Section 10.2 of this chapter, the high variation of material mechanical characteristics with environmental conditions (temperature and hygrometry) leads to a lack of reproducibility. As it is sketched by the experimental study conducted by (Champiré et al., 2014), one way to overcome this problem is to conduct all tests at controlled temperature and hygrometry, and to interpret the results accounting for this hydromechanical coupling.

However, this recommendation is not sufficient. Indeed, as reviewed by (Aubert et al., 2013), a significant increase in compressive strength is commonly observed with a decreasing aspect ratio of the tested sample (ie, length versus radius for a cylindrical sample or height versus width for a prismatic sample). A conventional way to account for this is to modify the measured value by a correction factor depending on the aspect ratio (eg, Eurocode 6: Design of Masonry Structures). Although, the use of a single correction factor implies that the processes which lead to the failure of the material remain similar whatever the shape of the sample. This postulate can be assumed only if the modification of the compressive stress with the aspect ratio remains moderate. This is not the case for adobes of heritage buildings or in the cases of some modern extruded bricks (themselves inspired by the heritage bricks) having a very low aspect ratio (close to 0.3), for which several studies refer to compressive strengths greater than 10 MPa and up to 45 MPa (Aubert et al., 2013), far above conventional compressive strength values of earthen brick (less than 5 MPa). Another possible bias of strength measurement is the nature of the sample interface with the test machine platens which often varies by procedure (addition of a rubber or wood layer, for instance). This impact has been studied by many authors (Olivier, 1994; Olivier et al., 1997; Ciblac and Morel, 2014); however, there remains no consensus on this issue. According to (Cianco and Gibbings, 2012), adding an interface component can improve the repeatability of tests, but (Hakimi et al., 1996) highlights the occurrence of localized damage next to the interfaces.

In general, however, the impacts of both the shape factor and the loading interfaces become negligible when the samples have an aspect ratio equal to or greater than 2 (Cianco and Gibbings, 2012). Consequently, one possible method to overcome these problems when determining the compressive strength of thin plates is to artificially increase their aspect ratio by testing a stack of elements. However, in such a case, the value obtained would be related to a structure made of this assemblage rather than to an intrinsic property of the material. To conserve information on the material, three-point bending tests, as described by (Morel and Pkla, 2002), may be a good alternative. Indeed, the results obtained using this methodology were found to be very close to those obtained from a conventional compressive strength test on the same materials.

Another mechanical parameter which is currently measured to evaluate the mechanical performance of earthen materials is the Young's modulus. To do so, in addition to the measurement of the axial stress, the axial strain over the middle third of the sample is measured. This measurement can be made using extensometers, spectroscopic cameras, or may be approximated from the displacement of the test machine piston or crosshead provided the machine compliance is known. The use of the Young's modulus and the compressive/tensile strengths to characterize the mechanical behavior of earthen materials assumes a linear elastic behavior with a constant Poisson's ratio, which is clearly not the case. Indeed, the use of only these parameters is found to be insufficient to properly model the complex behavior of earthen walls (Olivier & Mesbah, 1995). This problem of representativeness of laboratory tests is not novel, and it has been highlighted by the first modern work on the behavior on earthen structures conducted by (Olivier, 1994) in the 1990s. It is therefore interesting to consider the real mechanical behavior of the material on site. As illustrated in Fig. 10.10, earthen materials exhibit a nonlinear and irreversible behavior closer to that of compacted clayey soils than to conventional building materials. More studies are necessary to quantify the impact of the activity of the clay on the overall macroscopic mechanical behavior of earthen materials. However, as emphasized by (Champiré et al., 2014), for a given water content, a linear behavior as a first approximation may give satisfactory results for stress levels below 60% of the compressive strength. This simple linear behavior, taking into account moduli variations with water content, can be sufficient to provide a first insight into the material behavior when the loading remains sufficiently far from failure.

10.4   Concluding remarks on earthen building sustainability

In the introduction of this chapter, the attractiveness of the use of raw earth as a building material has been underlined. Indeed, earth is an abundant resource, which does not need any industrial processing to be used. Consequently, the use of a local earth as a building material results in significant reduction of the amount of embodied energy of a house and reduced impact associated with transportation. In compensation, earthen construction requires a larger and more qualified workforce than common construction techniques. In other words, the fossil energy used for earthen building construction is partly replaced by human energy.

The sole consideration of the embodied energy is, however, not sufficient to conclude on the sustainability of a construction method. It is necessary to consider at least the energy consumed in operation (heating, ventilation, air-conditioning, lighting), maintenance and, possibly, rehabilitation or dismantling (demolition, treatment, or recycling of waste) the structure.

Concerning the operation costs, the second section of this chapter underlines that the microstructure of earth enables hydric exchanges between the environment and in-pore water molecules through condensation/evaporation and sorption/desorption phenomena. This ability is known to have moisture-buffering and temperature-controlling properties. This undoubtedly influences the evaluation of benefits on the energy consumed for heating, cooling, and air-conditioning. Moreover, as it is shown in Fig. 10.7, the thermal benefits of an earth wall will be significantly enhanced by its ability to absorb solar energy. Nevertheless, a limitation of earthen constructions is their incompatibility with traditional waterproof insulation systems. Indeed, by preventing the hydric exchange with the outside environment, the use of these materials will induce an increase of the wall water content which will ultimately lead to a risk of mechanical failure. To meet modern standards in terms of energy consumption, it is therefore necessary to consider alternatives solutions, which are based on optimized design. Such designs enable the greatest utilization of the hydrothermal and solar radiation absorption properties of the wall, and, when appropriate, may consider the use of permeable insulation systems such as hemp concrete. The additional costs resulting from the use of these bio-based insulating materials would be balanced by the reduction of surface area which needs to be insulated. Concerning the maintenance, rehabilitation, and dismantling, the environmental cost of earthen buildings is most advantageous since earth is a totally recyclable material. Indeed, provided it is not stabilized, it can be reused for construction or discharged into nature without any risk to the environment.

Finally, from a socioeconomic point of view, the costs associated with the construction material supply and transportation are particularly attractive in the case of earthen buildings. Conversely, an additional cost of about 10−20% is commonly observed for new earthen construction in industrial countries. This is mainly due to the proportionally high price of labor and a tax system which encourages the use of fossil energy rather than human power. This point may however be tempered since a case study conducted by a French architect (Rigassi, 2012) suggests that a part of this additional cost should be amortized by the energy savings associated with the regulation of the temperature and of hygrometry inside the building. Moreover as global warming progresses, tax policies of developed countries must change to mitigate such dangerous phenomenon. Regardless of the case, the use of earth as a building material remains highly attractive, for both environmental and economic reasons, in countries where the labor cost remains low.

Acknowledgments

The authors would like to thank their former PhD students Pierre Antoine Chabriac, Lucile Soudani, and Florian Champiré. They are also indebted to Stephane Cointet and Nicolas Meunier for the earth supply and their technical support. The present work has been supported by the French Research National Agency (ANR) through the "Villes et Bâtiments Durables" program (Project Primaterre no. ANR-12-VBDU-0001). This work is also cofunded by the French Environmental agency ADEME, Lafarge (TEZ10-44), the Région Rhône-Alpes (11024835 01 — ENT012v de la Direction du Climat, Environnement, Santé et Énergie, and a Fellowship CMIRA for JC Morel).

References

Allinson, D., Hall, M., 2010. Hygrothermal analysis of a stabilised rammed earth test building in the UK. Energy and Buildings 42, 845—852.

Aubert, J., Fabbri, A., Morel, J., Maillard, P., 2013. An earth block with a compressive strength higher than 45 MPa! Construction and Building Materials 47, 366—369.

Aubert, J., Marcom, A., Olivia, P., Segui, P., 2014. Chequered earth construction in south-western France. Journal of Cultural Heritage 16 (3).

Azeredo, G., Morel, J., Lamarque, C., 2008. Applicability of rheometers to characterizing earth mortar behavior. Part I : experimental device and validation. Materials and Structures 41, 1465—1472.

Barden, L., Madedor, A., Sides, G., 1969. Volume change characteristics of unsaturated clays. Journal of the Soil Mechanics and Foundations Division 95 (SM1), 35—52.

Barrett, E., Joyner, L., Halenda, P., 1951. The determination of pore volume and area distributions in porous substances. I. Computations from nitrogen isotherms. Journal of the American Chemical Society 73, 373—380.

Basma, A., Al-Homoud, A., Husein Malkawi, A., Al-Bashabsheh, M., 1996. Swelling-shrinkage behavior of natural expensive clays. Applied Clay Science 11, 211—227.

British Standards Institute, 1990. BS 1377-2: 1990 Soils for Civil Engineering Purposes — Part 2: Classification. BSI, London.

Brooks, R., Corey, A., 1964. Hydraulic Properties of Porous Media. s.l. Colorado State University.

Bui, Q., Hans, S., Morel, J., Do, A.-P., 2011. First exploratory study on dynamic characteristics of rammed earth buildings. Engineering Structures 33, 3690—3695.

Bui, Q.-B., Morel, J.-C., 2009. Assessing the anisotropy of rammed earth. Construction and Buildings Material 23, 3005—3011.

Bui, Q., Morel, J., Hans, S., Meunier, N., 2009a. Compression behaviour of non-industrial materials in civil engineering by three scale experiments: the case of rammed earth. Materials and Structures 42, 1101—1112.

Bui, Q., Morel, J., Venkatarama Reddy, B., Ghayad, W., 2009b. Durability of rammed earth walls exposed for 20 years to natural weathering. Building and Environment 44, 912—919.

Chabriac, P., 2014. Mesure du comportement hygrothermique du pisé (Measurment of the hygrothermal behavior of rammed earth) (Ph.D. thesis, s.l.: Ecole Nationale des Travaux Publics de l'Etat).

Chabriac, P., et al., 2014. A procedure to measure the in-situ hygrothermal behavior of earth Walls. Materials 7, 3002—3020.

Champiré, F., Fabbri, A., Morel, J.C., Wong, H., 2014. Experimental identification of key mechanical parameters of earth materials, 40th IAHS World Congress - Funchal. Portugal 16-19/12/2014.

Cianco, D., Gibbings, J., 2012. Experimental investigation on the compressive strength of cored and molded cement-stabilized rammed earth samples. Construction and building materials (28), 294–304.

Ciblac, T., Morel, J., 2014. Sustainable Masonry. John Wiley & Sons, Hoboken, NJ.

Delage, P., Cui, Y.-J., 2000. L'eau dans les sols non saturés (Water in unsaturated soils). Techniques de l'ingénieur.

Denoth, A., 1982. The pendular-funicular liquid transition and snow metamorphism. Journal of Glaciology 28, 357–364.

Dixit, M.K., Fernandez-Solis, J., Lavy, S., Culp, C.H., 2010. Identification of parameters for embodied energy measurement: a literature review. Energy and Buildings 42, 1238–1247.

Doumenc, F., 2009. Eléments de Thermodynamique et Thermique (Fundamentals of Thermodynamics and Heat). Université Pierre et Marie Curie, Paris.

Fraunhofer, n.d. IBP/WUFI. Available at: http://www.wufi.de. (accessed 08.02.13.).

Garrison, J., Ruffner, E., 1983. Adobe: pratical and technical aspects of adobe conservations. The Journal of Arizona History 448–501.

Habert, G., Castillo, E., Vincens, E., Morel, J., 2012. Power: a new paradigm for energy use in sustainable construction. Ecological Indicators 23, 109–115.

Hagymassy, J., Brunauer, S., Mikhail, R., 1968. Pore structure analysis by water vapour adsorption. Journal of Colloid and Interface Science 29, 381–390.

Hakimi, A., Yamani, N., Ouissi, H., 1996. Report: results of mechanical strength tests on samples of compressed earth. Material and Structure 29, 600–608.

Hall, M., Allinson, D., 2009a. Analysis of the hygrothermal functional properties of stabilised rammed earth materials. Building and Environment 44, 1935–1942.

Hall, M., Allinson, D., 2009b. Assessing the effects of soil grading on the moisture content-dependent thermal conductivity of stabilised rammed earth materials. Applied Thermal Engineering 29, 740–747.

Hall, M., Djerbib, Y., 2006. Moisture ingress in rammed earth: Part 3-The sorptivity and the surface inflow velocity. Construction and Building Materials 20, 384–395.

Hall, M.R., Lindsay, R., Krayenhoff, M., 2012. Modern Earth Buildings. Woodhead Publishing, Cambridge, UK.

Hamard, E., Morel, J., 2013. A procedure to assess suitability of plaster to protect vernacular earthen architecture. Journal of Cultural Heritage 14, 109–115.

Hamilton, H.-R., McBride, J., Grill, J., 2006. Cyclic testing of rammed earth walls containing post-tensioned reinforcement. Earthquake Spectra 22 (4), 937–959.

Harris, D., 1999. A quantitative approach to the assessment of the environmental impact of building materials. Building and Environment 34, 751–758.

Houben, H., Guillaud, H., 1994. Earth Construction: A Comprehensive Guide. IT Publications, London.

Jaquin, P., 2008. Analysis of Historic Rammed Earth Construction. s.l. (Ph.D. thesis of Durham University).

Jaquin, P., Augarde, C., Gallipoli, D., Toll, D., 2009. The strength of unstabilised rammed earth materials. Géotechnique 59, 487–490.

Kouakou, C., Morel, J., 2009. Strength and elasto-plastic properties of non-industrial building materials manufactured with clay as a natural binder. Applied Clay Science 44, 27–34.

Künzel, M., 1995. Simultaneous Heat and Moisture Transport in Building Components One- and Two-dimensional Calculation Using Simple Parameters. s.l.. Fraunhofer Institute of Building Physics (Ph.D. thesis).

Lai, B.T., Wong, H., Fabbri, A., Branque, D., accepted in 2015. A new constitutive model of unsaturated soils using bounding surface plasticity (BSP) and a non-associative flow rule. Innovative Infrastructure Solutions.

Lei, X., et al., 2014. A thermo-chemo-electro-mechanical framework of unsaturated expansive clays. Computer and Geotechnics 62, 175–192.

Liuzzi, S., Hall, M.R., Stefanizzi, P., Casey, S.P., 2013. Hygrothermal behaviour and relative humidity buffering of unfired and hydrated lime-stabilised clay composites in a Mediter-ranean climate. Buildings and Environment 61, 82–92.

Maskell, D., Heath, A., Walker, P., 2014a. Inorganic stabilisation methods for extruded earth masonry units. Construction and Building Materials 71, 602–609.

Maskell, D., Heath, A., Walker, P., 2014b. Comparing the environmental impact of stabilisers for unfired earth construction. Key Engineering Materials 132–143, 600.

Miccoli, L., Müller, U., Fontana, P., 2014. Mechanical behaviour of earthen materials: a com-parison between earth block masonry, rammed earth and cob. Construction and Buildings Materials 61, 327–339.

Millogo, Y., Morel, J., 2012. Microstructural characterization and mechanical properties of cement stabilised adobes. Materials and Structures 45, 1311–1318.

Minke, G., 2001. Construction Manual for Earthquake-resistant Houses Built of Earth. GATE-BASIN at GTZ GmbH, Eschborn.

Mollion, V., 2009. Etude du comportement mecanique du pisé: étude de l'influence de l'eau sur les propirétés mécanique du matériau (Study of the mechanical behavior of rammed earth: a study of the influence of water on the mechanical properties of the material). ENTPE, Vaulx-en-Velin.

Morel, J., Mesbah, A., Oggero, M., Walker, P., 2001. Building houses with local materials: means to drastically reduce the environmental impact of construction. Building and Environment 36, 1119–1126.

Morel, J., Pkla, A., 2002. A model to measure compressive strength of compressed earth blocks with the '3 points blending test'. Construction and Building Materials 16, 303–310.

Morel, J., P'Kla, A., Di Benedetto, H., 2003. Essai in situ sur blocs de terre comprimée. Interprétation en compression ou traction de l'essai de flexion en trois points (In situ test on compressed earth blocks. Interpretation of compression or tension from three-point flexure test). European Journal of Environmental and Civil Engineering 7, 221–237.

Morel, J., Pkla, A., Walker, P., 2007. Compressive strength testing of compressed earth blocks. Construction and Building Materials 21, 303–309.

Nowamooz, H., Masrouri, F., 2008. Hydromechanical behavior of an expansive bentonite/silt mixture in cyclic succion-controlled drying and wetting tests. Engineering Geology 101, 154–164.

Olivier, M., 1994. Thèse: Le matériau terre, compactage, comportement, application aux structures en blocs de terre (Earth material: compaction, behavior, application to Compacted Earth Blocks structures). s.n, Villeurbanne.

Olivier, M., Mesbah, A., 1986. Le matériau terre: essai de compactage statique pour la fabrication de briques de terres compressées (Earth material: static compaction test for the manufacture of compressed earth bricks). Bulelin del Liaison du Laboratoire des Ponts et Chaussées 37–43.

Olivier, M., Mesbah, A., September 6–8, 1995. Modèle de comportement pour sols compactés (Model of behavior for compacted soils). In: Proceedings of the First International Conference on Unsaturated Soils, France.

Olivier, M., Mesbah, A., El Gharbi, Z., Morel, J., 1997. Test method for strength tests on blocks of compressed earth. Materials and Structures 30 (203), 515−517.

Piattoni, Q., Quagliarni, E., Lenci, S., 2011. Experimental analysis and modelling of the mechanical behavior of earthen bricks. Construction and Building Materials 25, 2067−2075.

Pkla, A., Mesbah, A., Rigassi, V., Morel, J., 2003. Comparaison de méthodes d'essais de mesures des caractéristiques mécaniques des mortiers de terre (Comparison of test procedures to measure the mechanical characteristics of earth mortar). Materials and Structures 36, 108−117.

Pusch, R., Yong, R.N., 2006. Microstructure of Smectite and Engineering Performance. Taylor & Francis Group, London.

Rigassi, V., 2012. Le point de vue d'un maître d'oeuvre (The point of view of a project manager). Vaulx-en-velin: Construction en pisé : où se situent les enjeux en 2012 (Rammed earth construction: where are the challenges in 2012).

Silveira, D., et al., 2012. Mechanical properties of adobe bricks in ancient constructions. Construction and Building Materials 28, 36−44.

Soudani, L., Fabbri, A., Chabriac, P.A., Woloszyn, M., Grillet, A.C., Morel, J.C., 2015. On the relevance of neglecting the mass vapor variation for modelling the hygrothermal behavior of rammed earth. First International Conference on Rammed Earth Construction (ICREC 2015), University of Western Australia, Perth.

Temimi, M., Ben Amor, K., Camps, J., 1998. Making building products by extrusion and cement stabilization: limits of the process with montmorillonite clay. Applied Clay Science 13, 245−253.

Van Damme, H., 2002. Colloidal Chemo-mechanics of Cement Hydrates and Smectite Clays : Cohesion vs Swelling. In: Encyclopedia of Surface and Colloid Science. Marcel Dekker, New York, pp. 1087−1103.

Van Genuchten, M., 1980. A closed form equation for predicting the hydraulic conductivity of unsaturated soils. Soil Science Society of America Journal 44, 892−898.

Venkatarama Reddy, B., Latha, M., 2014. Retrieving clay minerals from stabilised soil compacts. Applied Clay Science 101, 362−368.

Venkatarama Reddy, B., Prasanna Kumar, P., 2010. Cement stabilised rammed earth. Part B: compressive strength and stress−strain characteristics. Materials and Structures 44, 695−707.

Vilane, B., 2010. Assessment of stabilisation of adobes by confined compression tests. Biosystems Engineering 106, 551−558.

Walker, P., Stace, T., 1997. Properties of some cement stabilized compressed earth blocks and mortars. Materials and Structures 30 (9), 545−551.

Ancient stone masonry constructions

L. Dipasquale
DIDA (Department of Architecture), University of Florence, Florence, Italy

L. Rovero
DIDA (Department of Architecture), University of Florence, Florence, Italy

F. Fratini
CNR ICVBC (National Council of Researches-Institute for Conservation and Promotion of Cultural Heritage), Sesto Fiorentino, Florence, Italy

11.1 Introduction

In much of the world, the largest part of the built heritage, both living and monumental, consists of masonry buildings. A feature of this heritage is the great variability of local architectural characteristics that resulted in a very wide range of structural diversity. In the various architectural styles, construction techniques have evolved with a strong link with the environment, cultural context, and available resources. Techniques evolve, respecting of the essential laws of mechanics, in almost Darwinian fashion. Building techniques must be preserved and enhanced, because they characterize the society and the identity of the very people, just as a spoken language (Fratini et al., 2011; Gamrani et al., 2011; Rovero and Tonietti, 2012; Sani et al., 2012).

The knowledge and study of building techniques form the basis of every operation of conservation, not only to preserve the building material heritage but also to preserve the body of knowledge, handed down orally or rediscovered, that underlie them. Such heritage of knowledge has produced sustainable building systems, in equilibrium with the environment, and therefore can suggest strategies in the modern context.

In addition to reintroducing masonry construction techniques for new construction thanks to the development of innovative systems for the extraction and cutting, the rich existing heritage of masonry buildings has shown great environmental adaptability, as demonstrated by the historical centers of many European cities that remain densely inhabited. Interesting challenges are therefore the reuse and rehabilitation, in respect of safety, linked to seismic risk and energy efficiency.

In this chapter, after an overview of historic vernacular applications of masonry buildings, the physical and mechanical properties of the different components of the masonries are discussed and the more common types of masonry walls are presented. Finally, some considerations are given to mechanical and environmental performance of masonry constructions and applications in modern architecture and innovative uses.

Nonconventional and Vernacular Construction Materials. http://dx.doi.org/10.1016/B978-0-08-100038-0.00011-1

11.2   Overview of ancient applications

The widespread use of stone in all historical periods is mainly due to its availability, strength, and durability, qualities that allow it to overshadow the difficulties of quarrying, processing, and transport. Since the dawn of human civilization, man has learned to exploit the rocky outcrops to make symbolic constructions and religious buildings. Tombs and religious constructions are made of large stones using interlocking systems without the use of mortars: megalithic structures, as *dolmen, menhir,* and temples, have been found in large areas of the Middle East but also in Asia and across Europe. Other wonderful examples of great skill in the use of stone in antiquity are pyramids (Egypt and Middle East) and step pyramids (in Central America), huge structures built of stone and sometimes mortar, some of which are among the world's largest constructions: the Great Pyramid of Giza (Egypt) is one of the largest in the world, with a base of over 52,600 m^2 in area; and the pyramid of Cholula, Puebla, Mexico, stands 55 m above the surrounding plain and in its final form measured 400×400 m.

The first man-made constructions are artificial caves carved into the rock. Splendid examples are found in many geographically and chronologically diverse civilizations: the great façades with superimposed orders of the Nabatean tombs of Petra in Jordan dug into the multicolored sandstone, the rock-hewn church in Lalibela, northern Ethiopia, the Byzantine churches of Cappadocia dug into volcanic formations, the rock architectures dug into limestone in many areas of the southern Mediterranean, including Spain (Andalusia) and Italy (the "sassi di Matera," the troglodyte settlements of Cava d'Ispica and Pantalica, inhabited in alternate phases until the second half of the nineteenth century, and the houses in the historical centers of Modica, Scicli, and Ragusa) (Sani et al., 2012).

Throughout history, rocks were used to build walls and define spaces, using both large blocks, more or less worked and built without mortar, and small blocks, processed and put in place with the aid of mortars.

In Ancient Rome, special masonry techniques, known as *opus*, have been developed thanks to the use of mortars with the addition of *pozzolana*. These walls were made of two external leafs with an internal core made of rock fragments bound by lime mixed with *pozzolana*. Significant among these are *opus quadratum* (stones cut into parallelepiped shape arranged in horizontal rows), *opus caementicium* (stone and mortar), *opus incertum* (stones randomly placed in the mortar, with the main face outward), *opus reticulatum* (stones with a square base arranged diagonally; this method is known to have excellent seismic performance), and *opus mixtum* (*opus reticulatum* with corner pieces in bricks).

Walls with large hewn stone in the Greek and Roman world were strengthened with lead clamps that were cast in their molten state into holes carved in the stones. This technique has been used in many temples of the Acropolis in Athens and in the Coliseum in Rome, as well as in many other important structures of the period.

In the medieval period, stone was the basic material for the construction of houses and castles. The masonry types are innumerable and adjusted according to the type of stone used and to the type of building constructed. A medieval stonemason would often carve a personal symbol onto their block to differentiate its work, providing at the same time a simple means of quality assurance and determining payment.

Starting from the Middle Ages, the use of brick coexists with the use of stone, becoming predominant in regions rich in clay soils. Stone remains the predominant building material for vernacular architecture in areas with abundant rocky outcrops but is also used for the construction of buildings of higher value, such as churches.

In the sixteenth century in Florence, and in the other cities that have been center stage for the Renaissance, the technique of stone working improves, creating façades decorated with precious ornamental stones (ie, marble, serpentinite) or with stones carved to create volume and play of shadows in perspective (ashlar stones, cornices, etc.).

In all ages, the realization of structural supports (pillars, columns, and arches) and the realization of coverings through vaulted systems led to the development of sophisticated techniques of stereotomy (descriptive geometry) for the design and cutting of the stone elements.

In the history of architecture, stone is also the main material used for cladding, flooring, decoration, and, together with tiles, roofing. Slabs used for roofs are documented in some villages of the northern Apennine (Italy), in the famous "trulli" of Apulia (southern Italy), and in many regions of France like Provence (Rovero and Tonietti, 2014; Dipasquale, 2012).

11.3   Stone masonry materials

Stone masonry is formed by stone blocks, of natural origin, that is, derived from the rocks, and by mortars of artificial origin, that is, man-made products. This section deals first with stone blocks, describing rock classification and a review of rock types used in architecture. Next, we address mortars, describing the raw materials, the processes of preparation, and the most representative types.

11.3.1   Natural forms of stone material: the rocks classification

Rocks constitute the outer part of the Earth crust and are natural associations of one or more types of minerals, being the final result of a rock-forming process. In their turn, the minerals are defined as natural materials, in the great majority of cases solid, inorganic, and homogeneous from chemical and physical points of view. When solid, they are in a so-called crystalline state. The crystalline state is characterized by a regular and ordered arrangement of atoms and/or molecular components. This ordered structure also determines the anisotropic properties of these natural bodies. Rocks are generally constituted by a limited number of mineral species. In some, only one mineral is present such as in limestone and marble, made up almost exclusively of calcite. In others, such as conglomerates, there are many types of minerals. In the following, a simplified classification is reported.

11.3.1.1   Magmatic rocks

Magmatic rocks are formed after the crystallization of a magma, a mixture of a liquid, gaseous, and crystalline phase at a temperature above $700°C$. The main components are: silica ($40-75\%$), alumina ($10-20\%$), iron oxides ($2-12\%$), calcium ($1-12\%$), magnesium (trace-12%), sodium ($1-8\%$), and potassium (trace-7%). In the gaseous phase, the main component is water and, to a lesser extent, carbon dioxide, hydrochloric acid, sulfur dioxide, etc.

Figure 11.1 Granite ashlars in a building of Giglio island (Tuscany, Italy).
Photograph: F. Fratini.

11.3.1.2 Intrusive rocks

Intrusive rocks are formed through crystallization of magma within the earth's crust. Due to the slow crystallization process, they are characterized by granular structures (Fig. 11.1). A typical intrusive rock is *granite*, widely used in the architecture of northern Portugal, Galicia, Brittany, northern Sardinia, and Corsica.

11.3.1.3 Volcanic rocks

Volcanic rocks are formed on the surface of the Earth; magma is brought to the surface through the phenomenon of volcanism (emission of lava). The rapid cooling of magma prevents the complete crystallization with the consequent formation of a porphyritic structure, characterized by a groundmass composed of small crystals in which a few large well-formed crystals (phenocrysts) are included. A typical volcanic rock is *basalt*, used, for example, in the architecture in the surrounding of Etna volcano (Sicily), in Ireland, and in northern Jordan. When the magma is viscous and rich in gas, the rocks will be particularly porous (vesicular structure) as in pumices and some kinds of rhyolites and trachytes (Fig. 11.2).

11.3.1.4 Sedimentary rocks

Sedimentary rocks are formed by granules (sediments) coming from the weathering of preexisting rocks. There are two distinct groups: clastic rocks derived from material transported in solid form and rocks of chemical and biochemical origin derived from material transported in solution. The processes that lead to the transformation in rock (diagenesis) are compaction, dissolution under pressure, and precipitation of new minerals (like calcite, silica, iron oxides), which leads to cementation of the sediment. Among the clastic rocks, the *conglomerates* are composed of detrital elements

Figure 11.2 The Zenòbito Tower (XVIth century) in the Capraia Island (Tuscany, Italy), built in trachyte blocks.
Photograph: F. Fratini.

(clasts) of coarse size (from 2 mm to 25.6 cm) derived from the transformation in rock of gravels (Fig. 11.3). The term *breccias* refers to those conglomerates whose clasts have not undergone transport and therefore have maintained sharp edges; breccias originate mainly from talus fans. The *sandstones* have clasts size between 62 μm and 2 mm and result from the transformation of sands into rock. The main components of the sandstones are: quartz, feldspars, mica, and clay minerals. When limestone clasts predominate, the rock is called *calcarenite*. The sandstones have been widely used in the architecture of central Italy (Florence, Cortona, Arezzo) (Fig. 11.4) as dressed/cut stones, decoration, architectural elements, and, in thin slabs, for roofing. The *shales*

Figure 11.3 A conglomerate block in "Fonte dell'Ovile" (XIIIth century, Siena, Italy).
Photograph: F. Fratini.

Figure 11.4 "Pietra Bigia" sandstone in a building of Arezzo (Tuscany, Italy). Photograph: F. Fratini.

have clast sizes of less than 62 μm and result from the transformation of clay muds into rock. They consist mainly of clay minerals which, due to their laminar structure, are deposited according to a preferred orientation parallel to the stratification, resulting in highly fissile rock. The shale cannot be used as a building material, but clay, mixed with water, is the raw material of much earthen architecture and of the production of bricks and terracotta. When the original mud is composed of tiny fragments of calcite, the rock that is formed is called *calcilutite* or *micritic limestone* (Fig. 11.5). There are also terms of transition to the shale (*marly limestones, calcareous marls, marls*). The *volcanic tuffs* can be included in the sedimentary rocks as they are the result of the

Figure 11.5 Micritic limestone (Pietra Alberese) in the masonries of Certosa di Firenze (XIVth century, Florence, Italy). Photograph: F. Fratini.

Figure 11.6 Tuff masonry in Pitigliano (Tuscany, Italy).
Photograph: F. Fratini.

deposition and subsequent transformation into rock materials (blocks, lapillus, ash) originated from explosive volcanic eruptions. The volcanic tuffs were widely used in the architecture of central and southern Italy (province of Viterbo, Rome, Naples area) (Fig. 11.6).

Rocks of biochemical origin
Rocks of biochemical origin derive from the removal of calcium and carbonate ions from seawater by organisms such as foraminifera and mollusks to form their own shell. The accumulation of the shells of these organisms produces the *organogenic limestones* that when poorly cemented are called *calcareous tufa. Dolomites, calcareous dolomites,* and *dolomitic limestones* contain, instead, major amounts of the mineral dolomite.

Rocks of chemical origin
Rocks of chemical origin derive from the process of precipitation of calcium carbonate from thermal springs that gives rise to *travertine* (Fig. 11.7) and *calcareous alabaster.* Evaporites are rocks formed after the chemical process of precipitation of calcium sulfate, sodium chloride, and other salts in lagoon basins characterized by hot and arid climates. Great importance is played by the deposits of calcium sulfate (gypsum and anhydrite) used to produce the gypsum binder. In architecture, carbonate rocks (limestones, dolomites, marly limestones, organogenic limestones, travertine) have been widely used as building materials both as dressed/cut stones, thin slabs for roofing and for the production of lime. In particular, the marly limestones have been widely used in the constructions of north−central Italy (ie, east Liguria, Florentine region) and the calcareous tufa in southern Italy (Sicily, Naples, Salento).

Figure 11.7 Travertine (whitish blocks) together with volcanic trachyte blocks in Castiglione d'Orcia (Tuscany, Italy).
Photograph: F. Fratini.

11.3.1.5 Metamorphic rocks

Metamorphic rocks have undergone changes in their mineralogical composition and texture due to increasing temperature and pressure (metamorphism) within the earth's crust. All rocks (magmatic, sedimentary, metamorphic) may be subject to metamorphism.

The typical appearance of the great majority of metamorphic rocks is the schistose texture; schistosity is the tendency of a rock to split into thin slabs according to sub parallel planes. Schistosity is the consequence of the oriented pressure sustained by the rock during the metamorphic process that caused an orientation of the minerals with elongated, fibrous, and lamellar shapes. Typical metamorphic rocks are *slates*, *schists*, and *gneiss*.

The *marbles* are a particular type of metamorphic rock, produced by metamorphism of carbonate rocks composed of calcite or dolomite. Generally, they show a low schistosity.

Metamorphic rocks have been widely used in the architecture of the Alpine countries, in the area of the Pyrenees, in northern Spain, Wales, and in the north of England, especially as thin slabs for roofing but also as dressed/cut stones. Famous is the "Ligurian slate," which, with its gray color, characterizes the roofs of Genoa and of many villages of Liguria (Italy) (Fig. 11.8).

11.3.2 Petrographic characteristics and use of rocks as building materials

The petrographic characteristics of the rock affect the properties of stone construction material. In particular, the mechanical properties, including compressive strength and the stiffness, and the physical properties, such as durability, workability, polishing, and

Figure 11.8 The Ligurian slate (ardesia) used as tiles for roofs and wall protection (Genoa, Italy).
Photograph: F. Fratini.

sculptability characterize the behavior of stone materials. The mechanical properties of stone blocks affect the global behavior and the bearing capacity of masonry structural elements. Rocks with good compressive strength (eg, granites, marbles, compact limestones) were used for columns, pillars, and cornerstones. Durability (ie, the resistance to decay caused by atmospheric agents) must be taken into account in relation to the intended use of the material. Rocks resistant to abrasion such as sandstones and porphyry were often used for flooring. Workability is better in soft stone (volcanic tuffs and tufa) that are split, sawn, and shaped more easily than sandstone, compact limestones, and granites. Polishing cannot be accomplished in rocks such as sandstones, because of the presence of minerals with very different levels of hardness such as quartz and clay minerals. Polishing is possible in compact and homogeneous rocks (eg, igneous rocks, compact limestones, marbles). Sculptability is favored in homogeneous, compact, and fine-grained rocks, free from veins and minerals that may be altered (pyrite). In fine-grained marbles, these conditions are well satisfied.

The majority of the rocks present in the Earth's surface can provide building stone. Those rocks that are not suitable for construction include the sedimentary rocks in which a strong clay component is present (eg, marls, marly clays, shales). These rocks are poorly coherent, laminated and fissile but, nevertheless, when the clay component is predominant, can be used to produce bricks, tiles, and earthenware in general. Other rocks unsuitable for use are very schistose metamorphic rocks (phyllites) and volcanites rich in a glass component (obsidian, pumice).

Vernacular architecture has the characteristic of using, in most cases, materials available in the immediate vicinity of the buildings. Therefore, these materials determine the character and uniqueness of each site, and technical and stylistic convergences between different territories may occur in instances of availability of similar materials.

The shape of stone blocks in buildings depends on the geological nature of stone; this also influences the type of processing and use. The following typologies can be distinguished: (i) river pebbles (unprocessed), (ii) unworked stones (shapeless or regular blocks), (iii) roughly worked stones (dressed stone), (iv) cut stone (worked on all sides) for special masonries, and (v) slabs for floors, walls, stairs, and roofs.

River pebbles may have more or less rounded surfaces as a function of the transport distance and a more or less spherical shape as a function of the textural homogeneity of the mother rock: rocks with homogeneous texture without preferential orientation of the constituent minerals, such as granites, basalts, and some types of limestones, give rise to spherical shapes, while those with preferential orientation, such as the metamorphic rocks and some types of sedimentary rocks, result in flat or elongated shapes.

Unworked stones can have an irregular shape if obtained by splitting rocks devoid of fractures and characterized by a homogeneous texture without cracking (eg, granites) or a regular shape when obtained by taking advantage of the presence of cracks (eg, columnar basalts), the thickness of layers (in the case of sedimentary rocks), or the schistosity (metamorphic rocks).

Dressed stones are obtained through rough working of hard rocks (using chisels of various shapes) and through cutting with axes, saws, etc. for soft rocks like volcanic tuffs and calcareous tufa.

Slabs can be obtained through splitting along planes of sedimentation for sedimentary rocks (limestones, sandstones) or through splitting along the planes of schistosity for metamorphic rocks (slates, schists, gneisses).

11.3.3 Mortars

Mortars are artificial materials produced by man since ancient times that are used for different functions including masonries, plasters, to adhere tiles or realize decorations (stuccos, etc.). Mortars with special uses (filling of holes, sealing, pointing of mortar joints etc.) or features (resistant to moisture, pigmented, etc.) also abound.

Mortars are obtained by mixing water and aggregate with different types of binders, capable of hardening the paste. Organic additives (egg yolk, egg white, casein, animal glue, "fig milk", walnut oil, linseed oil, blood, resins, etc.) and natural fibers (straw, animal hair, iron filings, etc.) can also be added, as well as synthetic fibers and pigments, to impart particular characteristics.

The aggregate is added to the mixture with the aim of reducing the shrinkage and enhancing the strength of the final product. Generally, aggregate is supplied from river beds, but it can also come from beach deposits and from unconsolidated sand outcrops. It can also be obtained through the mechanical crushing of rocks.

Among the binders used to make the mortar, the following categories can be distinguished: (i) clay binders, (ii) gypsum binders, and (iii) lime-based binders. *Clay-based binders* are probably the earliest binders used by man, who realized how the dried clay develops a remarkable cohesion and is able to bind together elements of masonry. Mixtures of earthy clay are normally used as bedding mortars for masonry in mud brick (adobe) and for plastering the same earthen walls, in this case often combined with plant fibers. To make the clay paste more resistant to moisture, the addition of lime is also documented. Lime is able both to form a calcitic binder and to react with the clay minerals, thereby inhibiting the absorption of water in the crystal lattice of these minerals (Gamrani et al., 2011).

With respect to *gypsum binders*, use probably began in Middle East, where large outcrops of sulfatic rocks are present, and spread, especially in Egypt at the time of the ancient dynasties. The binder is obtained by burning evaporitic rocks of sulfatic composition at relatively low temperatures. At 130°C, these rocks undergo a chemical transformation, due to partial dehydration, giving rise to hemihydrate calcium sulfate called fast-setting plaster or plaster of Paris. Kneading with water quickly rehydrates and hardens the mixture. By burning the gypsum stone at 160–180°C, "gypsum for plasterers" or anhydrous gypsum is obtained, which on very fine grinding, is suitable for bas-reliefs, statues, etc., realized through molds or shaping in paste.

Gypsum binder has been widely used in historic buildings, especially for its capability to set and harden quickly. It can be used without aggregate because during setting it does not undergo shrinkage. Nevertheless, knowledge of its actual use in historical architecture has been forgotten because the gypsum mortars produced at present, obtained from the hemihydrate gypsum or from anhydrous gypsum, are hygroscopic, absorb a lot of water, and tend to pulverize. Therefore, today, gypsum binders are used mainly indoors to realize plasters, stuccos and decorations, as well as in the form of a fluid paste injected into masonry for consolidation operations. In some regions of Europe rich in sulfate rocks, like the Paris basin, some parts of Germany, and eastern Spain, gypsum mortars have been used extensively, with good results of durability, for exterior plasters, bedding mortar and even for realization of floors (Vegas et al., 2010; La Spina et al., 2013; Sanz and Villanueva, 2004). The good behavior of these mortars is particularly evident when sulfatic rocks, rich in carbonate and clay impurities, have been used as a raw material. In fact, traditional burning in furnaces where the temperature is not controlled causes the formation of calcium oxide and calcium silicates that give to the gypsum binder its particular resistance to decay.

Among the *lime binders*, *air-hardening calcic lime* is obtained by burning of pure limestones at a temperature of 850–900°C. At first, quicklime (calcium oxide) is obtained that, mixed with water, gives rise to slaked lime or lime putty (calcium hydroxide). The lime putty mixed with aggregate, without the necessity of adding water, hardens through the process of setting (evaporation of water and fixation of carbon dioxide [carbonation]) with the consequent transformation of the calcium hydroxide into calcium carbonate (Fig. 11.9).

From the burning of dolomite or dolomitic limestones, *air-hardening magnesian lime* is obtained that, on setting, produces mortars with mechanical properties higher than those from the air-hardening calcic lime. Regarding these two types of limes, and in particular for the calcic lime, experience has taught that the quality of the lime putty (plasticity, workability, carbonation, mechanical characteristics of the finished product) improves with aging, that is, with storage in an anaerobic environment (under a layer of water, covered with moist sand or in airtight containers) for as long as possible. In this regard, the recommendations of Pliny the Elder (1952)

Figure 11.9 Air hardening calcic lime mortar from a Florentine building of the XIXth century observed in thin section under transmitted light at the optical microscope. In the center an underburnt fragment of the limestone used to produce the lime.
Photograph: F. Fratini.

in "Naturalis Historia" are well known, suggesting a prolonged maturation. Early Roman provisions required curing in water for at least 3 years before use. The reasons that the curing improves the quality of the lime putty have been investigated by some scholars (Rodriguez Navarro et al., 1998; Hansen et al., 1999), who found that with aging, the crystals of calcium hydroxide undergo major changes in shape (from prismatic to tabular) and in size (from micron to submicron). These changes induce a significant improvement in the physical properties of putty, which becomes more plastic.

Nevertheless, there is a traditional technique of slaking and use of lime called "dry slaked lime." This technique does not require curing and it is used in the production of "hot lime mortars."

It is a technique common especially in Northern Europe, reported in treatises (Kraus et al., 1989), and represented in paintings. It was used both to produce bedding mortars and to fill the cores of masonry construction. In Northern Europe, this technology is still used (British Standard Code of Practice, 1951), in particular for restoration and as a traditional building technology, thanks to its good workability, durability, affordability of the final product, and suitability for winter use.

Usually, the mortars made with this technology are characterized by (i) a high amount of binder (ratios binder:aggregate 1:1 to 1:2), (ii) presence of numerous whitish inclusions consisting of lime not well mixed in the paste (dusty lumps, lumps of putty, remnants of under burnt lime fragments) (Kraus et al., 1989; Middendorf and Knöfel, 1991; Forster, 2004), and (iii) excellent mechanical properties with good adhesion to masonry and high resistance to frost.

There are various methods of making these mortars (Doglioni et al., 1986, Foster, 2004), but in summary we can say that fragments of quick lime are covered with sand, wet several times, and mixed before use. Doglioni et al. (1986) reports that in the case of use as bedding mortar, the mixture was sieved while, when used to fill the core of masonries, once cast with stone chips and pressed, was sprayed with water from above. It is this procedure, washing away the lime, that resulted in the precipitation of calcite concretions in the areas below.

With regard to the excellent mechanical properties and high durability of the "hot lime mortars," the heat developed in the mixture during slaking can have a positive effect in improving the binding between lime and aggregate because a basic attack on the surface of the silicatic aggregate is favored, making the granules rougher and more reactive (Jedrzejewska, 1967; Gibbons, 1993). Additionally, the steam that is produced can aid the formation of large pores, making the mortar more resistant to frost action.

The firing of marly limestones (with a percentage of clay from 5% to 20%) results in *hydraulic lime*. The slaking of this lime is done by sprinkling the clods of quicklime with water until they disaggregate to powder. This powder must then be stored in dry conditions; otherwise, it hardens and can no longer be used. Unlike air-hardening lime, the setting of hydraulic lime can take place in very humid environments and, indeed, under water. In the past, hydraulic lime was realized by adding to the lime putty additives such as pyroclastic material (volcanic ash, pumice) and ground earthenware. These substances, rich in active silica and alumina, react with the lime-forming calcium silicates and aluminates, which make the mortar capable to set in very humid conditions. Therefore, these types of mortars were often used for flooring and as tank and aqueduct coatings.

Organic additives are all those organic substances (milk, egg, beeswax, oils, starches, natural resins or acrylic, etc.) that, when mixed with lime, form binders with particular characteristics (adhesion, water repellence, etc.) (Arcolao, 2001). Some of these binders can be called mastics and be used for bonding, to fill cracks or fissures, or to form protective layers for floors, balconies, etc. Generally, the mixtures of lime, brick dust, and linseed oil or litharge were most frequently reported in the texts of the eighteenth and nineteenth centuries, while those based on pitch date to more ancient times. Pitch, defined by sources as colophony, is a natural resin produced in the form of secretion from some resinous plants. Recipes with pitch-based binders are numerous and testify to its widespread use, especially for its protective, adhesive, and waterproof properties. Often, the pitch-based compound was added with oil, animal fat, or wood ash to increase the workability and water resistance and with ox blood, wax, or to increase toughness, to accelerate the setting, reduce shrinkage, and to create less-viscous mixtures able to better penetrate the masonry.

In the first half of the nineteenth century, a new type of binder was developed, *Portland cement*, obtained by burning at very high temperature (1450°C) marly limestone or mixtures of limestone and clay. This revolutionary material was an important turning point in the history of architecture and resulted in the gradual abandoning of traditional mortars. Portland cement, compared to traditional binders, is characterized by a quick set and higher mechanical strength, allowing a wider use in the structural field. Cement mortars, however, have demonstrated poor durability in many applications, demonstrating the continued value of traditional binders.

11.4   Masonry constructions

The use of stone blocks assembled by mortar has characterized the architecture of all periods, creating a wide repertoire of technological and typological forms, which differ based on the specific characteristics of the stone and the construction techniques used. Different levels of performance and mechanical behavior correspond to different levels of development of the construction techniques used. These result from the quarrying capacity and processing of the stone and the technological methods adopted in the construction phase. Masonry buildings are continuous, consisting of bearing walls rather than framed structures. When subject to vertical loads, the walls demonstrate good mechanical behavior and typically have excellent resistance to sustained (creep) or accidental loads. In fact, in masonry buildings, all walls play a structural function, the transverse dimensions are consistent, and applied compressive stresses are typically low.

In the case of horizontal actions, such as those induced by the earthquakes, masonry structures exhibit an intrinsic weakness due to the low tensile strength of the material. Masonry consisting of stone elements assembled with or without mortar have poor adhesion between the stone blocks. In constructions without mortar, only gravity and friction resist tension forces, whereas in mortared construction, adhesion contributes to the capacity. Dangerous crack patterns or catastrophic collapses demonstrate the vulnerability of masonry construction to seismic action; this is mostly attributed to the lack of tensile strength of the walls. In particular, crack patterns highlight the separation between orthogonal walls, damages due to shear actions in the plain of the walls, and the initiation of wall overturning (Fig. 11.10).

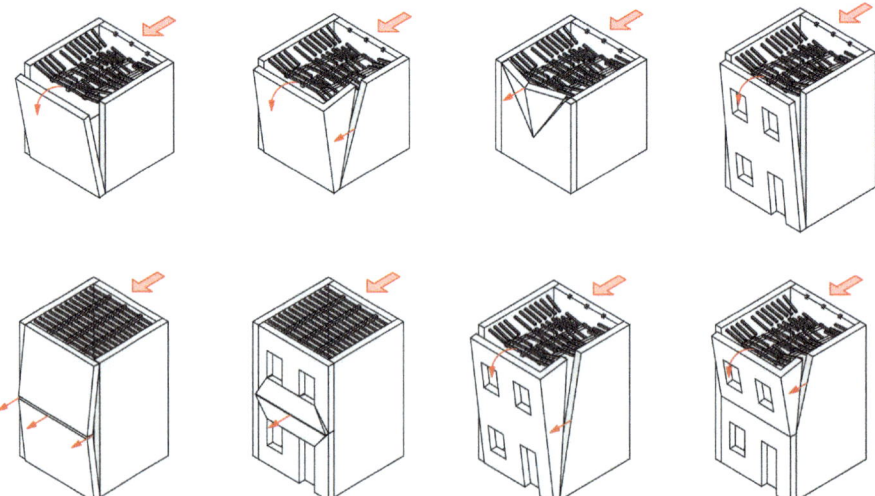

Figure 11.10 Collapse mechanisms of masonry walls subjected to horizontally thrust. Drawings: D. Omar Sidik.

To evaluate the mechanical efficiency of a masonry building in the event of horizontal actions, the main factors to consider are the ability of the building to behave like a box and the degree of monolithic behavior of the walls. The box behavior depends on the efficiency of the connections between orthogonal walls and between walls and floors; in the case of good connections, the seismic action is absorbed by the walls with greater stiffness, the overturning of the individual walls is mitigated, and only shear fracture in transversal walls appears.

The degree of monolithic behavior of a masonry wall made up of separate blocks is different than the theoretical model of wall consisting of a single block, which responds to the vertical loads with homogeneous stress and resists the seismic actions with rocking behavior. A wall consisting of small and irregular pieces, assembled in a chaotic way, cannot behave as a monolith and, when affected by the earthquake, disintegrates during low values of the seismic action. Only if the wall has a certain degree of compactness does the rocking mechanism engage, allowing greater resistance to horizontal loads.

11.4.1   The box behavior

Masonry buildings have a box behavior when the interlocking between orthogonal walls is effective. Construction techniques have resulted in different approaches for the realization of interlocking. The most common technique uses "cornerstones," stones of a larger size than those used in the walls and typically of regular shape. In this manner, an effective coupling of orthogonal directions through the thickness of the walls is possible.

Efficient systems of restraining (hooping) building walls to maintain the box were made with wooden or iron chaining. The use of wood for hooping is widespread in countries in which the earthquake risk is high and wood is abundant. Very interesting are examples in the Balkans, Anatolia, and sub-Himalayan regions (Dipasquale et al., 2014a,b). An evolution of wooden hooping is the iron chaining that begins to be used in the thirteenth century, developing along with the techniques of iron working, first in France and then across all the northern Mediterranean basin.

11.4.2   The masonry wall

The primary factors affecting the mechanical quality of a masonry wall are the shape and size of the stone blocks, the quality of the mortar, and the masonry texture (Dipasquale and Volpi, 2010). With regard to the size of blocks, a wall shows good performances if larger elements represent the majority of the section or if the size of the elements is homogeneous. The mortar makes the contact and friction between the stone elements relatively uniform in order to avoid concentrations of load when the surfaces are not regular. In the case of walls built with elements of small dimension, in which the meshing of the stones is not sufficient to ensure the monolithic behavior of the wall, the mortar should compensate providing transverse interlocking. The low tensile strength of the mortar, however, prevents it from completely fulfilling this task. Nevertheless, when the laying of stones is well organized, the cohesion of the mortar is less important: even a poor mortar accomplishes the function of making the contact uniform.

In terms of wall texture, a good mechanical solution must prevent any vertical separation of the wall into leafs exhibiting independent behavior. However, since the size of the stone blocks is typically smaller than the wall thickness, it is inevitable that the wall tends to behave as two vertical elements. For this reason, connections are needed between the external and internal wall elements to ensure a monolithic behavior, particularly when subject to seismic loads. This connection is obtained through the meshing of overlapping stones, which are crossing stones that bind the two opposite leafs. The resistance of the wall is ensured when the meshing between stones is sufficient to provide an equilibrium path for the transmission of loads through contacts between stone blocks.

Additionally, horizontally oriented "bonding courses," realized through the use of small stones, stone flakes, or brick layers, are useful to improve the mechanical behavior of the wall (Fig. 11.11) (Dipasquale and Volpi, 2010; Rovero and Fratini, 2013). The horizontal courses allow the formation of linear hinges in the event of horizontal loads. These trigger a reversible tipping mechanism and eventual self-righting stability of the wall, when the horizontal excitation ceases. In the instance of irregular masonry courses, the rotation triggered by an earthquake would result in irregular portions of the masonry and large variations in contact loads between the stones. When the rotation initiates, the combined action of the weight of the wall and the horizontal force (which should be dissipated along the edge of the wall) runs obliquely along the thickness of the wall, affecting the transverse monolithic

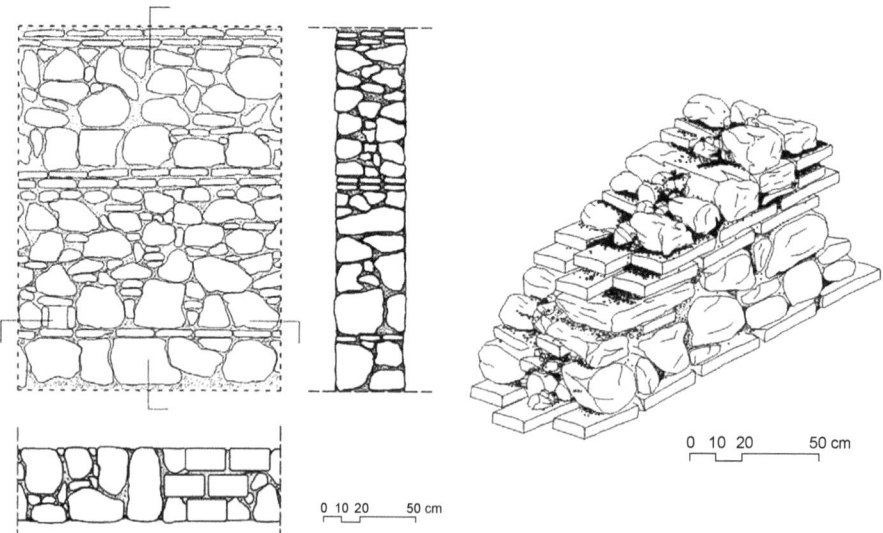

Figure 11.11 Representation of masonry walls of good quality, with bricks to improve the horizontality of the laying planes and diatones to connect the opposite leafs of the wall. Surveyed in Chefchaouen, Morocco.
Drawings: L. Dipasquale & V. Volpi.

character of the wall and potentially resulting in failure of the wall. Bonding courses help to make these forces more uniform through the wall thickness.

11.4.3   Masonry wall types

According to the shape of the stone elements and their assembly, different types of masonry walls can be distinguished.

11.4.3.1   Dry stone walls

Dry stone walls are composed of blocks of stones that are laid down without any mortar to bind them together. The structural integrity of dry stone walls arises from compression forces and the interlocking of the stones. The wall thickness depends on the nature and size of the stones and boulders available; these often come from nearby fields during preparation for agriculture.

The tools used for cutting and roughing of the stones are elementary: hammer, pickaxe, and shovel for soil preparation and the mallet and chisel to make a flat surface. The walls are built up by course, and at intervals, large tie-stones are placed that span both faces of the wall and increase the strength and integrity of the wall. In many cases, the last row is realized by placing larger pieces, that span the entire width of the wall, to connect the two leafs and close the wall.

Such walls have been traditionally used in building field boundaries and retaining walls for terracing, but dry stone buildings, bridges, and other structures also exist. In Europe, dry stone structures can be found mostly in regions having natural rock outcrops or large stones exist in quantity in the soil; they are abundant in the upland areas of Britain and Ireland (particularly Connemara) and throughout the Mediterranean, mostly in southern Italy (Apulia and Sicily) (Fig. 11.12(a−c)), France, Greece, and Spain.

11.4.3.2   Rubble masonry

Rubble masonry has been extensively used for vernacular building worldwide where a local source of stone is readily available. This type of masonry consists of blocks of undressed or rough stones placed with mortar. Depending on the characteristics of the local rocks, the dimension of the stones, and their arrangement in the wall, a large variety of subtypes can be distinguished: *Random rubble masonry* consists of not-squared-off stones of different sizes and shapes arranged in courses of equal (*coursed random rubble masonry*) or different height (*uncoursed random rubble masonry*) (Fig. 11.13(a) and (b)). In both cases, the larger stones are laid down first and the spaces between them are filled up with chips of stone. The more regular the courses, the better the mechanical performance of the wall will be. When the rock is very compact and hard, generally the blocks of stone are irregular in shape and dimension. For this reason, the organization of the blocks for the stability of the wall requires a considerable amount of binding mortar, capable of ensuring the assembly of the various elements, as well as the success of the masonry which depends on the strength of the mortar itself (Fig. 11.13(c)). *Squared rubble masonry* consists of face stones that

(a) (b)

(c)

Figure 11.12 (a, b) Dry stone construction near Ragusa (Sicily, Italy). *(Photograph: L. Dipasquale.)* (c) Dry stone construction in a "trullo" of Alberobello (Apulia, Italy) *(Photograph: L. Dipasquale.)*

(a)

(b) (c)

Figure 11.13 (a) Random rubble masonry with a little amount of bedding mortar in Lunigiana (Tuscany, Italy). *(Photograph: L. Dipasquale.)* (b) Random rubble masonry in Alberobello (Apulia, Italy). *(Photograph: L. Dipasquale.)* (c) Random rubble masonry with a lot of bedding mortar in Cortona (Tuscany, Italy). *(Photograph: L. Rovero.)*

are squared and brought to hammer-dressed or straight cut finish before being laid. Stones can to be laid in courses of equal layers with uniform joints (*coursed square rubble masonry*) (Fig. 11.14(a)) or can be arranged in several irregular patterns or courses of different size (*uncoursed square rubble masonry*) (Fig. 11.14(b) and (c)).

Together with quarried stone, river stone (pebbles) were sometimes employed in the construction of rubble walls. Such alluvial soil provides high amounts of such material in suitable sizes. In addition to the blocks, stone flakes can be used, together with

Figure 11.14 (a) Coursed square rubble masonry in Ragusa (Sicily, Italy). *(Photograph: L. Dipasquale.)* (b) Uncoursed square rubble masonry in Apulia, Italy. *(Photograph: L. Dipasquale.)* (c) Uncoursed square rubble masonry in Cusco, Perù. *(Photograph: L. Rovero.)*

pieces of brick and roof tiles, in order to fill the uneven gaps remaining among the stones and to obtain a kind of uniformity. Generally, large stones are employed at quoins and jambs to increase the strength of the masonry. Bricks can be also used with the aim of creating horizontal layers so as to improve effectiveness in the irregular stone work (such as in the Roman *opus listatum* shown in Fig. 11.15).

The evaluation of the quality of the rubble masonry structure depends also on the disposition of large stones, or through stones used as internal cross-connections, whose dimensions exceed half the thickness of the wall. In the absence of large stone blocks connecting the outer leafs of the wall, the masonry arrangement needs a high use of mortar in order to allow the transfer of loads and to ensure the integrity of the whole, although this alone cannot guarantee a monolithic behavior and resistance against out of plane actions. Rubble masonry is often covered by a plaster, which plays the role of protection and counteracts the dangerous phenomenon of washing out due to rain.

11.4.3.3   Ashlar masonry

Ashlar masonry is composed of regularly shaped stone blocks, with a dressed exposed face, which may feature a variety of treatments: tooled, smoothly polished, or rendered with another material for decorative effect. All the faces of the block are adjacent to those of other blocks, and joints can be very thin. The height of each course is kept

Figure 11.15 Random rubble masonry with horizontal layers of bricks in Lamezia (Calabria, Italy).
Photo: L. Dipasquale.

uniform, and all the joints are uniform (Fig. 11.16(a) and (b)). As the dressing of stones requires heavy labor and wastage of material, soft rocks are generally used. It is common, therefore, to find dressed stones used only as a facing (while the backing is rubble or brickwork) or just in the most important parts of the building (angle quoins, lintels, beams, arches, etc.), except in works of great importance and solidity. Ashlar masonry has been used by many ancient cultures (early examples can be found in the Knossos palace in Crete, in the step pyramid of Djosar in Egypt, and in Macchu Picchu in Peru) and is used worldwide for public or high-quality historical buildings.

Ashlar blocks were also used to create domed or arched structures in vernacular buildings (Fig. 11.17(a)). The superposition of successively smaller rings of ashlar stone was used also to construct beehive-shaped domes, used, for example, in the corbelled tombs, known as *tholoi*, which are commonly found throughout ruins of ancient civilizations in the Mediterranean and also in the vernacular architecture of the same region

(a)   (b)

Figure 11.16 (a) Ashlar masonry in Baku, Azerbaijan *(Photograph: L. Dipasquale.)* (b) Ashlar masonry in Matera (Italy). *(Photograph: L. Rovero.)*

(a)   (b)

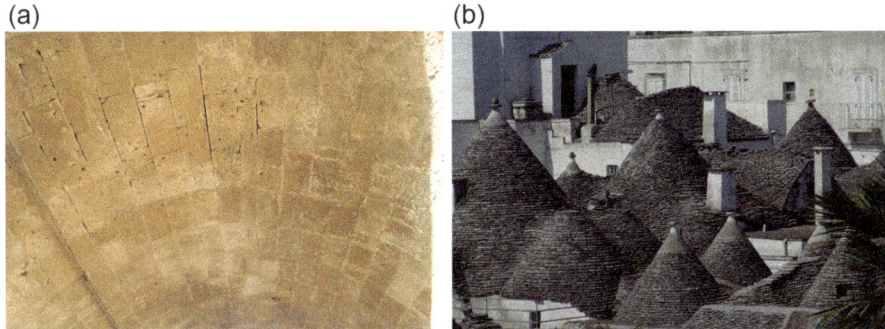

Figure 11.17 (a) Ashlar masonry used in a barrel vault in Ragusa (Sicily, Italy). *(Photograph: L. Dipasquale.)* (b) Corbelled domes in Alberobello (Apulia, Italy) *(Photograph: L. Dipasquale.)*

(talayots and ponts in Balearic islands, cabanes in Provence, pagghjari in Corse, mitati in Crete, chozos-barracas-pozos-bombos in Spain, nuraghi and pinnéttas in Sardinia, trulli-furneddi, specchie in Apulia) (Fig. 11.17(b)) (Dipasquale and Jorquera, 2009).

11.4.3.4   Masonry with wood reinforcing

Masonry with wood reinforcing is a very common in seismic-prone areas. The tensile strength of wood offers reinforcement against horizontal loads and enables the dissipation of substantial amounts of energy in the case of earthquakes. Moreover, timber elements divide the structure into sections, preventing the spread of cracks occurring in portions of the masonry. Two main categories of wood reinforcement can be identified: the hooping and frame systems. Hooping arranges the wooden beams horizontally within the load-bearing masonry during the construction process. The empty spaces between the beams are filled with fragments of brick or stone. The resulting ring beams can be inserted at the floor levels, at openings and lintels, or regularly distributed along the height of the construction. This system can be found in seismic regions of the Mediterranean from the Balkans to Turkey, Maghreb, Greece, and Italy (Dipasquale et al., 2014a,b) (Fig. 11.18).

The second category includes wooden-frame systems, which are articulated in round or square section beams and pillars, and frequently include diagonal bracing elements. The empty spaces defined by the frame are filled with locally available materials (earth, stone, or brick). One of the most ancient examples in Italy of timber-frame buildings techniques is the *opus craticium* by Vitruvius, today visible in some of the surviving houses of the archeological sites of Herculaneum and Pompeii (Fig. 13.19). In the Mediterranean area, relevant traditional examples of timber-frame structures together with masonry can be found in Turkey, Greece, and parts of Eastern Europe. In these countries, common traditional buildings techniques are based on the use of masonry-laced bearing wall constructions on the ground floor level and lighter infill-frames for the upper floors. The ground floor masonry walls are often laced with horizontal timbers; these elements can be thin timber boards laid into the wall so that they overlap at the corners or squared wooden beams.

Figure 11.18 Masonry with hooping wood reinforcement in Elbasan (Albania).
Photograph: L. Rovero.

Figure 11.19 *Opus craticium* in Pompei, Regio I Insula XII.
Photograph: L. Giacopelli.

11.5   Mechanical performance

11.5.1   Assessment of the safety

The problem of safety assessment of existing masonry buildings is very complex
because the wall structures, unlike those made of steel or reinforced concrete, do
not allow the definition of a single mechanical model capable of realistically modeling
any situation. Masonry buildings, in fact, show a great variability in both materials and
construction techniques, being the result of local cultures and traditions. Many studies
have been devoted to understanding the structural behavior of masonry solids — from

the elastic theory to limit analysis approaches, from micromechanical modeling to distinct elements analysis. The approaches are differentiated by purpose, level of approximation, and field of application.

The Italian Technical Standards (NTC, 2008) is a leading code with performance-based approaches for the safety assessment of existing masonry buildings, subject to seismic actions. According to NTC (2008), safety is evaluated using a limit states approach with reference to the ultimate limit states carrying out global analyses and checks against local mechanisms. The standards requires that the structural analysis considers displacement demands and that is carried out assuming that the structure can deform beyond its elastic limit through the opening of fractures, that dissipate energy but do not compromise the bearing capacity for vertical loads. This approach allows imparting upon masonry structures (although consisting essentially of brittle materials (stones, bricks, and mortar)), a "pseudo ductile" structural behavior with considerable capacity beyond the elastic phase, a behavior critical to sound seismic behavior.

To assess global behavior, it is necessary to conduct nonlinear analyses considering the building in its entirety. Masonry buildings are conventionally modeled using a geometrically equivalent frame whose pillars and beams represent the walls. The application of nonlinear static analysis to historic buildings is not always possible and has many unconvincing aspects. The representation of the structure as an equivalent frame is possible only for isolated regular structures (both in plan and in elevation) with aligned windows. Moreover, typically, the mechanical model provides that the interlocking between the walls and between walls and floors is perfect. Essentially, the procedure of calculation assumes that the building to be verified exhibits box-like behavior with effective interlocking, a condition that is unlikely in many historic structures. The resulting analyses are likely to overestimate the in situ structural capacity.

Greater criticism can be made of modeling aggregates of buildings for which standards introduce the concept of structural units that are analyzed separately taking into account the actions that result from contiguous structural units.

In addition to global behavior, safety assessment of masonry buildings subjected to seismic action must be conducted considering the possible local collapse mechanisms. This requirement more realistically accounts for the mechanical behavior of masonry structures characterized by the absence of effective linkages between orthogonal walls and between walls and floors and therefore characterized by high vulnerability to out-of-plane actions (as postseismic scenarios have unfortunately highlighted). The local mechanism approach is based both on the pioneering studies of Giuffré (1991, 1993) concerning the behavior of walls and their damage mechanisms and on the theories of Heyman (1995, 1996) that extended the application of limit analyses to masonry buildings. The method is based on assuming that the masonry behaves like a rigid material, having no tension strength but infinite compressive strength. Therefore, the more "macroscopic" characteristics of the masonry are considered and this fact allows one to carry out the assessment within the conditions of the limit analysis, without the need to identify elastic material properties, whose definition, in the case of heterogeneous materials such as masonry, is not easy. The walls are represented as rigid blocks subject to their own weight, to possible loads transmitted from floors, and to horizontal loads defined, through a multiplier, as function of vertical ones. Based on the observation of post-seismic damage and on the specific local building conditions, a kinematically congruent mechanism can be assumed through the introduction of plastic hinges. Using the kinematic theorem of limit analysis makes it possible to identify the upper limit of the actual multiplier for the horizontal loads at collapse, from the lowest of all multipliers determined for the hypothesized mechanisms. The identification of mechanisms is facilitated by a survey of similar buildings already damaged by seismic activity. From this, the fracture patterns can guide the selection of the macroelements involved in the mechanism and the subsequent selection of the position of the hinges.

The assessment procedure based on local mechanisms provides a very simple tool, capable of capturing the essence of the behavior of masonry buildings and highlighting the most vulnerable parts. The "simplification" taken may lead to an underestimation of the mechanical behavior under earthquake, especially when using the linear kinematics; nevertheless, the approach allows one to obtain the limiting value of the seismic action that can be withstood.

The detailed study of the building, considering the type of the masonry walls and efficiency of the interlocking, allows an enrichment of the approach based on local mechanisms, thus being an improved approximation of behavior. It is possible, for instance, to introduce resistive forces produced by the friction between the blocks and/or the cohesion produced by the mortar. If the walls are made without an adequate cross-connection, taking the form of two essentially separate leafs, the reduced ability to withstand out-of-plane actions should be considered. In such cases, mechanisms that enforce the presence of two leafs and the possible cohesive capacity or interaction between them must be considered.

11.5.2   Estimation of mechanical parameters

To perform an assessment of global behavior, it is necessary to make an analysis of the whole building, specifying the actual mechanical properties of the different materials constituting the masonry. The identification of mechanical parameters of masonry structures is complex. In fact, strengths and stiffness of masonries depend on many factors, such as strengths of component blocks and mortar, blocks shape, volumetric ratio between components, and wall texture. Taking into account the complexity resulting from the great number of variables, a fair assessment of the load-carrying capacity of masonry can be made only by in situ test, as highlighted in Binda et al. (2000).

In situ tests with semidestructive methods may not always be performed. In these cases, an assessment of the compressive strength of masonry walls can be carried out on the basis of a qualitative criteria evaluation, as proposed in Borri and De Maria (2009). The method, the Masonry Quality Index (MQI), consists of the evaluation of the presence, the partial presence, or the absence of certain parameters that define the "rule of the art," namely a set of construction techniques that, if executed during the construction of a wall, provide a good mechanical behavior and ensure the compactness and monolithic nature of the structure. A synthetic evaluation of the wall quality is obtained through three overall scores, the MQIs, that define the quality of masonry in relation to three actions: vertical actions, out-of-plane actions, and in-plane actions. An estimation of the mechanical parameters (compressive strength, shear strength, and Young's modulus) of masonry can be obtained through correlation curves, obtained from experimental data (Borri and De Maria, 2009).

11.6 Applications in modern architecture and innovative uses

While in traditional architecture stone plays a dominantly structural role, with the emergence of industrial techniques in the building sector and the structural use of reinforced concrete and steel, the use of stone has become mostly limited to veneers and claddings, becoming the skin of the building only. In modern architecture, stone is first reinterpreted by the Modern Movement, which puts aside the structural function of the material and gives it a new role, more related to its texture. Examples of this trend can be seen in the works of Ludwig Mies van der Rohe and Frank Lloyd Wright in the United States and, in Italy, in the buildings of Giuseppe Terragni, Adalberto Libera, and Giovanni Michelucci. In Late Modernism, stone is used on façades with the innovative system of "ventilated walls," which uses thin industrial processed slabs detached from the main structure by means of a metallic shelf, leaving an air chamber between the slab and structure. This gap provides an important function in regulating the interior thermal environment. During mid 1980s, some architects, such as Arata Isozaki, Aldo Rossi, James Stirling, and others, used stones with different colors and textures to emphasize the aesthetics of the surfaces.

The use of bearing masonry in modern and contemporary architecture is rather limited. It should be highlighted that the work of Antoni Gaudi (Crypt of the Guell colony, Barcelona, 1898–1917), that at the time seemed simply a reminder of the building traditions of the past and today seems paradoxically innovative. In recent years, there are some interesting examples from international architects who have enhanced stone in their projects, using different language and renewing its stereotomic conception: from Jorn Utzon (Can Lis House in Maiorca, 1973) to Aldo Grassi (Alessi House in Lago Maggiore, 1989) to Kengo Kuma (Stone Museum in Tochigi, 2000).

Many cases of contemporary architecture use reinforced masonry. Stone is reinforced with concrete, for example, in the St. John the Baptiste Church (Florence, 1960) of Giovanni Michelucci or in the Therme Vals (1991) of Peter Zumthor (Fig. 11.20). In both cases, walls are made by two stone exterior layers used as formwork for the subsequent casting of reinforced concrete.

The modernization of the production cycle of quarries by means of computer numerical control (CNC) machine tools has led to innovative technologies for reinforced stone. As examples, both the Expo pavilion in Seville (1992) of Peter Rice and the Liturgical Hall "Padre Pio" in San Giovanni Rotondo of Renzo Piano (2008) (Fig. 11.21) are based on the assembly of blocks of stone, individually designed, cut with high precision with use of the CNC machine, and stacked using posttensioned metallic cables to realize loading arches (Salerno et al., 2010).

Figure 11.20 Therme Vals wall structure of Peter Zumthor, Vals, Graubünden (Switzerland) (cc by 2.5).

Figure 11.21 Liturgical Hall "Padre Pio" of Renzo Piano in San Giovanni Rotondo (Apulia, Italy).
Photograph: F. Fratini.

The use of stone in massive form reappears in recent years in the works of the French architect Gilles Perraudin, who, for 15 years, has been designing contemporary buildings in which the supporting walls are made entirely of blocks of stone (Fig. 11.22). Blocks of local calcareous stone of large dimension have been used to build wine cellars (including the cellar of Vauvert in the Gard region, 1998) and buildings for cultural activities (Center for Training and Apprenticeship in Marguerittes, 1998, Wine Museum in Patrimonio, Corsica, 2011). Recently, stone construction was efficiently used in a social housing project (Cornebarrieu, suburb of Toulouse). The use of stone responded to a test of its feasibility with respect to very high requirements in terms of energy, environmental impact, and cost—having been constructed on a budget of about 1000 euros per square meter. The technique used by G. Perraudin in this project was based on placing one piece on the other and holding the large blocks

Figure 11.22 Chai viticole Vauvert by Gilles Perraudin.
Credits: Perraudin architectes.

in place by their sheer mass, without the use of cement or binding elements. A lime mortar mixed with stone sawdust from the quarry was used to prevent wind blowing through the joints. Despite the traditional building site, new technology and machinery allowed the building to be constructed in just few months, despite the weight of the stone blocks.

Another innovative use of stone in contemporary architecture can be seen in gabion walls. This technique consists of cages—generally made of galvanized steel wire or mesh—filled with rocks without mortar. The technique has been employed from the late nineteenth century to build retaining walls, and stabilize shorelines, stream banks, or slopes against erosion. In recent years, gabion walls have also been used in contemporary buildings, since many architects have recognized the effectiveness of such structures: gabions can be made using local stone, including small pieces of (otherwise waste) quarry material, the construction process is simple and low cost. Gabions are modular, can be constructed off site and stacked in various shapes; they are also resistant to being washed away by moving water. Ian Ritchie Architects used gabion structures to build the cultural Centre of Terrasson, France, while Herzong & de Meuron explored the environmental potential of this technique in the Dominus Winery in Napa Valley, California (Fig. 11.23). Here, the gabions are used to moderate the extreme temperatures of the Napa Valley, forming a *brise soleil* and thereby reducing thermal gain during the day, while during the night the mass of the stone acts as a thermal insulator (Dernie, 2003).

11.7 Role of masonry in sustainable construction

Stone as a building material demonstrates a number of important inherent qualities, which are well known to those whose work focuses on conservation in historic buildings. These same qualities can and should be leveraged in innovative construction methods and applications for contemporary architecture.

Figure 11.23 Dominus Winery by Herzog & de Meuron realized with gabions.

Masonry constructions have contributed to define the cultural identity of many areas worldwide, and they still are, thanks to their great durability, the principal constituent part of the current architectural heritage, despite the changing environment of cities and environmental context in which they reside. Great durability and low maintenance requirements of masonry construction mean important benefits also for contemporary buildings, in terms of sustainability from both environmental and socioeconomic respects. Of course, certain environmental conditions—such as temperature, precipitation, wind, water, and air pollution—can cause of degradation of stone materials and influence the physical aspect of the building surfaces and their durability (Fig. 11.24). Moreover, factors such as poor design of details, poorly executed repairs, and inadequate maintenance of the building can accelerate the processes of decay to both stone and the building in general. The reaction to exposure to their environment

Figure 11.24 Degradation of a calcareous masonry in Ragusa (Italy).
Photograph: L. Dipasquale.

varies from stone to stone, which is the reason why durability testing is required in modern construction. Such testing provides product performance over its lifetime and helps the designer to choose the appropriate type of stone depending on its use. Internationally recognized standards for assessing the environmental durability of stone products are promulgated by ASTM International and in Europe by the Construction Products Regulation 305/2011. In European Union countries, durability testing on building stone are required to obtain the CE marking, which is the manufacturer's Declaration of Conformity affixed to a product to indicate that it conforms to certain safety requirements, during both initial type testing and factory production (Dipasquale, 2012).

Today, stone materials can gain an advantageous position compared to other building materials, if they demonstrate their competitive qualities. Among these qualities, reduced environmental impact is a key strength. In terms of embodied energy and embodied carbon—such as many studies comparing natural stone with other building materials have demonstrated—the total energy inputs consumed throughout a stone construction life cycle and the carbon footprints of the entire process are lower than those of other building materials, such as concrete, bricks, and concrete block (Table 11.1) (Crishnaa et al., 2011; Garzonio et al., 2010; Morela et al., 2001; Urquhart, 2008; Stone Federation Great Britain, 2011). Although energy used for the extraction of indigenous natural stone, transportation to the factory, processing, and manufacturing bypass the initial carbon impact associated with the production of materials such as concrete, steel, and brick, the transportation of stone over vast distances around the world is a negative

Table 11.1 Embodied carbon of common construction materials

Building materials	kg CO_2/ton
Sandstone	64
Granite	93
Marble	112
General concrete	130
General clay bricks	220
Slate	232
Timber	450
Facing bricks	520
General building cement	830
Steel: bar and rod	1710
Steel: galvanized sheet	2820

Data for granite, sandstone, and slate are from Stone Federation Great Britain, 2011. Natural Stone the Oldest Sustainable Material, SFGB; the others are from University of Bath (Inventory of Carbon & Energy).

factor in terms of environmental impact. For this reason, the most sustainable approach to rock selection will be when it is sourced from a local quarry or reclaimed from demolished buildings (a practice that has gone on for millennia). Once the masonry is installed, recurring embodied energy, represented by the energy used to maintain, replace, and recycle materials and components of a building throughout its life, is lower than that of other materials (Hammond and Jones, 2006).

Further, stone is an abundant and practically inexhaustible resource. Because of its longevity, stone blocks can be either reused in the original structure or repurposed for use in another structure built with compatible material. Even if stone cannot be reused, it can be crushed for use as aggregate.

In terms of energy efficiency, stone masonry can play an important role by virtue of its high thermal inertia, which is the property that enables building materials to slowly absorb, store, and later release significant amounts of heat, contributing to indoor temperature stabilization against outdoor variations. The most energy is saved in summer, in geographical zones in which significant reversals in heat flow occur within a wall between night and day: the heat absorbed during the day by the mass of the wall can be cooled by natural ventilation during the night.

Masonry buildings are a green and a healthy solution also because they are not harmful. There are no toxic materials used in processing, nor are there direct greenhouse gas emissions during processing. Masonry building materials do not contain chemical additives or involve risks to health.

In conclusion, stone masonry can continue to be regarded as a holistic construction technique. The scientific community must invest a deeper knowledge of the material towards a progressive enhancement of its potential performance, by the considering and learning from practices well established over the centuries. Innovation in terms of sustainability is only now becoming sufficiently understood.

References

Arcolao, C., 2001. Le ricette del restauro. Malte, intonaci, stucchi dal XV al XIX secolo (The Recipes of the Restoration. Mortars, Plasters, Stuccos From the Fifteenth to the Nineteenth Century). Ed. Marsilio, Venezia.

Binda, L., Saisi, A., Tiraboschi, A., 2000. Investigation procedures for the diagnosis of historic masonries. Construction and Building Materials 14 (4), 199–233.

Borri, A., De Maria, A., 2009. L'indice di Qualità Muraria (IQM): evoluzione ed applicazione nell'ambito delle Norme Tecniche per le Costruzioni del 2008 (The index of Masonry Quality (IQM): evolution and application in the context of the Technical Standards for Construction of 2008). In: Proceedings of 13th Italian National Conference for Earthquake Engineering, Bologna, Italy.

British Standard Code of Practice, 1951. Masonry Walls Ashlared with National Stones or with Cast Stones, 121.201, London.

Crishnaa, N., Banfillb, P.F.G., Goodsira, S., October 2011. Embodied energy and CO_2 in UK dimension stone. Resources, Conservation and Recycling 55 (12), 1265–1273.

Dernie, D., 2003. New Stone Architecture. Laurence King Publishing, London.

Dipasquale, L., 2012. The Enhancement of Stone Material for the Local Development in the Iblean Area. Innovation, Sustainability and Competitiveness. University of Florence (Ph.D. thesis).

Dipasquale, L., Megna, V., Prescia, R., 2013. Dry Stone Buildings in Sicily, Italy: An Environmental and Territorial Resource. Vernacular Heritage and Earthen Architecture Contributions for Sustainable Development, Taylor & Francis Group, London, pp. 489—494.

Dipasquale, L., Jorquera, N., 2009. Corbelles domes of Apulia. In: Mecca, S., Dipasquale, L. (Eds.), Earthen Domes and Habitats. The Villages of Northern Syria. An Architectural Tradition Shared by East and West. ETS, Pisa, pp. 123—142.

Dipasquale, L., Omar Sidik, D., Mecca, S., 2014a. Earthquake resistant structures. In: Correia, M., Dipasquale, L., Mecca, S. (Eds.), Versus: Heritage for Tomorrow: Vernacular Knowledge for Sustainable Architecture. Firenze University Press, Firenze, pp. 232—244.

Dipasquale, L., Mecca, S., Omar Sidik, D., 2014b. Local seismic culture and earthquake-resistant devices: case study of Casa Baraccata. In: Mileto, C., Vegas, F., García Soriano, L., Cristini, V. (Eds.), Vernacular Architecture: Towards a Sustainable Future. CRC Press, Taylor & Francis Group, London, pp. 255—260.

Dipasquale, L., Volpi, V., 2010. "Masonry Walls", Chefchaouen, Architettura e cultura costruttiva. ETS, Pisa, pp. 127—138.

Doglioni, F., Bellina, A., Bona, A., Biscontin, G., Cusinato, G., Volpin, S., Driussi, G., 1986. Ricerca sulle tecnologie storiche di costruzione e manutenzione del Duomo di Sant'Andrea a Venzone (UD): le malte da sacco murario (Research on the historical building and maintenance technologies of the Saint Andrew Dome of Venzone (UD): the mortars constituting the nucleus of the of the masonries). Proceedings of the Congress "Scienza e Beni Culturali: manutenzione e conservazione del costruito tra tradizione e innovazione" (Science and Cultural Heritage: maintenance and conservation of the building between tradition and innovation), Bressanone 1986, Ed. Libreria Progetto Padova, pp. 571—595.

Foster, A., 2004. Hot lime mortars: a current perspective. Journal of Architectural Conservation 10 (3), 7—27 (Donhead Publishing Ltd).

Fratini, F., Pecchioni, E., Rovero, L., Tonietti, U., 2011. The earth in the architecture of the historical centre of Lamezia Terme (Italy): characterization for restoration. Applied Clay Science 53 (3), 509—516.

Gamrani, N., R'khaChaham, K., Ibnoussina, M., Fratini, F., Rovero, L., Tonietti, U., Mansori, M., Daoudi, L., Favotto, C., Youbi, N., 2011. The particular "rammed earth" of the Saadian sugar refinery of Chichaoua (IVth century, Morocco): mineralogical, chemical and mechanical characteristics. Environmental Earth Science 66 (1), 129—140. http://dx.doi.org/10.1007/s12665-011-1214-6. Published on line 29/07/2011.

Garzonio, C.A., Montanari, F., Torricelli, M.C., 2010. Pietra Serena: Product Quality and Environmental Sustainability. Editore Libria, Melfi.

Gibbons, P., 1993. Hot lime technique: some preliminary investigations. In: 2nd International EUROLIME Meeting, Copenhagen 1993, pp. 119—123.

Giuffré, A., 1991. Lettura Sulla Meccanica Delle Murature Storiche (Survey on the Mechanic of Historical Masonries). Kappa, Roma.

Giuffré, A., 1993. Sicurezza e conservazione dei centri storici. Il caso Ortigia (Safety and preservation of historic centers. The case of Ortigia-Syracuse, Italy). Laterza, Bari.

Hammond, G., Jones, C., 2006. Inventory of Carbon and Energy. Department of Mechanical Engineering, University of Bath. Available at: http://www.ecocem.ie/downloads/Inventory_of_Carbon_and_Energy.pdf.

Hansen, E., Tagle, A., Erder, E., Baron, S., Rodriguez Navarro, C., Van Balen, K., 1999. Effect of aging of lime putty. In: Proceedings of International RILEM Workshop "Historic Mortars: Characteristics and Tests", Paisely (UK), 12—14 May, 1999, p. 10.

Heyman, J., 1995. The Stone Skeleton: Structural Engineering of Masonry Architecture. Cambridge University Press.

Heyman, J., 1996. The stone skeleton. International Journal of Solids and Structures 2, 249–279.

Jedrzejewska, H., 1967. New methods in investigation of ancient mortars. Archaeological chemistry. In: Symposium on Archaeological Chemistry. American Ceramic Society, Washington DC, pp. 156–157.

Kraus, K., Wisser, S., Knöfel, D., 1989. Über das Löschen von Kalk in der 18 Jharunderts: Literaturswertung und Laborversuche (About the slaking of lime in the nineteenth century: literatures evaluation and laboratory tests). Arbeitsblätter für Restauratoren [Worksheets for Restorers] 1, 206–221.

La Spina, V., Fratini, F., Cantisani, E., Mileto, C., Vegas López-Manzanares, F., 2013. The ancient gypsum mortars of the historical façades in the city centre of Valencia (Spain). Periodico di Mineralogia 82 (3), 443–457. ISSN:2239-1002. http://dx.doi.org/10.2451/2013PM0026.

Middendorf, B., Knöfel, D., 1991. Investigations of mortars from medieval brick buildings in Germany. In: Proceedings of the 13th International Conference on Cement Microscopy. EDS Gouda – Nisperos and Bayles, Tampa, pp. 118–123.

Morela, J.C., Mesbaha, A., Oggerob, M., Walkerc, P., December 2001. Building houses with local materials: means to drastically reduce the environmental impact of construction. Building and Environment 36 (10), 1119–1126.

New Technical Codes for Construction, (NTC). January 14, 2008. Ministry Decree, January 14. Gazzetta Ufficiale, February 4, 2008.

Pliny the Elder, 1952. In: Warmington, E.H. (Ed.), Naturalis Historia. De Loeb Classical Library, Cambridge University Press.

Rodriguez Navarro, C., Hansen, E., Ginell, W.S., 1998. Calcium hydroxide crystals evolution upon aging of lime putty. Journal of the America Ceramic Society 81, 3032–3034 (Ed. Blackwell Publishing Oxford).

Rovero, L., Fratini, F., 2013. The Medina of Chefchaouen (Morocco): a survey on morphological and mechanical features of the masonries. Construction and Building Materials 47, 465–479.

Rovero, L., Tonietti, U., 2012. Structural behavior of earthen corbelled domes in the Aleppo's region. Materials and Structures 45, 171–184.

Rovero, L., Tonietti, U., 2014. A modified corbelling theory for domes with horizontal layers. Construction and Building Materials 50, 50–61.

Salerno, G., Formica, G., Gabriele, S., Varano, V., 2010. Stone-masonry new constructions: science and history in the service of beauty and environment. In: Cruz, P.J.S. (Ed.), Structures and Architecture. Taylor & Francis Group, London.

Sani, F., Moratti, G., Coli, M., Laureano, P., Rovero, L., Tonietti, U., Coli, N., 2012. Integrated geological-architectural pilot study of the Biet Gabriel-Rufael rock hewn church in Lalibela, northern Ethiopia. Italian Journal of Geosciences 131 (2), 171–186.

Stone Federation Great Britain, 2011. Natural Stone the Oldest Sustainable Material. SFGB.

Sanz Arauz, D., Villanueva Domínguez, L., 2004. Albarracín y el yeso rojo (Albarracín and the red gypsum). Informes de la Construcción 56 (493), 47–52.

Urquhart, D., 2008. Natural Stone Masonry in Modern Scottish Construction: A Guide for Designers and Constructors. Scottish Stone Liaison Group, Charlestown.

Vegas, F., Mileto, C., Fratini, F., Rescic, S., 2010. May a building stand upon gypsum structural walls and pillars? the use of masonry made of gypsum in traditional architecture in Spain. In: 8th International Masonry Conference, July 4–7, 2010, Dresden Germany, pp. 2183–2192.

Part Four

Timber, bamboo and paper

Nonconventional timber construction

K.I. Crews
University of Technology Sydney, Ultimo, NSW, Australia

12.1 Natural forms of timber materials

12.1.1 Introduction

Timber has been described as the 'Gift of God and friend of man' (Hudson et al., 1986). It is a truly amazing and diverse material. Each tree that produces wood is unique, and each type or species has distinctive qualities and characteristics that distinguish and differentiate it from other species. When a tree is felled, the wood is processed into what we refer to as timber or lumber.

12.1.2 Forests — the production factory

Most of the structural materials that we use today are processed from finite resources that consume large amounts of energy and contribute significantly to the release of carbon dioxide and greenhouse emissions into the atmosphere. In contrast with this, wood literally grows on trees. The production factory for wood is a forest, where each tree takes in water, nutrients and carbon dioxide and uses solar energy to produce oxygen and timber. Consequently, timber is one of the only structural engineering materials that can be totally renewed, provided replanting occurs either through plantations or through managed regeneration after harvesting or felling has occurred.

The uniqueness of each tree means that timber tends to be a variable material that results in its structural behaviour being relatively complex. However, today it is scientifically well understood and essentially behaves as a natural composite material.

12.1.3 Timber — natural rounds

The simplest form of processed wood is to use the tree in its natural, round state. For this to occur, the tree needs to have grown essentially vertically so that the trunk of the tree is relatively straight. The processing after felling usually requires removal of branches and in some cases shaping of the trunk to remove any abnormalities that would prevent the log from being incorporated or used as intended. Typical applications of round timbers include girders for simple bridges or roof structures, wall elements in log houses and utility poles to support electrical distribution and telecommunication services.

Nonconventional and Vernacular Construction Materials. http://dx.doi.org/10.1016/B978-0-08-100038-0.00012-3

12.1.4 Timber − sawn members

Perhaps the most common form of processing wood is to saw the log into individual pieces of timber that are sized and shaped to provide structural elements that may be used individually or combined into a more complex structural system, using fasteners or connectors to join the pieces of timber together. Beams and planks are sawn into rectangular sections, whilst columns are usually square in shape. Felling of old-growth forests has, in the past, often produced long-spanning, large rectangular beams − made possible by the sheer diameter of the tree (Fig. 12.1).

Over the past 100 years, the nature of forests and timber resources has changed significantly, primarily due to the fact that many virgin forests consisting of large-diameter, old trees have been logged. When trees have been replanted, the regrowth has consisted of either plantation species, which tend to be faster growing and are intensively managed for harvesting, or, in the case of native forests, are selectively logged on a predetermined rotation basis. The result of this change of available resource is that the nature of wood products in regrowth or plantation forests tends to be different from that of old-growth forests. One of the key differences is the size of the log when the trees are felled for processing, which in turn restricts the commonly available section size and length of sawn timbers to smaller than 300 × 90 mm and less than 6 m, respectively.

12.1.5 Timber − processed elements and products

The demand for wood products means that trees are felled at an earlier age and consequently the diameter of the log is significantly less than that which was previously available from old-growth forests. This has led to the development of more complex and sophisticated processing techniques, whereby the log is either peeled, chipped or stranded into veneers and/or fibre that is in turn reconstituted by gluing the wood material into what is referred to as an engineered wood product.

Figure 12.1 A 1.8-m-diameter Douglas Fir log, 460 years old when felled. Pacific Forestry Centre, Natural Resources Canada − Victoria, BC.

This processing of wood has a number of significant advantages. First, it permits fabrication of panel elements, such as plywood and oriented strand board (OSB) or larger section timber elements, such as glued laminated (glulam) or laminated veneer lumber (LVL) members, capable of spanning in excess of 25 m as simply supported beams and even further when combined into truss or portal frame structural systems. Second, processing reduces the influence of natural growth characteristics on the properties of structural elements, usually resulting in an increases to strength and stiffness, accompanied by a reduction in material variability. Third, processing permits enhanced levels of quality control and dimensional precision that is not as readily available for round and sawn timber products. The 'down side' with processing is that it is by nature more energy intensive and necessitates the use of various types of glue to fabricate the final product.

12.2   Traditional and historic applications (round and sawn timber)

12.2.1   Simple structures

Historically, wood was often used in its natural form to provide shelter from the elements (Fig. 12.2) or to bridge an obstacle, such as a stream or gully (Fig. 12.3). Preparation was minimal — the tree was felled, branches removed, then the logs were installed into position. Where required, logs were connected together using twine or natural rope. Whilst examples of this basic structural form are still used today, the applications are generally for temporary structures, which have a relatively short service life.

The simplest application of round timber construction that is prevalent today — is the timber pole (Fig. 12.4), which is frequently used to support utilities such as power and telecommunication infrastructure. In these applications, the pole behaves as a cantilever, resisting a load applied at the top of the pole by designing the pole to have adequate capacity at ground level, in order to resist the bending moment induced by the load.

Processing requirements vary depending on the timber species used but generally include de-barking the pole, shaping it by shaving irregularities to make the pole appear straighter, partial drying of the pole (known as seasoning) and impregnating the pole with some form of chemical preservative to enhance the durability of the pole against fungal and insect attack.

Another common application is for fencing, where the timber is used either in round or rectangular sections as both posts and beams.

12.2.2   Bridges

Timber has been used as a structural material to build bridges for hundreds of years. Simply supported girder (or stringer) bridges, with timber decking planks and running boards, are still operational in many parts of the world, often spanning 10 m or more. Multiple span bridges, consisting of two or more noncontinuous spans, often include

Figure 12.2 A simple shelter constructed of logs (only the branches have been removed).

corbel members to increase the effective span of the primary elements and provide some partial load transfer over the headstocks, as can be seen in Fig. 12.5.

Longer spans of 30 m or greater require timber to be used in truss and/or arch forms. The design of timber truss bridges developed rapidly during the latter part of the 19th century, particularly in North America and Australia, in order to provide road and rail crossings to support transport infrastructure to access remote areas and exploit natural resources. Unfortunately, many of these bridges were not designed for durability, were not well maintained and have deteriorated due to exposure to the elements and other hazards.

Whilst many timber truss bridges are no longer in service for road or rail transport, there are some notable exceptions, where heritage bridges have been rehabilitated and upgraded to carry current design loads and where modern timber truss bridges have been built to show case timber as a viable structural material for the 21st century.

12.2.3 Buildings

Variations on the ways in which timber has been historically used in structural forms for buildings are almost endless. Post and beam construction, arch forms and roof

Figure 12.3 A simply supported log bridge crossing a small river.

structures, for buildings such domestic dwellings, industrial factories and multistorey warehouses, rural buildings and storage facilities, churches and large public meeting places (Fig. 12.6), are just some of the numerous applications of timber in building applications. Traditional construction practices in Northern Europe often built buildings entirely from timber, whilst in other parts of Europe timber was often used for roof structures supported on masonry walls.

Before the development of mechanical connectors, such as bolts and nails, larger timber buildings tended to be post and beam construction, designed to avoid tensile forces by structural forms that keep the timber in compression or bending and use wooden dowels, where necessary, to prevent relative movement by securing members together. The development of techniques to bend timber members by steaming, led to widespread use of timber compression arches, particularly for roof structures (Fig. 12.7).

These traditional forms for timber find applications today in wall frames, roof trusses and post and beam construction used in modern timber houses, as seen in Fig. 12.8.

Figure 12.4 Typical timber utility pole used for power transmission.

(a)   (b)

Figure 12.5 An 80-m-long multispan timber girder bridge in Australia.

(a) (b)

Figure 12.6 Historical market place building in France, using post and beam construction.

Figure 12.7 Church roof in southern France.

Figure 12.8 Example of a modern house, using post and beam construction.

12.3   Material properties

12.3.1   Understanding the nature of wood as a natural composite

Wood is an organic material that is created as the process of photosynthesis converts carbon dioxide, water and nutrients into a network of complex chains of cellulose molecules, which in turn combine together to form cells. These cells are bonded together with lignin, which acts as a gluing agent to form long, slender fibres that are aligned with the long axis of the trunk. The fibres become what is known as the 'grain' in the wood, which not only creates a distinctive texture and appearance but also contributes to the inherent material properties of a piece of timber.

Many of the physical properties of timber members that interest structural designers are functions of the fibrous microstructure of the wood material:

- Density: Cell structure and size, moisture content
- Strength: Density, moisture content, cell size
- Shrinkage: Cell structure and size, moisture content
- Stiffness: Density, cell structure and size, moisture content
- Colour: Extractives
- Fire resistance: Density, extractives
- Termite resistance: Cell size, extractives
- Electrical resistance: Moisture content, cell size
- Mechanical damping: Cell structure, moisture content

12.3.2   Structural properties

As a consequence of the directional fibres, the properties of wood are anisotropic with much higher stiffness and strength parallel to the grain than across the grain. To illustrate this simply, we can liken the structure of wood to a bunch of parallel drinking straws (representing the fibres or grain of the wood), which are bonded together using a weak glue.

When load is applied parallel to the axis of the straws (Fig. 12.9(a)), they are very strong in tension (for some species, >200 MPa) and have reasonably good compressive strength until they start to buckle. The compression strength of wood fibre is usually <100 MPa, but the failure stress can be sustained over relatively large deformations of the sample. Bending can be thought of as a force couple that induces tension on one edge and compression on the other, and because of the high tensile strength of clear wood, the bending failure of small clear specimens is almost always limited by compression strength and is consequently ductile.

However, if the load is applied perpendicular to the axis of the straws (Fig. 12.9(b)), they will tend to crush under compression (exhibiting considerable ductility) and are weakest in tension, where the loose bonds between cells fails and the straws literally tear apart (Boughton and Crews, 2012). It is important for engineers and architects to develop an understanding of these properties that reflect

(a)

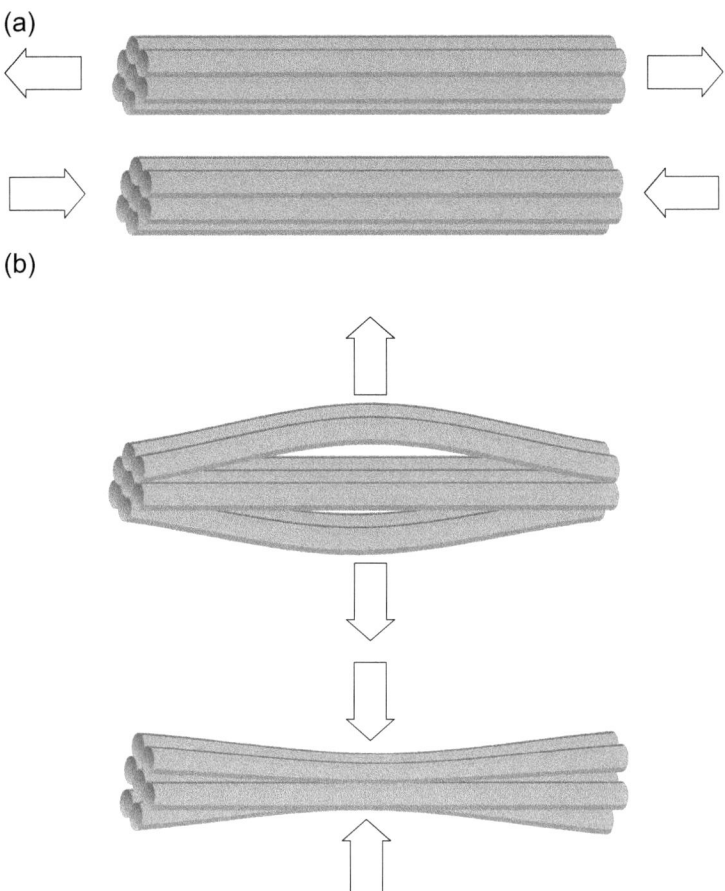

(b)

Figure 12.9 Orthotropic nature of wood fibres. (a) Strong in loading parallel to grain. (b) Weak in loading perpendicular to grain. (Arrows indicate direction of applied force.)

the different ways that timber behaves when loaded in different directions to the wood fibres, in order to be competent to design and detail safe and durable structures.

12.3.3 Wood and timber

From an engineering point of view, wood is as different from timber as cement is from concrete. It is therefore important to recognise that wood and timber are in essence two different materials, which is based on the fact that there is a fundamental difference in their respective modes of failure (Madsen, 1992). Clear wood (ie, free from any natural characteristics such as knots, gum veins, sloping grain, splits or checks) is very strong in tension and, when subjected to bending, tends to fail in the compression zone, resulting in essentially ductile behaviour. Structural timber on the other hand, contains clear wood but also has knots and other growth characteristics that interrupt the continuity of the wood fibres or create stress concentrations, which in turn reduce

Figure 12.10 Flexural failure on tension side of hardwood beam.

its strength when compared with that of clear wood, particularly when subjected to tension or bending.

Failure of structural timber members under tension or bending is normally brittle and tends to be initiated on the tension edge in the vicinity of knots or where relatively steep localised slope of grain occurs (Fig. 12.10). Both of these growth characteristics interrupt the continuity of the wood fibres and induce tension perpendicular to the grain, which we have already noted is the weakest property of wood material.

12.3.4   Effects of moisture

The cells in wood are very porous by nature and serve to transport waterborne nutrients from the ground to the leaves in a tree; for example, the cells in a living tree are full of water. This produces a very high moisture content.

Moisture content (mc) is measured as:

$$mc = (weight\ of\ water)/weight\ of\ dry\ wood) \times 100\% \qquad [12.1]$$

In a growing or newly felled tree, the moisture content measured in this way is frequently >100%. However, once a tree dies or the wood is cut from the trunk, the moisture begins to evaporate and will eventually dry out to be in equilibrium with atmospheric moisture. The drying process involves two basic stages — removal of 'free' moisture that is within the cell cavity, followed by removal of 'bound' moisture that is combined within the cell walls. The transition point at which time bound moisture begins to evaporate is known as the fibre saturation point (FSP), which is regarded as between 25% and 30% moisture content, depending on which part of the world you happen to be located.

Unseasoned (green) timber members will slowly lose moisture to reach the equilibrium moisture content for the environment in which the timber has been placed. The loss of moisture below the fibre saturation point means that the timber tends to shrink during the drying process, as bound water is removed from the cell walls. This effect is illustrated in Fig. 12.11. Seasoned timber will lose little additional moisture in most environments and, therefore, any shrinkage in service tends to be minimal.

Equilibrium moisture content (*EMC*) in most environments is normally within the range of moisture contents to which timber is seasoned. Seasoned timber tends to have superior dimensional stability and is much less prone to warping and splitting in service than unseasoned timber. In higher grades of timber, particularly hardwoods, the process of seasoning can enhance the characteristic properties of timber, increasing

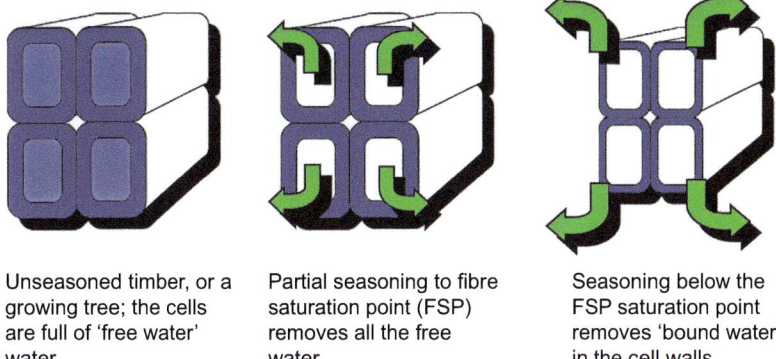

| Unseasoned timber, or a growing tree; the cells are full of 'free water' water | Partial seasoning to fibre saturation point (FSP) removes all the free water | Seasoning below the FSP saturation point removes 'bound water' in the cell walls |

Figure 12.11 The effect of seasoning on moisture.

stiffness by up to 25%. Testing of small clear specimens shows that bending strength can increase up to 60% and compression strength by up to 90% (Boughton and Crews, 2012).

Provided the *EMC* of the environment in which the timber is used is <18%, seasoned timber is also less vulnerable to fungal decay, since the reduction in moisture creates an 'environment' in the timber that is much less attractive to fungal colonisation. Drying of timber to remove both the free moisture and the bound moisture can be accomplished in a number of ways. The two most common methods, which are often used in combination, are air drying and kiln drying; the former normally dries timber to about 20% to 25%, whilst the latter is used to dry timber to 10% to 15%. The aim is to remove the moisture at a uniform rate through the piece, so that internal growth stresses are carefully relieved to avoid splitting and there is minimal moisture content differential within the piece. Where a large moisture content differential exists, then degradation of the timber (seasoning or drying degrade) can occur if the seasoning process is not carefully controlled.

12.3.5   Time-dependent behaviour

As noted previously, wood can be understood as a naturally occurring composite material, and because of this fact, there have been many different studies undertaken around the world on both full-scale timber members and small clear specimens, to investigate and quantify the effects of sustained (time dependent) loading on both deformation or creep and strength.

From a design point of view, these two effects need to be understood as duration of load in stiffness and duration of load on strength. Because the individual wood fibrils are only loosely bonded to form wood fibres, the fibrils can slide over the top of each other under stress. This material behaviour leads to two related effects:

- Creep under long-term loads, in which deflection of beams and elongation or shortening of axially loaded members may increase under constant load
- Reduction in strength under long-term loads

Both of these effects are exponential. Creep deformations increase significantly with changes in moisture content, particularly when the moisture content is cycled. The mechanism by which this occurs is complex and is generally referred to as mechanosorptive behaviour. Under normal use conditions, the ambient humidity changes, sometimes throughout the day but certainly between seasons. The changing moisture loading of the air means that moisture will move into or out of the timber. This can lead to substantial increase in creep or a significant loss of strength under long-duration loads.

Even under constant relative humidity conditions, unseasoned timber will gradually lose moisture over time. The moisture movement through the timber will contribute to higher creep deflections, as illustrated in Fig. 12.12(a). The elastic deformation will occur instantaneously as soon as the load is applied, the value of which is predicted by statics as a function of the section properties and the modulus of elasticity. However, if the load remains, the deformation will slowly continue to increase even though the load remains constant. Fig. 12.12(a) shows two idealised deformation curves − one for seasoned timber and one that has more creep deformation and applies to unseasoned timber.

Fig. 12.12(b) shows the reduction in strength that is due to microscopic damage to the wood matrix caused by the duration of load effects. It too is an exponential

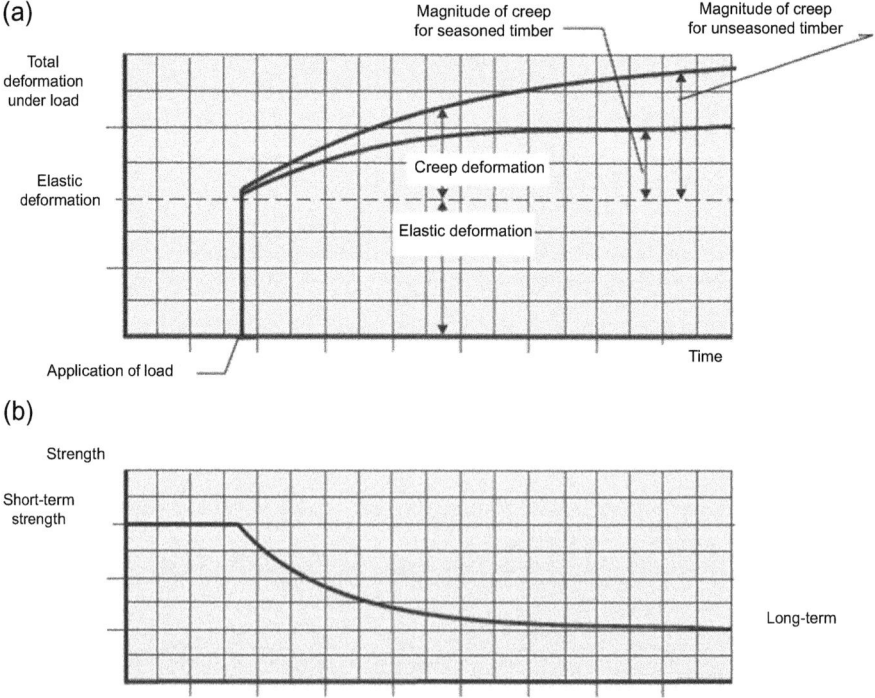

Figure 12.12 Time-dependent effects of load on timber. (a) Creep deformations under load, (b) loss of strength under load.

relationship and can be described as a cumulative damage effect. Both duration of load and creep effects are modelled in timber codes throughout the world by the use of correction factors, which reduce the short term strength and increase the initial deflection, as the duration of load increases.

12.3.6 Durability

Integral to any valid limit states design philosophy is the fact that durability needs to be considered during the design process, irrespective of the material type, if a structure is to perform its intended function for the intended service life. Durability can be defined as 'the capacity of a product, component, system building or structure to perform the function for which it was intended, be it aesthetic, structural or amenity for a specified period of time' (Boughton and Crews, 2012).

In most parts of the world today, most new buildings are designed for an intended life of 50 to 100 years, and because all building materials are subject to degradation, durability considerations are very important to ensure an appropriate service life for the structure being designed. Unfortunately, many designers have been trained to only consider the performance of a structure in its 'as new' construction condition; the need for the structure to be durable is often addressed as something we 'add on', by specifying some sort of protective system (eg, painting) rather than being an integral part of the design process. The result is the problem of ageing infrastructure: structures that deteriorate and need significant maintenance or rehabilitation well before the service life is reached. By contrast, in many parts of Europe we can find numerous examples of buildings and structures still functioning after hundreds of years of service — and in some cases there are timber structures that are >1000 years old. Therefore, the perception of timber buildings and structures having a life of only a decade or so can be erroneous, and it is important to study these buildings to gain insights into how to design timber structures to be durable.

Because timber is a cellulose-based material, it is subject to biological hazards, in addition to the types of degradation agents applicable to other materials such as steel and concrete. It is therefore important to identify all the hazards that will create degradation, if we are to be able to design for durability. The performance of a timber member could be compromised by any of the following:

- The wood is attacked by insects or fungi (rot).
- Adhesives break down over time, particularly if exposed to ultraviolet and/or moisture.
- Connectors corrode.
- Fire causes loss of cross section in a member.

Durability-based design will take these factors into account and specify:

- detailing that will minimise the factors that lead to degradation, in particular how the timber members are isolated or protected from any identified hazards,
- appropriate timber species, considering natural durability,
- sizes that allow for satisfactory performance after some loss of section or
- appropriate chemical treatment to give performance over the required lifetime.

It is important to recognise that durability considerations will affect other limit states such as strength and serviceability. The process of design for durability includes the following steps:

1. Determine the specific application – linked to the functional requirements of the structure/ building.
2. Estimate the performance requirements and determine the various 'limit states' of strength, serviceability and stability. The performance requirements will need to consider the following factors, each of which can determine the extent to which durability affects the design:
   * required service life
   * required reliability for both functionality and safety
   * initial versus ongoing costs for acceptable maintenance and rehabilitation
3. Determine what, if any, environmental hazards are present (eg, the exposure class or hazard level including the severity and type of the hazard).
4. Develop appropriate design details (including flashings, physical barriers, moisture isolation details, sacrificial elements) and materials specifications (eg, species, preservative treatment(s) and surface finishes) for members and connections.
5. Check that the functional and performance requirements will be met by the proposed design and specifications and refine the design details if necessary.
6. Define inspection and maintenance procedures to ensure that the functional requirements of the structure/building continue to be satisfied for its service life.

In order to make appropriate design decisions about durability, it is first necessary for a designer to understand the nature of the hazard to which the timber will be exposed. In nearly all cases, this is not straightforward. However, one simple and basic 'rule' for timber design always provides a good starting point: 'Keep it dry'.

Obviously, there are many design situations where timber is required to be used in wet or moist environments, but the principle of designing details so as to not allow moisture pockets or create moisture traps is a good one to adhere to wherever possible. In this way, the detailing can be used to reduce the hazard. The 'keep it dry' principle leads to the following:

* Arrange upper surfaces of exposed timber members to have a slope so that they shed water.
* Eliminate moisture traps where possible.
* Avoid having timber in contact with ponded water (use concrete plinths).
* Apply and maintain protective coatings.

Biological degradation (both termite attack and fungal attack) will be enhanced if the wood is damp. Both are also enhanced if the timber is in contact with the ground so that the ground-dwelling termites or fungi have access to the wood, but we must not forget that airborne fungal spores mean that timber that is well above ground level can also rot.

Designing timber structures to be durable embodies both the art and science of engineering; it is not simply specifying a durable timber or a level of preservative treatment. Rather it requires a fundamental understanding of timber as a material and an approach that 'designs durability' into the details of a structure, whereby the use of modelling tools and material specifications is part, not all, of the final solution. However, experience in many places of the world has clearly demonstrated that in the hands of creative and suitably trained designers, this renewable engineering material is aesthetically pleasing, structurally reliable, cost efficient and durable (Crews, 2003).

12.4 Modern engineering applications

When timber products are used for structural applications, the design properties are selected and specified on the basis that timber product has been graded or manufactured and assigned a set of structural properties. In some cases, structural timber might also be used where it is visible or touched or for a specific architectural feature. In these cases, it has both structural and appearance functions. Structural timber and timber products can be used in a variety of applications, including:

- Domestic construction, eg, housing or multiresidential complexes
- Large-span structures, such as public auditoriums
- Commercial buildings
- Bridges

12.4.1 Grading

Timber grading is the process by which timber is sorted into groups with similar structural properties in each group. Inevitably, there is a range of properties within a group and some overlap in properties between the groups. Grading of structural timber can be performed in a number of ways, including:

- Visual stress grading
- Machine stress grading using a variety of scanning technologies
- Machine proof-grading

The two most commonly used methods are visual stress grading and machine stress grading. In theory, any method that sorts timber into groups of material with similar structural properties could be used. However, in order to be valid, the sorting methods must produce results that are both consistent and repeatable. To achieve this level of reliability, rules for using acceptable sorting methods have been developed. In most countries, these rules are generally either industry product standards or national product standards.

These standards are used in production as the basis of applying a grade stamp to each piece of timber or each timber product, for example, plywood or OSB. Design professionals do not normally have to be expert in the use of these standards but should be aware of their existence as they are often referred to, or used in, materials specifications. A basic knowledge of the principles of visual stress grading is certainly useful for engineers and architects undertaking site inspections and building certification.

12.4.2   Reconstituted or manufactured timber products

A large range of different types of reconstituted timber products has been developed, tested and commonly used in construction. They are manufactured to minimise the effect of strength-reducing characteristics, and their properties have lower coefficients of variation than those of sawn timber. These products are generally manufactured under a national or international product standard or an industry standard and have stringent industry supervised quality control programs in place.

Manufactured timber products are often referred to as engineered wood products (EWP's) and can be broadly classified into three categories:

1. Sheet products, such as plywood and OSB
2. Beam products, such as glulam, strand lumber and LVL
3. Massive timber products, such as CLT (cross-laminated timber)

12.4.2.1   Sheet products

Plywood is made by slicing wood into thin veneers and gluing the veneers together to form a sheet. Each veneer is laid up at right angles to the previous one giving a sheet with just more than 50% of the fibres in the main direction (the face grain direction) and just less than 50% at right angles to that direction.

Consequently, plywood sheets have properties that are nearly the same in both directions in the plane of the sheet, making it ideal as a membrane. Plywood is used in nailing gussets in trusses, webs in I-beams or box beams and bracing in buildings. Thicker sheets can function well as slab elements and are often used in formwork or floors. Fig. 12.13 shows plywood gussets connected to LVL webs, as part of a 6-m-span composite floor beam being tested.

OSB is manufactured from large flakes of wood fibre that are glued and pressed into large, flat sheets, which can be used in some of the same applications as plywood, eg, I-beam webs or bracing panels. Properties and design methods for European or North American OSB products can be obtained from the manufacturers.

Figure 12.13 LVL composite beam with plywood gussets.

Strand lumber products (such as LSL and ESL) are manufactured from flakes or strands of wood fibre that are glued and pressed into rectangular shapes as a substitute for sawn timber. Few standards for the manufacture of these products exist, and designers must rely on manufacturer information regarding properties and design methods.

12.4.2.2   Beam products

Glulam is manufactured by a process in which large structural members can be made from a large number of smaller sawn pieces. It is common to face laminate the pieces to produce a deep section with the laminates horizontal, although in Europe, glulam has also been manufactured with face and edge lamination to produce deep sections with vertical members.

By staggering the butt joints in the laminates, it is possible to manufacture very long glulam beams. Simple butt joints create discontinuities in the fibres of the beams, and in order to overcome this problem and provide some level of continuity, a process known as finger jointing is implemented, which enables a large surface area of glue to transmit the force across the discontinuity at the end of one lamination to the next. The glue in finger joints transmits load in shear across the glue line, which creates a much stronger joint than a butt joint and provides performance close to that of continuous laminations from a single piece of timber.

Because the beams are fabricated from smaller pieces of timber that can be easily bent, it is possible to manufacture curved (see Fig. 12.14), tapered and cambered glulam beams at a cost only slightly higher than that of straight beams.

It is common practice to form a slight upward camber (hog) in nominally straight glulam beams so that under permanent actions they appear flat. Glulam can be manufactured to almost any depth, length or shape; as such, it is suitable for use in large span structures or for unusual configurations.

Manufacture of LVL involves peeling off plies from good quality logs and vertically laminating them with the grain on most plies running in the same direction. This gives orthotropic properties in a similar way to the properties of sawn timber. There are no generic grades for LVL, since each manufactured product has its own grade and properties. As such, a designer needs to use design characteristic properties obtained from a manufacturer and specifies the product by its commercial name.

The effect of a single growth characteristic (or defect) in an LVL laminate is minimal, since the laminates are so thin compared with the thickness of the whole member. The properties of LVL show much less variation than those of both sawn timber and glulam. In fact, LVL has a coefficient of variation in structural properties that is similar to hot rolled steel sections produced in many countries. Since LVL is a manufactured product, it can be made to very large dimensions, with most modern plants involving continuous-length production. Large sheets of LVL can be ripped or crosscut to give a very large range of product sizes. LVL can be used wherever sawn timber is used, but because of the long continuous lengths and the high structural reliability of the product, LVL is often used in

Figure 12.14 Curved glulam beam.

high-strength applications such as the flanges in timber I-beams and ply-webbed box beams, or portal frames and other large span structures. LVL is also commonly used in formwork bearers.

12.4.2.3 Massive timber products

CLT is a large-dimension manufactured timber product, sometimes referred to as 'massive timber'. Panels, as shown in Fig. 12.15(a), are typically 100−300 mm thick and the other dimensions are typically 3 m or more in either direction. Lamination boards used in CLT are usually 90 mm wide and 20 mm thick (Fig. 12.15(b)) and are laid out side-by-side to form large plates of sawn wood. Each plate is glued to other plates to form a product that has been described as 'plywood on steroids'.

CLT is still regarded as a relatively new timber product, and although North American and European design guidelines have been published, CLT design and construction remains a specialist area which builds on normal structural design principles for timber. Fig. 12.15(a) illustrates the arrangement of boards in CLT panels.

(a)   (b)

Figure 12.15 (a) CLT panel (five ply) and (b) CLT laminates prior to gluing.

In some respects, CLT is a 'hybrid' between glulam and plywood in that the laminates are similar to glulam in the way the adjacent layers of boards are face glued to form a deeper section, and similar to plywood in the way that the grain orientation of each layer is at 90° to the layer below it. This means that within the plane of each CLT panel, there are fibres running in both principal directions, which makes CLT a timber panel product; although recent research in Europe has also been using CLT in beam applications.

Due to its prefabricated manufacturing as a panel product, CLT is sometimes referred to as 'tilt-up timber'. Because of this, it can be used in a similar way to precast concrete panels, which makes it well suited for apartments and schools where there are many walls in the structure.

- Where the spans are appropriate, it can be used to form floor slabs (usually five layers). In some cases, a topping is required to minimise vibration and noise transmission.
- CLT is commonly used for wall panels (usually three to five layers). These can be lifted in by crane and fastened directly to the floors underneath. Fixing directly to other wall panels provides a strong and effective floor support for the next level and minimises the need for false work or propping. Doors and windows can be pre-cut if necessary and chases can be designed in to accommodate plumbing and wiring.
- CLT panels can also be used as roof panels and because of the lower loads, three to five layers are often sufficient.

Because the timber fibre alignment in CLT panels has similar characteristics to plywood, many of the techniques used for design of plywood can also be applied to CLT. For example:

- Out-of-plane loading of CLT panels (floor applications)
- In-plane loading (wall applications in resisting racking forces)
- Axial compression (wall applications in resisting gravity loads)

CLT panels are lighter and easier to erect than those made of concrete, making CLT an attractive and promising option for design. The speed of erection and the absence of wet trades make it possible to achieve very speedy construction times (Boughton and Crews, 2012).

12.5   Significant case studies

The development of EWPs has been accompanied by somewhat of a renaissance of timber design for modern buildings and structures, over the past 20 to 30 years. There are many excellent examples of iconic timber structures throughout Europe, North America and, more recently, Australia and New Zealand. An overview of a select few of these is presented in the following section.

Case study 1: Canada − Credit Valley Hospital, Mississauga, Ontario.
Architects: Farrow Partnership.
Engineering and Construction: Timber Systems Limited (Fig. 12.16).
Credit Valley Hospital is a fully integrated medical complex that with a particular focus on ambulatory recovery and cancer treatment. The design challenge for both the architect and engineer/contractor was to create a relaxed, open atmosphere, using natural warm materials to create the sense of a healing environment among patients and staff. Glulam timber was used as the main structural material, fabricated using sophisticated connection systems to create tree-like structural forms. The end result is that visitors to the hospital feel that they have entered an open forest, rather than a clinically 'cold' health building. It is an excellent example of how timber can be used to create a 'human' space in a modern medical facility. (Farrow Partnership, 2015).

Case study 2a: Norway.
Tynset Bridge, Tynset.
Design: Norconsult AS; Plan Arkitekter; Moelven Limtre AS and Public Roads Adminstraton Vegdirektoratet.
Construction: Moelven Limtre AS and Public Roads Administration, Hedmark (Fig. 12.17).
The Tinset Bridge was constructed in 2001 as an outstanding demonstration of technology developed during the third phase of an initiative known as the Nordic Timber Bridge Program. It was the first large timber bridge in Norway designed to support full highway loadings, with a design life of 100 years. The main structural system consists of three, two hinged truss glulam arches (2×27 m and 1×70 m spans), supporting a

(a)   (b)

Figure 12.16 (a) and (b) Glulam primary structure in the hospital foyer.

Figure 12.17 Side view Tynset Bridge.

stress laminated timber deck with galvanised steel tension ties and crossbeams. The timber arches are double impregnated with CCA and creosote, protected by a sophisticated system of copper flashing; whilst the deck planks are creosote treated. The final design was cost competitive compared to steel and concrete bridge alternatives, providing an outstanding example of how timber can be successfully used to design an attractive, durable modern bridge structure (Aasheim, 2002) (Fig. 12.18).

Case study 2b: Norway.
Rena River Military Bridge.
Design: SWECO Structural Engineers.
The bridge across the Rena River (known as 'Kjøllsæterbrua') was built in 2005 as part of the road network within a Norwegian military training facility. The two lane timber truss bridge is 158 m long (Fig. 12.19) incorporating a continuous truss superstructure with six spans, where the longest span is 45 m. The truss is manufactured from some 250 m³ of high-grade glulam, assembled using elements with lengths of up to 30 m — which is the maximum length that can be transported using public roads in Norway. The service life of the bridge is 100 years.

Figure 12.18 Connection and flashing details — Tynset Bridge.

Figure 12.19 A 158-m-long bridge across Rena River, Norway.

Due to it use by military vehicles, the bridge was required to meet both civil and NATO design loading requirements. The Norwegian Public Roads Administration requires capacity for two 65-ton trailers (one in each lane), whilst the Military Load Class 100 (MLC 100) specifies that the bridge had to be designed for a convoy of vehicles 30 m apart and weighing up to 109 tons each. The dynamic load allowance increases these basic design loads by 15%. Such heavy live loads also necessitated construction of a 310-mm-thick concrete deck as the wearing course, with a self-weight of 8.5 kPa. The deck is noncomposite and 'floats' on top of the timber trusses to allow for temperature and shrinkage movements. Consequently, it is believed to be the 'World's strongest timber bridge' (Abrahamsen, 2008).

Case study 3: Australia.
Dr Chau Chak Wing Building − Sydney.
Architects: Gehry Partners.
Engineers: ARUP.
Opened in January 2015, the Dr. Chau Chak Wing Building is the spectacular new home to the University of Technology Sydney Business School. Inspired by the idea of a treehouse, the building was designed by architect Frank Gehry as 'a growing learning organism with many branches of thought' to create not only a new Sydney landmark but also internal spaces that inspire real and relevant research and learning outcomes, interdisciplinary collaboration and the cross-pollination of ideas (UTS, 2015).

Whilst most of the building is constructed from masonry and concrete, Gehry's love of timber is reflected in two oval classrooms that have been constructed (one above the other) from around 150 large laminated timber beams within the ground-floor atrium void, which is framed using glulam timber supporting the glass façade. The massive timber beams (Fig. 12.20) each weigh up to two tonnes, with the longest measuring 12 m. The geometry of the oval rooms necessitated floor spans of up to 8 m with a relatively low profile, which was ideally suited to a wide-beam timber concrete composite

Figure 12.20 External view, oval classrooms.
Photograph: Andrew Worssam.

Figure 12.21 TCC floor over one of the oval seminar rooms.
Photograph: Andrew Worssam.

(TCC) system using inclined screw fasteners as the connection system, as indicated in Fig. 12.21. When the building was opened, this was the longest spanning TCC floor in Australia and New Zealand.

Case study 4: Australia.
CLT Forté Building, Melbourne.
Design and Construction: Lend Lease.
The Forté building in Melbourne's Docklands area is Australia's first timber high rise apartment building and, when completed late in 2012, was the tallest timber building of its type in the world. The residential complex is 10 storeys (approx. 32.2 m high) and houses 23 apartments with retail premises at ground level. It is also the first major building in Australia to be constructed using CLT (Fig. 12.22) and the first residential building to achieve an As Built Five Star Green-Star certification (Victor Harbour, 2013).

The decision by Lend Lease to build in CLT has been informed by a number of drivers. According to company estimates, the use of CLT provided similar levels of

Figure 12.22 Prefabricated panels — Forté building (Emma Cross photographer).

structural integrity as traditional concrete buildings whilst delivering better thermal performance and reducing the building's projected life-cycle CO_2 emissions by around 1400 tonnes. The building was almost entirely prefabricated, and subsequently the CLT construction process involved less material on site and was cleaner, simpler and faster than a reinforced concrete alternative. This also led to improvements in safety and on site quality control (Lend Lease, 2012).

The building comprises 759 CLT panels of European spruce (*Picea abies*), weighing a total of 485 metric tons. The spruce for the CLT panels was grown and harvested in Austria; the panels were manufactured and then shipped to Australia in 25 shipping containers. The building arrived on-site like flat pack furniture, including the 5500 angle brackets and 34,550 screws (Figs. 12.23 and 12.24) required for erection (Woodsolutions, 2015a).

12.6   Codes and standards

Codes and Standards for the use of timber throughout the world can be broadly categorised as belonging to either 'design standards' or 'product standards'. Whilst there is

Figure 12.23 Connecting panels — Forté building (Emma Cross photographer).

Figure 12.24 CLT construction in Austria similar to that used in the Forté building.

no internationally accepted standard for design of timber, the need to be able to trade timber products between countries has been an incentive to develop 'international' standards, primarily through ISO – the International Standards Organisation. These standards focus on test methods and determination of characteristic properties that can then be used for design of timber members in a particular country.

Current design standards for structural timber in North America, Europe, Australia and many parts of the world are all published in a Limit States or Load and Resistance Factor Design format. These types of standards are based on a philosophy of identifying the 'limits' of acceptable structural behaviour for function and appearance; strength or damage; robustness and public safety; whilst resisting various combinations of anticipated 'load events'. This is represented by Eq. [12.2], where R_d is the design capacity of the member and S^* is the relevant load event.

$$R_d \geq S^* \tag{12.2}$$

The design capacity is based on the geometry of the member (eg, area, section modulus); its characteristic material properties (eg, bending, tension, compression or shear); modification factors to account for the behaviour of timber in specific situations; and a capacity reduction factor to account for uncertainty and the desired level of reliability.

North American and Australian codes (such as CSA 086 and AS1720.1 respectively) have the form of Eq. [12.3a], whilst the Eurocode (EC5) uses the form of Eq. [12.3b]. However, despite the slightly different format, the approach is essentially the same.

$$R_d = \Phi k_{mod} \cdot R_k \tag{12.3a}$$

$$R_d = k_{mod} \cdot R_k / \gamma_M \tag{12.3b}$$

where R_k is the nominal capacity (characteristic material strength $\times$ section property), R_d is the design capacity.

$$k_{mod} = k_{dol} * k_{env} * k_{geom}$$

and

k_{dol} = time-dependent effects (duration of load)
k_{env} = environmental effects (temperature and moisture)
k_{geom} = geometric size (or volume) and stability effects
Φ = capacity reduction factor
γ_M = material adjustment factor

Both of these latter two terms account for uncertainty in the material properties and essentially 'adjust' the level of reliability and probability of 'failure' to that required for the structural element being designed.

Despite the apparent differences between current design codes, ongoing development of standards over recent years have seen a convergence in the philosophy underpinning design procedures for timber in use throughout the world, and it is now not difficult to perform 'equivalency' designs using different standards to arrive at similar solutions for a given set of loading events.

12.7   Conclusions

Timber is a remarkable material and since the early 2000s, there has been a resurgence in its use in new buildings and a growing awareness among design professionals of the potential role timber has to play in the 21st century as a modern building material. Because it is inherently a 'green' construction material, the environmental and sustainability credentials of timber are beyond dispute and superior to the 'normal' building materials such as concrete, steel and masonry.

So how will timber be used in the future?

As at 2015, there is now a worldwide interest in multistorey timber buildings. This is not just because of the environmental advantages of timber construction when compared to buildings constructed from concrete and steel; it also relates to efficiencies in design, prefabrication, constructability and quality of the finished product (Chapman, 2014).

Yet, it should be noted that tall timber buildings are not 'new'. Large wooden temples, such as the five-storey (32.45-m-high) Hōryū-ji pagoda were built in Japan some 1400 years back, still stand today despite being located in a moist environment and an area of high seismicity. Both Australia and Canada have a history of constructing tall timber buildings including examples in Vancouver of up to nine stories. Such heavy timber post and beam structures have been functioning effectively and safely for >100 years (Green, 2012). Many of these buildings have been 'recycled' from their initial lives as storage facilities, into new multistorey office and residential complexes.

In 2008, the Stadthaus (Murray Grove) CLT project in London was the impetus for putting CLT squarely 'on the map' and has stimulated continued innovation in what is referred to as 'mass timber' building — evident in current proposals for bigger and taller buildings in wood up to 30 stories, where 'mass timber' includes products such as LVL, CLT and LSL (Green, 2012).

The recently completed 14-storey apartment block in central Bergen, Norway, at 49 m high has pushed the boundaries further, and if the proposed 18-storey (53-m-high) tower on the University of British Columbia campus goes ahead, it will be, at least for a short time, the tallest building of its kind in the world. By the beginning of 2015, there were some 30 massive timber buildings around the world greater than five stories in height that had been constructed in the previous decade (Woodsolutions, 2015b).

This increasing interest in multistorey timber buildings has also brought into focus some pressing research needs: high performance connections and connection systems, long-term behaviour of CLT (partially rolling shear), dynamics and means of reliably quantifying human comfort/response factors for long-span floors, development of 'deemed to comply' provisions for safety based on robust fire engineering and optimal use of alternative materials such as concrete and steel with timber to produce structurally efficient timber hybrids — particularly for buildings in excess of 20 stories in height.

A number of research and development programs have already commenced to address these needs, and there is an air of excitement and expectation among architects, engineers and researchers in the timber engineering community to provide the enabling technologies for the new generation of timber buildings to become a reality. The sky really is the limit.

References

Aasheim, E., 2002. Nordic Timber Bridges. Nordic Timber Council, Stockholm, Drottning Kristinas väg71, SE-114 28 Stockholm, Sweden.

Abrahamsen, R., 2008. Bridge across rena river - "World's strongest timber bridge", Miyazaki, Paper 001. In: Proceedings - World Conference on Timber Engineering, Miyazaki, Japan.

Boughton, G.N., Crews, K.I., 2012. Timber Design Handbook. Standards Australia HB108, Sydney. ISBN: 0 7337 2057 9.

Chapman, J., 2014. Integrating cross-laminated timber panels to construct buildings to 20 levels, (Paper 392) WCTE 2014 Proceedings. In: Salenikovich, A. (Ed.), World Conference on Timber Engineering, Quebec City, Canada, August 10—14, 2014.

Crews, K.I., 2003. Development of durability design processes for timber structures. Australian Journal of Structural Engineering 4 (3), 157—167. ISSN: 1328-7982.

Farrow Partnership, 2015. http://www.farrowpartnership.com/index.php/health-education/credit-valley-hospital (accessed 29.01.15.).

Green, M., 2012. The Case for Tall Wood Buildings. MGA - 57 E Cordova St, Vancouver BC V6A 1K3 Canada.

Hudson, Henningham, 1986. A Story of the Timber Industry in NSW: 1788—1986. Aust Forest Industries P/L 1986. ISBN: 1 86252 678 8.

Lend Lease, 2012. Forte Fact Sheet (accessed 03.06.12.). http://www.lendlease.com/australia/projects/forte.aspx#.

Madsen, B., 1992. Structural Behaviour of Timber, Vancouver Canada. Timber Engineering Ltd. ISBN:0 9696162 0 1.

UTS, 2015. Media Toolkit 2015 (accessed 05.02.15). http://www.uts.edu.au/partners-and-community/initiatives/city-campus-master-plan/completed-projects/dr-chau-chak-wing-media.

Victor Harbour, 2013. http://www.victoriaharbour.com.au/live-here/forte-living (accessed 05.02.15.).

Woodsolutions, 2015a. http://www.woodsolutions.com.au/Inspiration-Case-Study/forte-living (accessed 09.02.15.).

Woodsolutions, 2015b. http://www.woodsolutions.com.au/Blog/World-Tall-Timber-Overview-2014 (accessed 10.02.15.).

Bamboo material characterisation

D. Trujillo
Coventry University, Coventry, United Kingdom

L.F. López
Independent Consultant, Bogotá, Colombia

13.1   Introduction

According to the Food and Agriculture Organisation of the United Nations (FAO, 2007), it is estimated that over one billion people worldwide live in traditional bamboo housing, and 2.5 billion people depend economically on bamboo. It has also been demonstrated that bamboo is a more sustainable material than steel, concrete and timber (Van Der Lugt et al., 2003) even when imported into Western Europe. Murphy et al. (2004) state that housing made with bamboo is both less expensive and more sustainable than reinforced masonry alternatives. As this chapter will demonstrate, some of bamboo's mechanical properties are remarkable. Yet engineers throughout the world very rarely consider using this structural material. This chapter aims to briefly explain some of the characteristics of bamboo, with the hope of contributing to its wider adoption. The chapter will start by explaining what bamboo is, and then it will present some of the efforts that have been made to characterise bamboo as a structural material. It will then move onto explaining the tests that are currently used to determine its physical and mechanical properties, and thereafter it will explain some of the current procedures used to establish design values from test data. The chapter concludes by listing a series of tasks that bamboo researchers should consider when working in future characterisation of bamboo.

Although characterisation can have a broad range of meanings in material science, in the context of this chapter, characterisation is assumed to refer to the process of studying geometric, physical, mechanical and elastic properties of bamboo at a macroscopic level, and mainly with the view of adopting it for structural applications. Examples from a range of authors' work are used to make observations about general trends in bamboo characterisation.

13.2   Bamboo as a plant

Bamboo is a giant grass (Liese, 1998), native to all continents except Europe and Antarctica. There are in excess of 1250 species worldwide, although fewer than 100 species have potential for structural use. The term bamboo as used throughout this chapter is a generalisation for those species having structural use with which the

Nonconventional and Vernacular Construction Materials. http://dx.doi.org/10.1016/B978-0-08-100038-0.00013-5

authors are familiar. These include the following: *Guadua angustifolia* Kunth; (the most valuable commercial species in South America), *Dendrocalamus asper*, *Phyllostachys edulis* (Moso; the most valuable commercial species in China) and *Bambusa blumeana*.

Bamboos, unlike trees, only exhibit primary growth and not secondary growth (Liese, 1998); this means they emerge from the ground having the diameter they will have throughout their life and will only grow vertically, not broaden, over the years. The culm, or aerial part, is hollow (though there are a few solid species), tapered and segmented. The anatomy of the culm can be divided into nodes and internodes. The internodes are fundamentally hollow tubes, with axially oriented cells. To the interior of the node a diaphragm is formed, and from the exterior the culm-sheath and branches form, as seen in Fig. 13.1.

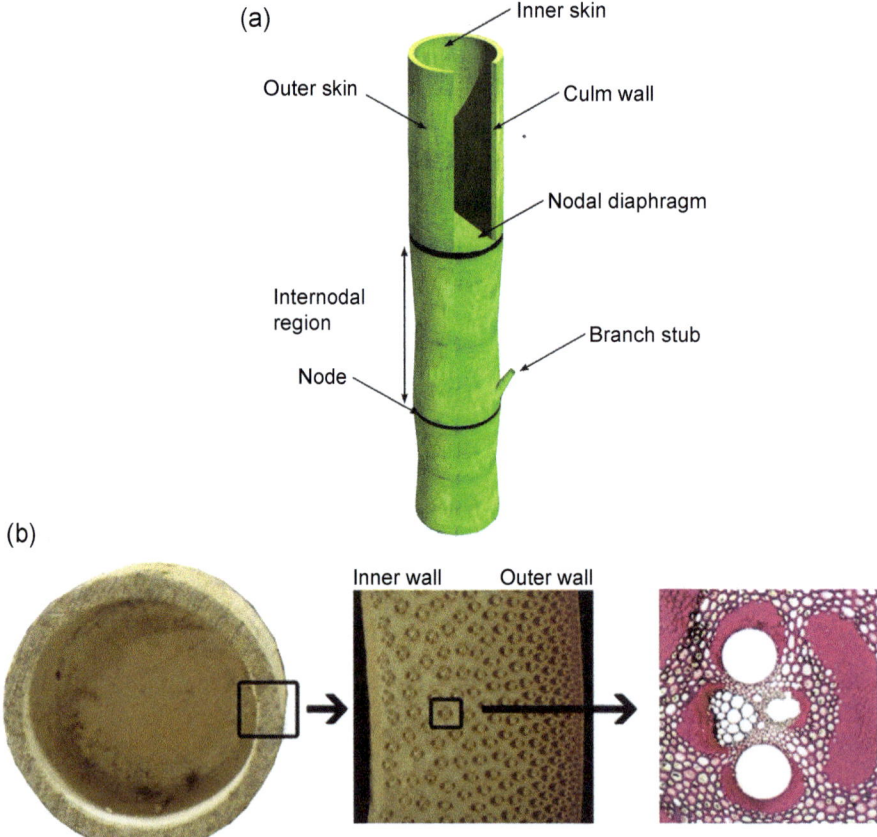

Figure 13.1 Anatomy of a bamboo culm: (a) schematic view of a culm segment, © Sebastian Kaminski, David Trujillo and Andrew Lawrence, (b) section of culm wall showing grading of vascular bundles (Gottron et al., 2014).

The wall of the culm is composed of a hard, shiny outer skin layer (the epidermis); a soft, light-coloured matrix material (the parenchyma) and the stronger and darker vascular bundles. The vascular bundles in turn are composed of metaxylem vessels, phloem and fibre sheaths (Liese, 1998) and impart the strength of the culm as fibres in a fibre-reinforced matrix. Vascular bundles are smaller and more numerous towards the exterior of the culm and larger and fewer towards the interior. As the culm tapers towards the top, the walls of the internode reduce in thickness, the number of vascular bundles decreases, but their density increases (Liese, 1998). In this manner, bamboo, a functionally graded material, has evolved to resist its primary loading in nature: its own self-weight and the lateral loading effects of wind. Typically the culm is constituted of approximately 52% parenchyma, 40% fibres and 8% vessels. A typical chemical constitution for bamboo would be 45% cellulose, 21% lignin, 32% soluble matters, 2% ash and less than 1% nitrogen (Amada et al., 1996).

Bamboo grows as a system of roots (technically, rhizomes) that produce culms. Regular extraction of some of the mature culms does not affect the health of the overall root network. Culms can reach full height (up to 20 m) in as little as 3 months. For some species, this is a rate of growth of 1 m a day (Kuehl and Yiping, 2012). However, culms will require a period of maturation before they have optimal strength for structural applications. This period ranges from 3 to 6 years. Their fast rate of growth signifies that they potentially can act as very effective carbon sinks, especially if regular, selective extraction is used (Kuehl and Yiping, 2012). If the extracted culms are used for durable products, such as housing, bamboo can act as a very sustainable and environmentally friendly material, sequestering on the order of 5 tonnes of carbon per hectare per year (Chinese Moso reported by Song et al., 2011).

13.3   Material properties of bamboos

Although bamboo has been used for millennia throughout the world for structural purposes, research into the determination of its structural properties is fairly recent and sparse. Although engineers identified the need to research the mechanical properties of bamboo from the late 19th century, for example Azuola-Guerra (1887), the oldest known published paper on mechanical properties of bamboo is Meyer and Ekelund (1923). Since then, numerous authors have contributed to the body of knowledge, many of which were compiled by Janssen (1991). In the last 20 years a vast contribution to the body of knowledge has been made, and yet, work into characterisation of bamboo species still needs to be furthered greatly. The following sections identify trends observed by diverse authors, which can be helpful in the process of characterisation of a bamboo species. They do not intend to be an exhaustive literature review or a summary of the state of the art.

13.3.1   Effect of density on mechanical properties

In a similar manner to other materials, in particular timber, the strength of bamboo correlates well with its density. This correlation has been observed to be linear and

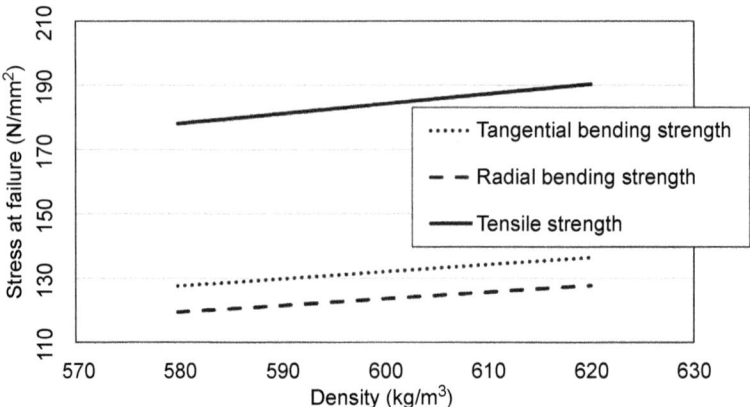

Figure 13.2 Relationship between density and ultimate stress (Zhou, 1981). Bending tests were carried out using strips cut tangentially or radially. Tensile tests were parallel to the fibres.

has been acknowledged for some time, for example, Ota (1950), Sekhar et al. (1962) and Zhou (1981) (see Figs. 13.2 and 13.3). Density values correlate well with compressive, bending and tensile strength; this is likely to be because denser samples have a higher content of cellulose. Janssen (1981) quotes Meyer and Ekelund (1923) by observing that shear strength also correlates to density, the explanation for this being related to the thickness of cell walls. However, observing other authors' results (eg, Abdul Latif et al., 1990), a clear correlation between density and shear strength is not discernible.

13.3.2   Variation of density and other properties in the culm

It has been observed that density increases from the bottom of the culm towards the top (eg, Zhou (1981), Fig. 13.4), and that density increases from the inside towards the outside of the culm wall at the internode (eg, Ota, 1950). Similarly the ratio of fibre

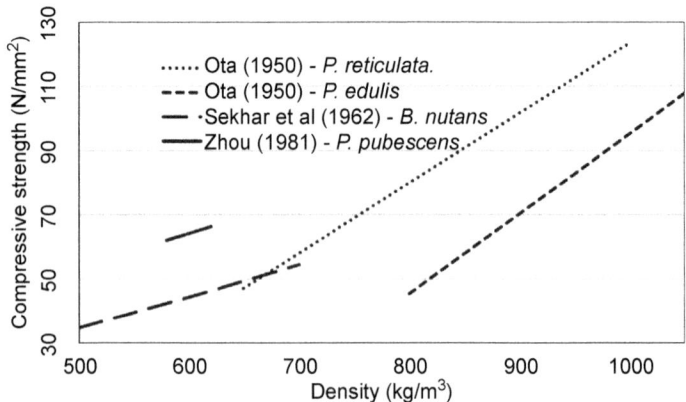

Figure 13.3 Relationship between compressive strength and density, several authors.

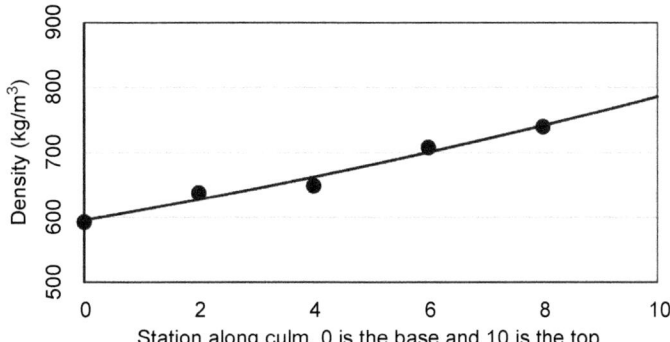

Figure 13.4 Variation of density along the culm, Zhou (1981).

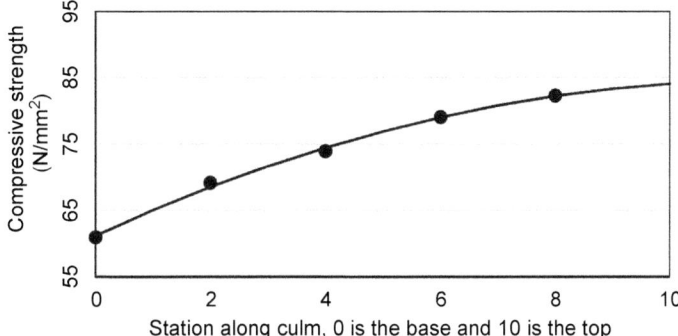

Figure 13.5 Variation of compressive strength along the culm, Zhou (1981).

to matrix material increases from the inside to the outside of the wall, and from the bottom of the culm to the top (eg, Amada et al., 1996). The increase in density from the bottom to the top of the culm is accompanied by an increase in compressive strength, as shown in Fig. 13.5 (eg, Ota (1950), Zhou (1981), Abdul Latif et al. (1990)) and an increase in tensile strength (eg, Amada et al., 1996). Zhou (1981) observed that the moisture content at harvest reduced from the bottom towards the top of the culm (Fig. 13.6); this trend was also observed by Abdul Latif et al. (1990).

13.3.3   Effect of age on other properties

Many properties of bamboo have been observed to be affected by the age of the culm at harvesting; these include density (eg, Fig. 13.7), moisture content at harvest (Fig. 13.8), strength (Fig. 13.9) and modulus of elasticity. Zhou (1981) and Lu et al. (1985) observed that many of these properties would peak at a particular age and then decrease again if the culms are not harvested. The optimal age for harvesting varies among species and location of the plantation. Correal and Arbeláez (2010) found a significant correlation between age and density, and age and compressive

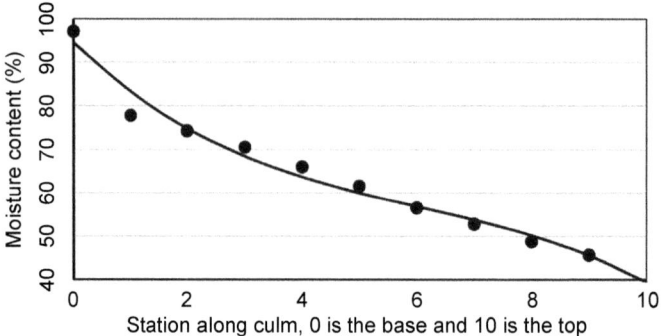

Figure 13.6 Variation of moisture content along the culm, Zhou (1981).

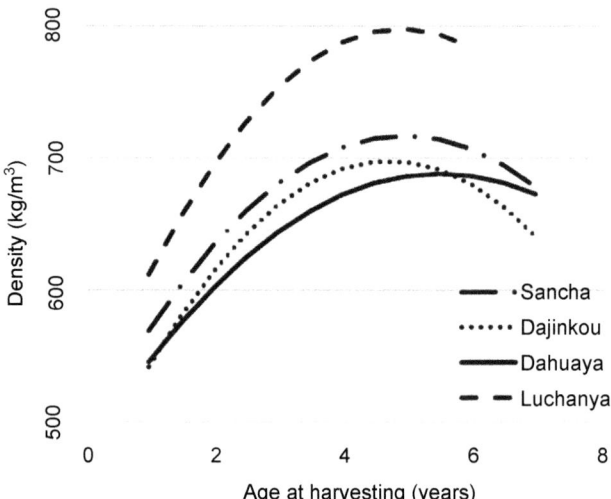

Figure 13.7 Effect of age at harvesting on density for *Phyllostachys glauca* from four different regions of Shandong province, Lu et al. (1985).

strength, but no relationship between shear strength and age or position along the culm. Increase in strength and stiffness has been associated with the thickening of the cell walls of fibre and parenchyma (Liese, 1998) which occurs with age.

Limaye (1952) also observed that the modulus of elasticity varied with age. The age range adopted in this study is quite limited. The variation to the modulus of elasticity is quite significant for green bamboo, but relatively small for dry bamboo in comparison to variations observed for other properties (Fig. 13.10). However, other authors arrive at conflicting conclusions. Low et al. (2006) find that both modulus of elasticity and strength are higher in *Sinocalamus affinis* in young specimens (1 year old) than in mature specimens (5 years old), although this could be explained by the rise and fall of these properties (ie, an optimal age between 1 and 5 years) as seen in Fig. 13.9.

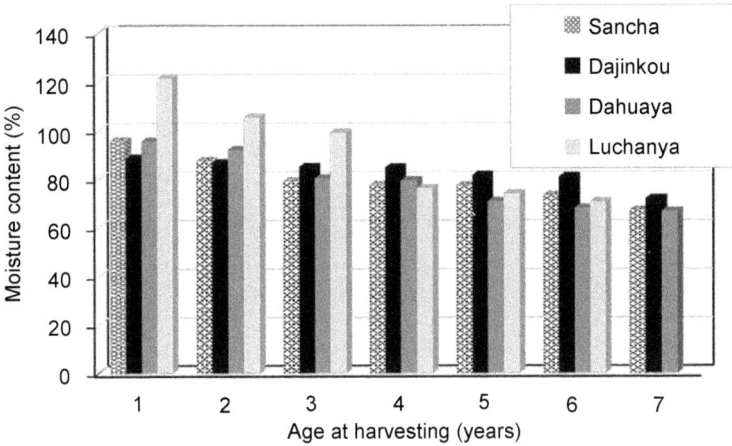

Figure 13.8 Effect of age at harvesting on moisture content for *Phyllostachys glauca* from four different regions of Shangdong province, Lu et al. (1985).

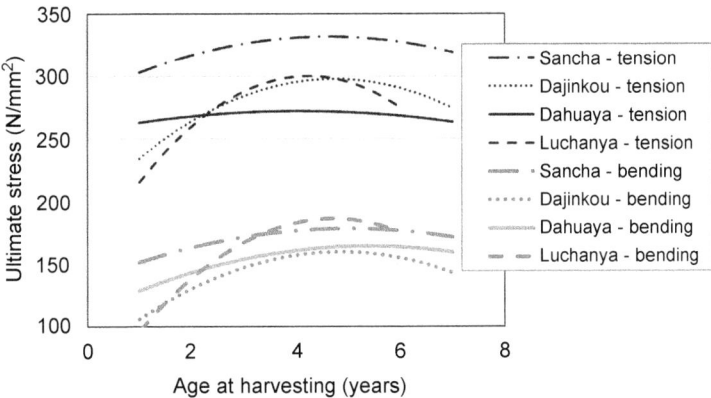

Figure 13.9 Effect of age at harvesting on strength for *Phyllostachys glauca* from four different regions of Shandong province, Lu et al. (1985).

Correal and Arbealáez (2010) found no significant correlation between modulus of elasticity in bending or compression and age for *G. angustifolia*.

13.3.4   Effect of moisture content

In a similar manner as timber, the strength of bamboo is affected by the moisture content of the specimen. Specimens exhibiting a moisture content greater than the fibre saturation point (FSP; the moisture content corresponding to no free water − only bound water − being present in the culm) exhibit fairly stable strengths, and those with moisture contents less than the FSP exhibit an increase in strength inversely proportional to the moisture content (Fig. 13.11). Ota (1953) established that the ratio between oven-dry and water-saturated (FSP) compression strengths is about 2.2. It would

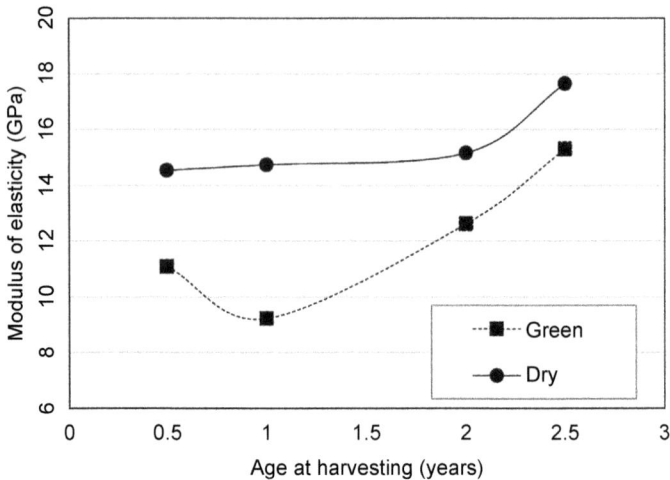

Figure 13.10 Effect of age on modulus of elasticity for *Dendrocalamus strictus* in green and dry condition, Limaye (1952).

seem that a ratio of around 2 is probably valid for shear strengths (Fig. 13.11). The effect of moisture content can be seen in the data presented by Limaye (1952) who tested both dry bamboo (typically at an MC = 12%) and green bamboo (with a moisture content greater than the FSP) (Fig. 13.12).

13.3.5 Dimensional properties of the culm

If bamboo is to be characterised with the aim to use it in its round form, ie, as a pole, its dimensional characteristics should also be considered in the process of

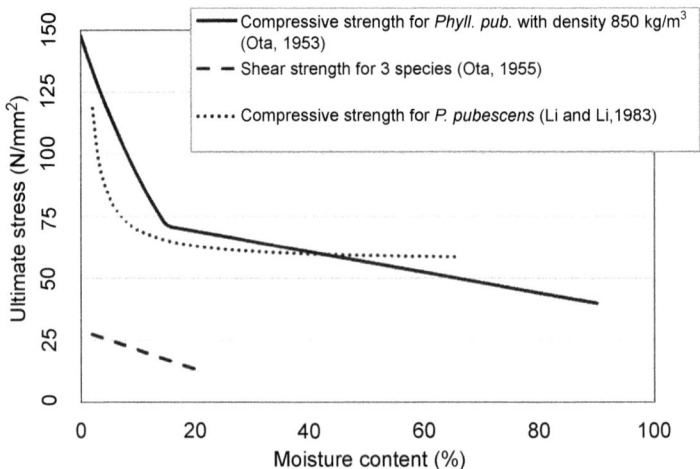

Figure 13.11 Effect of moisture content on strength, several authors.

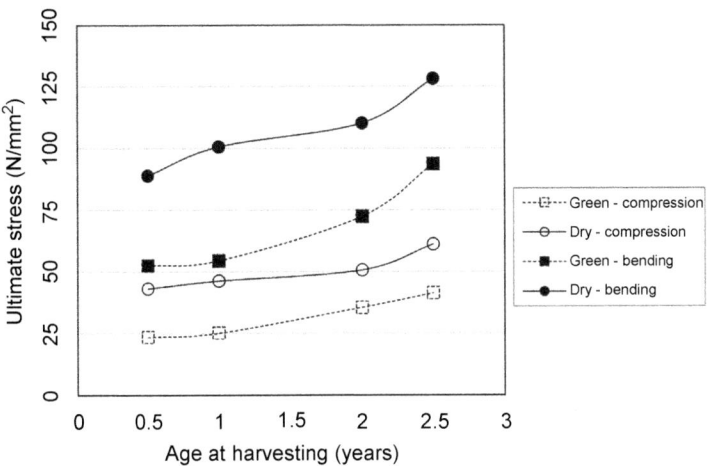

Figure 13.12 Effect of age on strength for *Dendrocalamus strictus* in green and dry condition, Limaye (1952).

characterisation. It is important to establish the typical diameters and wall thicknesses for a given species, although reporting of this data is less common. Length of internodes and taper also vary considerably among species. To this effect, the work of Shigematsu (1958), who studied the form of the culm of 15 different Japanese species, provides an excellent example of what can be achieved (Fig. 13.13). Shigematsu (1958) introduces the 'wall thickness index', which is a percentage of 2 times the wall thickness over the diameter to normalise dimensions. As Fig. 13.13 represents the dimensional characteristics of 15 species, it is possible to assume, in the opinion of the authors of this chapter, that some observations are fairly universal: (1) the length

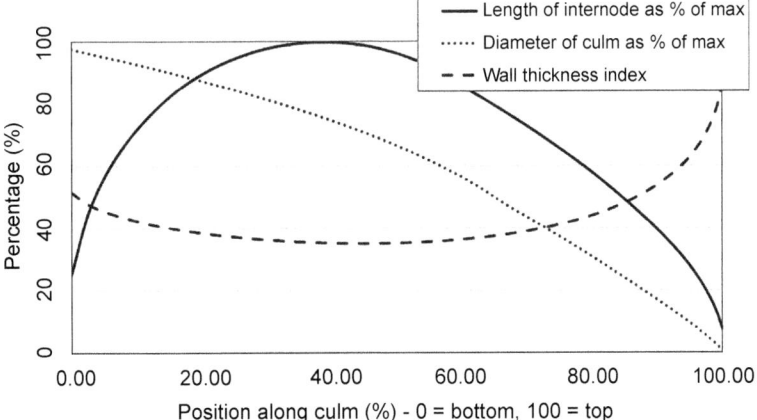

Figure 13.13 Dimensional properties of the culm for 15 Japanese species. Adapted from Shigematsu (1958).

of the internode peaks somewhere between a third and the middle of the culm; (2) the form of the taper of the culm is not linear; and (3) for a given species, a single wall thickness index is likely to be representative for the lower 70% of the culm (the part of most structural interest).

Amada et al. (1996) carried out a similar analysis for the geometric characteristics of *P. edulis*. They observed that diameter and wall thickness decreased along the culm in nearly a linear manner, provided the reference for position along the culm was not stated in units of length, but as node number. They also observed that the elastic section modulus varied along the length of the culm in a proportional manner to the bending moment caused by wind.

13.3.6 Functionally graded material

Amada et al. (1996) concluded that bamboos have evolved to be highly efficient composite materials or 'smart structures'. For example, the reduction in both the diameter and wall thickness along the culm also closely matches the reduction in bending moment along the culm. This results in bamboo having an almost constant maximum surface stress along its length when subjected to wind forces. Another interesting result is the near constant radius of gyration throughout the culm, which ensures the buckling capacity of the culm does not vary significantly over the height of the culm. Similarly, the percentage of area occupied by vascular bundles — a term the authors call volume fraction of the [vascular] bundle sheaths — increases with height to compensate for the loss of stiffness resultant from the loss of diameter and wall thickness. The distribution of the vascular bundles across the wall thickness mirrors the bending stresses caused by wind, ie, the more highly stressed perimeter has a greater concentration of vascular bundle sheaths and therefore fibre.

Though many of the observations made by Amada et al. (1996) were known, the elegance of their analysis was novel at the time. The authors did not attempt to separate strength and stiffness properties from the geometrical characteristics of the section or the location in the culm, this approach of considering both geometric and material properties and their variation along the culm is an important lesson provided by this paper for anyone intending to characterise bamboo.

13.4 Tests for material and physical property determination

As mentioned earlier, research into the physical and mechanical properties of bamboo has a history in excess of 90 years. During this period, different approaches have been used, sometimes with not entirely satisfactory results. Tests depend, to an extent, on the intended end use of the material. Tests carried out on laminated bamboo products (also referred to as engineered bamboo) can, in most instances, adopt timber standards (eg, Correal and López, 2008). Some specific differences do exist between engineered timber and engineered bamboo, but these are beyond

Figure 13.14 Bamboo rarely is absolutely round. *Note also the fissures at the southeast quadrant and northwest quadrant* (scale shown is in cm).
© David Trujillo.

the scope of this chapter. However, when the intention is to understand the properties of pole or round culms, simple adoption of timber testing procedures is not appropriate in most instances. Bamboo is hollow, round (though frequently more oval shaped, refer to Fig. 13.14) and tapered. All tests present some difficulty, and a direct extrapolation from timber, or other construction materials for that matter, is not possible.

For instance, in bending tests for *G. angustifolia*, Gnanaharan et al. (1994) determined that short specimens (using midpoint flexure specimens having a span of 700 mm and a shear span-to-culm diameter ratio of approximately 4.4D) do not provide reliable bending strength values as failure will occur through either shear or crushing prior to bending. A similar conclusion was arrived at, independently, by Sánchez and Prieto (2001). Gnanaharan et al. (1994) also observed that using strips of bamboo in lieu of whole round sections is not a reliable alternative. Their conclusion was that long specimens (in this instance, tested in third-point flexure (ie, four-point bending) over a span of 3000 mm, resulting in shear span-to-culm diameter ratio of approximately 12.5D) should be used. Load was applied by means of wooden saddles similar to the ones in Fig. 13.15(a). A variation to the wooden saddle system was presented by Correal and Arbeláez (2010) using straps; this allows for a more even distribution of loads, avoiding localised crushing, and permits rotation of the individual load point. An adaptation to the strap system was developed at Coventry University (Fig. 13.15(b)). Nugroho and Bahtiar (2013) observe that to obtain accurate values for the modulus of rupture in a four-point bending tests (such as that for ISO 22157-1 (ISO, 2004a)), culm taper should also be considered.

Round bamboo members can present a whole range of failure modes when subject to bending tests, which can affect interpretation of results. ISO 22157-1 (ISO, 2004a) recommends using specimens of a length at least equal to 30 times the diameter in a third-point flexure arrangement. This ratio was determined by Vaessen and

(a)   (b)

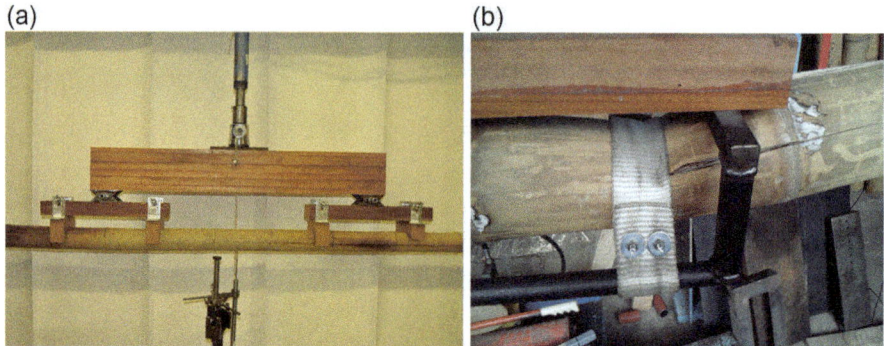

Figure 13.15 (a) Wooden saddle for bending test, (b) straps allow a more even distribution of stress at the supports or points of load application.
© David Trujillo.

Janssen (1997) to be in fact 26.3, which results in a shear span-to-diameter ratio of 10. If an alternative four-point bending arrangement is used, the shear span-to-diameter must still be observed, and even then, shear failures may still be observed. These occur in one of the shear spans between a point of load application and the nearest support. Failures that occur between the central points of load application are undoubtedly bending failures, but these failures occur in a range of manners, ranging from crushing of fibres in compression to collapse of culm in the compression side of the culm (as seen in Fig. 13.16).

Numerous types of shear tests have been proposed over the years as summarised by Janssen (1981), who proposes the currently widespread 'bow-tie' test method (as presented in Fig. 13.17), which has been incorporated into ISO 22157-1 (ISO, 2004a). This test places a segment of culm into longitudinal compression between two 90 degree quadrant plates (arranged like a bow-tie). The plates are rotated 90 degrees relative to each other at the ends of the specimens, resulting in four longitudinal shear planes along the length of the specimen along which failure occurs. Once again, adoption of timber test methods for round bamboo is not possible. Mitch (2009) suggests that a relationship between tension perpendicular to fibres and shear resistance may be derived, since both values are a function of the matrix material.

Compression parallel to the fibre tests also require consideration, as interaction with the loading platens may occur affecting results, a phenomenon identified by Arce (1993). Essentially, the friction between the culm plates effectively restrains the lateral expansion of the culm at the platen. Janssen (2000) recommends the use of Teflon plates or some other means to minimise friction between the platens and the culm segment. This recommendation has been incorporated in to ISO 22157-1 (ISO, 2004a). Harries et al. (2012) found that using sulphur capping compound (as is done for concrete cylinders) works well and has the additional advantage of ensuring that the culm ends are parallel and uniformly loaded.

Testing the tensile strength of the whole round culm is not easily achieved, and as such is not covered in standards. Instead the tensile strength of strips extracted from the

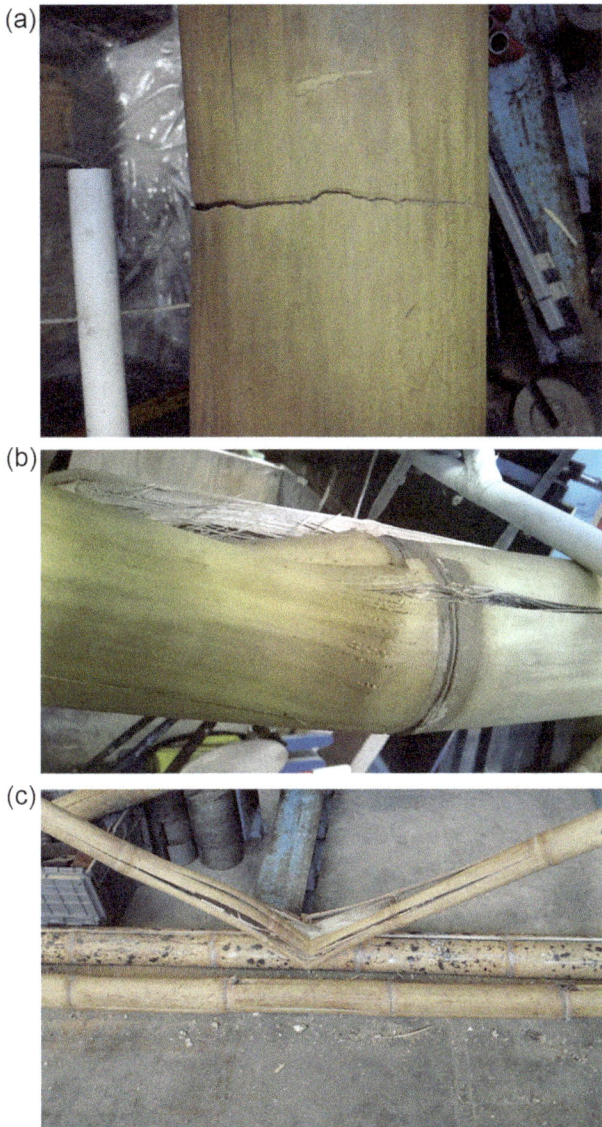

Figure 13.16 Bending failure modes in bamboo: (a) crushing of fibres in compression, (b) collapse and buckling of fibres in compression, and (c) collapse of culm.
© David Trujillo.

culm wall are the conventional way to calculate the strength of bamboo in tension. The strip is tabbed or cut as a 'dog bone' to ensure failure does not occur at the grips (in a manner similar to steel). Richard and Harries (2015) observed a range of failure modes for this test; these involved failure at the clamps, splitting, 'brooming' and pure tensile rupture. Reported tensile strength values for radially oriented *Bambusa stenostachya*

Figure 13.17 Bow-tie-shaped platens used in shear tests for bamboo.
© David Trujillo.

specimens ranging from 132 N/mm^2 to 141 N/mm^2, dependent on the fixity conditions of the tension grips used. They reported that testing tangentially oriented specimens extracted from near the middle of the culm wall resulted in slightly more uniform characteristic values. Tension tests that include nodal regions exhibit capacities approximately half of those that do not.

A test that is even less common is that for the determination of the tensile strength perpendicular to the fibres. The first reported attempt was by Arce (1993), with fairly unsatisfactory results; yet he rightly identified the importance of this property when he stated *Bamboo culms do not fail in compression, in bending or shear, but do fail when a maximum tangential tensile stress is reached.* A different test procedure was proposed by Castrillón and Malaver (2004) and later used by Pacheco (2006), consisting in pulling a length of internode transversely apart in direct tension. Mitch (2009) critiques this method, pointing out that although it requires a simple calculation, it is impractical, as the two semicircular wedges that are placed inside the culm need to fit exactly inside the section of culm, and as every culm is of a different size and shape, a range of wedges would be needed to achieve a tight fit. If a perfect fit is not achieved, this will result in bending stresses perpendicular to the fibre being exerted with the tensile stresses. Mitch (2009) instead proposed a test procedure using a split pin (Fig. 13.18), which has the benefit of determining both the tensile strength perpendicular to the fibre and the mode I stress intensity factor, K_1. This procedure can accommodate diverse diameters of specimens without major difficulty, although preparation of the test specimen is laborious. This test procedure has the disadvantage of inducing a failure mode more akin with prying than pure tension, although this results in a lower-bound solution that may be appropriate for establishing parameters for design.

Work investigating compression perpendicular to the axis dates back to Atrops (1969), who reported resistance values in units of force and not in terms of stress. Presumably, this was done because the researcher acknowledged the complexity of

Figure 13.18 Split-pin test as proposed by Mitch (2009).
© David Trujillo.

the induced stresses, as round bamboo specimens that are loaded perpendicular to their axis crush in a 'three-pinned arch' failure mode that induces flexural stresses (tension and compression) through the culm wall thickness (refer to Trujillo (2007) and Sharma et al. (2013)). Other authors (eg, Lozano (2010)) have mistakenly called this failure mode compression perpendicular to the fibre. Sharma et al. (2013) provide a rigorous analysis of the stresses induced in a more correctly termed 'edge-bearing test' (Fig. 13.19), and they suggest that results would be of value to derive a correlation between tension strength perpendicular to the fibre and bending strength (modulus of rupture) perpendicular to the fibre. This correlation has yet to be established, but it is worthwhile noting that these values vary by an order of magnitude. Mitch (2009)

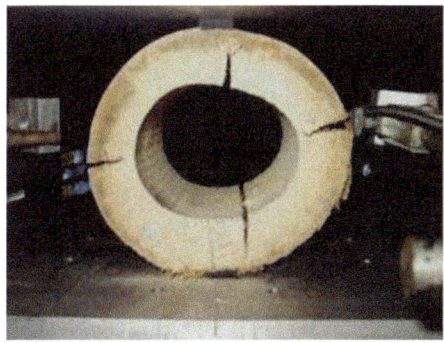

Figure 13.19 Edge-bearing test.
© Bhavna Sharma.

reported an average ultimate stress in tension perpendicular to the fibres using the split-pin method of 1.0 N/mm^2 for *B. stenostachya*. Whereas, Sharma et al. (2013) obtain values ranging from 3.3 to 5.8 N/mm^2 for bending perpendicular to the fibres (dependent on whether the wall was bending inwards or outwards) for the same species. Due to the nature of the edge-bearing test, the behaviour varies for thin- and thick-walled specimens. A *D/t* ratio of 10 is proposed as the threshold between behaviours (Sharma et al., 2013).

13.4.1 Development of test standards

The first bamboo testing standard in the world was introduced in 1973 by the Bureau of Indian Standards (BIS, 1973) under the name 'IS:6874 — Methods of tests for round bamboos'. From the late 1990s, Dr Jules Janssen, through the International Network for Bamboo and Rattan (INBAR), led an initiative to develop international standards for bamboo. This effort culminated in 2004 with the publication by the International Standards Organisation (ISO) of 'ISO 22157-1: Bamboo — Determination of physical and mechanical properties — Part 1: requirements' (ISO, 2004a), 'ISO 22157-2: Bamboo — Determination of physical and mechanical properties — Part 2: laboratory manual' (ISO, 2004b) and 'ISO 22156: Bamboo — Structural design' (ISO, 2004c). ISO 22157-1 (ISO, 2004a) contains the following test procedures: moisture content, mass by volume (ie, density), shrinkage, compression [parallel to fibres], bending, shear and tension [parallel to fibres]. ISO 22157-2 (ISO, 2004b) explains how to use these standards and provides useful advice for those undertaking experimental work. ISO 22156 (ISO, 2004c) presents, in the form of a model code, a design philosophy for bamboo and some outline guidance for structural design, although it contains little of the practical guidance that would be expected in a design standard. It does, however, contain guidance with regards to the derivation of characteristic and allowable stresses. This particular aspect will be discussed in greater depth later.

ISO 22157-1 (ISO, 2004a) has been widely adopted throughout the world; countries that have either directly adopted or adapted this standard include Colombia, Ecuador, Peru, Jamaica, Ethiopia and India. As such, they have proven invaluable as a basis to ensure that test results between researchers are comparable. They have also created a basis upon which to develop further standardisation. However, these standards are not without fault, as will be discussed hereafter.

ISO 22157-1 (ISO, 2004a) does not include tests to determine tensile strength perpendicular to the fibre. Because, as has been argued, this is such an important property in bamboo, this seems to the authors of this chapter a serious omission. The shear test contained in ISO 22157-1 (ISO, 2004a) has also been critiqued. Mitch (2009) points out that it rarely triggers more than one, and at the most three, shear planes (this is also the experience of the authors), yet ISO 22157-1 (ISO, 2004a) requires the derivation of shear strength to be based on the assumption that four planes will be triggered. This is not necessarily a problem from the design point of view, as assuming four shear planes results in a lower-bound value for shear capacity, but can be unhelpful if the desire is to determine the real shear strength. Deriving the shear strength based on only the shear planes that were triggered, ie, dividing the total force by the area of the triggered shear planes, is a common error that violates the principles of equilibrium. Mitch (2009) also points out that the requirement that ends of the culm specimen be cut perfectly parallel is very difficult to achieve (particularly in the field). Janssen (1981) also noted that shear strengths obtained through the bow-tie test seem to be larger than those derived from shear failures in bending tests. This calls into question the utility of this test, although it can be explained: the bow-tie test results in pure mode II shear failures. Flexural tests introduce a component of mode I behaviour whose interaction naturally 'weakens' the mode II capacity (Richard, 2013).

As previously discussed, bending tests require further adaptations such as straps to avoid crushing. Richard and Harries (2015) have identified a number of practical issues associated with testing tension parallel to the fibres that can result in inconsistent results (including grip fixity, specimen orientation). Thus, while a good basis, the ISO 22157 documents are presently being considered for revision to account for advances made since they were first approved.

13.4.2   Case study: Guadua angustifolia

Guadua angustifolia Kunth is one of the most studied bamboo species in the world. Work investigating this South American species dates back from the 1980s under the tutelage of Oscar Hidalgo López. Most characterisation discussed in this section will relate to the physical and mechanical properties observed, although it is worth noting that many other authors (eg, Londoño et al. (2002)) have characterised other aspects of the anatomy of its culm including its fibre content, cellular structure, etc. Londoño et al. (2002) also report average geometric properties, listed in Table 13.1. Interestingly, these results do not seem to fit into the equations derived by Shigematsu (1958), with the exception of length of internode, which warrants further research for this species.

As mentioned, determination of the mechanical properties of *G. angustifolia* date back to the 1980s. The resulting large database of test results (an example of the then extant database is shown in Fig. 13.20, adapted from López and Trujillo (2002)), alongside the broad spectrum of structural experience (Trujillo, 2007, Trujillo et al., 2013) has built considerable confidence in the properties of *G. angustifolia*, which has resulted in arguably the most comprehensive code for structural use of bamboo in the world: the Colombian Seismic Standard NSR-10 (AIS, 2010).

Table 13.1 **Average dimensions for the culms in study, Londoño et al. (2002)**

	Average diameter (mm)	Average wall thickness (mm)	Length of internode
Bottom segment (node 0−16)	111.5	18.2	200 (min)
Middle segment (node 16−40)	110.5	13.2	350 (max)
Top segment (node 40−72)	58.4	8.9	Not reported

Table 13.2 Mechanical properties parallel to the fibres for *Guadua angustifolia* for three distinct regions of Colombia, Lozano (2010)

Condition	Bending					Compression					Shear					Tension			
	Green				Dry[a]	Green				Dry[a]	Green				Dry[a]	Green			
Region	A	B	C	Mean		A	B	C	Mean		A	B	C	Mean		A	B	C	Mean
Sample size (n)	61	12	11	84		76	70	74	220		91	75	80	246		61	49	51	161
Mean (N/mm²)	69.0	65.0	68.8	68.4		33.9	32.3	32.4	32.87	54.8–56.2	5.66	5.52	6.44	5.87	7.10–7.47	61.5	55.5	66.9	61.4
Coefficient of variation (%)	26.0	39.1	48.2	31.1		30.0	27.0	17.6	25.6		24.3	18.9	17.0	21.3		21.7	18.5	22.2	22.3
Moisture content (%)				71					87					98					93

[a]Dry values are quoted by Lozano (2010), but correspond to other researchers.

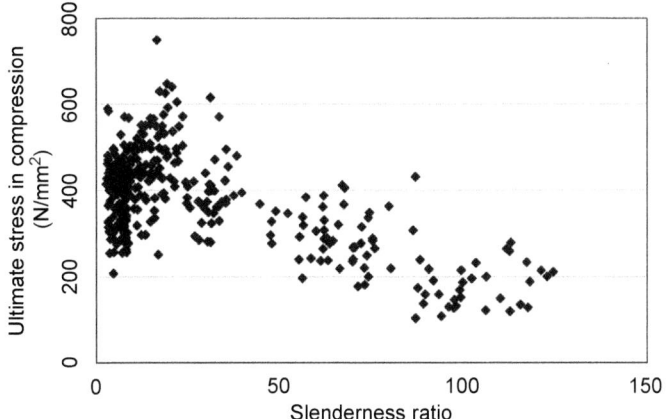

Figure 13.20 Stress at failure versus slenderness for *Guadua angustifolia.*
Adapted from López and Trujillo (2002).

One very extensive investigation into the properties of *G. angustifolia* is Lozano (2010). This report compiles data from 246 samples from three distinct regions of Colombia: Quindío, Cundinamarca and Valle del Cauca (Tables 13.2–13.4). Note that results are reported under the generic names of zone A, B and C to avoid generating the perception that a certain region produces better bamboo. The value of this work is that it provides insight into intraspecies variability. It is interesting to note that variation between regions is not remarkably large. Another very valuable investigation into the properties of *G. angustifolia* is that by Correal and Arbeláez (2010), which studied the effect of age at harvesting on several mechanical properties. It is interesting to note that both aforementioned studies tested bamboo in green conditions

Table 13.3 Mechanical properties perpendicular to the fibres for Colombian *Guadua angustifolia*, several authors

| | Mechanical properties perpendicular to fibres | | |
	Bending[a]	Tension	Compression[b]
Source	Lozano (2010)	Pacheco (2006)	Correal and López (2008)
Condition	Green	Green and dry	Dry
Sample size (*n*)	150		29
Mean (N/mm²)	6.59	0.73	5.4
Coefficient of variation	43.65%		14.66%
Moisture content	81%		10%

[a]Lozano (2010) calls this failure mode compression perpendicular to fibres.
[b]These values were obtained for laminated bamboo; however, they are quoted here for reference.

Table 13.4 Average elastic properties for *Guadua angustifolia* for three distinct regions of Colombia, Lozano (2010)

| | Modulus of elasticity | | | | Modulus of rigidity | Poisson's ratio |
	Bending	Compression parallel to fibres	Tension parallel to fibres	Circumferential		
	Green	Green	Green	Green	Green	Green
Sample size (*n*)	86	136	211	142	34	N/A
Mean (N/mm^2)	13,904[a]	9082[a]	6542	564	634	0.35
Coefficient of variation	36.06	63.68	43.70	77	32.45%	19.97%
Moisture content	71%	87%	93%	81%	68%	N/A

[a]Correal and Arbeláez (2010) also use *Guadua angustifolia* in green condition, yet obtain a modulus of elasticity in bending of approximately 17,200 N/mm^2 and in compression of 16,900 N/mm^2.

for their characterisation work to control for effects in variation of strength/stiffness arising with working with bamboo with a moisture content below the FSP.

13.5   Derivation of design values

As mentioned earlier, ISO 22156 (ISO, 2004c) provides a means to derive characteristic values and allowable stresses. Although derivation of design values is arguably beyond the scope of the process of characterisation of a material, this procedure is presented here due to its practical benefit to engineers intending to use bamboo. Clause 7.2.1 of ISO 22156 (ISO, 2004c) proposes the following equation (Eq. [13.1]) corresponding to the fifth percentile value defined with 75% confidence:

$$R_k = R_{0,05} \left(1 - \frac{2,7 \frac{s}{m}}{\sqrt{n}} \right) \tag{13.1}$$

where R_k is the characteristic value; $R_{0,05}$ is the fifth percentile from the test data; m is the mean value from the test data; s is the standard deviation of the test data; n is the number of tests (at least 10) (ISO, 2004c)

Clause 7.4 complements it with equation (Eq. [13.2]):

$$\sigma_{all} = R_k \times G \times \frac{D}{S} \tag{13.2}$$

where σ_{all} is the allowable stress, in N/mm^2; R_k is the characteristic value; G is the modification for the difference between laboratory quality and practice, default value 0.5; D is the modification value for duration of load: 1.0 for permanent load, 1.25 for permanent plus temporary load, 1.5 for the above plus wind-load; S is the factor of safety, default value 2.25 (ISO, 2004c)

This procedure has been adopted and adapted in the Colombian Legislation for Earthquake Resistant Construction − NSR-10 (*Reglamento Colombiano de Construcción Sismo Resistente*), in chapter G.12 − Guadua (bamboo) structures. As the name suggests, this procedure is only appropriate for one species of bamboo: *G. angustifolia* Kunth. Equation (G.12.7-2) of NSR-10 (AIS, 2010) states:

$$F_i = \frac{FC}{F_s \times FDC} f_{ki} \tag{13.3}$$

where F_i is the allowable stress; f_{ki} is the characteristic value (derived exactly as R_k in Eq. (13.1)); FC is the modification for the difference between laboratory quality and practice, taken as 0.5 in tension, 0.6 in shear and 1.0 for compression and bending; F_s is the factor of safety, taken as 2.0 for bending and tension, 1.8 for shear and compression perpendicular to the axis and 1.5 for compression parallel to the fibres; and FDC is the modification value for load duration in accordance to the load type: taken as 1.5 for bending and tension, 1.2 for compression and 1.1 for shear. This

Table 13.5 Allowable stress for *Guadua angustifolia* at a moisture content = 12% (AIS, 2010)

Type of stress	Allowable stress value
Bending	15
Tension	18
Compression parallel to fibres	14
Compression perpendicular to axis (assumes internode has been filled with cement mortar)	1.4
Shear	1.2
Moduli of elasticity	
Mean modulus of elasticity, $E_{0.5}$	9500
Modulus of elasticity at 5th percentile, $E_{0.05}$	7500
Minimum modulus of elasticity, E_{min}	4000

differentiation based on load type is a legacy from the timber code, as explained in López and Trujillo (2002). Allowable stresses are modified further to account for load duration (to reduce the overall factor of safety when load duration is short), moisture content, operational temperature, buckling of beams, presence of shear and buckling of columns.

The allowable stress values and moduli of elasticity for *G. angustifolia*, as stated in NSR-10, are presented in tables G.12.7-1 and G.12.7-2 of the standard and are summarised in Table 13.5 (AIS, 2010). The stated modulus of elasticity is surprisingly low, when compared to the results of aforementioned authors with green bamboo. The reason for this low value is that the code writers adopted an 'apparent modulus of elasticity' as observed in whole frames and not individual elements. Their concern was that the state of the art in connection design did not allow for reliable predictions of slippage, so a low modulus of elasticity (about 50% of experimentally observed values in dry conditions) would result in satisfactory designs for frames and trusses. As this point is not caveated in the code, designers will tend to overdesign simply supported members such as beams. According to the code, three values of modulus of elasticity are provided $E_{0.5}$, to be used for general structural analysis of elements, $E_{0.05}$, for stability calculations (ie, buckling) and E_{min}, for safety critical elements. Examples of safety critical members are not provided.

13.6   Further work and future developments

As argued earlier, if the characterisation of bamboo is carried out with the aim of adopting it for structural purposes in its pole, or round form, direct adoption of timber tests is not possible. Similarly, characterisation of bamboo species should not be limited to the determination of physical and mechanical properties. Since bamboo in pole form has not undergone any transformation other than cutting to length, it will not have a standard diameter, wall thickness, internode length, cross-sectional shape or taper. It is equally important to characterise these geometric properties for a given species, as to characterise the mechanical properties, since they will also affect the resulting strength characteristics of pole bamboo. Future characterisation work should try to arrive at generalisations about these characteristics in a similar manner as was done by Shigematsu (1958) for 15 species of Japanese bamboo.

The process of characterisation of a bamboo species should also take into account the phenomenon of splitting. Do all species exhibit the same sort of behaviour? It is important to note that if a single split occurs in a member subject to bending, and it were in the proximity of the neutral axis of the member, it is bound to reduce the shear strength of the member by half. If the split were to occur in a slender member subject to compression, the effect on the resistance to hoop stresses is likely to compromise the resistance to buckling of the member. The severity and gravity of this phenomenon has not been adequately studied to the authors' knowledge. The mechanics of splitting in bamboo needs to be better understood, including the factors that induce it, the frequency of the phenomenon, the consequences in terms of strength reduction and means to control or prevent it. The recent growth in interest in characterising the tensile strength perpendicular to the fibres will make a significant contribution to this process.

The destructive testing procedures contained in ISO 22157-1 (ISO, 2004a) have proven useful to the process of characterisation of bamboo species; however, more guidance is needed with regard to sampling within a batch. For example, with what frequency should sampling be done? Should a sample be subjected to all the tests contained in the standard, or can a single test be used as a reference or perhaps stand surrogate for the others, in a similar manner as to concrete's compression strength is used for the determination of its shear and tensile strengths and modulus of elasticity? Which test would be the best such reference? Alongside this guidance, more economical and field-ready means to infer strength of bamboo need to be considered, as access to the specialist laboratories required to carry out all ISO 22157-1 (ISO, 2004a) tests is impractical or prohibitively expensive in many parts of the world where bamboo is being used structurally. Simple field tests that correlate well to all other mechanical properties should be developed, as proposed by Sharma et al. (2013). More importantly, can nondestructive tests be used to develop a grading system, such as that used for timber? Lin et al. (2006) demonstrated that nondestructive tests could be used to correlate density and modulus of elasticity. It should be possible to develop visual or mechanical grading (through nondestructive means) for bamboo, as proposed by Trujillo (2013). These different initiatives are clearly complementary.

ISO 22156 states "strength and stiffness parameters shall be determined on the basis of tests for the type of action effects to which the material will be subjected (…), or on the basis of comparisons with similar bamboo species (…) or on well-established relations between the properties" (ISO, 2004c). It might be possible to combine or select procedures according to the context of application. As well as the preceding, if the current freedom of methods as outlined in ISO 22156 is to be maintained, to avoid burdening users of bamboo species that do not have access to the costly process of characterisation of a species or a species within a region, a range of factors of safety could be adopted. Species that are poorly understood and not subject to any grading methods, having higher factors of safety than those that are subjected to a rigorous mechanical strength grading regime, reflecting the confidence level generated by the process.

Factors of safety and serviceability limits need to be subjected to greater scrutiny and discussion. Not all failure modes in bamboo are as potentially catastrophic as the others. Compression strength parallel to the fibres exhibits a relatively ductile failure mode, while tension perpendicular to the fibres is brittle. Situations where shear failure could occur, for example members subject to bending or connections, need to account for the risk of splitting, which could reduce the resistance by half if it coincided with a dowel or with the neutral axis of the flexural member. The effect of splitting on long members subject to compression should be considered too, as it is likely to affect the member's resistance to both local and global (Euler) buckling.

It might be of value to separate tests that have more academic value to the process of characterisation from tests that can be readily used in the field and are more useful from the structural point of view. This approach is analogous to the separation of clear specimen tests from structural-size tests in timber (Dinwoodie, 2000). An example is the tension parallel to fibres test, which is of little value for most structural applications, as it is a failure mode that is extremely uncommon in large-scale bamboo structures. Introducing tests aimed at characterising the behaviour of connections might be more beneficial, for example, an adaptation of EN 383 (BSI, 2007) or ASTM D 5764 − 97a(2013) (ASTM, 2013) to bamboo.

13.7   Concluding remarks

Although our understanding of bamboo has progressed immensely over the last 90 years, a great deal still needs to be done to place it on par with timber. Like timber, bamboo is a renewable natural resource that has many associated environmental services during its period of growth, including carbon sequestration, erosion control, water cycle regulation, etc. Additionally, it is a more rapidly renewable resource − reaching full height in less than 6 months and maturity in 3−5 years. This is an important quality, as there is a smaller period of return on investment compared to timber. Bearing in mind that bamboo grows primarily in developing economies, where long-term capital investments are riskier and less common, its adoption as a mainstream structural material would seem more appealing as a business model to a small-scale farmer.

Pole-form bamboo is generally an inexpensive commodity when purchased close to its source; presumably this is due to its rapid growth, although it is also possibly due to the low regard in which bamboo is often held (poor man's timber). Although low-cost is an appealing quality, it is bamboo's remarkable strength and good environmental credentials which should serve to change perceptions. In many parts of the world where bamboo grows, sustainably grown timber of structural quality is simply not available. Therefore, bamboo's real competitors are reinforced concrete, masonry and steel: artificial materials that are perceived to be of superior quality. Bearing in mind that the many regions of the world where most urbanisation is going to take place in the 21st century are precisely those where bamboo grows, changing perceptions and enabling its use could make a significant difference to the carbon emissions, safety in the built environment and the overall quality of life these countries will have in the future. Using bamboo in place of its artificial competitors could potentially improve environmental conditions, reduce exploitation of nonrenewable resources, generate employment among rural workers, reduce dependence on imported materials and exposure to commodity price fluctuations, just to mention a few. Hopefully, this chapter makes a small contribution to that change.

Acknowledgements

The authors of this chapter would like to thank Mutsuko Grant for translating Shigematsu (1958), and Caori Takeuchi for generously sharing her wealth of knowledge.

References

Abdul Latif, M., Wan, T., Fauzidah, A., 1990. Anatomical features and mechanical properties of three Malaysian bamboos. Journal of Tropical Forest Science 2 (3), 227—334.

AIS (Asociación Colombiana de Ingeniería Sísmica), 2010. NSR-10-Reglamento colombiano de construcción sismo resistente - Capítulo G.12: Estructuras de guadua. AIS, Bogotá, Colombia.

Amada, S., Munekata, T., Nagase, Y., Ichikawa, Y., Kirigai, A., Zhifei, Y., 1996. The mechanical structures of bamboos in viewpoint of functionally gradient and composite materials. Journal of Composite Materials 30 (7), 800—819.

Arce, O., 1993. Fundamentals of the Design of Bamboo Structures. Thesis. Eindhoven University, Eindhoven, The Netherlands, p. 281.

ASTM International (ASTM), 2013. ASTM D 5764 — 97a (2013) Standard Test Method for Evaluating Dowel-bearing Strength of Wood and Wood-Based Products, p. 5.

Atrops, J.L., 1969. Elastizitaet und Festigkeit von Bambusrohen [Elasticity and strength of Green bamboo]. Der Bauingenieur 44 (6), 220—225. Quoted in Janssen (1991).

Azuola-Guerra, 1887. La Guadua. Revista Anales de Ingeniería 1 (3). Bogotá, Colombia.

BIS (Bureau of Indian Standards), 1973. Methods of Tests for Round Bamboos. IS:6874. Bureau of Indian Standards, New Delhi.

British Standards (BSI), 2007. BS EN 383:2007 Timber Structures — Test Methods — Determination of Embedment Strength and Foundation Values for Dowel Type Fasteners.

Castrillón, B.M., Malaver, D.M., 2004. Procedimientos de ensayo para la determinación de las propiedades físico-mecánicas de la guadua [Test procedures for the determination of the physical and mechanical properties of guadua Bamboo]. Dissertation. Universidad Nacional de Colombia, sede Bogotá, p. 92.

Correal, J., Arbeláez, J., 2010. Influence of age and height position of Colombian Guadua-bamboo mechanical properties. Maderas: Ciencia y Tecnología 12 (2), 105−113 (Chile).

Correal, J., López, L., 2008. Mechanical properties of Colombian glued laminated bamboo. In: Xiao, et al. (Eds.), Proceedings of the First International Conference on Modern Bamboo Structures. Taylor and Francis, London.

Dinwoodie, J.M., 2000. Timber: Its Nature and Behaviour. E & FN Spon, London.

Food and Agriculture Organisation of the United Nations−FAO, 2007. World Bamboo Resources. Non-Wood Forest Products, Rome.

Gnanaharan, R., Janssen, J., Arce, O., 1994. Bending Strength of Bamboo: Comparison of Different Testing Procedures with a View to Standardization. INBAR Working Paper No. 3. Kerala Forest Research Institute and International Development Research Centre, Peechi, India.

Gottron, J., Harries, K., Xu, Q., 2014. Creep behaviour of bamboo. Journal of Construction and Building Materials 66, 79−88. http://dx.doi.org/10.1016/j.conbuildmat.2014.05.024.

Harries, K.A., Sharma, B., Richard, M.J., 2012. Structural use of full culm bamboo: the path to standardisation. International Journal of Architecture, Engineering and Construction 1 (2), 66−75.

International Standards Organization (ISO), 2004a. 22157-1 Bamboo−Determination of Physical and Mechanical Properties−Part 1: Requirements, p. 17.

International Standards Organization (ISO), 2004b. ISO 22157-2 Bamboo−Determination of Physical and Mechanical Properties−Part 2: Laboratory Manual, p. 20.

International Standards Organization (ISO), 2004c. IS0 22156 Bamboo−Structural Design.

Janssen, J., 1981. Bamboo in Building Structures. Thesis. Eindhoven University, Eindhoven, The Netherlands, p. 253.

Janssen, J., 1991. Mechanical Properties of Bamboo. Kluwer Academic Publishers, Dordecht, The Netherlands.

Janssen, J.J., 2000. Designing and building with bamboo - Technical Report 20. International Network for Bamboo and Rattan, Beijing, China.

Kuehl, L., Yiping, K., 2012. Carbon Off-setting with Bamboo. INBAR Working Paper 71. International Network for Bamboo and Rattan−INBAR, Beijing, P.R. China.

Liese, W., 1998. The Anatomy of Bamboo Culms. INBAR Technical Report 18. International Network for Bamboo and Rattan - INBAR, Beijing, P.R. China.

Limaye, V.D., 1952. Strength of bamboo (*Dendr. strictus*). Indian Forest Records, New Series, Timber Mechanics 1 (1), 17. Quoted in Janssen (1991).

Lin, C., Tsai, M., Wang, S., 2006. Nondestructive evaluation techniques for assessing dynamic modulus of elasticity of moso bamboo (*Phyllosachys edulis*) lamina. Journal of Wood Science 52, 342−347.

Londoño, X., Camayo, G.C., Riaño, N.M., López, Y., 2002. Caracterización anatómica del culmo de *Guadua angustifolia* Kunth [Anatomical characterization of *Guadua angustifolia* Kunth]. Seminario−Taller Avances en la investigación sobre Guadua Pereira.

López, L., Trujillo, D., 2002. Diseño de uniones y elementos en estructuras de guadua [Design of connections and elements in guadua Bamboo structures]. Seminario - Taller Avances en la investigación sobre Guadua, Pereira, Colombia.

Low, I.M., Che, Z.Y., Latella, B.A., Sim, K.S., 2006. Mechanical and fracture properties of bamboo. Key Engineering Materials 312, 15−20.

Lozano, J.E., 2010. Validación de la *Guadua angustifolia* como material estructural para diseño por el método de los esfuerzos admisibles [Validation of *Guadua angustifolia* as a structural material for design by the method of allowable stresses]. Produced by Universidad Nacional de Colombia sede Bogotá.

Lu, X., Wang, K., Yi, X., Liou, J., He, J., 1985. A study on the physic-mechanical properties of culmwood of Phyllostachys Glauca of Shandong. Journal of Bamboo Research, Zhejiang Forestry Science Research Institute, Hangzhou, China 4 (2), 98−106. Quoted in Janssen (1991).

Meyer, H.F., Ekelund, B., 1923. Tests on the mechanical properties of bamboo. Session 1922−1923 The Engineering Society of China 22 (7), 141−169. Quoted in Janssen (1991).

Mitch, D., 2009. Splitting Capacity Characterization of Bamboo Culms. Honors College thesis. University of Pittsburgh, p. 89.

Murphy, R., Trujillo, D., Londoño, X., 2004. Life cycle assessment (LCA) of a Guadua house. In: Simposio Internacional Guadua 2004, Pereira, Colombia, 27 September−2 October, pp. 235−244.

Nugroho, N., Bahtiar, E., 2013. Bamboo taper effect on third point loading bending test. International Journal of Engineering and Technology 5 (3).

Ota, M., 1950. Studies on the properties of bamboo stem (part 7). The influence of the percentage of structural elements on the specific gravity and compressive strength of bamboo splint. Bulletin of Kyushu University of Forestry 19, 25−47. Quoted in Janssen (1991).

Ota, M., 1953. Studies on the properties of bamboo stem (part 9). On the relation between compressive strength parallel to grain and moisture content of bamboo splint. Bulletin of Kyushu University of Forestry, Vol. 22, pp. 87−108. Quoted in Janssen (1991).

Pacheco-Puentes, C.A., 2006. Resistencia a la tracción perpendicular a la fibra de la *Guadua angustifolia* [Tensile strength perpendicular to the grain of bamboo angustifolia]. Dissertation. Universidad Nacional de Colombia, Sede Bogotá.

Richard, M., 2013. Assessing the Performance of Bamboo Structural Components. Doctoral dissertation. University of Pittsburgh.

Richard, M., Harries, K., 2015. On inherent bending in tension tests of bamboo. Wood Science and Technology 49 (1), 99−119.

Sánchez, J., Prieto, E., 2001. Comportamiento de la *Guadua angustifolia* sometida a flexión [*Guadua angustifolia* behavior subject to bending]. Dissertation. Universidad Nacional de Colombia, sede Bogotá, p. 101.

Sekhar, A.C., Rawat, B.S., Bhartari, R.K., 1962. Strength of bamboos: *Bambusa nutans*. Indian Forester 67−73. Quoted in Janssen (1991).

Sharma, B., Harries, K., Ghavami, K., 2013. Methods of determining transverse mechanical properties of full-culm bamboo. Construction and Building Materials 38, 627−637.

Shigematsu, Y., 1958. Analytical investigation of the stem form of the important species of bamboo. Bulletin if the Faculty of Agriculture, University of Miyazaki 3, 124−135. Quoted in Janssen (1991).

Song, X., Zhou, G., Jiang, H., Yu, S., Fu, J., Li, W., Wang, W., Ma, Z., Peng, C., 2011. Carbon sequestration by Chinese bamboo forests and their ecological benefits: assessment of potential, problems, and future challenges. Environmental Reviews 19, 418−428.

Trujillo, D., 2007. Bamboo structures in Colombia. The Structural Engineer 85 (6), 25−30.

Trujillo, D., 2013. Prospects for a method to infer non-destructively the strength of bamboo: a research proposal. Proceedings of Third International Conference on Sustainable Construction Materials and Technologies, 18−21 Aug 2013 in Kyoto, Japan.

Trujillo, D., Ramage, M., Chang, W.S., 2013. Lightly modified bamboo for structural applications, construction materials. Proceedings of the ICE. Construction Materials 166 (4), 238−247. Available from http://dx.doi.org/10.1680/coma.12.00038.

Vaessen, M.J., Janssen, J.J., 1997. Analysis of the critical length of culms of bamboo in four-point bending tests. Heron 42 (2), 113−124.

Van Der Lugt, P., Van Den Dobbelsteen, A., Abrahams, R., 2003. Bamboo as a building material alternative for Western Europe? A study of the environmental performance, costs and bottlenecks of the use of bamboo in Western Europe. Journal of Bamboo and Rattan 2 (3), 205–223.

Zhou, F., 1981. Studies on physical and mechanical properties of bamboo wood. Journal of Nanjing Technical College, Forestry Products (2), 1–32. Quoted in Janssen (1991).

Bamboo design and construction

14

J.F. Correal
Universidad de los Andes, Bogotá, Colombia

14.1 Introduction

Bamboo, which is giant grass, plays an important economic and cultural role in various regions around the world. It has been used as an edible vegetable, cosmetic and fertilizer; in crafts, pulp, paper and fabrics; and as fuel as an oil, gas and charcoal. In addition, bamboo is one of the oldest construction materials and has been used in the fabrication of bridges, rural houses and water channels in some Asian and Latin American countries. Particularly in the construction of buildings, and despite its high strength:weight ratio, bamboo has been mostly restricted to rural houses, mainly due to the lack of building codes. This issue, amongst others, has been a barrier to the use of bamboo compared with concrete, steel, masonry and wood. Nonetheless, increasing knowledge of the behaviour of bamboo coming from research, emergence of new products (laminated products and fibre composites) and development of testing and structural standards (building codes) will influence the use of bamboo in the near future. Key sustainability characteristics of bamboo, such as having a fast growth rate, being renewable and biodegradable, as well as the benefit of sequestering carbon, suggest that the material can play an important role in the construction industry. In fact, it has been estimated that some species of bamboo could have a utilization rate four times greater than that of wood (De Flander and Rovers, 2009).

This chapter reviews the bamboo species suitable for construction purposes, the development of bamboo-based products and the type of structures that are typically designed and built with bamboo. Further, the anatomy of bamboo culms, physical and mechanical properties, some preservation methods and fire performance based on the latest research will be covered. In addition, design considerations such as bamboo testing and structural design standards and building codes will be presented. Part of this chapter will focus on the design procedure of members subjected to bending, axial and combined forces, as well as the design of shear walls based in the Colombian Building Code, NSR-10 (AIS, 2010). An overview of different types of connections between bamboo members and dowel-type connection capacity will be presented. Finally, fabrication and construction of bamboo structures will be covered, taking into account local experiences in some countries such as reported in the NSR-10 (AIS, 2010).

14.1.1 Bamboo species

Woody and herbaceous bamboo belongs to the Poaceae family, Bambusoideae subfamily, having about 90 genera and more than 1200 species. Bamboos have a growth

Nonconventional and Vernacular Construction Materials. http://dx.doi.org/10.1016/B978-0-08-100038-0.00014-7

area of almost 38,000 million hectares, approximately 3.2% of the total forest area around the world. Bamboo is a diverse plant that grows naturally in tropical and subtropical regions and is commonly found in Asia, with 65% of the total world bamboo resources, followed by Central and South America with 28% and Africa with 7% (Lobovikov et al., 2007). Bamboo resource statistics vary significantly amongst countries, since bamboo is not routinely included in forest inventories. In addition, there is no standard methodology developed for bamboo resource reporting with clear definitions, formulas, guidelines, or reclassification approaches. Based on available data, China has the highest bamboo biodiversity in Asia (500 native species), followed by Japan (139 species), India (119 species), Indonesia (118 species) and Myanmar (97 species). With the exception of Brazil, which has 232 bamboo species, most of the Central and South American countries have fewer species than Asian countries. Colombia and Venezuela have around 50 species, whilst Costa Rica, Peru and Mexico have about 30 species of bamboo. African countries have the lowest diversity of bamboo species, with four species reported for the United Republic of Tanzania, followed by three species for Togo and two species for Uganda and Ethiopia.

Not all species of bamboo are suitable for application in construction. According to McLure (1953), the following species are the most versatile and useful for building construction: *Arundanaria, Bambusa, Cephalostachyum, Dendrocalamus, Gigantochloa, Melocanna, Phyllostachys, Schizostachyum, Guadua* and *Chusqea*. Some of the species used in several applications in the construction of buildings are presented by Jayanetti and Follet (1998) based on the data compiled by McLure (1953), as reproduced in Table 14.1.

It is important to highlight that suitability of bamboo species for construction depends on the specific application in relation with the physical and mechanical properties of each species. Further, the size of the bamboo culm is relevant, especially in the construction of certain structural elements subjected to bending and axial forces (see Sections 14.4 and 14.5), since it is important in the design of these elements to achieve not only strength but also adequate stiffness.

14.1.2 Bamboo-based products

Bamboo is also very versatile, achieving 100% material utilization in most cases (Fig. 14.1). Bamboo shoots and sculpture crafts are examples of products obtained from the rhizome. Products such as bamboo beverages, medicine, pigment, fertilizer and fodder are acquired from its leaves. Culms are the most broadly used part of bamboo with a wide and diverse range of uses, from bamboo chopsticks, furniture and crafts to raw bamboo culms, bamboo-based panels and bamboo fibre composites. Processed waste from the culms is used for different products such as woven strand, charcoal, activated carbon and bamboo powder, which is raw material for composite decking and paper production.

Engineered bamboo panels are one of the types of products available for use in building construction. There are many varieties of bamboo-based panels with thicknesses varying from 2 to 40 mm and dimensions, depending on the specification of the manufacturing equipment or requirement of the project. The most common

Table 14.1 Bamboo species and applications in the construction of buildings

Application in building construction		Scientific name of most suitable bamboo species
Framing		*Bambusa balcooa, B. nutans, B. polymorpha, Gigantochloa apus, Guadua angustifolia, Phyllostachys edulis*
Walls	Wattling	*Bambusa balcooa, B. nutans, B. polymorpha, Gigantochloa apus, Guadua angustifolia, Phyllostachys edulis*
	Whole or half culms	*Bambusa balcooa, B. nutans, B. polymorpha, Gigantochloa apus, Guadua angustifolia, Phyllostachys edulis*
Roof	Tiles	*Bambusa balcooa, B. nutans, B. polymorpha, Gigantochloa apus, Guadua angustifolia, Phyllostachys edulis*
	Shingles	*Schizostachyurn lima, S. lumatnpao*
Floor	Boards	*Bambusa polymorpha, Guadua angustifolia, Phyllostachys edulis*
	Strips	*Bambusa nutans, B. polymorpha, Gigantochloa apus, Guadua angustifolia, Phyllostachys edulis*
Sheathing	Whole culms	*Bambusa nutans, B. polymorpha*
	Strips	*Bambusa balcooa, B. nutans, B. polymorpha, Gigantochloa apus, Guadua angustifolia, Phyllostachys edulis*
Matting		*Bambusa polymorpha, Gigantochloa apus*
Lashings		*Gigantochloa apus*
Concrete formwork	Boards	*Bambusa polymorpha, Guadua angustifolia*
	Shoring	*Bambusa balcooa, B. nutans, B. polymorpha, Guadua angustifolia*
Scaffolding		*Bambusa balcooa, B. nutans, B. polymorpha, Gigantochloa apus, Guadua angustifolia, Phyllostachys edulis*
Pipes and troughs		*Bambusa balcooa, B. nutans, B. polymorpha, Gigantochloa apus, Guadua angustifolia*

Adapted from Jayanetti, D.L., Follet, P.R., 1998. Bamboo in Construction: An Introduction. INBAR Technical Report No. 16, International Network for Bamboo and Rattan.

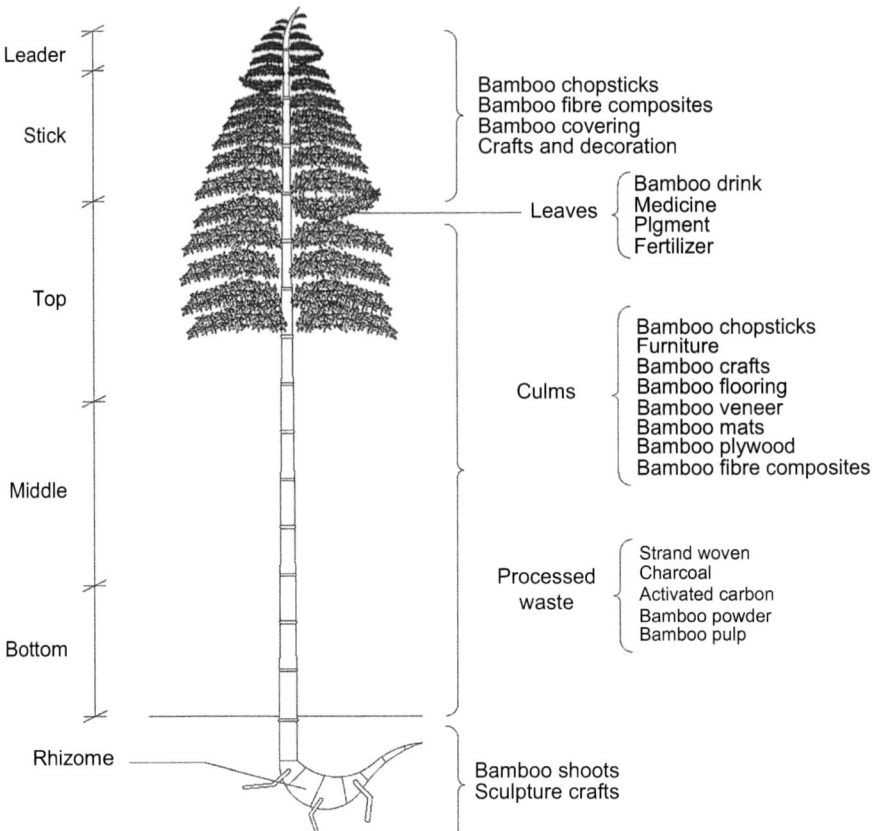

Figure 14.1 Parts of the bamboo plant and its utilization.

bamboo-based panels are plybamboo, laminated bamboo, mat plybamboo, curtain plybamboo, laminated bamboo strips, mat−curtain plybamboo, bamboo chipboard and bamboo−wood composite. More detailed information about engineered bamboo panels can be found in Chapter Engineered bamboo and in Qisheng et al. (2001).

Pole bamboo (ie, that is used in its native form) is another product frequently used in low-rise building construction. The middle and top parts of the culm (Fig. 14.1) are usually used as structural elements such as columns and beams, since these parts exhibit less tapering over their length. The remaining parts of the culm like the stick are used for some laminated products, whereas the leader is used as nonstructural material for decoration of buildings such as wall covering.

14.1.3   Types of structures

Although bamboo has been used for different applications relevant to construction such as scaffoldings (widely used in Asia up to 30−40 stories in height) and water piping, low-rise buildings and pedestrian bridges (Fig. 14.2) are the most common

Figure 14.2 Common structures constructed with bamboo.
Courtesy of Simón Vélez.

types of structures constructed with bamboo. Recent development of laminated products may increase the use of bamboo in the construction industry, since these products may be a substitute for laminated wood products not only for heavy- and light-frame building construction but also for the construction of vehicular bridges (Xiao et al., 2010).

Currently, the majority of pole bamboo construction is focused on low-rise buildings, which have a structural system similar to light-frame (platform) wood construction (Fig. 14.3). Roof and floor systems consist of single or double bamboo joists that are supported on the top of walls and sheathing composed of split bamboo mats covered with cement mortar to establish the diaphragm action necessary for the building's lateral force resisting system. Culm internodes in contact with bamboo joists need to be filled with mortar to avoid local crushing. It is important to ensure that all the elements composing the roof and floor systems are designed to resist the load path in the plane of the elements. A classic approach to considering diaphragm action is to consider the floor as a structural unit that behaves as a deep beam that transfers lateral loads to supporting walls based on the relative stiffness of the diaphragm and walls. Thus, the horizontal load path involves different elements of the diaphragm such as sheathing, framing, chords and struts. Finally, architectural details of the roof, such as the length and slope of overhangs, need to be taken into account to provide adequate protection against environmental exposure, since bamboo is susceptible to deterioration from weather conditions such as rain and sunlight.

Other main components of the bamboo structural system are the bearing frame composed of vertical (studs) and horizontal (sill and top plate) bamboo elements (Fig. 14.4). This type of wall is referred to as bahareque wall in some Latin American countries. 'Fish-mouth' cuts at the ends of the bamboo studs are made by a hand tool or cylindrical drill to connect to the sill and top plate. Even though nails and lashing are traditionally used in bamboo connections, threaded bars are recommended for connecting studs to sill and top plates (Fig. 14.5(a)). A good construction practice for bamboo is to fill the internodes involved in the connection with fluid cement mortar and provide additional metal clamps or bands around the bamboo, to reduce the possibility of

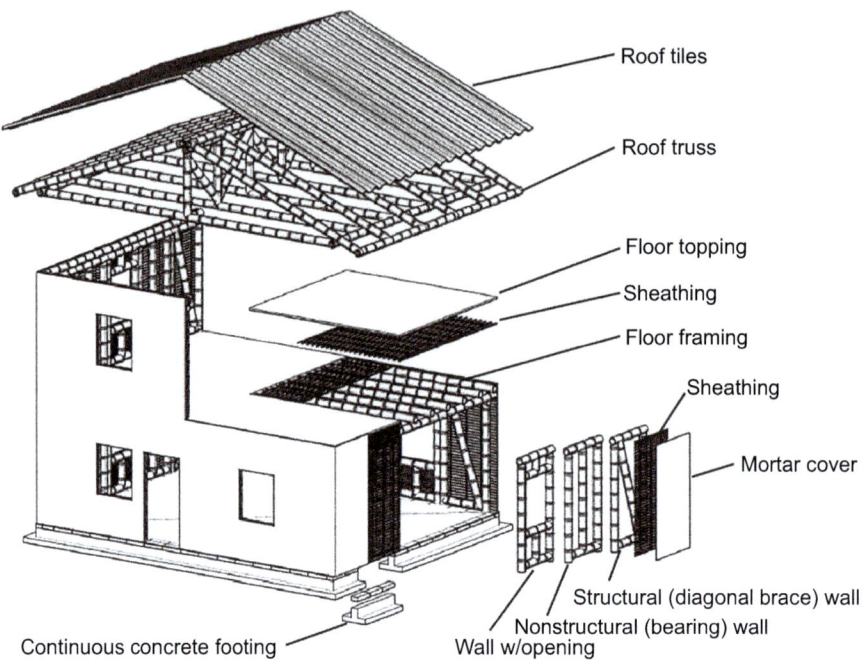

Figure 14.3 Bamboo light-frame system for low-rise buildings.

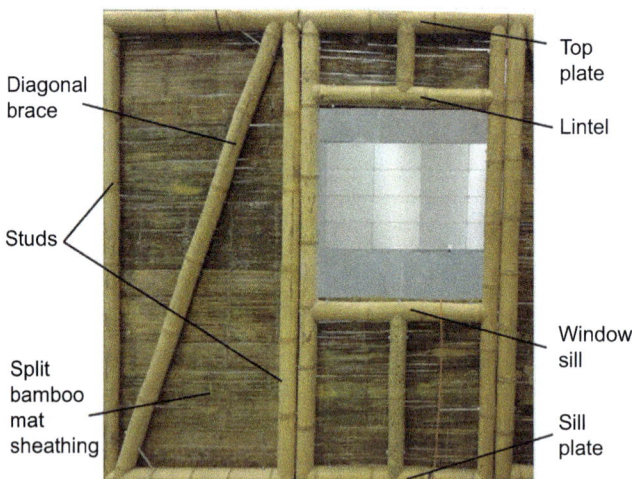

Figure 14.4 Typical bahareque framed system.

crushing and splitting failures, respectively (Fig. 14.5(a)). The bamboo frame is sheathed with either split bamboo or horizontal wood strips and a steel mesh cover. In both cases, it is recommended to use screws instead of nails for attaching the split bamboo or wood strips to the bamboo frame. Mortar is used for protection purposes

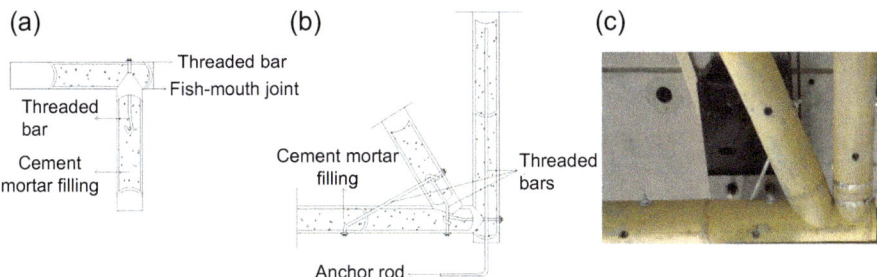

Figure 14.5 Typical connections in bamboo frame: (a) stud-to-plate connection, (b) diagonal brace connection and (c) connection with metallic straps.

and as finishing material. Braced bamboo walls are part of the lateral force resisting system in bamboo buildings. The construction of these walls is the same as for bearing walls, but the connection at the corner of the frame is usually more complex (Fig. 14.5(b)). Recent research data have shown that bearing walls with proper connections between the sheathing and the bamboo frame can achieve the same performance as braced walls (Arbeláez and Correal, 2012), but more research is needed to fully understand the structural behaviour of bamboo walls subjected to lateral loads. Walls that are part of the lateral force resisting system must be continuous along the entire height of the buildings, from the foundation level to the roof.

Continuous and isolated column concrete footings are the most common type of foundations used in bamboo building structures. Regardless of the type of foundation used, and taking into account that bamboo may decay in a matter of months, direct contact with soil and/or water must be avoided. Therefore, it is not only recommended to elevate bamboo elements above the ground but also considering preservation treatments for bamboo, as well as corrosion-resistant fasteners for connections to the foundation.

14.2 Bamboo as a material

To understand the potential of bamboo as a construction material, it is imperative to comprehend the structure of the bamboo culm. The anatomical structure of bamboo has a direct relation to its physical and mechanical properties. In addition, it is very important to preserve bamboo to improve durability, which is a primary concern for natural-origin materials such as bamboo. On the other hand, the knowledge of the fire behaviour of bamboo is also imperative to determine levels of safety for structures as required by building codes (see Section 14.2.4).

14.2.1 Anatomy of bamboo culms

Most bamboo species have an essentially circular cross section, but it is also possible to find square (*Chimonobambusa quadrangularis*) cross sections, and even man-made

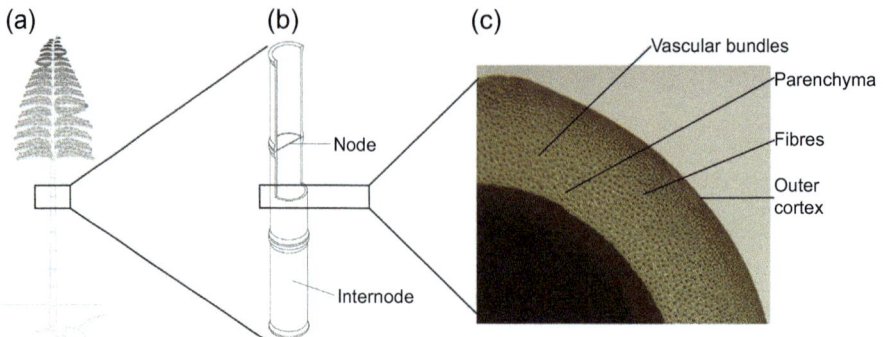

Figure 14.6 Structure of the bamboo culm: (a) bamboo plant, (b) nodes and internodes and (c) microstructure of culm wall.

triangular cross sections. The culm is the above-ground part of bamboo (Fig. 14.6(a)) and is composed of nodes and internodes (Fig. 14.6(b)). The number, spacing and shape of the nodes vary between species. The nodes or diaphragms provide transverse interconnection of the culm wall, which in turn strengthens the culm. Radially oriented cells like those found in wood are present in the nodes, whereas the cells in the internodes are longitudinally oriented. The internodes are hollow tubes and have a varying wall thickness, which forms the bamboo cavities.

The cortex of the culm contains axially elongated cells, shorter cork and silica cells and the stomata (Liese, 1998). The cells in the internodes run parallel to the height of the culm and they are composed of approximately 50% parenchyma, 40% fibres and 10% conducting tissue (Liese, 1985). A higher concentration of fibres is located at the outer face of the culm wall and decreases towards the inner face (Fig. 14.6(c)). Although the amount of fibres is reduced from bottom to top of the culm, bamboo fibre density increases with height, since bamboo culms are tapered from base to top. The fibres' length:width ratio varies between 150 and 250 and depends not only on the species but also on its location along the culm (Liese, 1985). From a structural point of view, bamboo is a naturally functionally graded material, with fibres embedded in the parenchyma matrix and set in the proper location (toward the outer side of the wall) to provide stiffness against bending, serving a purpose analogous to reinforcing steel in concrete.

14.2.2 *Physical and mechanical properties*

Taking into account the anatomy of the culm, bamboo is a natural material that exhibits viscoelastic and anisotropic behaviour but is generally considered in design as an orthotropic material. Physical and mechanical properties are different along three orthogonal axes. Nonetheless, variations of bamboo properties are more important along the bamboo culm length, since density increases with height due to the tapered shape of the culm, as discussed earlier.

Additionally, age at harvest is a key factor that influences the physical and mechanical properties of bamboo. Optimal strength properties in relation with the age of bamboo vary between species, but it is a general assumption that bamboo matures after 3 years (Liese, 1985). On the other hand, properties of bamboo can decay with increased age; therefore, it is important to select, according to the species, the most appropriate age at which to harvest bamboo (Correal and Arbeláez, 2010) to ensure a better performance from the point of view of strength and stiffness.

Moisture content affects the long-term performance of bamboo and must be considered when designing bamboo structures. Similar to wood, green (freshly harvested) bamboo has lower strength than dry bamboo. The effect of moisture content is determined by the fibre saturation point, which varies within culms and amongst species but generally ranges between 20% and 30%. Moisture content can also affect the dimensional stability, bending strength and load duration (creep) effect.

Testing standards, discussed further in this chapter, are used to determine physical and mechanical properties of pole bamboo. From the structural point of view, the following physical and mechanical properties are the minimum required for the design of buildings using bamboo: moisture content, density, shrinkage, compression, tension, shear and bending strengths and modulus of elasticity. Compression strength is determined both parallel- and perpendicular-to-grain, whereas tension and shear strength are obtained for the parallel-to-grain direction only. Typical values of physical and mechanical properties for selected species are shown in Table 14.2. Once the strength and stiffness properties have been determined, a statistical analysis is conducted over a large sample (at least 20) of strength values. The fifth percentile exclusion value serves as the starting point for the development of allowable stresses for each property. The minimum, fifth percentile, and mean values for modulus of elasticity are determined to calculate beam or column stability factors or deflections. Further details about the allowable strength design (ASD) method applied to bamboo construction are provided in Section 14.3.

14.2.3 Bamboo preservation

One of the main concerns for the use of bamboo in construction is durability. Although natural durability depends on the species, the expected service life of untreated bamboo is 1–3 years when it is in contact with soil, 4–5 years for elements protected from exterior exposure, and >15 years under very favourable conditions such as internal framing or when rendered. Treatment of bamboo is needed to improve durability. Whilst traditional (nonchemical) methods have been used, chemical methods are more appropriate for bamboo preservation in large-scale building projects. Penetration of chemicals into the bamboo culm tissue is more difficult than in wood, due to the absence of ray cells forming a radial transportation system that permits easy penetration through the entire wall thickness. Effective and safe preservation methods have been developed using chemicals such as boric acid, borax and boron. Open tank for cold soaking, modified Boucherie and pressure treatment are the most commonly used preservation methods against biological attack. The open tank method is simple and provides good protection for bamboo culms. The culms are immersed in a tank with a water-soluble preservative for several days so that a slow penetration process takes place. It is important to drill a hole into each internode, to allow the chemical

Table 14.2 **Mechanical properties of common bamboo species used in construction**

Species	Common name	Specific gravity	Bending strength (MPa)	Modulus of elasticity (MPa)	Tension strength (MPa)	Compression strength (MPa)
Dendrocalamus strictus[a]	Calcutta bamboo	0.64	137	9790	156	—
Guadua angustifolia[b]	Guadua	0.68	107	21,530	127	64
Phyllostachys edulis[c]	Moso bamboo	0.79	88	11,400	—	75

[a]Ahmad and Kamke (2005).
[b]CIMOC (2011).
[c]Chung and Yu (2002).

solution to penetrate the inside cavity of the bamboo. The modified Boucherie method uses a pressure system that is connected to one end of the culm, and the preservative is forced axially through the culm vessels until it comes out from the other end of the culm. This method reduces the time of treatment from days to hours. The chemical solution residue may be recycled once it is cleaned, adding more chemical solution to achieve the original concentration. Pressure treatment is one of the best preservative methods for bamboo but is also expensive. It requires specialized equipment (an auto-clave) that applies pressure ranging around $0.5-1.5 \text{ N/mm}^2$. Preservatives such as creosote or waterborne preservatives are typically used. For additional information regarding preservative methods, the reader is directed to Jayanetti and Follet (1998) and Janssen (2000).

14.2.4   Fire resistance

Bamboo is a combustible material like wood, which, when exposed to a certain amount of heat from a fire, presents both a mass loss and a reduction in strength. Under such conditions, if bamboo is used as a structural material for buildings, it is imperative to know how it behaves under fire. Fire reaction and resistance are properties of the material behaviour that allow the understanding of the capacity of structural elements during fire exposure. Fire reaction is associated with the ease with which a material ignites and the spread of fire under specific conditions, whereas fire resistance is the ability to maintain its sealing and isolation capacity from heat transfer and structural resistance under fire exposure. Some species of bamboo have shown favourable fire behaviour (Mena et al., 2012; Xiao and Ma, 2012), and in certain cases bamboo exhibits a better performance than some wood species. Particularly, the research con-ducted by Mena et al. (2012) indicates that from the fire reaction point of view, pole and laminated *Guadua angustifolia* Kunth (referred throughout this chapter as *Guadua a.* K.) has a higher ignition critical flux and requires a longer exposure time to achieve ignition under a constant flux than plywood made from Radiata pine (Fig. 14.7(a)). In addition, *Guadua a.* K. exhibited a higher ignition critical flux than traditional refer-ence values for wood species. This might be explained by the microstructure of bamboo, since outer cortex is composed of silica cells that may provide protection against fire. Regarding fire resistance, similar or lower charring rates than plywood were found. A lower decrease in flexural strength as temperature increased was found for pole and laminated *Guadua a.* K. relative to plywood (Fig. 14.7(b)).

   Although the fire behaviour of materials is important to determine levels of safety in buildings, the expected performance of the entire building under fire exposure is essen-tial to establish the risk. Building codes prescribe fire protection requirements appli-cable to all buildings to perform at the same level of safety, regardless of the material (CEN, 2002; ICC, 2012). Currently, building fire safety is incorporated in building codes through a combination of passive and active features. Whilst passive features are focused on prescriptive measures of fire-rated building elements and/or appropriate means of egress, active features are concentrated on automatic fire detec-tion and alarm or systems to suppress fire spread. Wood buildings have been success-fully designed to meet the requirements established by building codes and it is possible

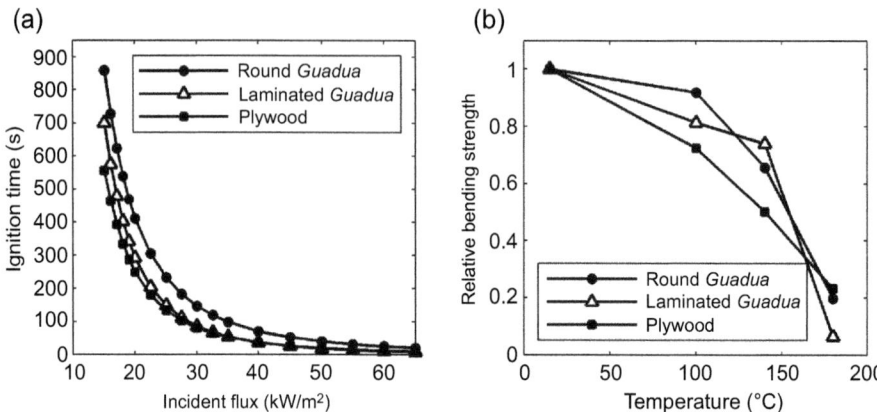

Figure 14.7 Fire behaviour of round and laminated *Guadua*: (a) heat flux versus ignition time and (b) relative flexure strength versus temperature.
Adapted from Mena, J., Vera, S., Correal, J.F., López, M., 2012. Assessment of fire reaction and fire resistance of *Guadua angustifolia* Kunth bamboo. Construction and Building Materials 27, 60−65.

to adapt or implement this building fire safety experience to bamboo buildings. Fire simulation tests carried out by Xiao and Ma (2012) showed that glued laminated bamboo frame buildings using similar design details as North American timber buildings (rock wool as thermal insulation and gypsum boards as cover for walls and ceilings) provide satisfactory fire performance. Even though it is possible to use successful design experience from timber structures, more research needs to be conducted on the fire behaviour of bamboo species.

14.3 Design considerations

Material behaviour and building performance are the main considerations in the structural design of buildings. Behaviour of the materials is based on testing standards as well as the design philosophy. Allowable stress design (ASD) and load and resistance factor design (LRFD) are the two design philosophies currently used in structural engineering practice. On the other hand, building codes provide performance requirements through minimum standards to safeguard life, health, property and public welfare by regulating the design, construction, quality of materials, use and occupancy, location and maintenance of buildings. In the following sections, a brief description of the current international testing and structural design standards, as well as national building codes applicable to the use of bamboo structures are presented.

14.3.1 Bamboo testing standards

Testing standards have been developed locally in different countries, but since 2001, International Organization for Standardization (ISO), with the cooperation of

International Network for Bamboo and Rattan (INBAR), assigned the Technical Committee ISO/TC 165 to develop an international standard for the determination of physical and mechanical properties of bamboo: ISO/DIS-22,157 (ISO, 1999). This standard is focused only on pole bamboo culms and it is used to obtain data to establish characteristic strength values for allowable stresses according to the ASD method. The ISO/DIS-22,157 testing standard is divided in two parts. The first part covers the test methods for the evaluation of moisture content, mass per volume, shrinkage, bending, compression, shear and tension strength. It is important to highlight that ISO/DIS-22,157 specifies compression, shear and tension tests that are performed parallel to the fibres. The second part is focused on providing methods and informative guidelines for laboratory staff to perform tests for the determination of the physical and mechanical properties of bamboo described earlier. Additional information regarding bamboo characterization is presented in Chapter Bamboo material characterisation.

14.3.2 Structural design standards and building codes

An international cooperation between ISO and INBAR was established to develop the first international standard for bamboo structural design [ISO/DIS-22,156 (ISO, 2001)] through Technical Committee ISO/TC 165. The scope of this standard applies to the use of bamboo structures made from pole bamboo, split bamboo and glued laminated bamboo. In addition, ISO/DIS 22,156 is based on the performance of the structure, and it covers both ASD and LRFD design philosophies (LRFD is focused solely on the definition of ultimate and serviceability limit states). Mechanical resistance, serviceability and durability are considered in this standard, whereas thermal and acoustic insulation are not.

A giant species of bamboo called *Guadua a.* K. grows naturally in Colombia, and people have used it as a structural material, based on the experience passed on from generation to generation. Moreover, *Guadua a.* K. is probably the most popular species of bamboo used as a construction material in Latin American countries. National building codes that incorporate bamboo as a structural material include the Indian (NBCI, 2005) and Colombian Building Codes (NSR-10; AIS, 2010). The Colombian Code is arguably more complete and robust and is the focus of the present discussion. The scope of Chapter G.12 of NSR-10 is focused on establishing requirements for the design of structures made from pole *Guadua a.* K. bamboo in both nonseismic and seismic regions. It is important to note that *Guadua a.* K. is a thick-walled bamboo species (typically having a diameter:wall thickness ratio <10) and that the design recommendations of NSR-10 are developed accordingly. A *Guadua a.* K. structure designed according to the requirements of the NSR-10 will have a safety level equivalent to that of structures designed with other materials. Also, requirements of the NSR-10 may be used for the design of elements in structures built entirely with *Guadua a.* K., or structures built with *Guadua a.* K. and other materials, like the bahareque system described in Chapter E.7 of the NSR-10. The design of residential buildings in Chapters G.12 and E.7 of NSR-10 is limited to two storeys, and no masonry or concrete walls in the upper storey are allowed. The requirements of NSR-10 are limited to those buildings whose main use is residential, commercial, industrial or educational. In addition, when *Guadua a.* K. structures with an area larger than 2000 m^2 are built, it is recommended to perform a loading test prior to occupancy.

14.3.2.1   Structural design basis

A structural design basis is required to provide adequate performance of buildings. In general, all bamboo elements in a structure must be designed, built and connected to withstand stresses produced by loading combinations and deflection limitations stipulated in building codes. Further, the structural design must reflect all the possible acting loads over the structure during the construction and service stages, as well as environmental conditions that may imply changes in the design assumptions or may affect the integrity of other structural components. Also, the analysis and design of bamboo structures must be based on the principles of engineering mechanics and strength of materials. From a practical point of view, bamboo elements are considered homogeneous and linear for the estimation of stresses due to applied loads.

An important consideration for the design of earthquake resisting buildings is the seismic response coefficient (R), which is used to reduce the linear elastic design seismic forces to account for the energy dissipation capacity of the structure. For example, the NSR-10 recommends a basic seismic design coefficient (R_0) for *Guadua a.* K. structures, whose lateral force resisting system is based on diagonally braced walls, to be equal to 2.0. For the case of a lateral force resisting system based on bahareque walls, the value of R_0 is 1.5.

14.3.2.2   Structural design method

The LRFD design philosophy requires developing load and resistance factors, which are based on statistical analysis and probability of failure to account for the variability of load conditions, materials and construction. ISO/DIS 22156 covers the definition of ultimate and serviceability limit states. Typically, material resistance factors are then calibrated using appropriate code-specified load factors and a target reliability, independent of the material considered. Whilst no code has yet to adopt an LRFD approach for bamboo, Harries et al. (2012) introduced the concept. Resistance factor determination will be an important task in the future for technical committees on bamboo.

On the other hand, the ASD design philosophy is focused on ensuring that stresses in the structure due to service loads do not exceed a limiting value within the elastic range of the material. This limit is called the allowable stress and is determined by factoring the characteristic value of the material property by a series of modification factors (which include such effects as the test load rate, load duration. and safety factor). The characteristic value of stress corresponding to the 95% tolerance limit (fifth percentile) obtained with 75% confidence level, f_{ki}, may be estimated from fifth percentile value obtained in laboratory tests, $f_{0.05i}$, as follows (AIS, 2010; ISO, 2001):

$$f_{ki} = f_{0.05i}\left(1 - \frac{2.7\frac{s}{m}}{\sqrt{n}}\right)$$

[14.1]

where m is the mean value of laboratory test data for stress i, s is the standard deviation of laboratory test data for stress i and n is the number of laboratory tests conducted (at least 10). The convention for the stress subscript i is typically b for bending, t for tension, c for compression-parallel-to-grain, p for compression-perpendicular-to-grain and v for shear.

Once the characteristic value is determined for each type of load action (bending, tension, compression and shear), allowable stresses are calculated as follows (NSR-10 Equation G.12.7-2):

$$F_i = \frac{FC}{F_S \cdot FDC} f_{ki}$$ [14.2]

where F_i is the allowable stress for action i, FC is the test load rate reduction factor, F_S is the safety factor and FDC is the load duration factor. Modifying factors used in the NSR-10 are presented as reference values in Table 14.3. Additional information about modifying factors may be found on ISO/DIS 22,156 (ISO, 2001). Further, allowable stresses used in NSR-10 for *Guadua a*. K. determined using Eq. [14.2] are shown in Table 14.4.

Stress values used in the design of different elements of structures, such as beams and columns, are also called adjusted design values, and are obtained according to the following equation (NSR-10 Equation G.12.7-3):

$$F_i' = F_i C_D C_m C_t C_L C_r C_c$$ [14.3]

where the adjustment factors are for load duration (C_D), moisture content (C_m), temperature (C_t), beam lateral stability (C_L), group action (C_r) and shear (C_c).

Adjustment factors for bamboo (NSR-10) are very similar to those used in the design of wood structures. These factors are not only applicable to stresses but also to moduli of elasticity. Chapter G.12.7 of the NSR-10 presents the values for adjustment factors for *Guadua a*. K. bamboo, which are reproduced in Table 14.5.

Table 14.3 Modifying factors for allowable strength estimation for *Guadua a*. K

Modifying factor	Bending	Tension	Compression-parallel-to-grain	Compression-perpendicular-to-grain	Shear
FC	1.0	0.5	1.0	1.0	0.6
F_S	2.0	2.0	1.5	1.8	1.8
FDC	1.5	1.5	1.2	1.2	1.1

Adapted from AIS, 2010. Reglamento Colombiano de Construcción Sismo Resistente NSR-10, Asociación Colombiana de Ingeniería Sísmica.

Table 14.4 **Allowable strength values for *Guadua a.* K**

Property	Notation	Allowable strength (MPa)
Bending	F_b	15
Tension	F_t	18
Compression-parallel-to-grain	F_c	14
Compression-perpendicular-to-grain	F_p	1.4
Shear	F_v	1.2
Modulus of elasticity (mean)	$E_{0.5}$	9500
Modulus of elasticity (5th percentile)	$E_{0.05}$	7500
Modulus of elasticity (minimum)	E_{min}	4000

Adapted from AIS, 2010. Reglamento Colombiano de Construcción Sismo Resistente NSR-10, Asociación Colombiana de Ingeniería Sísmica.

14.4   Bending members

The design of pole bamboo elements subjected to bending follow the same basic procedure used in the design of beams in other structural materials. Because of the high ratio of modulus of rupture (MOR) to modulus of elasticity (MOE), bamboo is a highly flexible material and bending design will typically be governed by allowable deflections. Nonetheless, the following effects must be analysed in the design of bamboo elements subjected to bending:

• Deflections
• Bending strength, including lateral stability for built-up beams
• Shear-parallel-to-grain strength
• Bearing (compression-perpendicular-to-grain)

The first three effects govern the resulting dimensions of the element, whereas the fourth effect must be considered in the design of the supports. The supports of a bamboo element subjected to bending should not fail due to bearing (compression-perpendicular-to-grain); therefore, it is common practice to fill the internode at the support with cement mortar as described previously. Concentrated loads must be applied, as much as possible, in the node of the bamboo element and all necessary actions should be conducted to avoid failure due to shear-parallel-to-grain or bearing at the point of loading. In this case, filling the adjacent internodes with cement mortar is recommended.

When using more than one culm for beam construction (built-up beams), these must be connected with bolts or threaded bars and metal clamps, to ensure the elements work in a composite manner. These connectors should be designed to withstand the shear forces acting along the element interfaces.

Table 14.5 Adjustment factors for *Guadua a*. K

		F_b	F_t	F_c	F_p	F_v	$E_{0.5}$	$E_{0.05}$	E_{min}
C_D	Permanent	0.90	0.90	0.90	0.90	0.90	—	—	—
	Ten years	1.00	1.00	1.00	0.90	1.00	—	—	—
	Two months	1.15	1.15	1.15	0.90	1.15	—	—	—
	Seven days	1.25	1.25	1.25	0.90	1.25	—	—	—
	Ten minutes	1.60	1.60	1.60	0.90	1.60	—	—	—
	Impact	2.00	2.00	2.00	0.90	2.00	—	—	—
C_m	MC $\leq$ 12%	1.00	1.00	1.00	1.00	1.00	1.00	1.00	1.00
	MC = 13%	0.96	0.97	0.96	0.97	0.97	0.99	0.99	0.99
	MC = 14%	0.91	0.94	0.91	0.94	0.94	0.97	0.97	0.97
	MC = 15%	0.87	0.91	0.87	0.91	0.91	0.96	0.96	0.96
	MC = 16%	0.83	0.89	0.83	0.89	0.89	0.94	0.94	0.94
	MC = 17%	0.79	0.86	0.79	0.86	0.86	0.93	0.93	0.93
	MC = 18%	0.74	0.83	0.74	0.83	0.83	0.91	0.91	0.91
	MC $\geq$ 19%	0.70	0.80	0.70	0.80	0.80	0.90	0.90	0.90
C_t	$T \leq 37°C$	1.00	1.00	1.00	1.00	1.00	1.00	1.00	1.00
	$37°C < T \leq 52°C$ MC $\geq$ 19%	0.60	0.85	0.65	0.80	0.65	0.80	0.80	0.80
	MC < 19%	0.85	0.90	0.80	0.90	0.80	0.90	0.90	0.90
	$52°C < T \leq 65°C$ MC $\geq$ 19%	0.40	0.80	0.40	0.50	0.40	0.80	0.80	0.80
	MC < 19%	0.60	0.80	0.60	0.70	0.60	0.80	0.80	0.80

Continued

Table 14.5 Continued

		F_b	F_t	F_c	F_p	F_v	$E_{0.5}$	$E_{0.05}$	E_{min}
C_L	$d/b = 2$	1.00	—	—	—	—	—	—	—
	$d/b = 3$	0.95	—	—	—	—	—	—	—
	$d/b = 4$	0.91	—	—	—	—	—	—	—
	$d/b = 5$	0.87	—	—	—	—	—	—	—
C_r	$s \leq 0.60$ m	1.10	1.10	1.10	1.10	1.10	—	—	—
C_c	$l/D = 5$	—	—	—	—	—	0.70	—	—
	$l/D = 7$	—	—	—	—	—	0.75	—	—
	$l/D = 9$	—	—	—	—	—	0.81	—	—
	$l/D = 11$	—	—	—	—	—	0.86	—	—
	$l/D = 13$	—	—	—	—	—	0.91	—	—
	$l/D = 15$	—	—	—	—	—	0.93	—	—

Adapted from AIS, 2010. Reglamento Colombiano de Construcción Sismo Resistente NSR-10, Asociación Colombiana de Ingeniería Sísmica.

14.4.1   Deflections

In general, bamboo has a high MOR:MOE ratio, which implies that the design of elements subjected to bending is governed by allowable deflection. The requirements and limits for allowable deflections are established in the design codes and are typically checked under live load alone and under total load (dead load plus live load) conditions.

Deflections for bamboo elements (beams, joists and floorings) must be calculated following the classic theory of elasticity and are usually estimated using the mean modulus of elasticity. However, if service conditions are severe or the required safety level is high, the fifth percentile modulus of elasticity or minimum modulus of elasticity may be used, although this is at the discretion of the structural engineer. In addition, deflections in short spans sometimes need to consider a correction of the modulus of elasticity to account for the effect of shear. For instance, Table 14.5 shows correction factors to the modulus of elasticity due to shear effects (C_c) for *Guadua a.* K. bamboo elements with span (l):outside diameter of the culm (D) ratios lower than 15 recommended by NSR-10.

Instantaneous deflections are predictable (with some degree of accuracy) provided the modulus of elasticity is known. On the other hand, long-term deflections resulting from creep are not predictable yet, and therefore these must be considered to be only estimates. For example, NSR-10 recommends for calculation of long-term deflections to increase the dead load (sustained loads) by a factor of 2.8 and live load by a factor of 1.3 for *Guadua a.* K. bamboo with moisture content (MC) lower than 19% and ambient temperature (*T*) lower than 37°C; if MC is higher than 19% and *T* is higher than 37°C, NSR-10 recommends increasing the dead load by a factor of 3.8 and live load by a factor of 1.4.

14.4.2   Flexure

The maximum tension and compression stresses due to bending are calculated for the maximum bending moment along the element. Principal bending stresses are parallel to the length of the element and thus parallel to the fibres of the bamboo. The design moment in bamboo is obtained using traditional elasticity theory and the corresponding stresses are calculated using the elastic section modulus, S (Eq. [14.4]). The moment of inertia of built-up beams of two or more bamboo culms should be calculated as the sum of the individual moments of inertia of each culm unless the builder or engineering professional ensures adequate composite action of the built-up beams (Sharma et al., 2011; Richard and Harries, 2012). The design bending stresses, f_b, should not be higher than the maximum allowable stress for bending, F'_b, as shown in Eq. [14.4], which are modified by appropriate adjustment factors (Section 14.3.2.)

$$f_b = \frac{M}{S} \leq F'_b \qquad\qquad [14.4]$$

where M is the flexural moment acting over the element and S is the elastic section modulus.

One of the adjustment factors that needs to be considered in bending is intended to account for the lateral stability of beams. This factor considers the reduction of loading capacity of an element subjected to bending due to the lateral instability or buckling of the portion of the cross section of the element stressed in compression (ie, lateral torsional buckling). Common framing conditions such as providing an effective connection of roof or floor diaphragms to the compression side of beam causes the unbraced length to be essentially zero; thus, lateral instability is mitigated. In the case of slender bamboo beams (beams composed of two or more bamboo culms), the need for lateral support of the compression side must be verified. Two methods may be used to determine the stability factor (C_L): a rule-of-thumb method, or tabulated lateral stability factors prescribed by NSR-10.

For the rule-of-thumb method, the lateral stability factor C_L is 1.0, provided that the following conditions are satisfied for each case of beams composed of two or more *Guadua a.* K. bamboo culms, based on the depth (d) to width (b) ratio of the beam:

- For $d/b = 2$, no lateral support is required.
- For $d/b = 3$, lateral displacement at the supports must be restrained.
- For $d/b = 4$, lateral displacement at the supports and the compression side of the beam must be restrained.
- For $d/b = 5$, lateral displacement at the supports must be restrained and continuous support for the compression side of the beam should be provided.

Alternatively, the tabulated lateral stability factor (C_L) used in NSR-10 for built-up beams of two or more *Guadua a.* K. bamboo culms are presented in Table 14.5.

14.4.3   Shear

The critical location for shear is calculated at a distance equal to the depth of the element measured from the face of the support. For beams of only one culm, this depth should be equal to the outside diameter of the culm, except for cantilever beams, where the maximum shear stress should be calculated at the face of the support. For built-up beams of two or more bamboo culms, the depth corresponds to the actual depth of the element. The maximum shear stress should be determined considering a nonuniform stress distribution along the cross section and must be lower than the maximum allowable shear-parallel-to-grain strength modified by appropriate adjustment factors (Table 14.5).

Special attention should be given to the design of transverse steel connectors in built-up bamboo beams composed of two or more culms. It is recommended to fill all internodes that are crossed by a connector with a fluid nonshrink cementitious mortar (Fig. 14.8). In addition, NSR-10 recommends that maximum spacing between steel connectors should be lower less than three times the depth of the beam or one-fourth of the span. NSR-10 further recommends that the first connector be located at a distance of 50 mm from the face of the support.

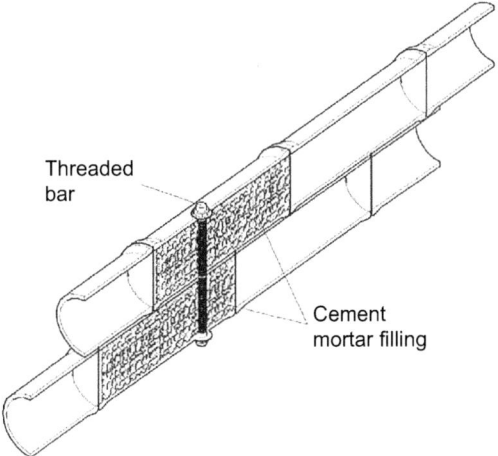

Figure 14.8 Internode filling with cement mortar.
Adapted from AIS, 2010. Reglamento Colombiano de Construcción Sismo Resistente NSR-10, Asociación Colombiana de Ingeniería Sísmica.

14.4.4 Bearing

Bearing stresses perpendicular to the fibres of bamboo occur especially at beam supports or where loads from other members frame into the beam as concentrated loads. The acting compression-perpendicular-to-grain stress must not exceed the allowable compression strength modified by the appropriate coefficients. Based on NSR-10, the acting compression-perpendicular-to-grain stress, f_p, should be calculated according to the following expression:

$$f_p = \frac{3RD}{2t^2l} \leq F_p' \qquad\qquad [14.5]$$

where R is the load acting over the element in the direction perpendicular to the fibres, D is the outside diameter of the culm, t is the culm wall thickness, l is the length of the support and F_p' is the allowable compression-perpendicular-to-grain strength.

It is recommended that internodes of the culm subjected to compression-perpendicular-to-grain stresses should be filled with fluid nonshrinkable cementitious mortar. Nonetheless, in the case that the internodes are not filled with mortar, a reduction of the allowable stresses should be applied. For instance, NSR-10 establishes that only one-fourth of the design allowable compression-perpendicular-to-grain strength should be considered if internodes are not filled with mortar. A further consideration is the wall thickness; NSR-10 is applicable to *Guadua a.* K., a thick-walled species ($D/t < 10$). The provisions of NSR-10 for bearing should be applied with caution to thinner-walled species (eg, *Phyllostachys edulis* [Moso]).

14.5   Axial force members

An axially loaded bamboo member has the force applied parallel to the longitudinal axis (same direction of the fibres) passing through the centroid of the cross section. These forces are either tension or compression. In bamboo structures, compression force members (like columns) are encountered more often than tension members due to the natural need of the structures to carry vertical gravity loads and because of the difficulty of fabricating efficient connections for tension members.

14.5.1   Compression elements

The most common axially loaded members in bamboo structures are columns. There are different types of columns, such as single culm columns, spaced columns and built-up columns. Spaced columns consist of two or more parallel culms that are separated by spacers with a rigid element, tied together at the ends (and occasionally along their height) of the columns. Built-up columns are made of two or more culms mechanically fastened together at their ends to provide continuity of each column. The main mechanisms of failure of compression elements are buckling and crushing. The slenderness ratio, λ, defines the primary measure of buckling:

$$\lambda = \frac{kl_u}{r} \qquad [14.6]$$

where l_u is the unbraced length of the column and k is the effective length coefficient dependent on the end support conditions of the column and whether side sway of the column is prevented. Typical theoretical and recommended design values of k are presented in Table 14.6; r is the radius of gyration, calculated as:

$$r = \sqrt{\frac{I}{A}} \qquad [14.7]$$

where I is the moment of inertia about the weak axis and A is the cross-sectional area. When two or more culms are used to compose an element subjected to compression,

Table 14.6 Column coefficient k for different end conditions

Support/end conditions		Coefficient k	
Support at element ends	Lateral restraint	Theoretical value	Empirical value
Fixed − pin	No restraint	2.00	2.40
Pin − fixed	No restraint	2.00	2.10
Fixed − fixed	No restraint	1.00	1.20
Pin − pin	Restrained	1.00	1.10
Pin − fixed	Restrained	0.70	0.80
Fixed − fixed	Restrained	0.50	0.65

the moment of inertia of the group should be calculated as the sum of the individual moments of inertia of each culm, unless the builder or engineering professional ensures adequate composite action may be achieved.

According to their slenderness ratio, *Guadua a.* K. bamboo columns are classified into short ($kl_u/r \leq 30$), intermediate ($30 < kl_u/r < C_k$) or long columns ($C_k \leq kl_u/r \leq 150$); the slenderness limit between intermediate and long columns, C_k, is given by the following expression:

$$C_k = 2.565 \sqrt{\frac{E_{0.05}}{F'_c}} \qquad [14.8]$$

where $E_{0.05}$ is the fifth percentile modulus of elasticity and F'_c is the allowable compression-parallel-to-grain strength modified by the appropriate adjustment factors. *Guadua a.* K. bamboo columns with slenderness ratios >150 are not allowed.

Shorter columns will be controlled by compression-parallel-to-grain strength, whereas larger columns will be controlled by global (Euler) buckling. The intermediate-length columns will be controlled by a combination of direct compressive stress and bending. For instance, NSR-10 defines that the acting compression stress, f_c, must not exceed the allowable compression-parallel-to-grain strength modified by the appropriate coefficients, F'_c, for short, intermediate and long columns, according to Eqs. [14.9]–[14.11], respectively.

$$f_c = \frac{N}{A_n} \leq F'_c \qquad [14.9]$$

$$f_c = \frac{N}{A_n \left[1 - \frac{2}{5}\left(\frac{\lambda}{C_k}\right)^3\right]} \leq F'_c \qquad [14.10]$$

$$f_c = 3.3 \frac{E_{0.05}}{\lambda^2} \leq F'_c \qquad [14.11]$$

where N is the load acting over the element in the direction parallel to the fibres, A_n is the net cross-sectional area of the element, λ is slenderness ratio of the element (Eq. [14.6]), C_k is the slenderness limit between intermediate and long columns (Eq. [14.8]). and $E_{0.05}$ is the fifth percentile modulus of elasticity.

14.5.2 Tension elements

The axial tension stress, f_t, for any bamboo cross section must not be greater than the allowable strength modified by the appropriate adjustment factors, F'_t. The equation for designing bamboo tension members is:

$$f_t = \frac{T}{A_n} \leq F'_t \qquad [14.12]$$

where T is the tension load acting over the element having a net cross-sectional area A_n.

The design of axial tension members needs to consider combined loading in tension and bending when an eccentricity between the connection centroid and the centroid of the member force exists. On the other hand, design of bamboo elements subjected to tension-perpendicular-to-grain should be avoided, due to its relatively low resistance in this direction. However, if these stresses are present in a structure, reinforcement in the areas subjected to such stresses should be provided, for example using external metal clamps (as shown in Fig. 14.5(c)).

14.6 Combined loading elements

Combined loading for bamboo elements consists of bending moment that occurs simultaneously with a tension or compression force. For combined loading, it is important to consider the net stresses at each face: for a member face subject to net compression, lateral stability must be considered. The combined stresses are analysed using the linear interaction equation presented in Eq. [14.13], which is applied to the stress condition at both extreme tension and compression faces of the section. For the case of net tensile stresses, the allowable bending stress used in the interaction equation, F_b', does not include the lateral stability factor ($C_L = 1.0$). On the other hand, if the combined stresses result in net compression, the lateral stability factor must be included in the calculation of F_b'.

$$\frac{f_t}{F_t'} + \frac{k_m f_b}{F_b'} \leq 1.0$$ [14.13]

where f_t and f_b correspond to the acting tension and bending stresses over the element, respectively, and F_t' and F_b' are the allowable tension and bending strengths, respectively. Combined compression force and bending moment is frequently observed in practical applications of bamboo structures. As discussed earlier, the lateral stability factor needs to be included in the calculation of the allowable bending. Also, for bending moment occurring simultaneously with compression force, a second-order bending stress, known as the $P-\Delta$ effect, is induced. In order to account for this effect, the actual bending stress is multiplied by an amplification factor, k_m given by Eq. [14.14]. For combined bending and tension, $k_m = 1.0$.

$$k_m = \frac{1}{1 - 1.5(N_a/N_{cr})}$$ [14.14]

where N_a is the acting compression force and N_{cr} is Euler critical buckling load calculated using the fifth percentile modulus of elasticity, $E_{0.05}$:

$$N_{cr} = \frac{\pi^2 E_{0.05} I}{(k l_u)^2}$$ [14.15]

where I is the moment of inertia of the cross section and kl_u is effective length of the element (Eq. [14.6]).

14.7 Shear walls

Shear walls in bamboo structures are typically used as the vertical elements in the lateral force resisting system. Lateral loads such as seismic and wind forces are distributed to the shear walls by the floor and roof diaphragms and then transfer these loads to the foundation system. Low-rise bamboo buildings with bahareque walls have exhibited good seismic performance in past earthquakes (EERI, 2000). In addition, bahareque walls have been adopted as building systems for one- to two-storey houses by the Colombian Building Code (NSR-10). Although bamboo laminated products allow the construction of light-frame building systems similar to those used in wood (Varela et al., 2013), the bahareque walls are one of the best solutions for construction of one- to two-storey houses with pole bamboo. A general description of the bahareque system (made of *Guadua a.* K. bamboo elements) based on NSR-10 is presented in this section.

Bahareque is a system composed of two main parts: the framing and the sheathing (Fig. 14.4). The framing is composed of two plates or horizontal elements at top and bottom, and studs or vertical elements, connected by nails, screws or other types of fasteners. The cross section of the top and bottom plates must have a minimum width equal to the diameter of the bamboo elements used as studs, and a depth >100 mm. In some cases, it is preferred to construct both top and bottom plates in sawn lumber since its connections provide greater stiffness and are less susceptible to crushing than bamboo elements. The studs may be constructed with dry bamboo with a diameter not less than 80 mm. The horizontal spacing between bamboo studs need not be less than 300 mm and should not exceed 600 mm on centre.

The bahareque wall sheathing can be either split bamboo mat (or wood panel) or horizontal wood strips, both covered with cement mortar applied over a steel mesh. In both cases, it is recommended to use screws to attach the split bamboo mat or wood strips to the bamboo frame instead of nails. Mortar covering should be applied over a steel wire mesh having a maximum diameter of 1.25 mm (BWG 18 gauge) and hexagonal openings not larger than 25.4 mm. The minimum required classification for the cement mortar should correspond to type N, with a volume ratio not exceeding four parts sand to one part cementitious material. Quality of cement and lime, when it is used, should be based on recognized standard specifications for the material.

Walls of cemented bahareque for low-rise bamboo buildings are classified into three groups:

1. Walls with diagonal elements are structural walls, or wall segments, composed of top and bottom plates, studs, diagonal bamboo elements and sheathing with split bamboo mat, wood panel or horizontal wood strips, covered with cement mortar applied over a steel mesh. These walls receive vertical loads and in-plane horizontal wind or seismic loads and are part of the lateral force resisting system in bamboo buildings. Structural walls with diagonal elements should be provided in the corners of buildings and at the ends of each group of structural wall segments.

2. Load-bearing walls without diagonal elements but otherwise the same as Item 1 may be used only for resisting vertical loads of the building.

3. Nonbearing walls are walls capable only of supporting their self-weight. Exterior walls are designed to resist the out-of-plane force of the wind blowing against them. These walls have no function other than separating spaces inside the household. Interior nonbearing walls must be connected to the top diaphragm by a connection that prevents overturning but does not allow the transmission of shear or vertical forces between the roof or floor and the wall. Nonbearing walls do not need to be continuous and are not required to be anchored to the foundation.

Structural diagonal bamboo walls and load bearing walls must be built over foundations beams. Structural walls must be continuous from the foundation level to the top diaphragm to which they are connected.

NSR-10 specifies a minimum length of bahareque walls in each principal direction of the building to provide a uniform resistance to seismic forces in the inelastic range (Eq. [14.16]). It is important to note that Eq. [14.16] is developed for bahareque walls sheathed on both sides. The effective length of a wall with one side sheathed is taken to be one-half of its actual length.

$$L_i \geq C_B A_p \qquad\qquad\qquad [14.16]$$

where L_i is the minimum total length of walls without openings in each direction, C_B is a wall density coefficient determined from Table 14.7 as a function of the effective peak acceleration A_a (based on seismic hazard maps) and A_p is the area tributary to the storey for which the walls are designed (ie, area of roof and second floor for first-storey

Table 14.7 Density coefficient values for bahareque walls

Effective peak acceleration, A_a (g)	Wall density coefficient, C_B
0.40	0.32
0.35	0.28
0.30	0.24
0.25	0.20
0.20	0.16
0.15	0.16
0.10	0.16
0.05	0.16

Adapted from AIS, 2010. Reglamento Colombiano de Construcción Sismo Resistente NSR-10, Asociación Colombiana de Ingeniería Sísmica.

walls; area of the roof for one-storey houses). For second-storey walls in two-storey houses, $0.66\,A_p$ may be used if lightweight materials are used for the roof, such as panels or metallic sheets without mortar covering. Bahareque walls must be distributed in an approximately symmetric manner in plan. Heavy plating or cladding should be avoided in facades. For bathrooms, the entire wet area must be made waterproof. All facades must be adequately fixed to avoid their collapse during an earthquake.

Cemented bahareque wall systems must be designed for the effects of combined loading. For the effects caused by seismic forces, the energy dissipation capacity of the structural system is taken into account, applying the seismic response modification factor (R). NSR-10 specifies an R-factor of 1.5 for cemented bahareque. Values obtained for stresses over each wall, due to vertical and lateral loads, must be less than the allowable strength. Table 14.8, adapted from NSR-10, establishes values for allowable design forces and stresses according to the structural composition and materials used for each type of bahareque wall. The values expressed in Table 14.8 assume that walls will be fully anchored to the foundation and between each other.

14.8 Connections

Connections are fundamental to maintain the integrity of any structure. Since bamboo cross sections are round, hollow and tapered, connections between bamboo elements are more difficult than other construction materials such as wood. Different kinds of bamboo connections have been developed. These depend mainly on the orientation of the joined elements and the type of fasteners used. Since it is very difficult in bamboo connections to ensure a moment-resisting capacity, it is recommended that bamboo connections in the structure are considered as pin-connected, and no moment transmission between the connected elements should be considered, unless one of these elements is continuous, in which case the transmission will only occur for the continuous element. Even though a general description of the different kinds of connections is presented in the next section, the dowel connection using steel bolts and filling the bamboo internode with cement mortar will be the focus of the discussion in this section.

14.8.1 Types of connections

Janssen (2000) and Jayanetti and Follet (1998) presented a comprehensive description of the different types of connections that can be made with pole bamboo. Some of these connections are shown on Figs. 14.9–14.13. Janssen (2000) introduced a classification based on three principles: (1) a joint between two bamboo culms can be either by contact between the full cross section or by collecting forces from the cross section to a jointing element; (2) collecting forces may occur from the inside (ie, within the hollow core of the bamboo) or outside of the cross section and (3) the jointing element can run parallel with the fibres or perpendicular to them.

Table 14.8 Allowable strength properties for cemented bahareque walls

Cross section for verification	Allowable shear strength, V (kN/m) Horizontal section of the lower end of the wall	Allowable compression strength, F_c (MPa) Effective cross section of end post or stud	Maximum vertical load (kN/m) Horizontal section of the lower end of the wall	Allowable tension load, T (kN) Anchor rod on end post or stud
Wall composition	Wood group classification			
Guadua and wood framing. Two diagonals, one in each end of the wall, in opposite directions. 40% Minimum of studs in sawn lumber. Top and bottom plate in sawn lumber. Sheathed with split *Guadua* mat, steel mesh and mortar cover on both sides of the wall	ES1 and ES2 — 15	15	39	15
	ES3 — 12	13	37	
	ES4 — 10	10	35	
	ES5 — 7	8	33	
	ES6 —	5	31	
Guadua framing. Two diagonals, one in each end of the wall, in opposite directions. Top and bottom plate in *Guadua*. Sheathed with split *Guadua* mat, steel mesh and mortar cover on both sides of the wall	8	8	41	10

Adapted from AIS, 2010. Reglamento Colombiano de Construcción Sismo Resistente NSR-10, Asociación Colombiana de Ingeniería Sísmica.

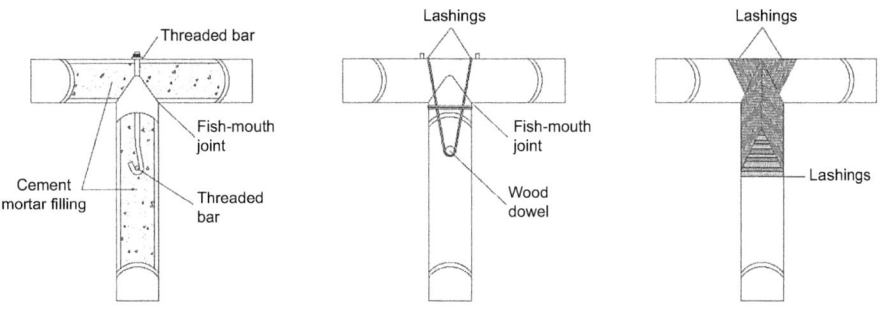

Figure 14.9 Typical orthogonal connections.

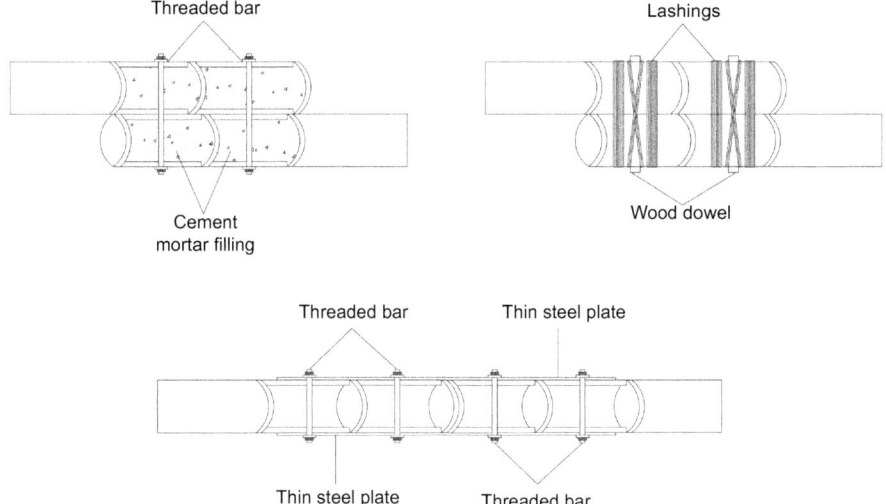

Figure 14.10 Typical longitudinal (sliced) connections.

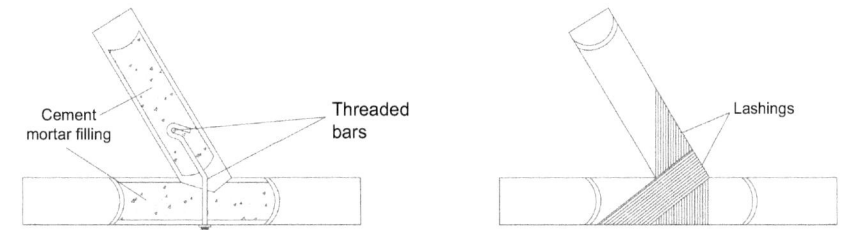

Figure 14.11 Typical angled connections.

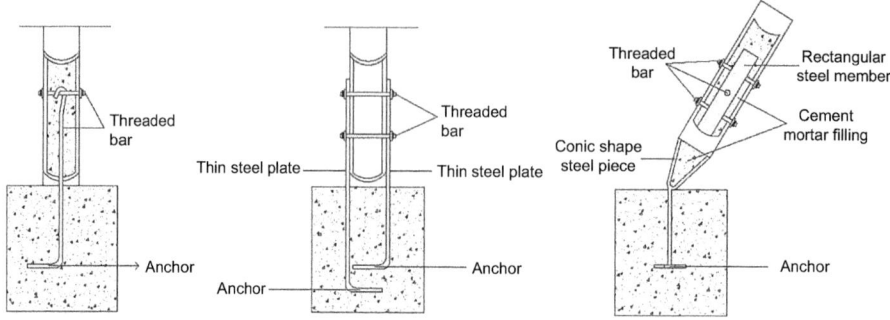

Figure 14.12 Typical post-to-foundation connections.

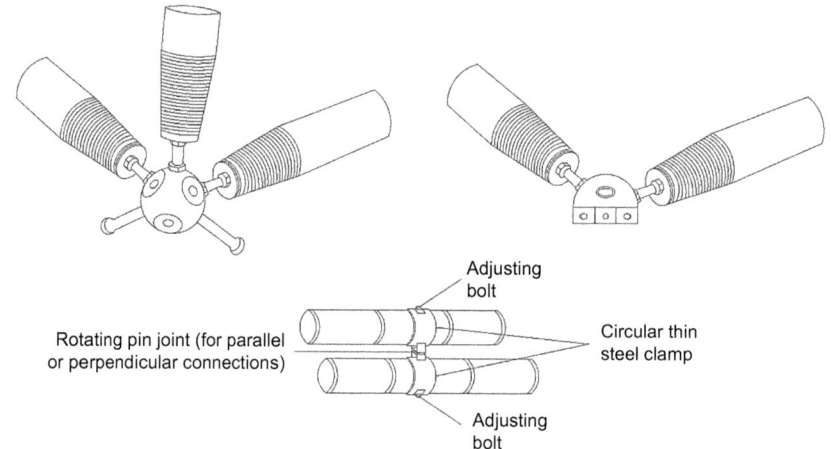

Figure 14.13 Other types of contemporary connections using metal end joints.

According to these principles, eight types of connections are presented. On the other hand, Jayanetti and Follet (1998) classified pole bamboo connections into four categories: (1) sliced, (2) orthogonal, (3) angled and (4) through. Both Janssen and Jayanetti and Follet emphasize the importance of taking into account three aspects in pole bamboo connections: (1) type of connector (ie, lashing ties, steel bolt, wood pins, steel clamps, galvanized steel gusset plates or prefabricated metal end joints); (2) the cut shape at the end of bamboo culms (ie, straight cut, fish-mouth or inclined fish-mouth) to fit elements at different angles and (3) the filling material within the bamboo internode in the connection, such as cement mortar or wood glued to the inside wall of the culm.

Steel bolts and clamps, galvanized steel gusset plates and prefabricated metal end joints are currently the most common type of connectors in pole bamboo construction. Since steel bolts and clamps are usually less expensive than other connectors, they are often used in the construction of low-rise bamboo buildings. On the other hand, the cut shapes at the joints of two or more elements of bamboo are very important to avoid crushing or splitting at the joint surfaces. Even though a handmade cut can be performed, it is recommended to perform a mechanical cut using a metal hole-saw and to make those cuts not farther than 100 mm from the nodes. Fig. 14.14 shows a fish-mouth cut and inclined fish-mouth cut used to join pole bamboo culms at 90 degree and an inclined angle orientation. Although wood-filling material can be used at the bamboo internode, cement mortar is the most common filling material in the connections of bamboo buildings, since cement mortar is easier to obtain and fabricate, and in some cases is less expensive than wood. NSR-10 recommends using a cement mortar with a minimum cement:sand ratio of 1:3 and to use plasticizer additive to ensure the fluidity of the mix. It is also recommended that the cement mortar should correspond to type M or S (compression strength between 12.5 and 17.5 MPa and 110−125% of fluidity). Even though the use of wood as a filling material requires extra work (turning the inside part of the culm and fastening it) for fabrication of joints, this type of filling has a similar specific weight as bamboo. Taking into account that cement mortar filling adds weight to the connection, additional research needs to be done regarding new filling materials that have adequate compression strength whilst remaining lightweight and easy to pour.

14.8.2 Dowel connection capacity

Dowel connections are one of the most widely used type of connections in bamboo construction. Bamboo dowel connections use steel bolts as a connector and usually the bamboo internodes are filled with cement mortar. Some of the recommendations given in the NSR-10 for *Guadua a.* K. bamboo dowel-type connections are presented here. Minimum bolt diameter is 9.5 mm, and bolts should be made of structural steel with yield strength >240 MPa. Drilled holes for bolts should have a diameter 1.5 mm larger than the bolt diameter. Drilled holes for filling the internodes should have a maximum diameter of 26 mm and must be filled with cement mortar to ensure the structural continuity of the element. For the case of bolted connections in the

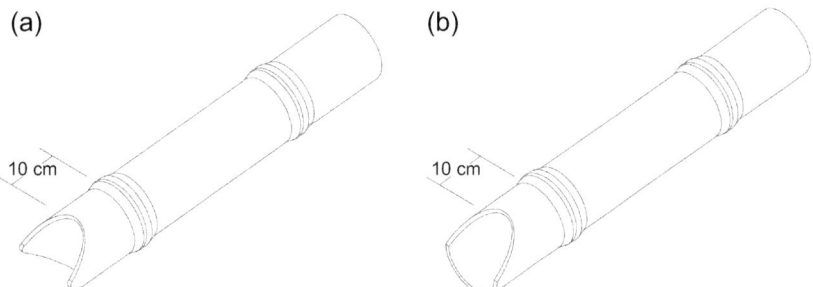

Figure 14.14 Typical cuts in bamboo: (a) fish-mouth cut and (b) inclined fish-mouth cut. Adapted from AIS, 2010. Reglamento Colombiano de Construcción Sismo Resistente NSR-10, Asociación Colombiana de Ingeniería Sísmica.

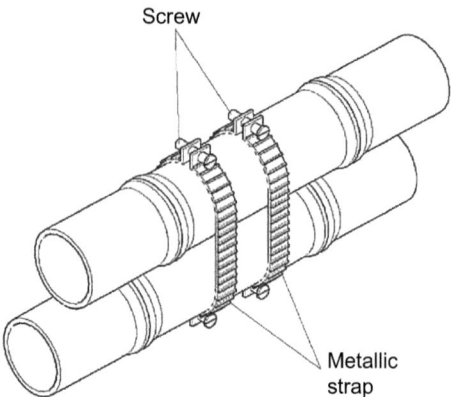

Figure 14.15 Connections using metallic straps.
Adapted from AIS, 2010. Reglamento Colombiano de Construcción Sismo Resistente NSR-10, Asociación Colombiana de Ingeniería Sísmica.

longitudinal direction of the bamboo, it must be ensured that failure does not occur in the partition diaphragms involved in the connection. In addition, all other metallic elements used in dowel-type connections that are exposed to unfavourable environmental conditions should be treated against corrosion.

The use of metallic straps or clamps as shown in Fig. 14.15 is permitted for the design of connections, provided appropriate precautions are taken to prevent crushing and compression-perpendicular-to-grain failure in individual culms, as well as slip between connected elements.

In the case of a dowel-type connections subjected to bearing or crushing loads, the bamboo internodes adjacent to the connection, or where bolts are located, must be filled with cement mortar. Metallic washers must be used between both the bolt head and nut and the bamboo element. Dowel-type connection capacity depends on the direction of the load at the main and side elements connected. For instance, NSR-10 defines three allowable loads for dowel-type connections: (1) when loading in the connection is parallel to the fibres, for the main element and the side elements (indicated as P in Fig. 14.16(a)); (2) when the force is parallel to the fibres of the main element, but perpendicular to the fibres of the side elements, or vice versa (indicated as Q in Fig. 14.16(b)) and (3) when the force is perpendicular to the fibres of one element and parallel to the fibres of the other element (indicated as T in Fig. 14.16(c)).

NSR-10 allowable loads for dowel-type connections subjected to double shear (P, Q and T directions) are shown in Table 14.9, as a function of the outside diameter of the culm (D) and the bolt diameter (d). Allowable loads listed in Table 14.9 are representative of *Guadua a.* K. bamboo culms with moisture content under 19%, used in a dry service environment. For connections of four or more members, each load plane should be evaluated as a single shear connection. The value for the connection should be estimated as the lowest nominal value obtained, multiplied by the number of shear planes.

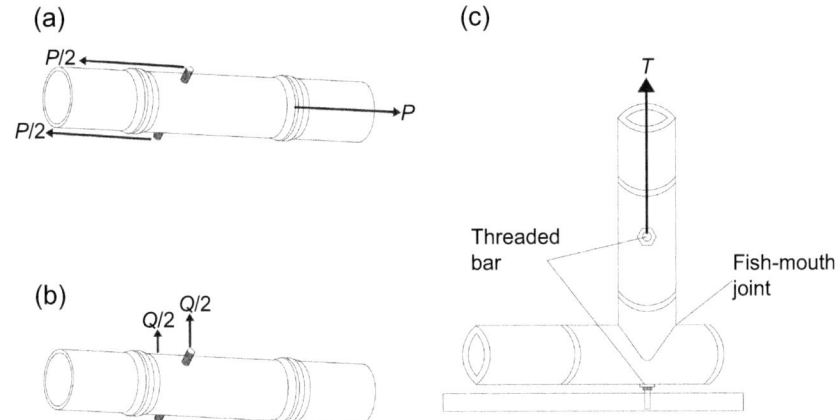

Figure 14.16 Directions of loading for dowel-type connections: (a) parallel to main and side members; (b) parallel to side members and perpendicular to main member, on perpendicular planes and (c) parallel to main member and perpendicular to side member, on the same plane. Adapted from AIS, 2010. Reglamento Colombiano de Construcción Sismo Resistente NSR-10, Asociación Colombiana de Ingeniería Sísmica.

Table 14.9 Allowable loads for dowel-type connections subjected to double shear

Bolt diameter, d (mm)	**Guadua** element diameter, D (mm)	P (N)	Q (N)	T (N)
9.5	80	7212	2885	2000
	90	8008	3203	2100
	100	8804	3522	2200
	110	9601	3840	2300
	115	10,041	4016	2400
	120	10,481	4193	2500
	125	10,922	4369	2600
	130	11,362	4545	2700
	135	11,802	4721	2800
	140	12,242	4897	2900
	150	13,039	5216	3000
12.7	80	9710	3884	2000
	90	9916	3966	2100
	100	10,943	4377	2200

Continued

Table 14.9 **Continued**

Bolt diameter, d (mm)	*Guadua* element diameter, D (mm)	P (N)	Q (N)	T (N)
	110	11,970	4788	2300
	115	12,521	5009	2400
	120	13,072	5229	2500
	125	13,623	5449	2600
	130	14,174	5670	2700
	135	14,725	5890	2800
	140	15,276	6110	2900
	150	16,303	6521	3000
15.9	80	11,540	4616	2000
	90	12,806	5122	2100
	100	13,250	5300	2200
	110	14,515	5806	2300
	115	15,185	6074	2400
	120	15,855	6342	2500
	125	16,525	6610	2600
	130	17,195	6878	2700
	135	17,865	7146	2800
	140	18,535	7414	2900
	150	19,800	7920	3000

Adapted from AIS, 2010. Reglamento Colombiano de Construcción Sismo Resistente NSR-10, Asociación Colombiana de Ingeniería Sísmica.

If the force in the connection follows the direction of the main element but forms an angle with the fibre direction of the side elements, or vice versa, the allowable load should be determined from the Hankinson equation:

$$N = \frac{PQ}{P \sin^2 \alpha + Q \cos^2 \alpha} \qquad [14.17]$$

where N is the allowable inclined load, P and Q are the allowable loads for the parallel and perpendicular directions from Table 14.9 and α is the angle between the main and side members of the connection.

**Table 14.10 Group reduction coefficient values
for connections with two or more bolts**

Type of connection	Number of bolts				
	2	3	4	5	6
Connections with *Guadua* elements	1.0	0.97	0.93	0.89	0.82
Connections with steel elements	1.0	0.98	0.95	0.92	0.90

Adapted from AIS, 2010. Reglamento Colombiano de Construcción Sismo Resistente NSR-10, Asociación Colombiana de Ingeniería Sísmica.

Allowable loads listed in Table 14.9 correspond to connections with only one bolt. The group reduction coefficient, C_g, should only be applied to P load when a connection requires more than two bolts in a line parallel to the loading direction. The C_g coefficient value for different numbers of bolts in a single gauge line is presented in Table 14.10.

Finally, NSR-10 specifies that the spacing between bolts should not be less than 150 mm or >250 mm, but in all cases an internode must fall between each bolt. The distance from a bolt to the free end of the element must be >150 mm in connections subjected to tension and >100 mm in connections subjected to compression.

14.9   Fabrication and construction

The entire process of material preparation, fabrication, construction and assembly should follow good engineering practices. In addition, the following recommendations (based on NSR-10) are presented for the preparation process of the bamboo culms:

- Bamboo culms used for structural applications must be selected at a proper mature age to obtain optimal strength properties.
- To prevent fungal attack, the maximum moisture content of bamboo culms used as structural elements should be 20%. In addition, drying and preservation of culms should be conducted.
- Bamboo culms should not have an initial deformation from its major axis $>0.33\%$ of its length and should not have a taper $>1.0\%$.
- Bamboo culms should not have any cracks around their perimeter, nodes or along the neutral axis of the element. In case of cracked elements, these must be located facing the upper or lower external fibre of the cross section.
- Bamboo culms with cracks $\geq 20\%$ of the length of the culm are not appropriate for structural use.
- Bamboo culms should not have any holes caused by xylophage insects prior to their use and should not present any degree of rot or decay.
- The overloading of bamboo culms should be avoided during transportation and storage. The maximum number of culms stacked one over the other should be seven.

- The storage of bamboo culms must be in a dry location, under cover, with proper ventilation and drainage. It is preferable to store culms in vertical position, isolated from the ground with no contact with organic matter.

NSR-10 specifies some recommendations regarding the fabrication and construction process of the bamboo buildings:

- No holes at the depth of the neutral axis are permitted for cross sections with concentrated loads or near the supports unless the internodes are filled with cement mortar. In all other cases, holes must be located at the depth of the neutral axis. Finally, holes at an element's tension face are not permitted in any cases.
- To determine the diameter of bamboo culms, the average of at least two perpendicular measurements needs to be performed at each end of the culm. The thickness of the culm wall culms is determined by the average of at least four measurements taken at each end of the culm at the same locations as the measurements taken for the determination of the diameter.
- Imperfections in cutting, assembly and cross section dimensions should not be >2% of the values specified in the drawings.
- All bamboo elements must have a visible and permanent identification mark that matches the engineering drawings or specifications.
- Bamboo structures must follow regulations established in construction building codes to have adequate performance against seismic and wind loads and under fire conditions.

14.10   Concluding remarks and future research

Key sustainability characteristics of bamboo such as having a fast growth rate, being renewable and biodegradable, as well as the benefit of sequestering carbon, make this material an excellent candidate for sustainable building design and construction. On the other hand, special attention needs to be paid to perform an adequate design with bamboo. Some of the important issues in design and construction with bamboo are (1) using bamboo at a proper mature age, (2) taking into account the effect of moisture content on the estimation of mechanical properties, (3) using treatment methods for preservation and (4) estimating fire reaction and fire resistance of bamboo structures and assemblies, amongst others.

Further, international ISO standards have been developed for the determination of physical and mechanical properties and structural design methods considering mechanical resistance, serviceability and durability, although are not implemented in most building codes worldwide. These standards provide design procedures for typical bamboo structural elements and assemblies such as (1) beams (considering deflections, bending strength and lateral stability, shear and bearing strength), (2) columns and other elements under compression (including buckling effects), (3) tension elements, (4) elements under combined loads (using interaction equations applicable to other materials), (5) lateral force resisting elements (such as shear walls) and (6) connections or joints (mainly dealing with dowel-type connections). Finally, some building codes (NBCI, 2005; AIS, 2010), which have implemented or adapted these standards as mandatory, have included not only more comprehensive design procedures but also several recommendations for the preparation, fabrication, construction and assembly of bamboo structures, regarding shape, tolerances, cutting and drilling provisions, transportation and storage.

Nevertheless, the need for technical and scientific knowledge around bamboo for structural use is a topic of great importance, to provide an adequate framework for including bamboo and bamboo structures in building codes. Major efforts have been carried out in recent years in terms of research around bamboo for construction. Some research topics worth noting include the characterization of physical and mechanical properties of several bamboo species (Correal and Arbeláez, 2010; Sharma et al., 2013; Lee et al., 2014; Xu et al., 2014), the creep behaviour of bamboo (Gottron et al., 2014) and the seismic behaviour of structural assemblies (Arbeláez and Correal, 2012; Correal et al., 2014).

Future research around bamboo is directed towards topics that are still unresolved or may be explored in depth, such as the application of bamboo structures for medium- and tall-rise buildings, the determination of load and resistance factors for LRFD, the more accurate estimation of long-term deflections, the effective moment of inertia for built-up beams and columns, the structural behaviour of walls subjected to lateral loads considering various sheathing and finishing materials or the characterization and design procedures of adequate mechanical connections (eg, considering the application of widely used theoretical approaches such as the European Yield Model to mortar- or wood-filled bamboo culms). Also, the determination of seismic response coefficients (R) for different bamboo structures should be addressed to provide more technical information for the inclusion of bamboo in building codes.

References

AIS, 2010. Reglamento Colombiano de Construcción Sismo Resistente NSR−10. Asociación Colombiana de Ingeniería Sísmica.

Ahmad, M., Kamke, F.A., 2005. Analysis of Calcutta bamboo for structural composite materials: physical and mechanical properties. Wood Science and Technology 39, 448−459.

Arbeláez, J., Correal, J.F., 2012. Racking performance of traditional and non-traditional engineered bamboo shear walls. In: 9th World Bamboo Congress, April, 10−15, Antwerp, Belgium.

CEN, 2002. Eurocode 1: Actions on Structures − Part 1−2: General Actions − Actions on Structures Exposed to Fire, EN 1991−1−2:2002. European Committee for Standardization.

Chung, K.F., Yu, W.K., 2002. Mechanical properties of structural bamboo for bamboo scaffoldings. Engineering Structures 24, 429−442.

CIMOC, 2011. Validación Tecnológica del Comportamiento de Estructuras de *Guadua* Rolliza Seca e Inmunizada, Informe Técnico Final, Centro de Investigaciones en Materiales y Obras Civil. Universidad de los Andes, Bogotá, Colombia.

Correal, J.F., Arbeláez, J., 2010. Influence of age and height position on Colombian *Guadua angustifolia* bamboo mechanical properties. Maderas: Ciencia Y Tecnología 12 (2), 105−113.

Correal, J.F., Echeverry, J.S., Arbeláez, J., 2014. Evaluación del comportamiento sísmico de estructuras de *Guadua rolliza*. In: XXXVI Jornadas Sudamericanas de Ingeniería Estructural, November, 19–21, Montevideo, Uruguay.

De Flander, K., Rovers, R., 2009. One laminated bamboo-frame house per hectare per year. Construction and Building Materials 23 (1), 210–218.

EERI, 2000. El Quindío, Colombia Earthquake. January 25, 1999, Reconnaissance Report. Earthquake Engineering Research Institute.

Gottron, J., Harries, K.A., Xu, Q., 2014. Creep behaviour of bamboo. Construction and Building Materials 66, 79–88.

Harries, K.A., Sharma, B., Richard, M.J., 2012. Structural use of full culm bamboo: the path to standardisation. International Journal of Architecture, Engineering and Construction 1 (2), 66–75.

ICC, 2012. International Building Code, 2012 IBC. International Code Council, Inc.

ISO, 1999. Determination of Physical and Mechanical Properties of Bamboo, ISO/DIS 22157. International Organization for Standardization.

ISO, 2001. Bamboo Structural Design, ISO/DIS 22156. International Organization for Standardization.

Janssen, J.J.A., 2000. Designing and Building with Bamboo. INBAR Technical Report No. 20. International Network for Bamboo and Rattan.

Jayanetti, D.L., Follet, P.R., 1998. Bamboo in Construction: An Introduction. INBAR Technical Report No. 16. International Network for Bamboo and Rattan.

Lee, P.-H., Odlin, M., Yin, H., 2014. Development of a hollow cylinder test for the elastic modulus distribution and the ultimate strength of bamboo. Construction and Building Materials 51, 235–243.

Liese, W., 1985. Bamboos – Biology, Silvics, Properties, Utilization. Deustche Gesellschaft für Technische Zusammenarbeit (GTZ), GmbH.

Liese, W., 1998. The Anatomy of Bamboo Culms. INBAR Technical Report No. 18. International Network for Bamboo and Rattan.

Lobovikov, M., Paudel, S., Piaza, M., Ren, H., Wu, J., 2007. World Bamboo Resources, a Thematic Study Presented in the Framework of the Global Forest Resources Assessment 2005, Non-wood Forest Products 18. Food and Agriculture Organization of the United Nations.

McLure, F.A., 1953. Bamboo as a Building Material, Peace Corps. Division of Volunteer Support, Washington, D.C.

Mena, J., Vera, S., Correal, J.F., López, M., 2012. Assessment of fire reaction and fire resistance of *Guadua angustifolia* Kunth bamboo. Construction and Building Materials 27, 60–65.

NBCI, 2005. National Building Code of India. Bureau of Indian Standards, India.

Qisheng, Z., Shenxue, J., Yongyu, T., 2001. Industrial Utilization on Bamboo. INBAR Technical Report No. 26. International Network for Bamboo and Rattan.

Richard, M.J., Harries, K.A., 2012. Experimental buckling capacity of multi-culm bamboo columns. Key Engineering Materials 517, 51–62.

Sharma, B., Mitch, D., Harries, K.A., Ghavami, K., Kharel, G., 2011. Pushover behavior of bamboo portal frame structure. International Wood Products Journal 2, 20–29.

Sharma, B., Harries, K.A., Ghavami, K., 2013. Methods of determining transverse mechanical properties of full-culm bamboo. Construction and Building Materials 38, 627–637.

Varela, S., Correal, J., Yamín, L., Ramírez, F., 2013. Cyclic performance of glued laminated *Guadua* bamboo-sheathed shear walls. Journal of Structural Engineering 139 (11), 2028–2037.

Xiao, Y., Zhou, Q., Shan, B., 2010. Design and construction of modern bamboo bridges. Journal of Bridge Engineering 15 (5), 533−541.

Xiao, Y., Ma, J., 2012. Fire simulation test and analysis of bamboo frame building. Construction and Building Materials 34, 257−266.

Xu, Q., Harries, K., Li, X., Liu, Q., Gottron, J., 2014. Mechanical properties of structural bamboo following immersion in water. Engineering Structures 81, 230−239.

Engineered Bamboo

Y. Xiao
College of Civil Engineering, Nanjing Tech University, Nanjing, China

15.1   Introduction

Bamboo is well known because of its characteristics of being fast-growing and eco-friendly and because of its high strength:weight ratio. In today's trend of sustainable development, there is renewed interest in the use of bamboo for modern building and bridge structures. Bamboo has been indispensable to human life, being used as building and bridge structural materials as long as wood. Its broad and time-honored apprlication in construction has an explanation. Bamboo is one of the fastest growing plants on Earth and can reach the full height of 15−30 m within a period of 2−4 months because of diurnal growth rates of about 20 cm up to 100 cm [1]. Its physical and mechanical properties are generally better than those of conventional timber species. Moso bamboo (*Phyllostachys pubescens*), for example, has a relative density ranging from 0.55 to 1.01 g/cm^3; the mean longitudinal tensile modulus of elasticity (MOE) ranges from 9 to 27 GPa, and the mean longitudinal tensile strength ranges from 115 to 309 MPa [2]. According to research by Kwan et al., bamboo is like a natural composite material and its natural structure is found to be similar to fiber-reinforced composites [3]. More importantly, mechanical properties including tensile strength, MOE, and bending strength are found to have approximately linear relationships with fiber density [3]. However, fiber content in bamboo is not uniformly distributed within its wall thickness.

Although the natural properties of bamboo are comparable to those of wood, the native geometrical shape of bamboo culms is difficult to use in modern construction. Using industrialized manufacturing process, several different types of bamboo composite materials, or engineered bamboo, have emerged in recent decades. Several researchers documented production and properties of laminated bamboo lumber (LBL), including Lee et al. [4], Nugroho and Ando [5], and Jiang et al. [6]. Recently, Mahdavi et al. [7] provided a thorough discussion and summary of LBL. Three different processing methods were introduced and their properties compared with wood. Results showed the capability of bamboo for replacing wood in structural applications. Shama et al. also contributed additional testing data to define the properties of LBL [8,9]. Bamboo composites discussed in these reports can all be classified as thick-layer laminated bamboo composites since the bamboo strips used in the lamination process are 3−6 mm in thickness. Typically, each layer of a thick-layer laminated bamboo veneer or LBL are close to the original thickness of bamboo culm wall, and usually three to five plies are included to make the veneers or LBL. Producing this type of bamboo composite involves comparatively high-level technology, quality control,

(a)   (b)

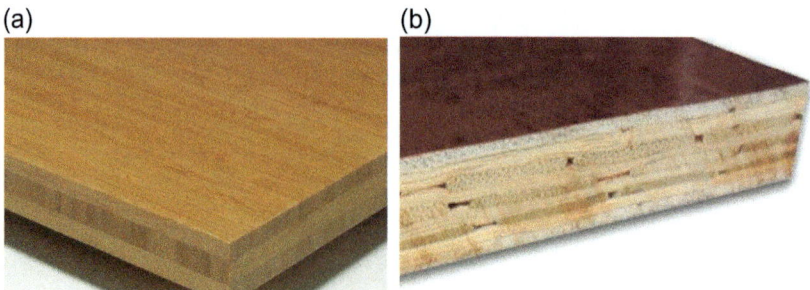

Figure 15.1 Laminated bamboos: (a) three-ply LBL having a total thickness of 24 mm and (b) glubam sheet having thickness of 28 mm.

and production cost. In a discussion report by Xiao and Yang [10], the costs are attributed to the irregularities of bamboo in the cross section and along the length of bamboo culm. Typically, these types of LBL are more suitable for high-end products such as flooring panels, as shown in Fig. 15.1(a).

Based on abundant material and existing production technology in China, a new type of laminated bamboo that can be cost-effectively used as structural elements was invented by Xiao and his research group: GluBam, imitating the well-known GluLam or glue laminated timber [11,12]. The structural glubam elements are made by laminating glubam sheets, which are typically 20–40 mm thick, produced by using a modified bamboo veneers (plybamboo) production process. As shown in Fig. 15.1(b), the glubam is based on a so-called thin-layer lamination of bamboo strips.

The glubam or plybamboo sheets follow the original technology of plywood production to laminate mats of thin-layer bamboo strips, each having a thickness of about 2 mm. Most of the mills in China are using a hot-pressing method, which is similar to that used for plywood and originally appeared in North American plants in the 1930s. Heating is usually from hot water or steam. In last two decades in China, production of plybamboo sheets matured in Hunan, Zhejiang, Jiangxi, and Fujian provinces, those with abundant bamboo forests. The cost of making plybamboo using thinner-layer bamboo strips is relatively low due to the ability to make fuller use of bamboo culms having a variety of geometries [10–12]. Plybamboo are used for concrete formwork and floors of shipping containers, although, typically not for structural applications. Glubam sheets are produced with more stringent specifications for the orientation of bamboo strips (and therefore the bamboo fibers) and sheet thickness, etc. in order to better control its mechanical properties and constructability.

15.2 Production of glubam

As an example of engineered bamboo, this chapter discusses glubam in more detail. The production of glubam has six steps: (a) raw bamboo selection, (b) splitting bamboo strips, (c) netting bamboo curtains or mats, (d) gluing with adhesives, (e) hot-pressing, and (f) postprocessing. Within the process, all steps but the last are aimed to produce standard glubam sheets with dimensions of 2440-mm long, 1220 mm wide, and 30 mm thick. Based on the design of the specific structure, the glubam sheets are shaped and laminated to form the final structural components.

15.2.1 Raw bamboo selection

First, bamboo materials that meet maturity requirements are selected and prepared for bamboo strips and then woven into mats. Moso bamboo aged 3−5 years is used as raw material for glubam. Usually, the average perimeter of bamboo culms of this age reaches 35 cm or larger. Fig. 15.2(a) shows a batch of newly harvested Moso bamboo at the author's production base in Hunan Province. The bamboo culms are then transported to a nearby mill.

15.2.2 Bamboo strips

The quality of bamboo strips is important for the resulting physical and mechanical properties of glubam. Uniformity of bamboo strips improves the performance of

Figure 15.2 Production of glubam: (a) raw bamboo culms, (b) boiling, (c) drying, (d) netting, (e) applying adhesive, (f) hot-pressing formation of sheets, (g) cold-pressing machine, and (h) glubam structural elements.

glubam. Compared with four processing techniques of LBL described by Nugroho and Ando [5], Rittironk and Elnieiri [13], Sulastiningsih and Nurwati [14], Lee et al. [4], and Mahdavi et al. [15], bamboo culms are split to more slender and thinner strips having 2−3 mm thickness and 20−30 mm width. During this process, wax on the outer skin and pitch on the inner surface of the culms are removed. Because bamboo strips obtained from this process are only about 2 mm thick, there is no need to flatten them. Another advantage of using thin-layer strips is that there is less constraint on suitable culm wall thickness or the length of bamboo culms used.

After the process of splitting bamboo strips, they are boiled in hot water to eliminate part of the organic material within these strips (see Fig. 15.2(b)) followed by air drying to achieve a moisture content (MC) of about 20% (see Fig. 15.2(c)). During this process, bamboo strips from different culms and from different parts of culms are randomly mixed.

15.2.3   Bamboo curtains

Air-dried bamboo strips are aligned parallel to each other and woven with strings to form a curtain layer. As shown in Fig. 15.2(d), the string-connected bamboo strips form curtains with length and width slightly larger than the final product of 2440 × 1220 mm. Depending on the orientation, the strips are cut either slightly longer than 2440 mm for longitudinal arrangement or slightly longer than 1220 mm for transverse orientation. The bamboo curtains are the basic component of the resulting glubam sheets. The process can be completed manually or partially automated. Due to the smaller thickness of these bamboo curtains, small gaps between bamboo strips are tolerated. The thin-layer bamboo curtains are more flexible than thicker ones used for LBL and therefore require less precision in their arrangement.

15.2.4   Gluing and pressing

The bamboo strip curtains are cleaned and dried in a kiln to an MC of 18% and are then saturated in phenol formaldehyde resin as shown in Fig. 15.2(e). The resin-saturated curtains are dried again and stored keeping the MC below approximately 15%. About 15 layers of resin-saturated bamboo strip curtains are stacked in parallel or orthogonally, depending on its design requirements. Finally, the stacked bamboo layers were pressed with a temperature of about 150°C, and pressure of 20 MPa for about 15 min, using a procedure similar to manufacturing plywood [16]. As shown in Fig. 15.2(f), several glubam sheets are typically pressure-formed at the same time using the pressure machine.

Like plywood production in the North America, phenol formaldehyde is used in making glubam sheets. Phenol formaldehyde is known to have certain toxicity; therefore, during production, workers are well protected from the raw material. When phenol formaldehyde solidifies, its toxicity is essentially controllable. The profession is always looking for other nontoxic or biodegradable alternatives, but, currently, phenol formaldehyde is the choice with wide availability, established and controlled properties and handling processes, good resistance to moisture, and acceptable cost.

15.2.5 Postprocessing

Typical glubam sheets are 30 mm (± 2 mm), 1220 mm wide, and 2440-mm long. Finger-jointing can be used to structural elements of different shapes and larger dimensions. Fig. 15.2(g) exhibits the cold-press machine. Fig. 15.2(h) shows some completed glubam columns with a length of 6 m for an actual construction project of 10-m span workshop building.

15.3 Material properties of engineered bamboo

Typical glubam structural components use sheets with bamboo strips arranged with 80% in the longitudinal direction and 20% in the transverse direction, referred to as 4:1 sheets. There are usually 15 or fewer layers of bamboo curtains, each having a thickness of about 2 mm.

Due to the fact that glubam is a new invention and no testing standard has been established, standards for timber structures are referenced in the study of the properties of glubam. In the Chinese timber structure design standard and similar ASTM standards [17,18], design values of mechanical properties of timber are from small clear specimen tests. Due to scarcity of timber material, little plywood is used in China as structural material except some thin wood boards used for decoration. In view of the nature of glubam production and design, it is reasonable to use the small clear specimen test method to obtain mechanical properties for glubam. A large number of specimens were tested, and the results are summarized in Table 15.1.

While it is important to define the basic mechanical properties of engineered bamboo such as glubam, studies on the combined stress states are also important. Yang et al. presented the failure analysis of glubam with bidirectional fibers using Hankinson formula and Tsai-Wu failure criterion [19]. Off-axis tension tests were performed on 4:1 glubam specimens. A revised Hankinson equation suitable for glubam was developed, yielding satisfactory agreement with the test results. It is found that the interaction coefficient in Tsai-Wu failure theory can only be established from the 15 degrees off-axis tension test. Also, three approximation methods, "Tsai-Hill," "Hoffman," and "Mises-Hencky", are applied to estimate the value of interaction. Four failure envelopes for glubam were presented and compared. By comparing off-axis test data, the failure envelope from Tsai-Wu theory provides an nonconservative prediction when in-plane shear is ignored. Even when in-plane shear is considered, the failure envelope obtained from Tsai-Wu theory is still deemed unsuitable to represent the behavior of glubam. Therefore, it seems that an empirical approach is currently appropriate to represent the behavior of 4:1 glubam subjected to a biaxial tensile stress state.

Table 15.1 Main mechanical properties of 4:1 glubam

Materials	Longitudinal in-plane tensile strength, f_{yt} (MPa)	Longitudinal elastic modulus, E_y (MPa)	Transverse elastic modulus, E_x (MPa)	G_{xy} (MPa)	G_{yx} (MPa)	Longitudinal in-plane compressive strength, f_{yc} (MPa)	Bending strength (MPa)	Density (kg/m^3)
Glubam	82	10,400	2600	4.6	7.2	51	99	800–880

15.4   Structural components

15.4.1   Columns

Five sets of small glubam columns were made with the same 56-mm^2 cross-section geometry and lengths, ranging from 402 to 1602 mm. Axial loading tests were conducted with the pinned end supports; thus, the effective length ratios of the specimens ranged from 24.7 to 98.9. The failure patterns of the specimens were strength failure (crushing) for shorter columns and elastic buckling failure for longer columns, while the columns with intermediate effective lengths fail in mixed failure modes. Fig. 15.3 shows the comparison of the test results with the Chinese and U.S. timber standards [17,18,20].

15.4.2   Beams and girders

Static tests of glubam girders (Fig. 15.4) show their excellent bearing capacity [12]. The use of fiber reinforced polymers (FRP) to reinforce the soffit of the girders can further enhance the load carrying capacity of glubam girders [12].

Fatigue conditioning of 2 million cycles at a stress range of 20% of the ultimate strength reduced the bearing capacity of glubam girders approximately 10% due to the development of weaknesses in finger-joints and at interfaces. When the maximum value of cyclic load does not surpass its design value, there was no distinct reduction in the stiffness of specimens compared with static tests. It is clear that excellent flexibility of bamboo contributes to the stability of glubam beams undergoing fatigue loading.

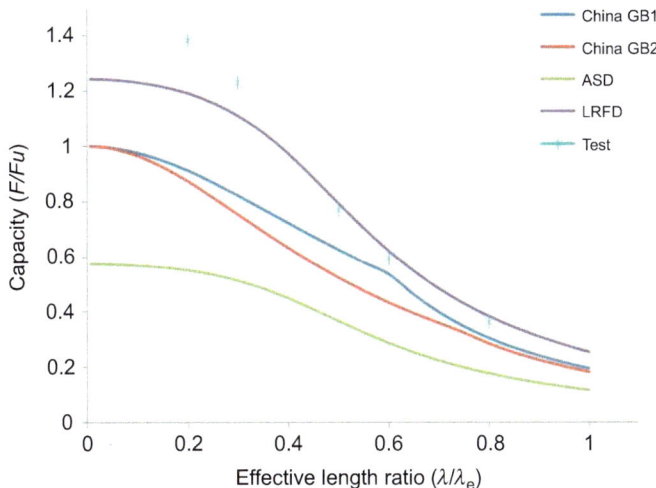

Figure 15.3 Comparison of column test results and design curves.

Figure 15.4 Static and fatigue loading test of full-scale glubam girder.

15.4.3   Shear walls

Significant numbers of experimental tests were conducted to study the lateral loading behavior of shear wall panels made with glubam sheets and glubam or wood frame studs [21]. Tests show that the lightweight wood shear walls with glubam sheathing panels (as shown in Fig. 15.5) have good seismic performance as well as the ability to meet the design requirements specified in most timber codes. The fabrication of this type of shear wall could be automated and made modular also affecting easy installation. Fig. 15.6 shows the comparison of monotonic and cyclic lateral force–deformation relationships of 2.44-m-long by 2.44-m-tall lightweight glubam frame shear wall specimens. Such lightweight glubam frame shear walls have comparable load and displacement characteristics and capacities as lightweight wood-frame shear walls reported in the literature.

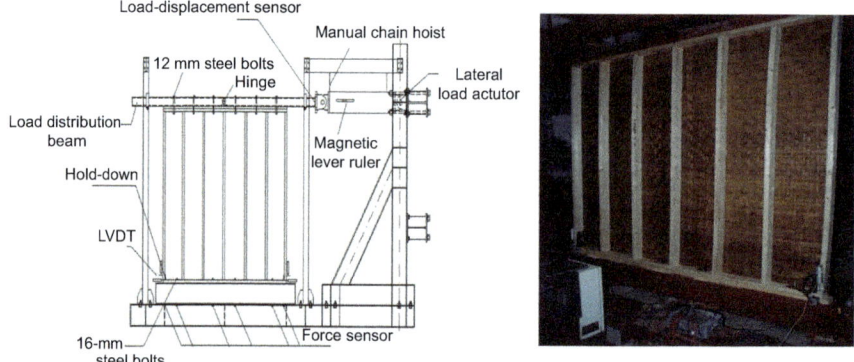

Figure 15.5 Test shear wall details.

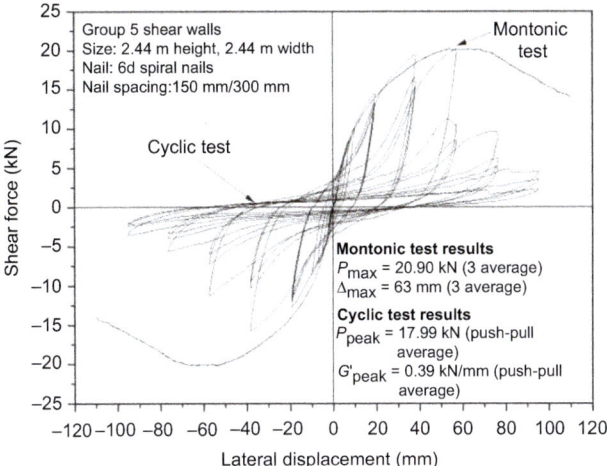

Figure 15.6 Test results of lightweight glubam shear walls.

15.5 Applications: buildings

15.5.1 Lightweight glubam-frame buildings

The author's research and development team has made efforts to build modern bamboo buildings using glubam and has achieved several breakthroughs. One of the targeted applications is the construction of permanent residence buildings. In 2008, the author's team designed and constructed a 250-m² two-story residential modern bamboo house using the newly invented technologies, as shown in Fig. 15.7.

The two-story demonstration house has 4 bedrooms, 3.5 bathrooms, 2-car attached garage, kitchen, dining area, and living room. Glubam studs of 35×80 mm, as shown in Fig. 15.7(a), were specially manufactured for vertical elements (studs) of the wall panels, similar to the so-called 2×4 stud-wall method. Each story was built using the platform framing method, with the first story erected on top of concrete strip foundations and concrete masonry block cripple walls and pedestals with a height of

Figure 15.7 Two-story glubam house construction completed in 2008: (a) installing wall frame, (b) roof frame work, and (c) completed glubam house.

800 mm. Several different forms of glubam trusses were studied experimentally prior to the construction. Fig. 15.7(b) shows the installation of the queen-post roof truss. Fig. 15.7(c) exhibits the completed building as of February 2009.

After the success of the world's first bamboo stud-wall residential building, the author's team designed and built other demonstration houses. One was at the famous Black Bamboo Park (Zizhu Yuan) in Beijing, and the other at the Cailun Forest in Leiyang, Hunan Province. Fig. 15.8 shows the two-story demonstration glubam houses constructed in 2009. Each of the two glubam houses has a floor area of about 100 m².

15.5.2   Glubam temporary shelters for disaster relief

In addition to permanent residential construction, the potential of glubam is also being explored to manufacture prefabricated mobile or modular buildings, such as disaster relief shelters, work or tool sheds, equipment sheds, etc. The author's team designed prefabricated glubam earthquake relief shelter units in the aftermath of the Great Wenchuan earthquake of May 12, 2008. The shelters adopted a modular design; therefore, the plan can be easily adjusted according to needs. The modular units are connected by bolts thus are easy to manufacture, assemble, disassemble, and reuse. Four to six workers can assemble the basic unit of 22.3 m² in about 4 h. Interior detailing is similar to wood-frame houses and air-quality tests indicating satisfactory performance according to standards. The seismic design was conducted following well-established design procedures. The basic relief house contains two windows, fans, locations for a liquid propane gas stove or bath unit, etc., satisfying basic shelter needs for a family of four. The basic unit for classrooms has an area of 50 m². Comparing with tents or light-gauge steel prefabricated shelters, the prefabricated glubam quake relief house has the advantages of improved heat and sound insulation, being more resistant to fire and, most importantly, being cost-effective. Fig. 15.9 shows examples of glubam classroom design and construction.

(a)   (b)

Figure 15.8 Completed glubam stud-wall houses: (a) 100-m² tea house built in Black Bamboo Park in Beijing in August 2009 and (b) 120-m² mountain house in Cailun Forest, Leiyang, Hunan Province, completed in November 2009.

(a) (b) (c)

Figure 15.9 Earthquake relief glubam prefabricated classrooms: (a) design, (b) completed glubam classroom, and (c) glubam classroom under construction.

15.5.3 Glubam heavy frame buildings

The author's research and development team also built several glubam special frame buildings; a typical example is shown in Fig. 15.10. This 350-m^2 building is located in a theme park in Changsha, China. In the heavy space frame building shown in Fig. 15.10(c), the largest girder had a length of 16.5 m, with a cantilever length of 7 m; the section was 800 mm deep and 120 mm wide.

15.5.4 Building combined with bamboo and solar energy

Collaborating with the Chinese Academy of Forestry and the International Center for Bamboo and Rattan, a new concept was recently developed for bamboo building design combining rooftop solar panels for partial energy supply to the demonstration house. In this demonstration project, glubam was only used for panels, whereas all the studs, structural elements for beams, and roof trusses were made with round bamboo culms or poles. Some of the structural bamboo poles were also used for routing the water heater pipes connected to the solar panels installed on the roof (Fig. 15.11).

Figure 15.10 Glubam heavy frame building.

Figure 15.11 Bamboo house equipped with solar energy near completion.

15.6   Applications: bridges

15.6.1   Glubam pedestrian bridges

In November 2006, the authors designed and built the first modern bamboo bridge as an initial trial of laminated bamboo technology. Based on experimental testing of proto type girder specimens, the authors adopted a modular design concept, enabling the efficiency of construction. Column and girder elements were all manufactured using bamboo veneer sheets. The 1.5-m-wide, 5-m-long pedestrian bridge was supported on six 300 × 84 mm laminated bamboo girders. Two stairways made with the same materials were attached to the pedestrian bridge. The surface of the bridge was covered with bamboo strip reinforced precast concrete panels. Fig. 15.12 shows the completed

Figure 15.12 The first glubam pedestrian bridge completed in late 2006.

modern bamboo pedestrian bridge. The bridge is functioning well since opening for use at the end of 2006.

The author's team also completed the design and construction of two pedestrian bridges in the Wanke Development Research Park in Dongguan, Guangdong Province. The Wanke 20-m-long bridge had four 5-m-long simply supported spans as shown in Fig. 15.13. The width of the bridge was 1.8 m. The bridge was designed with five 250 × 84 mm glubam girders for each span. The completed bridge is shown in Fig. 15.13.

The Wanke 40-m-long bridge has six simply supported spans with different lengths, and an irregular shape, where two approaches from the two abutments intersect with an angle of approximately 9 degrees in the fourth span, as shown in Fig. 15.14(a). The major challenge in design was concentrated in the fourth span, which not only was the longest with a span length of 10 m but also had an irregular plan layout.

Figure 15.13 The Wanke 20-m-long bridge near completion.

(a)   (b)

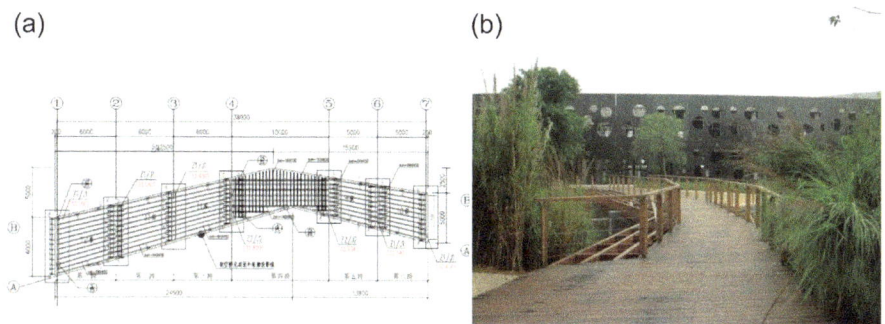

Figure 15.14 The Wanke 40-m-long bridge: (a) plan layout of glubam girders and (b) completed bridge.

This span was essentially designed with a glubam beam grids system with 600 × 112 mm main girders in the longitudinal direction and smaller 100 × 56 mm beams in the transverse direction. On the concave side, three girders integrated together forming a K shape in the plan.

15.6.2 Truck load roadway bridge

In 2007, the author's team was given the opportunity to design and construct a truck-carrying 10-m-long bridge in the village of Daozi, Leiyang, Hunan Province. The bridge is a single-lane bridge crossing the Xunjiang River and connects the rural roadway network, as a part of the government's agriculture development [12].

An existing 1.5-m-wide old stone bridge had been at the site for a long time; however, it was not capable of carrying modern transportation, nor was it safe for pedestrians, particularly during rainy days. The requirement for the new bridge was to enlarge the width to 3.5 m in order to allow single-lane truck traffic. A decision was made to build the new bridge utilizing the existing stone piers. The length of the new bridge is 22.8 m consisting four simple spans. The middle 10-m-long span was designed and constructed with laminated bamboo girders reinforced with carbon fiber−reinforced polymer (CFRP) materials. The remaining shorter spans were constructed using reinforced concrete beams and slabs.

The design truckload was 2-axle 8-ton truck. Based on analysis, nine 600-mm-deep × 100-mm-wide glubam girders were used to carry the 10-m-long span. The girders were interconnected by bolting bamboo veneer plates between and on top of the girders to provide the overall integrity. Waterproof asphalt sheets were laid on top of the cover plates before placing the 200-mm-thick, 3.5-m-long, and 1.0-m-wide precast concrete pavement panels. For simplicity, the authors designed the bridge assuming a 16-ton truckload applied at the mid-span of the bridge. The final design was based on controlling the mid-span deflection to be below the limiting value of 1/600th of the span length, per the Chinese bridge design code [22]. Also, for simplicity and safety, the CFRP layers were neglected in the design. Based on the test results of the laminated bamboo girders and neglecting shear lag effect, the bridge could carry an ultimate load of more than 80 tons applied at its mid-span.

Vertical bars were anchored in drilled holes on top surfaces of the existing stone piers to form the reinforcement for a 400-mm-thick concrete layer with a design strength of 30 MPa. Fiber-reinforced rubber pads with a thickness of 20 mm were placed between the glubam girder ends and the piers in order to reduce the effects of impact under traffic loads. Components were installed on the piers and fastened using bolts. Without the need of heavy equipment, the installation of the bamboo bridge was completed within 10 days by a team of four to eight workers daily. The two exterior girders were wrapped in waterproof rubber sheets. Two layers of waterproof asphalt sheets were also applied between the precast concrete surface panels and the veneer deck plates on top of the girders. The joints between the precast concrete panels were sealed by waterproof sealant. Finally, steel railings were installed.

Figure 15.15 Glubam roadway bridge under truckload testing.

With the purpose study, the performance of the world's first truck-safe modern bamboo bridge, field tests were carried out on the main span, 30 days after completion of the bridge. As shown in Fig. 15.15, the load test was carried out using a truck loaded with stones, representing the possible overload truck in the local area. The wheel-lane distance was 1.5 m and the distance between the front and the rear wheel axles was 3.0 m. The weight of the overloaded truck was about 8.6 tons, distributed approximately 1.6 tons to the front axle and 7 tons to the rear. Dial gauges were installed under each of the nine girders at the mid-span for measuring their deflections. Testing results confirmed that the deflection of the bridge was below the prescribed code limitation [12].

15.7   Analysis of carbon footprints

In recent years, it has been well recognized that a major greenhouse gas (GHG) is carbon dioxide (CO_2) and that the major cause of human activity—generated carbon (C) in the atmospheric is the burning of fossil fuels for energy.

Like wood, bamboo is also an organic material, composed of about 50% of carbon. As a natural carbon sink, CO_2 is entrapped in bamboo or wood until the final end of the material's use when it is burned or permitted to decompose. Thus, compared with steel, concrete, and masonry, the production (or growth) of wood and bamboo is a carbon-sequestering process. During its growth, bamboo can absorb even more CO_2 than trees. Based on a study conducted in China, the annual carbon fixation of bamboo forest is 5.1 t/ha^2/year, which is 1.46 times that of Chinese fir at the fast-growing stage and 1.33 times that of tropical mountain rain forest [23]. Based on a study by Xiao et al. [19], the glubam is verified as essentially a carbon negative material, outperforming other conventional materials, such as steel and concrete.

15.8   Future research needs

Engineered bamboo is actively investigated by several research groups, including the author's. Limited research on fire resistance and durability has been reported [24,25] and thus remains an urgent need. The potential for hybrid uses of bamboo and other materials, such as wood, steel, FRP, etc. should also be explored. Despite increased research to define the material and component behaviors of engineered bamboo and several successful applications, it is necessary to establish design guidelines and specifications for widespread use of engineered bamboo structures. New engineered bamboo products should also be developed and investigated. Recently, several Chinese manufacturers developed a new engineered bamboo product, named restructured bamboo or heavy bamboo having high density, strength, and MOE. These new products are promising for outdoor use; however, their performance and suitability for structural design should be addressed in greater detail.

References

[1] Liese W. Research on bamboo. Wood Science and Technology 1987;21(3):189–209.
[2] Yu HQ, Jiang ZH, Hse CY, Shupe TF. Selected physical and mechanical properties of moso bamboo (*Phyllostachys pubescens*). Journal of Tropical Forest Science 2008;20(4): 258–63.
[3] Kwan SH, Shin FG, Yipp MW. Consideration of bamboo as a natural composite material. Acta Materiae Compositae Sinica 1987;4(4):79–83.
[4] Lee AWC, Bai X, Bangi P. Selected properties of laboratory-made laminated bamboo lumber. Holzforschung 1998;52(2):207–10.
[5] Nugroho N, Ando N. Development of structural composite products made from bamboo II: fundamental properties of laminated bamboo lumber. Journal of Wood Science 2001; 47(3):237–42.
[6] Jiang SX, Zhang QS, Jiang SH. On Structure, production, and market of bamboo-based panels in China. Journal of Forestry Research 2002;13(2):151–6.
[7] Mahdavi M, Clouston P, Arwade S. Development of laminated bamboo lumber: review of processing, performance, and economical considerations. Journal of Materials in Civil Engineering 2011;23(7):1036–42.
[8] Sharma B, Gatoo A, Rock M, Ramage M. Engineered bamboo for structural applications. Construction and Building Materials 2015;81:66–73.
[9] Sharma B, Gatoo A, Ramage M. Effect of processing methods on the mechanical properties of engineered bamboo. Construction and Building Materials 2015;83:95–101.
[10] Xiao Y, Yang RZ. Discussion of Development of laminated bamboo lumber: review of processing, performance, and economical considerations by M. Mahdavi, P.L. Clouston, and S. R. Arwade. Journal of Materials in Civil Engineering July 1, 2011;23(7).
[11] Xiao Y, Shan B, Chen G, Zhou Q, She LY. Development of a new type of glulam— glubam. In: Xiao Y, editor. Modern bamboo structures. Changsha, China: CRC; 2008.
[12] Xiao Y, Zhou Q, Shan B. Design and construction of modern bamboo bridges. Journal of Bridge Engineering 2010;15(Compendex):533–41.
[13] Rittironk S, Elnieiri M. Investigating laminated bamboo lumber as an alternate to wood lumber in residential construction in the United States. In: Proc., 1st int. conf. on modern bamboo structures. Abingdon, U.K: Taylor & Francis; 2007. p. 83–6.

[14] Sulastiningsih IM, Nurwati. Physical and mechanical properties of laminated bamboo board. Journal of Tropical Forest Science 2009;21(3):246—51.

[15] Mahdavi M, Clouston PL, Arwade SR. A low-technology approach toward fabrication of laminated bamboo lumber. Construction and Building Materials 2012;29:257—62.

[16] Laboratory FP. In: USDA FS, editor. Wood handbook—Wood as an engineering material; 1999. Madison, Wis.

[17] GB/T50329-2002. Standard for testing methods of timber structures. Beijing: China National Standard; 2002.

[18] West Conshohocken, PA ASTM. D143—94 Standard test methods for small clear specimens of timber; 2007.

[19] Xiao Y, Yang RZ, Shan B. Failure analysis of typical glubam with bidirectional fibers by off-axis tension tests. Journal of Construction and Building Materials 2014;58:9—15.

[20] Feng L, Lv XH, Xiao Y, She LY. Experimental studies on glubam columns under axial load (in Chinese). Journal of Industrial Buildings 2015;45(4):13—7.

[21] Xiao Y, Li Z, Wang R. Lateral loading behaviors of lightweight woodframe shear walls with ply-bamboo sheathing panels. ASCE Journal of Structural Engineering March 2015; 141(3).

[22] JTG D62. Design guidelines for reinforced and prestressed concrete highway bridges (in Chinese). People's Transportation Press; 2004.

[23] Zhou G, Jiang P. Density, storage and spatial distributbion of carbon in phyllostachy pubescens forest. Scientia Silvae Sinicae 2004;40(6):20—4 [in Chinese with English abstract].

[24] Xiao Y, Ma J. Fire simulation test and analysis of laminated bamboo frame building. Journal of Construction and Building Materials September 2012;34:257—66.

[25] Xiao Y, Li L, Yang RZ. Long-term loading behavior of a full-scale glubam bridge model. ASCE Journal of Bridge Engineering September 2014;19(9).

Engineered bamboo for structural applications

B. Sharma

University of Bath, Bath, United Kingdom

Engineered bamboo utilizes the base bamboo material processed and manufactured into a variety of composites materials. The following summary expands upon the preceding chapter and further explores the development and emerging research on two categories of engineered bamboo product intended for primary structural applications: bamboo scrimber and laminated bamboo. Bamboo scrimber, also known as strand woven or parallel strand bamboo, is manufactured from crushed sections of the bamboo culm, which are coated in a resin, compressed, and cured. The term scrimber originates from a timber-based product that was developed in Australia to utilize small-diameter material that maintained the longitudinal direction of the fibers [1]. In comparison, laminated bamboo is manufactured with strips of bamboo that are processed to form rectangular sections, which are subsequently laminated to form a board. Both products are typically formed into boards for flooring and surface applications; however, the mechanical properties are comparable to other structural materials [2]. Use of bamboo board and panel products, however, is currently limited to secondary structural applications such as flooring for shipping containers and trucks. Products such as bamboo oriented strand board (OSB) are under development for a variety of applications. Bamboo OSB is manufactured by processing the bamboo culm into strands which are then coated in a resin and pressed into the final product [3–7]. Bamboo OSB has also been used to form the webs of engineered bamboo joists [8].

Global commercial production methods of laminated bamboo products are diverse, with variation in the species of bamboo, type, and size of the laminates, resins and adhesives used, and processing methods [9]. Methods used to obtain the base bamboo material for the laminates include splitting and flattening culms [10–13], as well as splitting and planing the bamboo into thin strips of bamboo [14,15]. In contrast, bamboo scrimber manufacturing utilizes a method in which the culm is split and crushed into bundles while maintaining the fiber direction [16,17]. The final scrimber product is manufactured by compression of the resin-coated fiber bundles into a structural section. Bamboo scrimber contains approximately 75% bamboo and 25% solid content of impregnating resin, whereas laminated products are greater than 85% bamboo. As a result, as shown in Table 15.2, the variation in manufacturing methods results in a range of product densities. Nonetheless, relatively little variation in mechanical properties is observed. Significantly, the bending modulus of rupture (MOR) and modulus of elasticity (E_L) for both the laminated bamboo and bamboo scrimber are generally greater than timber, effectively overcoming the generally poor flexural performance of bamboo culms.

As described in the previous chapter in reference primarily to "Glubam" laminated bamboo, the combination of engineered bamboo board and panel products can form

Table 15.2 **Typical mechanical properties of engineered bamboo materials [2,15]**

Materials	Density (kg/m^3)	Longitudinal in-plane tensile strength, f_{yLt} (MPa)	Longitudinal elastic modulus, E_L (GPa)	Shear parallel to fibers, τ_{xy} (MPa)	Longitudinal in-plane compressive strength, f_{yLc} (MPa)	Modulus of rupture (MOR) (MPa)
Laminated bamboo	~510–980	82–144	6–15	9–16	36–66	39–145
Scrimber	~600–1240	138	7–14	15	62	54–266

the basis for a sustainable structural system. With increased research and development, engineered bamboo can serve as a competitive and sustainable alternative to conventional structural materials.

References

[1] Bowden J. The great 'scrimber' mystery. Inwood Magazine 2007;74:20−1.
[2] Sharma B, Gatóo A, Bock M, Mulligan H, Ramage M. Engineered bamboo: state of the art. Proceedings of the Institution of Civil Engineering − Construction Materials 2014;168(2): 57−67.
[3] Lee AN, Bai X, Peralta PN. Physical and mechanical properties of strand board made from Moso bamboo. Forest Product Journal 1996;46(11/12):84−8.
[4] Sumardi I, Ono K, Suzuki S. Effect of board density and layer structure on the mechanical properties of bamboo oriented strandboard. Journal of Wood Science 2007;53:510−5.
[5] Febrianto F, Sahroni, Hidayat W, Bakar ES, Kwon GJ. Properties of oriented strand board made from Betung bamboo (*Dendrocalamus asper* (Schultesf.) Backer ex Heyne). Wood Science Technology 2012;46:53−62.
[6] Semple KE, Zhang PK, Smith GD. Stranding Moso and Guadua bamboo. Part I. Strand production and size classification. Bioresources 2015;10(3):4048−64.
[7] Semple KE, Zhang PK, Smith GD. Stranding Moso and Guadua bamboo. Part II. Strand surface roughness and classification. Bioresources 2015;10(3):4599−612.
[8] Aschheim M, Gil-Martín L, Hernández-Montes E. Engineered bamboo I-joists. Journal of Structural Engineering 2010;136(12):1619−24.
[9] Mahdavi M, Clouston PL, Arwade SR. Development of laminated bamboo lumber: review of processing, performance, and economical considerations. Journal of Materials in Civil Engineering 2011;23(7):1036−42.
[10] Lee AWC, Bai X, Bangi AP. Selected properties of laboratory-made laminated-bamboo lumber. Holzforschung 1998;52:207−10.
[11] Mahdavi M, Clouston PL, Arwade SR. A low-technology approach toward fabrication of laminated bamboo lumber. Construction and Building Materials 2012;29(4):257−62.
[12] Nugroho N, Ando N. Development of structural composite products made from bamboo I: fundamental properties of bamboo zephyr board. Journal of Wood Science 2000;46: 68−74.
[13] Nugroho N, Ando N. Development of structural composite products made from bamboo II: fundamental properties of laminated bamboo lumber. Journal of Wood Science 2001;47: 237−42.
[14] Xiao Y, Yang RZ, Shan B. Production, environmental impact and mechanical properties of glubam. Construction and Building Materials 2013;44:765−73.
[15] Sharma B, Gatóo A, Ramage MH. Effect of processing methods on the mechanical properties of engineered bamboo. Construction and Building Materials 2015;83:95−101.
[16] Huang D, Zhou A, Bian Y. Experimental and analytical study on the nonlinear bending of parallel strand bamboo beams. Construction and Building Materials 2013;44:585−92.
[17] Zhou A, Bian Y. Experimental study on the flexural performance of parallel strand bamboo beams. The Scientific World Journal 2014, Article ID 181627. http://dx.doi.org/10.1155/ 2014/181627.

Paperboard tubes in structural and construction engineering

L.C. Bank
The City College of New York, New York, NY, United States

T.D. Gerhardt
Sonoco Products Company, Hartsville, SC, United States

16.1 Introduction

Paperboard tubes are ubiquitous and are found in every aspect of modern-day life. In the industry, the terms paperboard "tube" and paperboard "core" are used interchangeably. They are produced for a myriad of industrial applications and consumer products and are sometimes considered to be commodity products, but most are actually highly engineered. On a global basis, the majority of paper tubes are used in industrial winding operations. In these applications, immediately after being manufactured, either paper, film, or textile yarn is wound directly onto paper tubes. The paper tube helps the wound material develop a stable roll structure and enables transportation of the manufactured material to a converting operation. These tubes range from small (10- to 25-mm diameter with 0.5- to 1.5-mm wall thickness) to wind paper, plastic, aluminum, cloth, and yarn consumer products, to large industrial cores (70- to 200-mm diameter having 10- to 25-mm wall thickness) to wind rolls of printing paper, plastic film, textiles, and specialty metals. Such tubes are designed to be used in interior environments for winding, transportation, and finishing operations; not exterior environments where they are exposed to rain or high humidity. Most tubes for industrial markets are structurally optimized and engineered products. This is achieved by selecting proper paperboard strength and stiffness, tube thickness, ply positioning in the wall, and adhesive type in the design process. For example, certain paper mill cores are designed both for fatigue strength to support a heavy cyclic load from winding a large roll of paper and for sufficient axial modulus to prevent core vibrations when unwinding the paper roll on high-speed rotogravure presses. Such cores generally require thick walls (10−16 mm) and fabrication from high strength paperboards. Many feature patented designs. Fig. 16.1(a) shows typical paper mill cores, and Fig. 16.1(b) shows typical concrete formwork tubes.

Paper strength is significantly reduced when it is exposed to elevated humidity environments, so the use of large paperboard tubes in structural engineering applications is uncommon. However, such applications are well known due to the unique and pioneering work of 2014 Pritzker Prize-winning Japanese architect Shigeru Ban (Anon, 2014a), who has designed and constructed many temporary or semipermanent paperboard tube structures for exhibition spaces, humanitarian emergency shelters,

Nonconventional and Vernacular Construction Materials. http://dx.doi.org/10.1016/B978-0-08-100038-0.00016-0

(a)   (b)

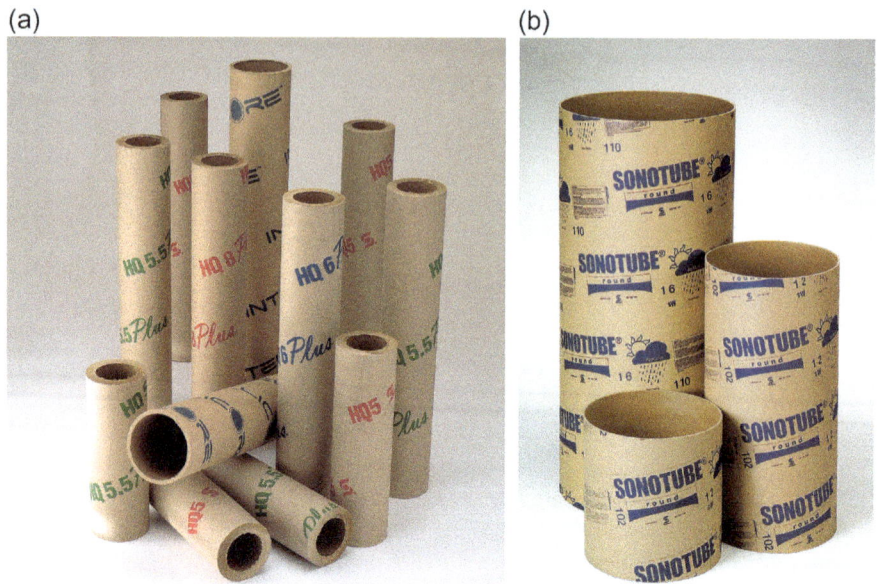

Figure 16.1 (a) High-strength paper mill cores and (b) concrete formwork tubes.
Photographs with permission from Sonoco Products Company.

single family houses, and even bridges since 1989 (McQuaid, 2003; Miyake, 2009; Jodidio, 2010). Ban's paperboard tube structures have been designed using a rigorous structural engineering approach that has included special treatments of the tubes to inhibit moisture penetration, material and structural testing of the structural components and joints, and structural analysis and design to ensure appropriate structural stability, safety and serviceability, and compliance with appropriate local permits (McQuaid, 2003; Correa, 2004). In these projects, Ban has collaborated with structural engineers (eg, Minoru Tezuka, Geno Matsui), major international structural engineering firms (eg, Buro Happold, ARUP, Terrell (Miyake, 2009)), and major universities in Europe and Japan (Tezuka et al., 1998). A number of other designers and engineers have experimented with paperboard tube structures and installations in recent years including a small school building by Buro Happold (Cripps, 2004), "Public Farm one" by WORK Architecture Company (Anon, 2014b) with LERA (structural engineers), "Portals to an Architecture" by Steve Preston (Preston, 2006; Preston and Bank, 2012), and "Brunkebergstorg 2014" by Guringo (Anon, 2014c). Much of the appeal of using paperboard for structures comes from their aesthetics, inherent recyclability, and potential for use as a sustainable building material. Paper and paperboard are some of the most highly recycled materials in the world (60% to 70% according to AFPA (1972)). The use of paperboard in temporary exhibitions is common, and life-cycle assessment (LCA) analyses have been conducted to evaluate the environmental impact of cardboard exhibition materials (Toniolo et al., 2013)

The use of paperboard tubes in construction engineering is much more well known than in structural engineering primarily due to the popularity of paperboard tubular forms for concrete columns for building frame construction. Often called "SONO-TUBES," they were originally trademarked by Sonoco Products Company in 1945 (SONOTUBE®), and a US patent was granted in 1954 patent for a design with an improved inside tube surface (polyethylene) (Copenhaver et al., 1954). Such tubes are now produced worldwide by many manufacturers. Forms are commercially available in standard diameters ranging from 50 to 1600 mm and in lengths up to 18 m. Paper tubes can also be used in the construction of voided-slabs. Fig. 16.2(a) shows a large diameter concrete form tubes in place for a column casting and Fig. 16.2(b) shows SONOVOID™ tubes in a concrete horizontal element prior to casting.

In a recent study, segments of large diameter (1600 mm) SONOTUBES were investigated for use as recyclable and inexpensive bridge deck formwork for concrete bridges having wide-flange prestressed girders with narrow gaps between the girders (Spottiswoode, 2007; Spottiswoode et al., 2012). Fig. 16.3 shows (a) the large-diameter tubes, (b) the small tube segments used as horizontal formwork, and (c) an experimental 200-mm-deep slab section cast on the paperboard segment.

16.2   Paper tube manufacturing and primary uses

Spiral paper tubes are manufactured using a sophisticated process as illustrated in Fig. 16.4. The green belts in Fig. 16.4 are stretched and rotated to drive the manufacturing process. The belts pull the paper plies onto a stationary winding

(a)   (b)

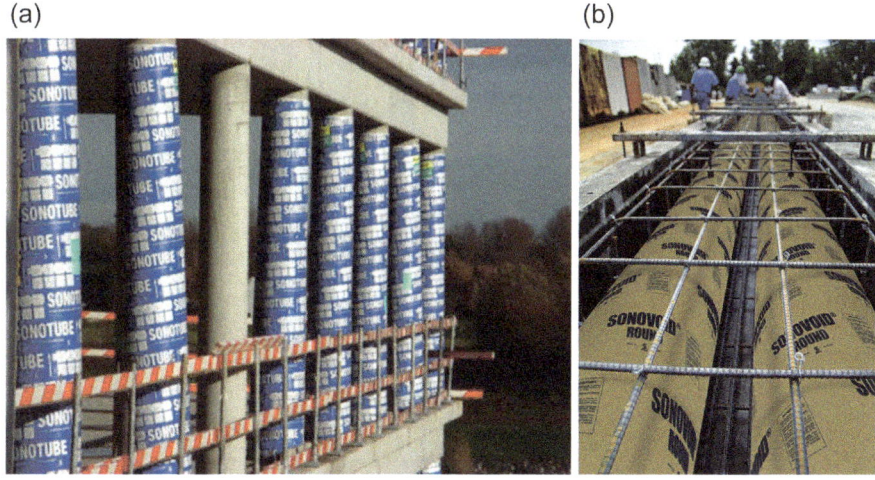

Figure 16.2 (a) Column forms and (b) voided slab tubes.
Photographs with permission from Sonoco Products Company.

(a)

(b)

(c)

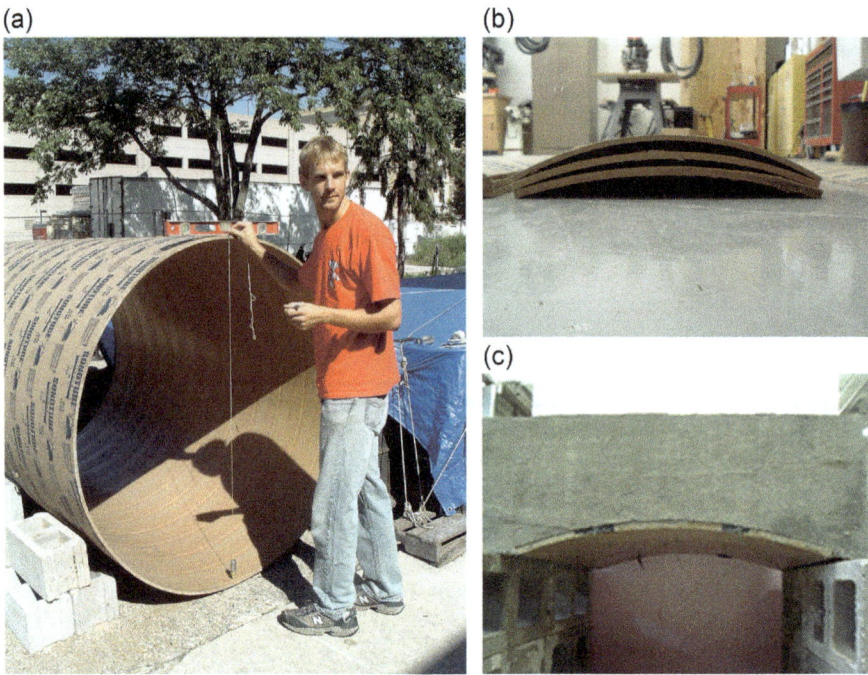

Figure 16.3 (a) A 160-cm-diameter tube, (b) tube segments (for 610-mm span), and (c) concrete cast on tube segment.
From Spottiswoode, A.J., 2007. An Investigation Into the Use of Paperboard Tube Segments for Bridge Deck Formwork (MS thesis), University of Wisconsin-Madison, with permission from Andrew Spottiswoode.

Figure 16.4 Paper tube manufacturing operation.
With permission from Pakea.

mandrel to form a continuous tube. Paper ply movement through the process is rather complex, but can be readily understood from Fig. 16.4. The ply entering from the right center of the image that comes into direct contact with the outside surface of the mandrel becomes the inside diameter of the paper tube. The subsequent layer (from the inside) is formed from the highest ply entering the mandrel from the left. The subsequent plies moving down the lattice form the sequential layers in the tube wall. Finally, plies that form the tube outside surface can be seen entering from the right in the lower portion of the image.

This can be a high-speed process as belt speeds as high as 160 m/min are possible. Tube production speed depends on the angle of wind, but for a belt speed of 160 m/min, the production rate is approximately 75 m/min for a typical 76-mm inside diameter (ID) core. This is a versatile process, as tubes can be fabricated using mandrels ranging from 20 to 1600 mm in diameter using anywhere from 2 to 40 plies of paper. Some manufacturing machines have been built to significantly increase the angle the plies make with the axis of the tube in order to achieve unique tube properties (Biggs and Dunlap, 1965; Jarvinen, 2003). At the front end of the process, large rolls of paper are hung on an array of unwind stands as shown in Fig. 16.5. The rotating belts unwind these rolls and pull the paper plies to the frame shown in Fig. 16.4. Typical unwind stands include an automatic mechanism that splices a new paper roll to an expiring roll at full speed, keeping the process continuous.

As noted earlier, each ply forms a layer in the tube that occupies a different radial position. Thus, the layer each ply forms has a slightly different diameter in the tube. To avoid excessive gaps in the wall of the tube, larger ply widths must be used for plies that form larger diameters in the tube. This geometric requirement complicates both the manufacturing operation and paper inventory requirements. For a single tube, as many as 10 different widths of paper may be required. Guidelines for

Figure 16.5 Unwind stands.
With permission from Pakea.

selecting a proper set of ply widths for a given tube construction are based on geometry and are well known (Akashiba, 1987).

Fig. 16.6 shows a view of the plies moving to the mandrel from the unwind stands. During this process, an adhesive is applied to each ply. Many different types of adhesives are used, and tube manufacturers typically use their own proprietary formulations. As tubes are designed to be used inside industrial facilities, adhesives are not generally designed for exterior use. However, for some concrete form applications, tubes have been designed to inhibit moisture penetration and strength degradation. These tubes use a polyvinyl cross-linked adhesive and are fabricated with highly sized paper (Kim et al., 2010). Paper sizing is a chemical treatment made during paper manufacture that reduces its tendency to absorb water (Chamberlain and Kirwan, 2013).

After formation, the tube moves past the winding mandrel and is then cut to the desired length as shown in Fig. 16.7(a). Adhesives are formulated so that bonds develop quickly enough to enable cutting immediately after the tube is formed.

Figure 16.6 Ply positioning and adhesive application.
With permission from Pakea.

(a)   (b)

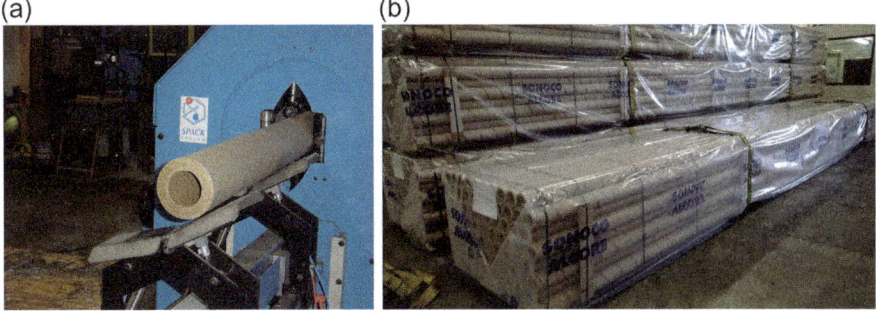

Figure 16.7 (a) Tube cutter *(With permission from Pakea.)* and (b) palletized tubes *(With permission from Sonoco Products Company.)*.

Various systems are available to cut tubes to length and also to automatically palletize the newly manufactured tubes as shown in Fig. 16.7(b).

Paper tubes are typically used to wind materials such as paper, film, metal sheet, or textile yarn into a roll immediately after they are manufactured. Typically, the wound rolls are then transported to a converting operation, such as a printing operation for paper or a coating operation for film. In some applications, the wound roll is shipped for direct use after winding, such as a film roll used in stretch wrapping operations. Paper tubes are also used in consumer markets to wind products like paper towels and bathroom tissue. These tubes are typically small diameter and manufactured using only a few plies of paper.

For these typical uses, paper tubes and the wound rolls are used inside manufacturing facilities or in the household. Paper tubes are not designed for exterior use. Exterior applications are challenging, and as is discussed next, paper tubes experience significant strength and stiffness loss when exposed to water or even high humidity conditions. Since paper tubes are typically not treated to inhibit moisture penetration, it is the paper's strength loss when combined with water that enables paper tubes to be easily recycled using standard pulping operations. Recycled paper tubes are an important raw material source for manufacturers of recycled paperboard.

16.3   Mechanics and properties of paperboard and tube materials

This section provides a brief summary of the mechanics of paper and paper tubes. Few common materials and structures exhibit such complex mechanical behavior. Paper consists mainly of cellulosic fibers held together by hydrogen bonds. An overview of the papermaking process and the properties of paper is provided in Chamberlain and Kirwan (2013). Paper is generally considered to be an orthotropic, nonlinear viscoelastic material that also exhibits an accelerated creep response when exposed to changing humidity conditions. Paper tubes are typically manufactured from recycled paperboard, an even more complex material since recycling paper impacts both fiber

length and bonding capability. When spirally wound, a paper tube becomes an anisotropic structure as the principal directions of the paper are not aligned with the principal directions of the tube. This anisotropy is evident as even subjecting a tube to
external pressure causes it to twist. Therefore, sophisticated methods are required to
design paper tubes for either traditional or nontraditional uses.

Generally, tubes are manufactured from paper that ranges in thickness from about
0.3 to 1.2 mm. Such paper is typically called paperboard; it is much thicker than linerboard, corrugating medium or paper used for printing and writing applications. In this
chapter, we use the terms "paper" and "paperboard" interchangeably. To manufacture
paper, a dilute mixture of fibers in water is distributed on a screen, pressed, and then
dried to form a paper sheet. This process creates a preferential alignment of fibers in the
direction the paper is manufactured, the so-called paper machine direction (MD). For
this reason, paper strength and stiffness is highest in its MD. The perpendicular direction across the width of a paper machine is called the cross machine direction (CD).
Typically, MD properties are 1.5—4 times higher than CD properties. However, it is
the thickness direction (ZD) properties that make paper an extremely unique material.
Paper's ZD modulus is typically 100—250 times less than its MD modulus. This very
low modulus gives paper tubes a unique structural advantage for winding materials
like film and paper, as will be discussed. The orthotropic modulus values for milk
carton stock have been reported as 7.44 GPa (MD), 3.47 GPa (CD), and 0.039 GPa
(ZD) (Mann et al., 1980). Such testing requires conditioning paper at standard conditions of relative humidity (RH) and temperature: 50% RH and 22°C. Such conditioning is required as all paper properties are sensitive to changes in moisture content and
temperature. For example, conditioning linerboard samples at 80% RH was shown to
reduce the modulus by 32% in the MD and 38% in the CD compared to conditioning
matched samples at 20% RH (Baum et al., 1981).

A material's stress—strain behavior is typically measured using quasi-static or ramp
loading. Even with such basic loading (and constant temperature and moisture content), paper's MD and CD stress-strain curves are highly nonlinear. In a pioneering
work to analyze paper strain energy, the stress—strain curves were successfully fit to
hyperbolic tangent functions (Johnson and Urbanik, 1984). This approach can also
reliably characterize the stress—strain behavior of the paper tube material itself, for
example, laminated recycled paper (Qiu et al., 1999). The authors machined blocks
from laminated paper to measure the orthotropic material properties. As shown in
Fig. 16.8, MD stress—strain data are accurately characterized with a hyperbolic tangent
function given as

$$\sigma_{MD} = a_1 \tanh(a_2 \varepsilon_{MD}) + a_3 \varepsilon_{MD} \qquad [16.1]$$

where a_1, a_2, and a_3 are constants fit to the data. When materials are wound around
paper tubes, pressure is transmitted into the tube through the paper ZD. Significantly,
Qiu et al. (1999) presents stress—strain data for ZD compression, which required
lamination of more than 100 plies of paper to form the blocks.

An elastic—plastic constitutive model was developed using a hyperbolic tangent
function to model anisotropic strain hardening in paper (Xia et al., 2002). In addition,

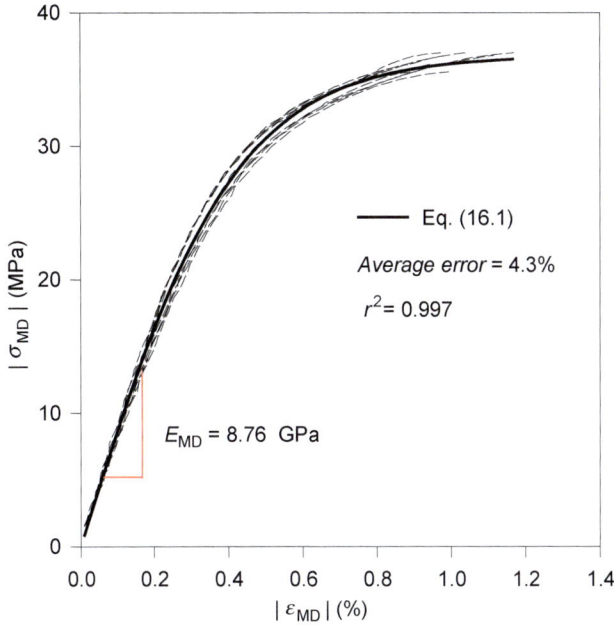

Figure 16.8 Measured MD stress—strain curves of recycled paper laminates.
From Qiu, Y.P., Millan, M., Lin, C.H., Gerhardt, T.D., 1999. Nonlinear properties of high
strength paperboards. ASME Journal of Engineering Materials and Technology 121, 272—277
with permission from the American Society of Mechanical Engineers (ASME), NY.

hyperbolic tangent functions were used to describe biaxial tensile data in the MD and
CD plane (Castro and Ostoja-Starzeski, 2003). All these studies quantified paper's
response to simple quasi-static loading at fixed temperature and humidity. Tensile tests
were recently conducted on paper at different, but fixed, moisture and temperature
levels (Linvill and Ostlund, 2014). For moisture contents ranging from 6.9% to
13.8% and temperatures from 23 to 168°C, the coefficients of the hyperbolic tangent
functions could be adjusted to fit the measured stress—strain data. Even for such basic
loading conditions, the stress—strain behavior of paper is complex.

Using paper in structural applications requires the ability to resist time-dependent
loads under conditions of varying humidity and temperature. For time-dependent
loading, paper is considered to be an orthotropic, nonlinear viscoelastic material. As
expected for such materials, the measured modulus depends strongly on test loading
rate (Gunderson et al., 1988). A nonlinear viscoelastic constitutive model that incorpo-
rates different, but constant, moisture levels was developed and validated using linerboard compression data by Pecht and Haslach (1991). Three viscoelastic theories were
compared to uniaxial paper data using both relaxation tests and stress—strain curves
measured at different loading rates by Joonas et al. (2007). As may be expected, higher
moisture levels cause higher creep rates, but paper also exhibits accelerated creep
when subjected to humidity changes. Remarkably, for variable humidity, the measured
creep rate is higher than even the rate measured when paper is held at the highest

humidity level (Von Byrd, 1984). Accelerated creep data are available for container-board (Fellers and Panek, 2007).

Fabricating paper into a spiral tube complicates the mechanics further as loading never coincides with the paper's MD or CD. Fig. 16.9 illustrates the geometric relationships in winding a paper ply to form a layer in a tube, where α is the angle between the paper MD and the tube circumference, W is the width of the ply, and D is the diameter of the layer the ply forms in the tube. The winding angle α can be easily determined from W and D. As was noted earlier, wider plies must be used for forming outer layers of the tube (Akashiba, 1987).

The paper ZD is perpendicular to the MD-CD plane and becomes aligned in the tube radial direction after winding. Paper's low ZD modulus, therefore, causes paper tubes to be extremely compliant in the radial direction. This compliance has a dramatic impact on hoop stress distributions when paper tubes are subject to pressure on their outside diameter, as during winding film around the tube (Gerhardt, 1990). Gerhardt derived a linear elasticity solution that was experimentally verified using strain gauge measurements. In Fig. 16.10, measured MD, CD normal strains, and in-plane shear strains on the tube outer surface are plotted until tube failure at an applied pressure of about 1800 kPa. The close agreement between experiment and theory in the linear range (less than about 600 kPa) validates the derived model. This agreement suggests that solid mechanics formulations can be used to develop structural models for paper tubes.

In addition to impacting hoop stresses, the radial compliance of paper tubes also provides an underappreciated benefit for winding materials. The compliance enables the outside surface of the tube to deform slightly even when relatively weak materials like paper or film are wound. Such small deformations are critical for starting the roll and developing good roll structure. Wound rolls are fragile structures as they are held together only by friction. The radial compliance of paper tubes make them very well suited for winding rolls of a wide variety of materials. In contrast, it is nearly impossible to wind a stable paper roll around a metal tube.

Paper tubes are designed differently for each market segment as winding operations differ significantly in speed, roll weight, roll length, chuck type, wound material properties, and stiffness requirements. For example, winding textile yarns can cause the tube inside diameter (ID) to shrink and stick on the support mandrel. The ID stiffness

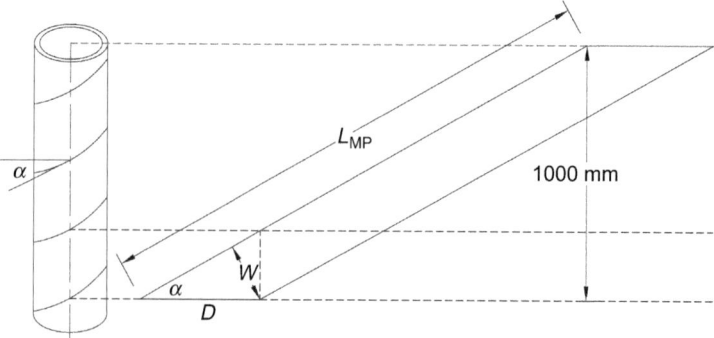

Figure 16.9 Inherent geometric relationships in winding a layer in a paper tube.

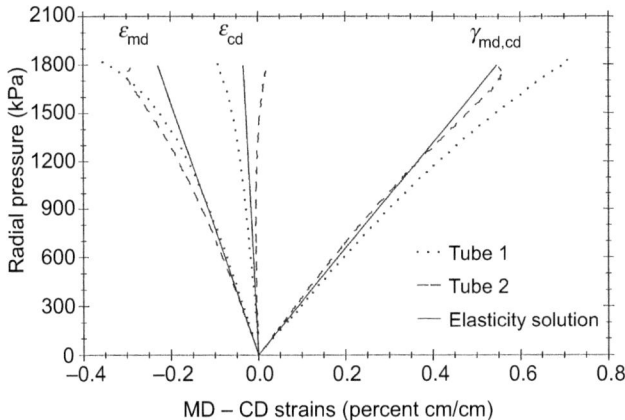

Figure 16.10 Measured strains on outer tube surface compared to elasticity solution. From Gerhardt, T.D., 1990. External pressure loading of spiral paper tubes: theory and experiment. ASME Journal of Engineering Materials and Technology 12 (4), 144–160 with permission from the American Society of Mechanical Engineers (ASME), NY.

of such tubes can be improved by using multigrade designs that position weaker paper in the middle of the tube wall (Qiu and Gerhardt, 1996). On the other hand, tube flat crush strength (T-108, 2008) can be optimized by positioning stronger paper in the middle of the tube wall (Wang et al., 1995).

Tubes used in the paper industry must meet demanding requirements both for fatigue and dynamic performance as they must resist both (1) cyclic loading during winding and unwinding heavy paper rolls and (2) vibrations during high speed unwinding in printing presses. One successful design approach is fabricating tubes with a higher winding angle α (Fig. 16.9) to both resist fatigue and increase tube bending modulus (Jarvinen, 2003). In practice, in a tube layer, adjacent plies are not perfectly butted together but have small gaps that become crack initiation sites during cyclic loading. Therefore, using a higher winding angle α improves fatigue strength by reducing the length of the gaps per unit length of tube (i.e., L_{MP} in Fig. 16.9).

Fig. 16.11 shows the dependency of tube bending modulus on winding angle α for tubes fabricated from four different types of paper (Haapaniemi and Jarvinen, 2005), where the theoretical spiral angle range of 0–90 degree is shown. Two of these paper grades were press-dried, a mechanical treatment made during paper manufacturing that significantly increases paper CD modulus (Gunderson, 1992). Typical paper tubes have winding angles ranging from 10 to 35 degrees depending on tube ID. But using a patented technology, 76-mm ID tubes can be manufactured with α greater than 60 degrees, thereby achieving a significant increase in tube modulus. Such high-angle tubes, both 76- and 160-mm IDs, are widely used in Europe for winding paper rolls that will be unwound on wide, high-speed rotogravure printing presses. Ultrasonic data measured on rotogravure presses confirmed that tube modulus was a key attribute for controlling vibrations as the roll unwinds (Gerhardt et al., 1993). However, to our knowledge, such high-angle tubes have never been used for

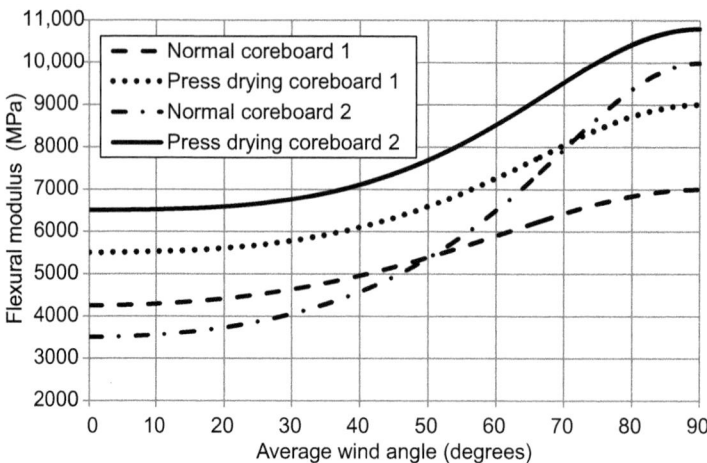

Figure 16.11 Tube-bending modulus versus winding angle.
With permission from Marko Ilomäki, Sonoco Products Company.

structures. Such designs could be considered for future structural projects as the improved bending modulus could offer design advantages. Also, using tubes fabricated from press-dried or Condebelt paper could offer a structural advantage from reduced hydroexpansivity along the length of the tube.

A variety of quality control tests have been developed to ensure paper tubes meet stiffness and strength requirements for the various winding operations. These tests include bending modulus (Bank et al., 1992; T-150, 2008; ISO 11093-7, 2011; ISO 11093-8, 2012), flat crush strength (T-108, 2008), radial crush strength (Rowlands et al., 1994), dynamic strength (Staples et al., 2000; Jarvinen et al., 1990), and axial strength (CT-107, 2009). Flat and radial crush quality control tests are commonly used in the industry (Gerhardt et al., 1999a).

Like paper, strength and stiffness of paper tubes are highly sensitive to moisture content. Testing procedures require conditioning at standard RH and temperature. Typically, flat and radial crush strength are both reduced about 50%, when tube moisture content increases from 5.5% to 13% (Gerhardt et al., 1999b). Dimensional stability is also significant, and for a 1% change in tube moisture, length changes of 0.12% and outside diameter changes of 0.09% are reported (TCR-2, 2008). When paper tubes are used in exterior environments with variable humidly conditions, significant changes in strength and dimension must be anticipated. This is clearly a challenge for structural applications.

As described, manufacturers have designed and optimized paper tubes for their key market applications (i.e., industrial winding operations). The comprehensive test methods, property data, and optimization strategies were all developed for these uses rather than structural applications. In fact, quality control tests do not exist to measure creep response of paper tubes, a key requirement for structural applications. So it is not surprising that the projects reviewed next have often relied on full-scale tests of tubes to determine the properties needed for structural design.

Figure 16.12 Paperboard tubes in different structural systems: (a) Japan Pavilion, Hannover Expo, 2000 *(From Correa, C., 2004. Designing with paper tubes. Structural Engineering International, IABSE, 4, 277–281 with permission from Structural Engineering, International IABSE, Switzerland.)*, (b) Museum of Modern Art, 2000 *(From Correa, C., 2004. Designing with paper tubes. Structural Engineering International, IABSE, 4, 277–281 with permission*

◀

from Structural Engineering, International IABSE, Switzerland.), (c) Portals to an Architecture, 2006 (From Preston, S.J., 2006. Portals to an Architecture (MS thesis), University of Wisconsin-Madison with permission from Steven Preston.), (d) Public Farm 1, PS1 - Museum of Modern Art, NY, 2008 *(Photograph by and with permission from Deniz Olcay, Work Architecture Company, NY.),* (e) Camper Traveling Race Pavilion 2011 *(Photograph by and with permission from Fredric Piau, Sonoco Products Company.),* (f) Musee Du Luxembourg, Paris, 2011 *(Photograph by and with permission from Fredric Piau, Sonoco Products Company.),* (g) Brunkebergstorg, 2014 *(Photograph by Erik Lindvall of Guringo, Sweden, with permission from Klara Börjeson.),* and (h) Singapore Biennale Pavilion, 2006 *(Photograph with permission from Eddie Chng, Sonoco Products Company.).*

16.4   Structural systems

Structural systems that use paperboard tubes as primary load-carrying structural members are not intrinsically different from those that would be constructed using tubular metallic members and connectors. However, due to their lower weight, strength, stiffness, and high strains to failure, structures built from paperboard tube often possess unique and surprising aesthetic elements and structural features.

Common structural systems used (see McQuaid, 2003; Miyake, 2009; Cripps, 2004; Preston and Bank, 2012) include barrel vaults and arches for large roof or enclosure spans, two- or three-dimensional arches, or pitched trusses for medium-span roofs, paperboard post and beam framing with vertical paperboard wall systems for small houses, and space frame skeletal structures. In all cases, the paperboard tubes are exposed to reveal both the structural systems and the unique use of the materials. Fig. 16.12(a–c) show arch-grid structures for roofing systems, Fig. 16.12(d–f) show vertical column structures used to form wall or frame systems, and Fig. 16.12(g–h) show three-dimensional space frame systems. While the primary structural members in these structures are paperboard tubes, the joints and structural connections use more conventional structural materials such as metals (eg, gusset plates, bearing plates, inserts, cables, bolts, nuts, and other hardware), timber (eg, baseboard, cap beams, small trusses and connectors, exterior sheathing), concrete (eg, foundations), natural and synthetic fabrics (eg, straps, ropes, cables, exterior sheathing, waterproofing membranes), and glazing materials (eg, windows).

16.5   Structural elements

16.5.1   Columns

Paperboard tubes are most commonly used as columns subjected to axial compressive loads as seen in the images in Fig. 16.12. Tube sizes ranging from nominal ID of 76 mm and wall thickness (t) of 12 mm to ID of 550 mm and wall thickness of 100 mm have been used as columns in paperboard structures. In cases where paperboard tubes are exposed to the environment for any length of time, they are typically coated with a polyurethane or other relatively impervious polymer coating to protect the tube from ambient environmental conditions (including rainfall).

The axial stiffness and axial strength of a tube vary as a function of the paper (paperboard) grade (type and specific weight) used to fabricate the tube and wind angle as discussed previously in Section 16.3. The axial load-carrying capacity varies as a function of the cross-sectional area. All properties are generally nonlinear functions of moisture content and duration of loading or imposed deformation (i.e., creep or relaxation) and, to a lesser extent, temperature. It is, therefore, common to perform simple full-scale axial tests of tubes in order to determine their mechanical properties for use in design (after preliminary estimates of loads are made to select the range of tube materials and geometries that will be appropriate). The Composite Can and Tube Institute (CCTI) provides a simple test method to measure full-section strength (CT-107, 2009), which can also be used for determining axial stiffness. Representative tube mechanical property data used in Ban's projects are provided in Table 16.1 (from Tezuka et al., 1998 and McQuaid, 2003). It is important to note the significant strength reduction is observed with both increased moisture content above 7−8% (typical as-received content in standard conditions at 21°C and 50% RH) and with sustained axial loading (creep rupture). Paper tube moisture content is typically measured using oven drying (destructive) test methods according to CT-111 (2008) and ISO 11093-3 (1994). CT-111 also allows for use of a hand-held digital nondestructive moisture meter. TCR-2 (2008) provides an in-depth discussion of the effects of moisture on the strength (and dimensional stability) of paperboard tubes.

Under axial compressive loading, the stress−strain curve of a paper tube is nonlinear and softens significantly as the ultimate load approaches. The modulus data are typically determined between 20% and 30% of the ultimate load (Tezuka et al., 1998). The ultimate failure is preceded by wrinkling of the outer linerboard layers at the ply gaps. This wrinkling tends to begin at between 85% and 90% of the ultimate compressive load capacity. Under sustained axial loading the tube axial modulus also decreases due to the nonlinear viscoelastic nature of the paperboard material and the adhesive. As noted in Section 16.3, data with respect to the extent of the axial modulus decrease with sustained load is lacking. Correa (2004) suggests a 30−50% decrease in short-term axial modulus under sustained long-term loads. Nevertheless, when tubes are subjected to moderate-term service or assembly loads, creep deformations are recoverable when the load is removed and do not lead to long-term strength or modulus losses (McQuaid, 2003; Dickson et al., 2001).

To use a paper tube as a column (or any other structural member), it must be able to transfer loads at its ends from other members framing into it (or bearing on it) into supporting foundations or other structural members. This is typically accomplished by designing a snug-fitting insert, usually made of wood, that fits into the end of the tube and transfers loads into or out of the tube shell. The tube is attached to the inserts using lag screws or steel bolts (when the insert is hollow and the interior is accessible). Bolt/tube connection shear capacity of the tube is obtained by testing for use in connection designs (see McQuaid (2003) for property data and test configurations used). It is typically assumed that the total load is shared equally between screws (or bolts) in a connection.

Table 16.1 Representative mechanical properties of paper tubes by static testing[a]

Project	Tube ID, t (mm, mm)	Load duration (s)	MC (%)	Axial strength, σ_a (MPa)	Axial modulus, E_a (GPa)	Flexural strength, σ_b (MPa)	Flexural modulus, E_b (GPa)
Library of a Poet (1991)	75.0, 12.5	Short	–	10.12	1.83	–	–
Paper House (1995)[b]	220, 30	Short	8.8	11.17	2.36	16.82	2.17
Paper Dome (1998)[c]	220, 30	Short	10.0	9.71	2.17	14.9	2.11
	75,10	Short	7.0	10.79	1.57	–	–
			13.0	5.39	0.54		
Japan Pavilion, Hannover Expo (2000)[d]	98, 22	Short	8.7	9.46 (100%)	1.57	14.5 (100%)	1.46
			10.1				
		110	8.7	8.07 (85%)			
		510	7.9	7.13 (75%)			
		5000	8.2	6.65 (70%)			
		36,000	6.8	5.70 (60%)			
		449	10.0	–	–	13.1 (90%)	–
		2461	10.0			11.6 (80%)	
		47,018	9.6			10.2 (70%)	
		1×10^6	9.2			8.7 (60%)	

[a] All tests conducted in quasi-static load–deformation conditions. Flexural test conducted in three-point or four-point bending configurations.
[b] Specific Gravity (S.G.) of 0.81–0.82 reported for these tubes.
[c] Tube spiral wind angle of 16.5 degree reported. See Figs. 16.4 and 16.5 in Tezuka et al. (1998) for more data on the effect of moisture content on tube strength and stiffness. In the absence of test data linear extrapolation using these data between seven% and 13% MC are recommended for axial strength and modulus predictions.
[d] Shear modulus (G) from full-section torsion test was also obtained in this series — 0.140 GPa (twisted with spiral winding direction) and 0.180 GPa (twisted against spiral winding direction).

16.5.2 Beams

Beams or flexural members are used to span horizontally and to carry transversely applied dead and live loads in addition to their self-weight. In most cases where beams have been used as structural members, superimposed dead and live loads are minimized, and simply carrying the tube self-weight without undergoing significant short-term and long-term deflections is a sufficient challenge. This is due to the low flexural modulus and, hence, flexural stiffness (EI) of tubular members (notwithstanding the fairly large second moment of area [I]) and the typical spans needed for building structures (4 m and greater). Therefore tubes are most effective as beam members when used in small-scale framing systems or small house—type structures (eg, Brunkebergstorg, Fig. 16.12(g), or Ban's emergency partitions, Anon, 2014a, or in rigid-like framing systems like the Vierendeel truss (frame) in the Singapore Biennale Pavilion, Fig. 16.12(h)). Connection details for beams also tend to be more complicated than for columns often requiring metallic, wooden, or molded polymeric inserts.

Fortunately, the flexural (or bending) stiffness and hence the flexural modulus of paperboard tubes are well-reported mechanical properties in the industry. The flexural stiffness (and first flexural mode of vibration) of a tube is of critical importance in paper winding, and unwinding operations as tube instability due to resonant frequency excitation must be prevented (Gerhardt et al., 1993). Both static (three-point bending test with special fixtures) and dynamic (modal testing) methods are provided by CCTI and ISO (T-150, 2008; ISO 11093-7, 2011; ISO 11093-8, 2012). Static testing is dependent on the testing fixture and beam theory used to determine the modulus and generally gives lower values than those obtained from dynamic testing (Bank et al., 1992). Dynamic modulus values (obtained from the first flexural mode of vibration) for a variety of tubes are provided in Table 16.2. The flexural strength of a paper tube is reported to be about 40—50% higher than its compressive strength, as seen in Table 16.1. The significant dependence on wind angle and specific weight of the paperboard used in the tube can be seen. The flexural modulus of a paper tube can be increased by adding a layer of stiffer material to the outside of the tube such as a thin layer of pultruded fiber-reinforced polymer (FRP) (Fink and Bank, 2006) as shown in Fig. 16.13.

The low modulus and relatively high strength (in relation to modulus) of paper tubes make them very "bendable." Fig. 16.14 shows a 12-m-long, 76-mm-ID tube slowly bent into a very small radius arc. Very interesting designs can be created with paper tubes in this fashion.

16.5.3 Arches, trusses, and frames

To span larger distances than possible with conventional paper tubes, arches, trusses, and frames are used as seen in Fig. 16.12.

Arches are particularly effective as the tubes can be prebent into curvilinear shapes and then placed in fixed positions prior to any creep recovery due to the very slow recovery rate. In addition, arches are primarily compression members, and vertical displacements are less pronounced to the eye. This technique is described in Preston (2006) and Preston and Bank (2012), where large spliced-tube arches were used in

Table 16.2 Representative mechanical properties of paper tubes via dynamic modal testing (as-received MC = ~8%)

Paper	Tube ID, t (mm, mm)	Wind angle[a] (degrees)	Specific gravity	Flexural modulus, E_b (GPa)
Bank et al. (1989)	76.5, 8.0	32	0.87	3.18
	76.4, 8.1	25	0.88	2.96
Bank et al. (1992)	77.8, 16.9	24	0.94	4.22
	78.0, 16.0	24	0.76	2.55
	76.8, 22.7	17	0.68	1.85
	76.6, 21.5	17	0.71	2.10
	76.8, 20.1	17	0.74	2.36
	76.6, 18.7	17	0.79	2.80
	77.4,16.3	17	0.88	3.16
	163.2, 6.86	12	0.53	1.73
	160.9, 13.0	12	0.94	4.12

[a]Wind angle, α, is shown in Fig. 16.9.

Figure 16.13 A 160-mm-ID, 16-mm-thick paper tube with an overmolded 3-mm pultruded outer layer.

Figure 16.14 A 12-m-long, 76-mm-ID tube bending into a very tight arc without breaking. From Preston, S.J., 2006. Portals to an Architecture (MS thesis), University of Wisconsin-Madison with permission from Steven Preston.

a temporary outdoor sculpture (Fig. 16.12(c)). Arches have been very effectively used in large barrel arch grid-structures by Shigeru Ban in the Japan Pavilion at the 2000 Hannover Expo and in the Museum of Modern Art (MOMA) canopy in New York, as seen in Fig. 16.12(a) and (b) (Dickson et al., 2001; Correa, 2004). An interesting arch—space—truss form was used by Ban to build a temporary footbridge in Remoulin, France, in 2007 (Anon, 2014a).

As with columns, it is important to provide appropriate constraints at the foundation. This is particularly important where the arch is required to be locked into a predetermined shape. This can be accomplished using metallic fixtures to ensure appropriate end-conditions or by embedding the arch into the soil. Care must be taken to ensure that moisture does not enter the tube at the foundation. Epoxy coating and capping the cut tube ends or encasing the paper tube in an impervious plastic lining are recommended. Fig. 16.15(a) shows the metallic "hinge" at the base of the MOMA double-layer grid canopy prior to assembly (Correa, 2004). Fig. 16.15(b) shows paper tubes embedded in an HDPE culvert pipe sleeve (Preston, 2006)

The choice of connections between tubes in overlapping arch-grid tube structures depends on their structural purpose. If the connection is only to keep the tubes in vertical contact, then polymer fabric and ropes straps have been used. These allow the tubes to move in-plane relative to one another and may be required to facilitate erection where the curvilinear arch shape is formed as the structure is lifted into its final position and then secured as in Ban's Japan Pavilion at the 2000 Hannover Expo. Such connections may also serve to relieve any built-up stresses due to long-term deformation and moisture-induced dimensional changes in the structure. However, if such connections are used, additional structural members such as restraining cables and end diaphragms are needed to ensure overall stability of the structure (Dickson et al., 2001). Alternatively and more commonly, the tubes are connected where they overlap to prevent such in-plane deformation. This increases the structural

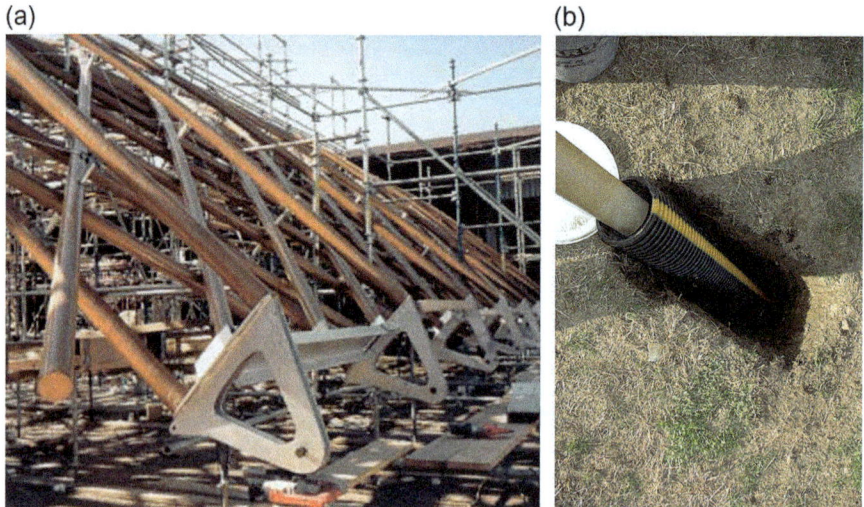

Figure 16.15 (a) Metallic end detail of MOMA arch grid *(From Correa, C., 2004. Designing with paper tubes. Structural Engineering International, IABSE 4, 277—281 with permission from Structural Engineering International, IABSE, Switzerland)* and (b) embedded ends for Portals arches *(From Preston, S.J., 2006. Portals to an Architecture (MS thesis), University of Wisconsin-Madison with permission from Steven Preston).*

rigidity and makes structural analysis more tractable. Connection details can be complicated, especially in the case of overlapping and multilayer systems. Fig. 16.16(a) shows a connection detail in the MOMA multilayer arch-grid (Correa, 2004). Preston used Fiberbolt® fiber-reinforced polymer threaded rods and nuts at the overlapping connections, which were bolted through the tubes in the Portals arches shown during assembly in Fig. 16.16(b).

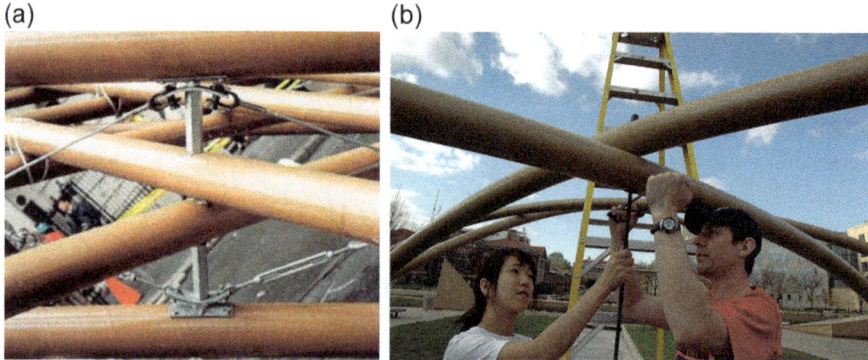

Figure 16.16 Connection between overlapping tube arches (a) MOMA, 2000 *(From Correa, C., 2004. Designing with paper tubes. Structural Engineering International, IABSE 4, 277—281 with permission from Structural Engineering International, IABSE, Switzerland)* and (b) Portals arches *(From Preston, S.J., 2006. Portals to an Architecture (MS thesis), University of Wisconsin-Madison with permission from Steven Preston).*

(a) (b)

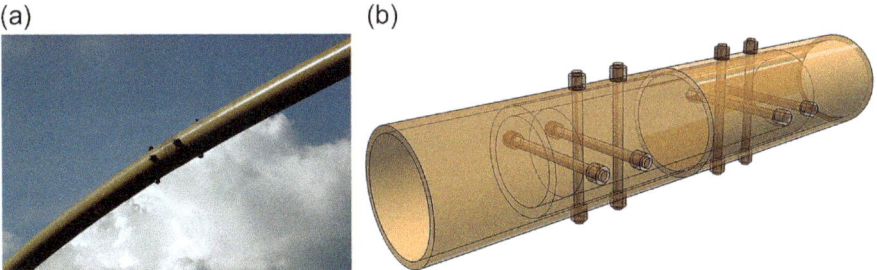

Figure 16.17 Tube longitudinal splice connection (a) as-built and (b) design sketch *(From Preston, S.J., 2006. Portals to an Architecture (MS thesis), University of Wisconsin-Madison with permission from Steven Preston).*

Paper tubes produced by spiral winding can be produced in any desired length. In general, it is preferable to have tubes produced in the full final length for the structural application to avoid lengthwise connections (splices). In many of Ban's larger projects (eg, Japan Pavilion and the MOMA canopy), tubes longer than 40 m were produced and shipped to the construction site. Shipping of such long tubes is not always possible or economically feasible, and in that case a longitudinal splice is required. Details of the Portals (Preston, 2006) splice connection using a smaller-diameter paper tube insert with FRP through-bolts is shown in Fig. 16.17.

Plane (two-dimensional) and space (three-dimensional) trusses have been used in a number of projects by Ban. In general, the paper tubes are only used for the compression members of the truss. The tension members are often metal cables or timber members (which may or may not be covered with paper tubes for aesthetic reasons). Trusses, as is typical, are loaded at their nodal points. However, secondary moments may be introduced depending on the fixity of the nodal connections and details of the joints. Much of the art (and cost) of designing paper tube trusses is in the design of the nodal joints. Complex joints (connectors) are typically made of wood with protruding cylindrical solid inserts, which are inserted into the paper tubes at the joint and then connected by through-bolts from the outside of the tube into the connection unit. Fig. 16.18 shows details of the truss system cables and molded nodal joints used by Ban in the Singapore Biennale Pavilion built in 2006.

For smaller structures, post-and-beam frame systems can be used. In these designs, the beam simply penetrates through the larger column tube section. A nice example of this is in Ban's 2013 Abu Dhabi Art Pavilion (Anon, 2014a). To maintain lateral stability of such structures, it is important to ensure a tight-fitting connection and bearing of the beam tube on both walls on the column tube (near and far ends) to transfer moment at the connection. If the frame is not braced, it will not be stable without moment resistance at the joints.

16.6 Structural analysis and design

Since the paperboard structures under discussion in this chapter are intended for use as building structures and may provide shelter or enclosure or be of sufficient size that

Figure 16.18 Details of space truss from the Singapore Biennial Pavilion, 2006. With permission from Eddie Chng, Sonoco Products Company.

their collapse could endanger those in and around them, they must be designed to ensure the safety of their occupants and bystanders (even if they are temporary structures or not fully enclosed). In most cases, a building permit will be required to build a structure of this sort, and a structural engineer should be engaged in the analysis and design.

The analysis (determination of the forces and displacements in the individual members and the entire structure under all possible load conditions) and the design (selection of the structural system (scheme), the materials, the member sizes and the connection types to ensure local and overall structural stability and serviceability) of a structure are typically the responsibility of the structural engineer. Structural engineers are well versed in performing these tasks when the materials (such as steel, reinforced concrete, and timber) are standardized and explicitly included in applicable structural design and building codes. When the structural materials are not explicitly included in applicable building codes, such as in the case of paperboard tubes, the structural engineer must establish an appropriate design basis for the structural elements and then perform an analysis and design according to this predeveloped design basis.

A design basis for a structure includes the choice of (1) the design philosophy (allowable stress design method or limit states design [applied in US design codes for strength only as load and resistance factor design or ultimate strength design]), (2) the loads and load combinations to be applied, (3) the resistance factors or safety factors to be used, (4) the ultimate member capacities or allowable stresses in the elements or materials, (5) the member effective lengths based on end-conditions, (6) the allowable member and global displacements (and natural frequencies), (7) the methods of structural analysis (linear, non-linear material and/or geometrical), and (8) the identification of material standards and test methods (such as ASTM, ISO,

CCTI). Eurocode 1990 provides details of the basis for structural design according to the Eurocodes (EN, 1990). A similar situation currently exists for the design of fiber-reinforced polymer pultruded structures, and this issue is discussed in greater detail in Bank (2006, chap. 12).

Typically, not all the steps listed above need to be developed from first principles. For example, often the load cases and combinations will be taken from an existing code of loads on structures (eg, ASCE 7 (ASCE, 2013), Eurocode 1 (EN, 1991)). In that case, the design basis simply needs to specify the exiting code document and its relevant sections. Such approaches are generally taken in the design of paperboard structures. A set of structural engineering contract documents for a building project will include drawings and specifications that contain the details of the elements and systems, and notes listing the appropriate codes and standards followed. However these are typically not made publically available. In addition, the structural analysis assumptions, models, and results are generally not provided by the structural engineers as part of the design package. It is therefore often not possible to ascertain how structures were designed — and this applies to paper tube structures as well.

Few cases have been documented in any detail. In the case of the Japan Pavilion, the structural engineers at Buro Happold designed the paper tubes based on test data obtained from testing of the tubes at Dortmund University (see McQuaid, 2003) "with partial safety factors similar to EC5 (Eurocode 5) for timber structures" (Dickson et al., 2001). Loads and load cases were as prescribed by the German standards. Elastic analysis was used but allowable material stresses were "reduced to account for long-term creep effects" (Dickson et al., 2001; Correa, 2004). It is assumed that a full computer structural analysis of the structure was conducted, but details are not available.

For the Portals to an Architecture project (Preston, 2006; Preston and Bank, 2012), an allowable stress design basis was selected because no probabilistic data for such structures could be obtained to develop resistance factors (or partial safety factors). For the type of unenclosed skeletal structure designed and the short duration of the exhibition (3 months), a safety factor of 1.5 was chosen. The allowable stresses and material properties were based on data provided by the paper tube supplier, which was Sonoco Products Company. Since the structure shared characteristics with highway sign structures, the loads and load combinations analyzed for the structure were taken from the Standard Specification for Structural Supports of Highway Signs, Luminaires and Traffic Signals (AASHTO, 2001). An elastic-material, geometrically nonlinear (P-Delta) analysis was conducted using the structural analysis (finite-element) software program Visual Analysis (IES, 2014). For the structural analysis, the ends of the arches were modeled as fixed connections, and the tube members were modeled as beam elements with pipe cross-sections. As an example of the output, Fig. 16.19 shows the bending moment diagram for the arches subjected to the dead plus wind load case. Full details of the modeling can be found in Preston (2006).

From the maximum moments the maximum bending stresses including biaxial bending effects were calculated at each node and compared with the allowable stresses and found to be satisfactory. Deflections were also checked (Preston, 2006).

Figure 16.19 Bending moment diagram (Z direction) looking west of Portals to an Architecture. Preston, S.J., 2006. Portals to an Architecture (MS thesis), University of Wisconsin-Madison with permission from Steven Preston.

16.6.1 Member design

Classic mechanics of materials equations are typically used to relate stress resultants (N, M, V) obtained from structural analysis to stresses for determination of member dimensions. The tube is assumed to be homogeneous (and also generally assumed to be linear elastic and isotropic notwithstanding the discussions of nonlinearity, viscoelasticity, and anisotropy in prior sections of this chapter) and full-section properties (cross-sectional area A, section modulus S, and second moment of area I) are used in design calculations.

For columns, the axial capacity (the lesser of the compressive capacity and the global (Euler) buckling capacity) and axial deformation are calculated from:

$$P_a = \sigma_a A \qquad [16.2]$$

$$P_{cr} = \frac{\pi^2 E_b I}{(kL)^2} \qquad [16.3]$$

$$\delta = \frac{PL}{AE_a} \qquad [16.4]$$

where P_a is the axial compressive capacity, σ_a is the compressive strength of the tube (see Table 16.1), P_{cr} is the axial (flexural) buckling capacity, E_b is the tube bending (flexural) modulus (see Table 16.1), k is the tube effective length factor, L is the tube nominal length, and δ is the axial length change when subjected to an axial load of P.

Local tube wall buckling is not likely to be a critical strength mode, as wall thicknesses are large relative to tube dimensions (Preston and Bank, 2012).

For beams, flexural capacity and deflection, Euler−Bernoulli beam theory is generally used and values are calculated from:

$$M = \sigma_b S \qquad [16.5]$$

$$y(x) = \iint \frac{M(x)}{E_b I} dx dx \qquad [16.6]$$

where σ_b is the bending (flexural) strength (see Table 16.1) and $y(x)$ is the beam deflection as a function of the location along the tube length, x, and the bending moment distribution along the tube length, $M(x)$.

Lateral torsional buckling is unlikely but can be checked using classical equations. Torsional buckling may need to be checked in frame structures.

Biaxial (or multiaxial) combined bending equations should be used when three-dimensional effects are present. It is important not to neglect axial deformations in trusses and frames. Interaction equations should be used for beam-columns when coupled axial and flexural loads exist. Bearing stresses at connections must be checked. Detailed three-dimensional stress analysis using a finite element method should be conducted to determine stresses in complex parts and connections. Full-scale testing of prototypes is recommended.

16.7 Conclusion

The chapter provides a comprehensive overview of paperboard tube manufacturing, properties of paper and tubes, structural applications, and analysis and design approaches that should enable architects and engineers to build structures using paperboard tubes. While paperboard tubes are readily available from many suppliers for potential use in structures, the authors stress the fact that currently available tubes are manufactured for the converting industries and are not intended for, nor optimized for, structural applications, and designers should proceed with caution. Close coordination with the tube manufacturer is needed to ensure that tubes produced for structural applications meet required structural and environmental demands for the intended duration of service.

Much of the interest in designing structures from paper tubes lies in the inherent recyclability of the paperboard materials and in the novelty of using materials that are generally perceived as being weak and not particularly durable. This chapter has demonstrated that highly engineered paperboard tubes can be used in large structures, many of which have endured over 20 years in service in the outdoor environment. Those that have been built for temporary exhibitions are typically recycled after their use. Paper tubes have also been used very effectively in humanitarian disaster relief efforts by Ban to build shelters and other enclosures. In these cases, the factors of being light weight, machinability with light hand-tools, ease of self-assembly and local availability, low price, and low environmental impact give paper tubes a significant advantage.

References

AASHTO, 2001. Standard Specification for Structural Supports of Highway Signs, Luminaires and Traffic Signals, fourth ed. American Association of State Highway and Transportation Officials, Washington, DC.

AFPA, 1972. American Forest & Paper Association. Washington, DC. http://www.paperrecycles. org/ (accessed 01.01.16.).

Akashiba, M., 1987. Spiral Wound Paper Pipe. National Association of Paper Pipe Industry, Japan.

ASCE 7, 2013. Minimum Design Loads for Buildings and Other Structures (ASCE/SEI 7-10). American Society of Civil Engineers, Reston, VA.

Anon, 2014a. Shigeru Ban Architects — Paper Tube Structures. http://www.shigerubanarchitects. com/works.html#paper-tube-structure (accessed 23.12.14.).

Anon, 2014b. PF1, Public Farm 1. http://www.publicfarm1.org/ (accessed 29.12.14.).

Anon, 2014c. Placemaking-brunkebergstorg 2014. www.guringo.com/project/brunkebergstorg-2014/ (accessed 23.12.14.).

Bank, L.C., 2006. Composites for Construction: Structural Design with FRP Materials. John Wiley & Sons, New York, NY, 551 pages.

Bank, L.C., Gerhardt, T.D., Gordis, J.H., 1989. Dynamic mechanical properties of spirally wound paper tubes. ASME Journal of Vibration, Acoustics, Stress and Reliability in Design 111, 489—490.

Bank, L.C., Cofie, E., Gerhardt, T.D., 1992. A new test method for the determination of the flexural modulus of spirally wound paper tubes. ASME Journal of Engineering Materials and Technology 114, 84—89.

Baum, G.A., Brennan, D.C., Habeger, C.C., 1981. Orthotropic elastic constants of paper. Tappi 64 (8), 97—101.

Biggs, W.A., Dunlap, C.K., (Sonoco Products Company), 1965. Spirally Wound Paper Tube. US Patent 3,194,275. July 3, 1965.

Byrd, V.L., 1984. Edgewise compressive creep of fiberboard components in a cyclic-relative-humidity environment. Tappi Journal 67 (7), 86—91.

Copenhaver, J.E., Clarkson, B.D., Woods Jr., (Sonoco Products Company), 1954. Concrete Form and Method of Molding Concrete Columns Therewith. US Patent 2,577,165. May 4, 1954.

Castro, J., Ostoja-Starzewski, M., 2003. Elasto-plasticity of paper. International Journal of Plasticity 19, 2083—2098.

CT-107, 2009. Axial (End-to-end) Compression Composite Cans, Tubes and Cores, May 1981 (Reviewed & Reapproved July 2009). Composite Can and Tube Institute, Alexandria, VA.

CT-111, 2008. Measuring Moisture Content Composite Cans, Tubes and Cores, March 1982 (Reviewed & Reapproved March 2008). Composite Can and Tube Institute, Alexandria, VA.

Chamberlain, D., Kirwan, M.J., 2013. Paper and paperboard — raw materials, processing and properties. In: Kirwan, M.J. (Ed.), Handbook of Paper and Paperboard Packaging Technology, second ed. John Wiley & Sons, Ltd, UK.

Correa, C., 2004. Designing with paper tubes. Structural Engineering International, IABSE 4, 277—281.

Cripps, A., 2004. Cardboard as a construction material: a case study. Building Research & Information 32 (3), 207—219.

Dickson, M., Hardie, G., Leiblein, K., Rogers, P., Wesbury, P., 2001. The Japan pavilion for the Hanover expo 2000. In: Theory, Design and Realization of Shell and Spatial Structures, International Symposium on Theory, Design and Realization of Shell and Spatial Structures, IASS 2001, 9—13 October, 2001, Nagoya, Japan, TP178, CD-ROM, 8 pages.

EN 1990, Eurocode — Basis of Structural Design, http://eurocodes.jrc.ec.europa.eu.

EN 1991, Eurocode 1: Actions on Structures, EN 1991-1-1:2002 to EN 1991-4: 2006, http://eurocodes.jrc.ec.europa.eu.

Fellers, C., Panek, J., 2007. Effect of relative humidity cycle start point and amplitude on the mechano-sorptive creep of containerboard. In: 61st Appita Annual Conference and Exhibition, 6—9 May, 2007, Gold Coast, Australia, pp. 291—296.

Fink, E., Bank, L.C., 2006. Pultruded glass fiber-reinforced plastic and paperboard composite tubes. Journal of Reinforced Plastics and Composites 25 (6), 601—616.

Gerhardt, T.D., 1990. External pressure loading of spiral paper tubes: theory and experiment. ASME Journal of Engineering Materials and Technology 12 (4), 144—160.

Gerhardt, T.D., Qiu, Y.P., Wang, Y., Johnson, C.G., McCarthy, M., Rhodes, D.E., 1999a. Engineering paper tubes to improve winding performance, Mechanics of cellulosic materials. In: ASME Joint Applied Mechanics and Materials Division Meeting Blacksburg, VA, USA, 27−30 June 1999a, pp. 169−176.

Gerhardt, T.D., Rhodes, D.E., Johnson, C.G., Wang, Y., McCarthy, M., 1999b. Performance of paper tubes. In: Proceedings of the 5th International Conference on Web Handling, 6−9 June, 1999b. Oklahoma State University, United States, pp. 265−277.

Gerhardt, T.D., Staples, J.F., Lucas, G.L., 1993. Vibrational characteristics of wound paper rolls: experiment and theory. Tappi 76 (6), 121−128.

Gunderson, D.E., Considine, J.M., Scott, C.T., 1988. The compressive load-strain curve of paperboard: rate of load and humidity effects. Journal of Pulp and Paper Science 14, 713−726.

Gunderson, D.E., 1992. An overview of press-drying, impulse-drying, and condebelt-drying concepts. Paperi Ja Puu − Paper and Timber 74 (5), 412−418.

Haapaniemi, J., Jarvinen, M., (Sonoco-Alcore Oy), 2005. Structural Ply of a Paperboard Core, a Paperboard Core Made Thereof, and a Method Improving the Stiffness of a Paperboard Core. US Patent US 6,962,736, November 8, 2005.

IES, 2014. Visual Analysis. IES (Integrated Engineering Software), Bozeman, MN.

ISO 11093-3, 1994. Paper and Board−testing of Cores−Part 3: Determination of Moisture Content Using the Oven Drying Method. International Organization for Standardization, Geneva, Switzerland.

ISO 11093-7, 2011. Paper and Board−testing of Cores−Part 7: Determination of Flexural Modulus by the Three-point Method. International Organization for Standardization, Geneva, Switzerland.

ISO 11093-8, 2012. Paper and Board −Testing of Cores−Part 8: Determination of Natural Frequency and Flexural Modulus by Experimental Modal Analysis. International Organization for Standardization, Geneva, Switzerland.

Jarvinen, M.S., Novaky, M.J.H., Rommy, H.P.J., (Ahlstrom), 1990. Method of Testing Cores. US Patent 4,942,769, July 24, 1990.

Jarvinen, M.S., (Sonoco-Alcore Oy), 2003. Paperboard Core With an Improved Chuck Strength, for the Paper Industry and a Method of Fabricating Such. US Patent 6,540,174, 1 April 2003.

Johnson Jr., M.W., Urbanik, T.J., 1984. A nonlinear theory for elastic plates with application to characterizing paper properties. ASME Journal of Applied Mechanics 51, 146−162.

Joonas, S., Kouko, J., Malinen, M., Kurki, M., Hamalainen, J., 2007. Paper as a viscoelastic material: comparison between different theories. In: 61st Appita Annual Conference and Exhibition, 6−9 May, 2007, Gold Coast, Australia, pp. 389−396.

Jodidio, P., 2010. Shigeru Ban: Complete Works 1985−2010. Taschen, Germany.

Kim, T.H., Lounsbury Jr., C.W., Rhodes, D.E., Smith, G.N., Van de Camp, J.W., (Sonoco products company), 2010. Water Resistant Wound Paperboard Tube. US Patent 7,842,362, November 30, 2010.

Linvill, E., Ostlund, S., 2014. The combined effects of moisture and temperature on the mechanical response of paper. Experimental Mechanics 54, 1329−1341.

Mann, R.W., Baum, G.A., Habeger, C.C., 1980. Determination of all nine orthotropic elastic constants for machine-made paper. Tappi 63 (2), 163−168.

McQuaid, M., 2003. Shigeru Ban. Phaidon Press, NY.

Miyake, R. (Ed.), 2009. Shigeru Ban: Paper in Architecture. Rizzoli, NY.

Pecht, M., Haslach, H.W., 1991. A viscoelastic constitutive model for constant rate loading at different relative humidities. ASME Mechanics of Materials 11 (4), 337−345.

Preston, S.J., 2006. Portals to an Architecture. University of Wisconsin-Madison (MS thesis).

Preston, S.J., Bank, L.C., 2012. Portals to an architecture: design of a temporary structure with paper tube arches. Construction and Building Materials 30, 657−666.

Qiu, Y., Gerhardt, T.D., (Sonoco Products Company), 1996. Multi-grade Paperboard Winding Cores for Yarns and Films Having Enhanced Resistance to Inside Diameter Reduction. US Patent 5,505,395, April 9, 1996.

Qiu, Y.P., Millan, M., Lin, C.H., Gerhardt, T.D., 1999. Nonlinear properties of high strength paperboards. ASME Journal of Engineering Materials and Technology 121, 272−277.

Rowlands, R.E., Saliklis, E.P., Wise, J.T., Gerhardt, T.D., Sonoco Products Company, 1994. Apparatus and Method for Testing Tubular Products. US Patent 5,339,693, August 23, 1994.

Spottiswoode, A.J., 2007. An Investigation into the Use of Paperboard Tube Segments for Bridge Deck Formwork. University of Wisconsin-Madison (MS thesis).

Spottiswoode, A.J., Bank, L.C., Shapira, A., 2012. Investigation of paperboard tubes as formwork for concrete bridge decks. Construction and Building Materials 30, 767−775.

Staples, J.F., Taylor, D., van de Camp, J.W., Lyles, S., Stevens, J.T., (Sonoco Products Company), 2000. Dynamic Strength Tester for Paper Cores. US Patent 6,050,148, April 18, 2000.

T-150, 2008. Part 1-Determination of the Bending Stiffness of a Fiber Core by: (A) Modal Analysis and (B) Three Point Bending (Formerly T-148) Part 2-method for Calculating the Critical Speed of Fiber Cores on a Printing Press or Other High Speed Application Composite Tubes and Cores, June 1998 (Reviewed & Reapproved March 2008). Composite Can and Tube Institute, Alexandria, VA.

T-108, 2008. Side-to-side (Flat) Crush, September 1985 (Reviewed & Reapproved March 2008). Composite Can and Tube Institute, Alexandria, VA.

TCR-2, 2008. The Effect of Moisture Content on the Dimensional Stability and Strength Properties of a Typical Paperboard Tube or Core, June 1989, (Reviewed & Reapproved March 2008). Composite Can and Tube Institute, Alexandria, VA.

Tezuka, M., Yamada, N., Ito, K., Ban, S., Tsuboi, Y., 1998. Application of paper tube to the member of building structure I. Mechanical properties of paper tube and the effect of water content. Mokuzai Gakkaishi 44 (5), 332−341 (in Japanese with English abstract and tables and figures), Japan Wood Res Soc, Tokyo, 113-0023, Japan.

Toniolo, S., Fedele, A., Manzardo, A., Scipioni, A., 2013. Recycling applied to temporary exhibition structures: investigation of the environmental impacts using a life cycle approach. In: 3rd International Energy, Life Cycle Assessment, and Sustainability Workshop & Symposium (ELCAS3), 7−9 July, 2013. Nisyros, Greece, pp. 1111−1119.

Wang, Y., McCarthy, M., Johnson, C.G., Gerhardt, T.D., Sonoco Products Company, 1995. Enhanced Crush Strength Construction Multi-grade Paperboard Tubes. US Patent 5,393,582, February 28, 1995.

Xia, Q.S., Boyce, M.C., Parks, D.M., 2002. A constitutive model for the anisotropic elastic-plastic deformation of paper and paperboard. International Journal of Solids and Structures 39, 4053−4071.

Index

A

Abrasion
  coefficient, 245—246
  test, 245—246
Accelerated weathering, 119
Acoustic properties, 141
Adjusted design values. *See* Stress values
Adobe, 44—46. *See also* Reinforced adobes
  blocks, 273—274
Adsorption, 131
Affinity with water, 275—280
Afghanistan, 26—27, 28f
Agave sisalana. See Sisal (*Agave sisalana*)
Age effect, 369—371
Agricultural additives, 263—264. *See also*
  Industrial additives
Air-cured hybrid composites, 85
Air-dried bamboo strips, 436
Air-hardening
  calcic lime, 311, 312f
  magnesian lime, 311—312
Alcock's linear shrinkage test, 220, 221f
Alkali-resistant synthetic fibers, 84
Allowable strength design method (ASD
  method), 401, 404
Allowable stress design. *See* Allowable
  strength design method (ASD
  method)
Alternative materials in modern bridge
  construction, 57—58
Amazonic fibers, 88—89
American Society of Civil Engineers
  (ASCE), 19
American Society of Mechanical Engineers
  (ASME), 463f
Ananas erectifloius. See Curauá plant
  (*Ananas erectifloius*)
Ancient stone masonry constructions, 301,
  314—321

applications in modern architecture and
  innovative uses, 325—327
box behavior, 315
collapse mechanisms of masonry walls,
  314f
degradation of calcareous masonry, 328f
embodied carbon of construction materials,
  329t
masonry wall, 315—317, 316f
mechanical performance
  mechanical parameters estimation,
  324—325
  safety assessment, 322—324
overview of ancient applications,
  302—303
role of masonry in sustainable construction,
  327—330
stone masonry materials, 303—313
Angled connections, 421f
Animal fibres, 113
Anji Bridge. *See* Zhao-Zhou Bridge
Arches, 469—473
Arg-e-Bam, 3, 3f, 7
  earthquake destruction, 4f
  evidence of termites in wall in, 6f
ASCE. *See* American Society of Civil
  Engineers (ASCE)
ASD method. *See* Allowable strength design
  method (ASD method)
Ashlar blocks, 320—321
Ashlar masonry, 319—321, 320f—321f
ASME. *See* American Society of
  Mechanical Engineers (ASME)
Atterberg limits, 254
Axial force members, 414. *See also* Bending
  members
  compression elements, 414—415
  tension elements, 415—416
Azar dry-stack blocks, 229

B

BA. *See* Bottom ash (BA)
Bacteria, 121
Bahareque, 417
    allowable strength properties for cemented,
        420t
    density coefficient values, 418t
    wall, 397–399
BaleHaus, 145f
Bam, 3–6
    archeological site in, 4f
    Arg-e-Bam, 3, 3f
    citadel, 44–45
Bamboo, 365, 393, 433. *See also*
        Engineered bamboo
    axial force members, 414–416
    bahareque framed system, 398f
    bamboo testing standards, 404–405
    bamboo-based products, 394–396
    bending failure modes in, 377f
    bending members, 408–413
    combined loading elements, 416–417
    composites reinforced with Bamboo
        cellulose fiber, 95–96, 96t
    connections, 419–427
    construction, 427–428
    Culm anatomy, 366f, 399–400
    curtains, 436
    derivation of design values, 385–386
    design considerations, 404
    fabrication, 427–428
    further work and developments, 386–388
    light-frame system for low-rise buildings,
        398f
    as material, 399
        bamboo culms anatomy, 399–400
        bamboo preservation, 401–403
        fire resistance, 403–404
        mechanical properties of bamboo
            species, 402t
        physical and mechanical properties,
            400–401
    material for 21st century, 48–57
        efficiency of bamboo, 49
        FGM, 50–53
        fibres distribution, 50f
        mapping of physical and mechanical
            properties of full-culm bamboo,
            53–57
    original location of samples in culm, 52f
    performance of bamboo and other
        materials, 49f
    scientific research of application of
        bamboo, 49
    variation of fibre distribution, 53f
material properties, 367
    effect of age, 369–371
    effect of density, 367–368
    dimensional properties of Culm,
        372–374
    functionally graded material, 374
    effect of moisture content,
        371–372
    variation of density and properties in
        Culm, 368–369
parts, 396f
as plant, 365–367
preservation, 401–403
shear walls, 417–419
and solar energy building combination,
    443
species, 393–394
    and applications, 395t
strips, 435–436
structural design standards and building
    codes, 405
    structural design basis, 406
    structural design method, 406–407
testing standards, 404–405
tests for material and physical property
    determination, 374–375
    development of test standards,
        380–381
    edge-bearing test, 379f
    G. angustifolia, 375
    Guadua angustifolia, 381–385
    shear tests, 376
    split-pin test, 379f
    tensile strength, 378
types of structures, 396–399
Bamboo oriented strand board (OSB), 450
Bamboo scrimber, 450
Banana (*Musa cavendishii*), 87–88
Barrier systems, 118
Basalt, 304
Beams, 439, 469
    products, 351–352
Bearing, 413

Bending members, 408. *See also* Axial force
    members
  bearing, 413
  deflections, 411
  flexure, 411−412
  shear, 412
Bhatar, 16−17, 30
Bhutan, 8f
Binders, 255−256, 259
Biocomposites, 111
  description, 111−117
    applications, 116−117
    biopolymers, 114−115
    composite forms, 115
    concerns about behaviour, 115−116
    natural fibres, 113−114
    natural forms of fibre/polymer,
        112−113
  durability concerns, 117−121
  managing durability, 121−122
Biodegradable polymers, 114
Biopolymers, 114−115, 115t
Blastfurnace, 259
Block production, 223−231
  batch quantities for CSEB production,
        225
  classification of usable compaction pressure
        CSEB production, 228t
  components for soil-cement production,
        225t
  curing of CSEBs, 229−231, 231f
  handling and temporary stacking of freshly
        produced block, 226f
  manual and mechanised production,
        227−229
  material mix proportion, 223t
  mixing of materials, 225−226
  raw materials, 223
  sieving process of soil, 224f
  soil preparation, 224−225
  soil-cement mix, 226f
Block units and masonry, strength
        evaluation of, 236
  construction using CSEBs, 232−233
    CSEB wall and foundation construction,
        234f
    dry-stack masonry, 232f
    elevation details of interlocking wall
        systems, 233f

hydraform dry-stack interlocking CSEB
        system, 233f
dry-stacking of masonry, 233−235
  construction of Great Zimbabwe ruins,
        235f
  dry-stack interlocking block masonry,
        235f
  effect of openings in, 236f
factors affecting strength in compression of
        CSEBs, 239−241
  cement content effect, 239−241
  clay content effect, 239
  28-day strength, 240f
general features of Hydraform interlocking
        blocks, 236
Hydraform blocks, 236−239
in-plane and out-of-plane tests, 241−244
  compression tests on dry-stack masonry
        wall, 241−242
  conventional masonry wall behaviour
        under lateral load, 242
  dry-stack masonry wall behaviour under
        lateral load, 242−244
  failure modes, 241f
  wall panel results in compression, 242f
key features of interlocking units, 231−232
Block-producing companies, 208
Blocks, 209
Bottom ash (BA), 261
'Bow-tie' test method, 376
BP6, 229
Breccias, 304−307
Bricks, 209
Bridges, 337−338
  materials
    glubam pedestrian bridges, 444−446
    truck load roadway bridge, 446−447
Brise soleil, 327
Brittle matrices, vegetable fibers as
        reinforcement for, 84−85
Buildings, 338−339
  materials
    bamboo and solar energy combination,
        443
    glubam heavy frame buildings,
        443, 443f
    glubam temporary shelters for disaster
        relief, 442
    lightweight glubam-frame, 441−442

C

Calcarenite, 304−307
Calcareous
  alabaster, 307
  dolomites, 307
  tufa, 307
Calcilutite, 304−307, 306f
Calcium carbonate (CaCO$_3$), 99
Calcium hydroxide (CH), 163
Calcium oxide (CaO), 160
Calcium silicate hydrate (CSH), 163
  gel, 258
Cannabis sativa L. See Industrial hemp core
  (*Cannabis sativa L.*)
Capillary domain, 280
Capillary suction. *See* Sorptivity
Capping arrangement, 237
Carbon (C), 447
Carbon dioxide (CO$_2$), 100, 447
Carbon fiber−reinforced polymer materials
  (CFRP materials), 446
Carbon footprints analysis, 447
Carbonation, 100
Carbonic acid (H$_2$CO$_3$), 100
CCTI. *See* Composite Can and Tube
  Institute (CCTI)
CD. *See* Cross machine direction (CD)
CEBs. *See* Compressed earth blocks (CEBs)
Cellulose nanocrystals, 92, 92f
Cellulosic fibers, 85
Cellulosic pulps, 90, 96−97
Cement, 217
Cement kiln dust (CKD), 160−170
  chemical properties, 161−162, 162t
  effect on properties of mortar and concrete,
    162−170
    compressive strength development,
      165t−166t
    durability properties, 168−170
    flexural strength development,
      167t−168t
    hydration properties, 163−164
    setting time, 162−163
    splitting tensile strength development,
      169t
    strength properties, 164−167
  effect of particle size on alkali content of,
    161t
  physical properties, 161, 161t

Cereal straw, 131
CFRP materials. *See* Carbon
  fiber−reinforced polymer materials
  (CFRP materials)
CH. *See* Calcium hydroxide (CH)
Chai viticole vauvert, 327f
Chemically bonded sand, 178, 178f
Cherrapunji root bridge, 43, 43f
Chicago, 20−21, 25f, 28
CINVA Ram press, 228, 228f
CKD. *See* Cement kiln dust (CKD)
Clay, 44−45, 219
  clay-based binders, 310−311
  clay-bonded sand, 178, 178f
  soils
    consistency and swelling properties, 256t
    structural forms in, 254f
    structure and properties, 252−255
    types and free swell values, 255t
Clay-based materials, 251−252, 252f
  without additives, 256−258
  construction using clay-based systems,
    264−268
    durability of unfired clay materials,
      267−268
    formation, 264−265
    storage, 265
Closed-loop manufacturing, 70
CLSM. *See* Controlled low-strength
  materials (CLSM)
Cob, 274
Codes and standards, 359−361
Coir binderless board, measuring impacts of,
  68b
Colombian Code, 405−406
Columns, 439, 466−467
Combined loading elements, 416−417
Combusted rice husk ash (CRHA), 189
Combustible volatiles, 119−120
Composite Can and Tube Institute (CCTI),
  467
Composites, 111−112
  forms, 115
Compressed earth blocks (CEBs), 214−215
Compressed stabilised earth blocks
  (CSEBs), 207−208
  average loading on typical single-storey
    wall, 219f
  block production, 223−231

comparative properties, 223t
cost comparison, 208–209
density and thermal properties of, 220–222
dimensional tolerance, 222
durability, 245–246
    abrasion test, 245–246
    oven-dried block and wire bristle brush, 245f
    sorptivity, 246, 246f, 246t
    of stabilised soil blocks, 245
limitations, 209
material properties of, 218–222
materials selection and block production, 209–222
    diagram of texture for soil for, 214f
    materials for CSEB production, 210
    settlement of soil materials, 212f
    soil identification and classification, 210–211
    soil identification and selection, 211–214
    soil textural triangle, 213f
nonstabilised soil construction, 216f
recommended limits, 222t
shrinkage limits, 221t
stabilisation of soil for, 214–217
stabilisers, 217–218
summary of advantages of, 208
zone of suitable plasticity, 215f
Compression
    elements, 414–415
    strength evaluation methods, 237
Compressive strength, 218–219
    evaluation, 236
Concrete, 159
    CKD effect on properties, 162–170
        compressive strength development, 165t–166t
        durability properties, 168–170
        flexural strength development, 167t–168t
        hydration properties, 163–164
        setting time, 162–163
        splitting tensile strength development, 169t
        strength properties, 164–167
    formwork, 453, 454f
    POFA effect on properties, 190–195
        compressive strength development, 194t
        durability properties, 193–195
        scanning electron microscopy images, 191f
        strength properties, 190–193
    RHA effect on properties, 184–189
        compressive strength development, 187t–188t
        durability properties, 189
        strength properties, 184–189
    SF effect on properties, 171–173
        compressive strength development, 174t–175t
        durability properties, 173
        hydration, 172
        setting time, 172
        splitting tensile and flexural strength development, 176t–177t
        strength properties, 172–173
        water demand and workability, 171–172
    WFS effect on properties, 181–182
        durability properties, 182
        strength properties, 181–182
Condensed SF. See Silica fume (SF)
Conglomerates, 304–307, 305f
Connections, 419
    allowable loads for dowel-type connections, 425t–426t
    angled connections, 421f
    dowel connection capacity, 423–427
    longitudinal connections, 421f
    orthogonal connections, 421f
    post-to-foundation connections, 422f
    types, 419–423
Construction, bamboo, 427–428
Contraflexure-based analysis, 24–25
Controlled low-strength materials (CLSM), 180–181
Conventional material, 63–64
Copper slag (CS), 164
Corn stalk (Zea mays L.), 88
Cost, 72
Coursed random rubble masonry, 317–318
Coursed square rubble masonry, 317–318, 319f
Crack growth, 46
Cradle-to-cradle analysis, 66–67
Cradle-to-gate analysis, 66–67
    scope, 68b
Cradle-to-grave analysis, 66–67

Credit Valley Hospital, 354
CRHA. *See* Combusted rice husk ash
    (CRHA)
Cross machine direction (CD), 460
Cross-laminated timber (CLT). *See* Massive
    timber products
CS. *See* Copper slag (CS)
CSEBs. *See* Compressed stabilised earth
    blocks (CSEBs)
CSH. *See* Calcium silicate hydrate (CSH)
Culm
    bamboo Culm anatomy, 366f
    dimensional properties, 372−374
    variation of density and properties,
      368−369
Cultural significance, 72−73
Curauá plant (*Ananas erectifloius*),
    88−89
CVBT block system, 229

D

Deciding between materials, 75−77
    intersectional innovation, 77
    Pareto optimality, 75−76, 75f
Decision making, 66, 76
Deflections, 411
Dendrocalamus giganteus (*D. giganteus*),
    51−53
Density effect, 367−368
Design decisions, performance
    consequences of, 121−122
Design values derivation, 385−386
Desmoncus polyacanthos Mart. *See* Jacitara
    palm (*Desmoncus polyacanthos*
    Mart.)
Desorption isotherm, 277, 277f
"*Dhajji dewari*" construction, 17, 17f−18f,
    32f
Digital image processing (DIP), 51
DIP. *See* Digital image processing (DIP)
Disassembly, 69−70
Dolomites, 307
Dolomitic limestones, 307
Dowel connection capacity,
    423−427
Dressed stones, 310
Dry slaked lime, 312
Dry stone walls, 317, 318f
Dry-stack system, 208

Dry-stacking of masonry, 233−235
    construction of Great Zimbabwe ruins, 235f
    dry-stack interlocking block masonry, 235f
    effect of openings in, 236f
Durability, 100, 132−134, 293−294,
    347−349
    concerns, 117−121
    managing, 121−122
      optimisation for tuned performance, 122
      performance consequences of design
        decisions, 121−122
Duralin process, 122
Duzce earthquake, 31f

E

E-glass fibers, 95, 96t
Earth stabilization, 289−290
Earth-based coatings, 136−137
Earthen materials and constructions,
    273−275
    assessing performance
      hygrothermal performance, 290−293
      mechanical stability and durability,
        293−294
    particularities, 275−290
      affinity with water, 275−280
      capillary pressure effect, 286f
      compressive strength in function of
        sample water content, 285t
      Earth stabilization, 289−290
      evidences of thermal buffering effect of
        rammed earth walls, 281f
      evolution of interface during drying
        process, 279f
      hygrothermal behavior, 280−283
      impact on building design, 288−289
      mechanical behavior, 283−288
      morphological description of earth, 276f
      stress−strain behavior comparison, 287f
    rammed earth house, 275f
Earthquake, 3−5, 14−16, 20−21
EC5. *See* Eurocode 5 (EC5)
Ecovative, 77
Edge-bearing test, 378−380, 379f
Elastic, 20
    analysis, 475
    elastic−plastic constitutive model,
      460−461
Embedded nanofibers, 98

EMC. *See* Equilibrium moisture content (EMC)
Emergency shelters, 453–455
Engineered bamboo, 374–375, 433.
  See also Bamboo
  bridges
    glubam pedestrian bridges, 444–446
    truck load roadway bridge, 446–447
  buildings
    bamboo and solar energy combination, 443
    glubam heavy frame buildings, 443, 443f
    glubam temporary shelters for disaster relief, 442
    lightweight glubam-frame, 441–442
  carbon footprints analysis, 447
  glubam production, 434–437
  laminated bamboos, 434f
  material properties, 437
  structural applications, 438t, 450–452
  structural components
    beams, 439
    columns, 439
    girders, 439
    shear walls, 440
    test shear wall details, 440f
Engineered composites with hybrid reinforcement, 98–100
Engineered wood products (EWPs).
  See Manufactured timber products
Enzymatic hydrolysis, 92
Equilibrium moisture content (EMC), 344–345
Eucalyptus pulps, 89–90
Eurocode 5 (EC5), 475
External absorbed water, 278–279

F
FA. *See* Fly ash (FA)
Fabrication, bamboo, 427–428
Feuillette house, 129f
FGM. *See* Functionally graded material (FGM)
Fiber-reinforced cement, 85
Fibre saturation point (FSP), 344, 371–372
Fibre-reinforced polymers (FRP), 111–112, 469

Ficus elastica (*F. elastica*), 43–44
Finger jointing, 351
Fique fibers (*Furcraea andina*), 88
Fire, 119–120
  reaction, 403
  resistance, 138, 403–404
Firing, 251–253, 264–265
Flexural members. *See* Beams
Flexure, 411–412
Fly ash (FA), 163, 261
4:1 glubam mechanical properties, 438t
4:1 sheets, 437
Frames, 26–28, 469–473
FRP. *See* Fibre-reinforced polymers (FRP)
FSP. *See* Fibre saturation point (FSP)
Full-culm bamboo, mapping of physical and mechanical properties of, 53–57
  coordination, positions of borings and device, 54f
  data for selection of bamboo for engineering projects, 56–57
  dimensions, 53–54
  mathematical equations, 56t
  nondimensional values, 57t
Functional unit for analysis, 68b
Functionally graded material (FGM), 50–53, 374
Fungus, 121
Furcraea andina. *See* Fique fibers (*Furcraea andina*)

G
Gabions, 327, 328f
GGBFS. *See* Ground granulated blast furnace slag (GGBFS)
GGBS. *See* Ground granulated blastfurnace slag (GGBS)
GHG. *See* Greenhouse gas (GHG)
Girders, 439
Glubam heavy frame buildings, 443, 443f
Glubam pedestrian bridges, 444–446
Glubam production, 434–435, 435f
  bamboo curtains, 436
  bamboo strips, 435–436
  gluing, 436–437
  postprocessing, 437
  pressing, 436–437
  raw bamboo selection, 435
Glubam sheets, 434

Glubam temporary shelters for disaster relief, 442
Gluing, 436—437
Glulam, 351
Gneiss, 308
GPOFA. *See* Ground POFA (GPOFA)
Grading, timber, 349
Granite ashlars, 304, 304f
Graphical isopleth system, 133
Gravity domain, 280
Greenhouse gas (GHG), 159, 447
GRHA. *See* Grounded rice husk ash (GRHA)
Grinding method, 93—94
Ground granulated blast furnace slag (GGBFS), 163
Ground granulated blastfurnace slag (GGBS), 259—261
    unconfined compressive strength against slag, 260f
    unfired building bricks, 261f
Ground POFA (GPOFA), 193—195
Grounded rice husk ash (GRHA), 185—186
Guadua a. K., 405—406
    adjustment factors for, 409t—410t
    allowable strength values for, 408t
    modifying factors for, 407t
    slenderness ratio, 415
Guadua angustifolia (*G. angustifolia*), 381—385
    allowable stress for, 386t
    average dimensions for culms, 381t
    average elastic properties for, 384t
    mechanical properties parallel to fibres for, 382t
    stress at failure versus slenderness for, 383f
Gypsum binders, 311
Gypsum plasters, 136

H
Half-timber, 17
Hankinson equation, 426, 437
Hatschek methods, 85, 86f
HDPE. *See* High-density polyethylene (HDPE)
Heat, 116
Hevea brasiliensis (*H. brasiliensis*), 43
High-calcium lime, 217—218
High-density polyethylene (HDPE), 122

High-magnesium lime, 217—218
Hımıs, 17, 31f, 33
Hollow-block systems, 208
Hot lime mortars, 312—313
Hybrid polymeric composites, 96
Hydraform
    blocks, 236—239, 238f
        capping arrangement, 237
        compression strength evaluation methods, 237
        direct flat loading on block, 239
        loading on block shoulders, 237
        testing of cubes cut from blocks, 239
    dry-stack block, 229
    dry-stack interlocking CSEB system, 233f
    interlocking block features, 236
Hydrated lime, 217—218
Hydraulic lime, 313
Hygroscopic domain, 279—280
Hygrothermal
    behavior, 280—283
    effects, 117—118
        moisture, 117—118
        temperature, 118
    performance, 290—293
    properties, 131—132, 139—140

I
ID. *See* Inside diameter (ID)
Illuviation, 210—211
Impact assessment, 68b
In situ straw bale construction, 146—149
    load-bearing straw bale construction, 146—149
    non-load-bearing applications, 149
In-pore condensation/evaporation processes, 278
INBAR. *See* International Network for Bamboo and Rattan (INBAR)
Inca Suspension Bridge, 41, 41f
India, 8
Industrial additives, 258—263
    GGBS, 259—261
    PFA, 261, 262f
    WSA, 262—263, 263f
Industrial by-products. *See* Supplementary cementitious materials (SCMs)
Industrial hemp core (*Cannabis sativa L.*), 88

Industrial-scale processes, 86b
Inside diameter (ID), 462–463
Interlamellar absorbed water, 278–279
Interlocking blocks, 230f, 231, 236
Interlocking systems, 208
Interlocking units, key features of,
    231–232
International Network for Bamboo and
    Rattan (INBAR), 38–39, 380,
    404–405
International Organization for
    Standardization (ISO).
    See International Standards
    Organisation (ISO)
International Standards Organisation (ISO),
    380, 404–405
Intersectional innovation, 77
Intrafiber adsorbed water, 278–279
Intrusive rocks, 304
Inventory, 68b
Iran, 3, 5
ISO. *See* International Standards
    Organisation (ISO)
Isopleth, 134
Italian Technical Standards, 323

J
Jacitara palm (*Desmoncus polyacanthos*
    Mart.), 88–89

K
Kah-gel, 45
Kashmir, 13–19
Kelvin's Law, 278
Kimmeridge clay, 259

L
Laminate composites, 115
Laminated bamboo, 450
Laminated bamboo lumber (LBL), 433–434
Laminated veneer lumber members (LVL
    members), 337
Latent heat of evaporation, 282–283
LBL. *See* Laminated bamboo lumber (LBL)
LCA. *See* Life cycle analysis (LCA)
LCI. *See* Life cycle inventory (LCI)
Life cycle analysis (LCA), 65–69,
    141–142, 453–455
    of straw bale construction, 141–142

Life cycle approach. *See* Life cycle analysis
    (LCA)
Life cycle assessment. *See* Life cycle
    analysis (LCA)
Life cycle inventory (LCI), 67
Lightweight glubam-frame buildings,
    441–442
Lignin, 121
Lignocellulosic fibers, 88–89
Ligurian slate, 308, 309f
Lime, 217–218
    binders, 311
    lime-based coatings, 136–137
    stabilization, 290
Limestone powder (LP), 185–186
Limit of proportionality (LOP), 97–98
Linoleum, 112–113
Linothorax, 112–113
Liturgical Hall "Padre Pio" of Renzo Piano,
    326f
Load and resistance factor design (LRFD),
    404, 406
Load-bearing straw bale construction,
    146–149
    load-bearing straw bale wall, 128f, 130f
    rendering of straw bale wall, 147f
    running bond coursing of straw bales, 146f
    temporary shelter for straw bale
        construction, 148f
Load-bearing walls, 418
LOI. *See* Loss on ignition (LOI)
Longitudinal connections, 421f
LOP. *See* Limit of proportionality (LOP)
Loss on ignition (LOI), 160
LP. *See* Limestone powder (LP)
LRFD. *See* Load and resistance factor
    design (LRFD)
LVL members. *See* Laminated veneer
    lumber members (LVL members)

M
Machine direction (MD), 460
Macrofibers, 85–89, 88t
Magmatic rocks, 303
Magnani methods, 85, 87f
Manufactured timber products, 336, 350,
    354
    beam products, 351–352
    curved glulam beam, 352f

Manufactured timber products (*Continued*)
  massive timber products, 352–354
  sheet products, 350–351
Marbles, 308
Masonry, 9–11, 14–17, 20
Masonry Quality Index (MQI), 324–325
Masonry wall, 315–321, 316f
  Ashlar masonry, 319–321, 320f–321f
  dry stone walls, 317, 318f
  masonry with wood reinforcing, 321
  rubble masonry, 317–319
Masonry with wood reinforcing, 321
Massive timber products, 350, 352–354
MC. *See* Moisture content (MC)
MCC. *See* Microcrystalline particles (MCC)
MD. *See* Machine direction (MD)
Mechanical behavior, 283–288
Mechanical stability, 293–294
Mechanosorptive behaviour, 346
Metakaolin (MK), 186
Metamorphic rocks, 308
Micritic limestone. *See* Calcilutite
Micro/nanofibrils, 93, 93f
Microcrystalline particles (MCC), 93
Microfibrils, 84–85
Microscopy techniques, 94
Microsilica. *See* Silica fume (SF)
Mineral stabilization, 289
MIPROMALO. *See* Mission de Promotion
    des Materiaux Local
    (MIPROMALO)
Miscanthus, 127–128
Mission de Promotion des Materiaux Local
    (MIPROMALO), 263–264
MK. *See* Metakaolin (MK)
Modern Movement, 325
Modulus of elasticity (MOE), 97–98, 186,
    408, 433
Modulus of rupture (MOR), 97–98, 408
MOE. *See* Modulus of elasticity (MOE)
Moisture, 117–118
  effects, 344–345
  seasoning on, 345f
  volumetric strain, 219–220
Moisture content (MC), 344, 401, 411, 436
  effect, 371–372
MOMA. *See* Museum of Modern Art
    (MOMA)
MOR. *See* Modulus of rupture (MOR)

Mortars, 159, 310–313
  CKD effect on properties, 162–170
    compressive strength development,
      165t–166t
    durability properties, 168–170
    flexural strength development,
      167t–168t
    hydration properties, 163–164
    setting time, 162–163
    splitting tensile strength development,
      169t
    strength properties, 164–167
  POFA effect on properties, 190–195
    compressive strength development,
      194t
    durability properties, 193–195
    scanning electron microscopy images,
      191f
    strength properties, 190–193
  RHA effect on properties, 184–189
    compressive strength development,
      187t–188t
    durability properties, 189
    strength properties, 184–189
  SF effect on properties, 171–173
    compressive strength development,
      174t–175t
    durability properties, 173
    hydration, 172
    setting time, 172
    splitting tensile and flexural strength
      development, 176t–177t
    strength properties, 172–173
    water demand and workability,
      171–172
  WFS effect on properties, 181–182
    durability properties, 182
    strength properties, 181–182
Moso (*Phyllostachys heterocycla
    pubescens*), 51, 51f–52f
Moso bamboo (*Phyllostachys pubescens*),
    433
Moulds, 267–268
MQI. *See* Masonry Quality Index (MQI)
Mud construction, 9–10
Musa cavendishii. *See* Banana (*Musa
    cavendishii*)
Museum of Modern Art (MOMA),
    469–471

N

Nanofibers, 91–93, 98
Nanofibrillated cellulose, 91–93
    grinding method, 93–94
    properties of, 94–95
Nanosilica (NS), 171
Natural ashes, 159–160
Natural fibres, 113–114, 113t
Natural forms
    of fibre/polymer biocomposites, 112–113
    of stone material, 303–308
        Granite ashlars, 304f
        intrusive rocks, 304
        magmatic rocks, 303
        metamorphic rocks, 308
        "Pietra Bigia" sandstone, 306f
        sedimentary rocks, 304–307
        volcanic rocks, 304
        Zenòbito Tower, 305f
Natural glass fibers, 95
Natural materials
    in historic construction, 39–44
        ancient suspension bridges, 42f
        Cherrapunji root bridge, 43, 43f
        depiction of Xerxes' Pontoon Bridge, 39f
        Inca Suspension Bridge, 41, 41f
        Zhao-Zhou Bridge, 40, 40f
    in modern bridge construction, 57–58
Natural rice husk ash (NRHA), 183
NOCMAT. *See* Nonconventional materials
        and technologies (NOCMAT)
Non-load-bearing applications, 149, 149f
Nonbearing walls, 418
Noncombustible volatiles, 119–120
Nonconventional materials, 38, 64
    alternative materials in modern bridge
        construction, 57–58
    bamboo material for 21st century, 48–57
    natural materials
        in historic construction, 39–44
        in modern bridge construction,
            57–58
    reinforced adobes, 44–48
Nonconventional materials and technologies
        (NOCMAT), 38
Nonconventional timber construction
    case studies, 354–359
    codes and standards, 359–361
    80-m-long multispan timber, 340f

material properties
    durability, 347–349
    effects of moisture, 344–345
    nature of wood, 342
    seasoning on moisture, 345f
    structural properties, 342–343
    time-dependent behaviour, 345–347
    wood and timber, 343–344
modern engineering applications, 349
    grading, 349
    manufactured timber products, 350–354
modern house, 341f
timber materials natural forms, 335
    1. 8-m-diameter Douglas Fir log, 336f
    forests, 335
    natural rounds, 335
    processed elements and products,
        336–337
    sawn members, 336
traditional and historic applications
    bridges, 337–338
    buildings, 338–339
    simple structures, 337
Tynset Bridge, 355f
Nonlinear viscoelastic constitutive model,
        461–462
Nonrenewable resources, 113
Nonstructural applications, 116–117
Nonwood pulps, 85
Nordic Timber Bridge Program,
        354–356
NRHA. *See* Natural rice husk ash (NRHA)
NS. *See* Nanosilica (NS)

O

OPC. *See* Ordinary Portland cement (OPC)
Optimisation for tuned performance, 122
Optimum water content, 227
Opus, 302
Opus craticium, 321, 322f
Ordinary Portland cement (OPC),
        160, 217
Organic additives, 313
Organic polymers, 118–119
Organosolv pulp production and properties,
        91
Oriented strand board (OSB), 337
Orthogonal connections, 421f
OSB. *See* Oriented strand board (OSB)

P

Pakistan, 32f, 33
Palm oil fuel ash (POFA), 190–195.
    See also Rice husk ash (RHA)
    chemical properties, 190, 192t
    effect on properties of mortar and concrete,
        190–195
        compressive strength development, 194t
        durability properties, 193–195
        scanning electron microscopy images,
            191f
        strength properties, 190–193
    physical properties, 190, 192t
Paper MD, 460
Paper tubes, 462–463
Paperboard, 460
Paperboard tubes, 453
    Ban's paperboard tube structures, 453–455
    manufacturing, 455–459, 456f
    mechanical properties of paper tubes, 468t
    mechanics and properties of paperboard
        and tube materials, 459–464
    paper strength, 453–455
    ply positioning and adhesive application,
        458f
    primary uses, 455–459
    structural analysis and design, 473–477
    structural elements, 466–473
    structural systems, 465f–466f, 466
    tube-bending modulus *vs.* winding angle,
        464f
    unwind stands, 457f
Parallel strand bamboo, 450
Pareto optimality, 75–76, 75f
Particles size analysis, 211–213
PC. *See* Portland cement (PC)
Peacock House, 28f
Permeability, 246
PFA. *See* Pulverized fuel ash (PFA)
PHA. *See* Polyhydroxyalkoanate (PHA)
PHB. *See* Polyhydroxybutyrate (PHB)
PHBV. *See* Polyhydroxybutyrate-
    cohydroxyvalerate (PHBV)
Phenol formaldehyde, 436–437
Phyllostachys heterocycla pubescens.
    See Moso (*Phyllostachys
    heterocycla pubescens*)
Phyllostachys pubescens. See Moso bamboo
    (*Phyllostachys pubescens*)

PI. *See* Plasticity indices (PI)
"Pietra Bigia" sandstone, 306f
Pinus pulps, 89–90
Pitch, 313
PLA. *See* Polylactic acid (PLA)
Plant growth, 130–131
Plasters, 136
Plasticity indices (PI), 255
Plybamboo sheets, 434
Plywood, 350
    on steroids, 352
POFA. *See* Palm oil fuel ash (POFA)
Pole bamboo, 396
Polyhydroxyalkoanate (PHA),
    114–115
Polyhydroxybutyrate (PHB),
    114–115
Polyhydroxybutyrate-cohydroxyvalerate
    (PHBV), 114–115
Polylactic acid (PLA), 114
Polymeric matrix, 96, 111–112
Polypropylene (PP), 84
Polyvinyl alcohol (PVA), 84
Portal analysis method, 28
Portland cement (PC), 162–163, 264, 313
Post-to-foundation connections, 422f
Pozzolans, 8–9, 218
PP. *See* Polypropylene (PP)
Prefabricated panel construction, 150–152,
    150f
    load-bearing straw bale panels, 151f
    straw bale cladding panels, 151f
Prefabrication, 136
Pressing, 436–437
Pressure, 116
Proctor compaction test, 227, 227f
Pulp fibers, 89–90, 90t
Pulp-reinforced cement-based composites,
    96–98
Pulverized fuel ash (PFA), 261, 262f
PVA. *See* Polyvinyl alcohol (PVA)
$P-\Delta$ effect, 416–417

R

R-curve. *See* Resistance-curve (R-curve)
Rana Plaza, 11, 12f
Random rubble masonry, 317–318, 318f,
    320f
Raw bamboo selection, 435

Reconstituted timber products.
 See Manufactured timber products
Red in colour and nonfibrous (RNF), 255
Reinforced adobes, 44–48. *See also* Adobe
 attributes and properties, 45–46
 comparative primary energy requirements
  of building materials, 46t
 effective use of adobes in building
  construction, 48
 improvement of physical and mechanical
  properties, 46–48
 influence of fibres, 47f
 interaction of natural reinforcing fibre and
  drying soil, 47f
Relative humidity (RH), 460
Renders, 136, 144
Renewable resources, 112, 116
Resilience, 71
Resin-saturated curtains, 436
Resistance-curve (R-curve), 87–88
RH. *See* Relative humidity (RH)
Rice husk ash (RHA), 182–189, 184f.
 See also Palm oil fuel ash (POFA)
 chemical properties, 183, 185t
 effect on properties of mortar and concrete,
  184–189
  compressive strength development,
   187t–188t
  durability properties, 189
  strength properties, 184–189
 physical properties, 183, 183t
Risk of decay, 139–140
River pebbles, 310
RNF. *See* Red in colour and nonfibrous
 (RNF)
Rubble masonry, 317–319
Rubble stone, 32f
Rule of art, 324–325
Rule-of-thumb method, 412

S
San Francisco earthquake and fire, 19–22,
 21f
Sandstones, 304–307
Saturation ratio, 276–277
SCBA. *See* Sugar cane bagasse ash (SCBA)
SCC. *See* Self-compacting concrete (SCC)
Schistosity, 308
Schists, 308

SCMs. *See* Supplementary cementitious
 materials (SCMs)
SE. *See* Specific energy (SE)
Seasoning, 337
Sedimentary rocks, 304–307
 biochemical origin, 307
 chemical origin, 307
Self-compacting concrete (SCC), 182
SF. *See* Silica fume (SF)
Shales, 304–307
Shear, 412
 tests, 376, 380–381
 walls, 417–419, 440
Sheet products, 350–351
Shell lime powder (SL powder), 186
Shelter, 63, 64f
Silica Dust. *See* Silica fume (SF)
Silica fume (SF), 170–173
 chemical properties, 171, 171t
 effect on properties of mortar and concrete,
  171–173
  compressive strength development,
   174t–175t
  durability properties, 173
  hydration, 172
  setting time, 172
  splitting tensile and flexural strength
   development, 176t–177t
  strength properties, 172–173
  water demand and workability,
   171–172
 physical properties, 170, 170t
Sinter, 251–252
Sisal (*Agave sisalana*), 85
Skeleton-frame
 construction, 22–26, 25f
 skyscraper, 21, 22f
Skyscraper, 21
SL powder. *See* Shell lime powder
 (SL powder)
Slabs, 310
Slag, 259
Slates, 308
Slenderness ratio, 414–415
Sliced connections. *See* Longitudinal
 connections
Soil
 burial tests, 121
 composites, 45

Soil (*Continued*)
  identification
    and classification, 210−211
    and selection, 211−214
  particle grading, 211t
  profile, 210−211, 210f
  stabilisation, 217
  texture, 211−212
  textural triangle, 213f
Soil-cement blocks. *See* Compressed
    stabilised earth blocks (CSEBs)
Sole evaporation/condensation processes,
    278−279
Solid interlocking blocks, 208
Solid walls, 26−28
Solid waste management, 159−160
"SONOTUBES", 455
Sorption isotherm, 277, 277f
Sorptivity, 246, 246f, 246t
Soy protein concentrate (SPC), 115
Specific energy (SE), 97−98
Spent foundry sand. *See* Waste foundry sand
    (WFS)
Spiral paper tubes, 455−457
Split-pin test, 379f
Squared rubble masonry, 317−318
Srinagar, Kashmir, India, 13−19, 30f
  canalside building of *taq* construction, 16f
  contemporary continuation of preindustrial
    technique, 15f
  four-and five-story buildings, 15f
  narrow lane with timber-laced masonry
    bearing wall buildings, 14f
  traditional houses, 13f
Stabilisers, 217−218
  cement, 217
  combination of lime and cement, 218
  combination of lime and pozzolans, 218
  lime, 217−218
Stoke's law, 212
Stone, 302−303
  masonry materials, 303−313
    mortars, 310−313
    natural forms, 303−308
    petrographic characteristics and use
      of rocks as building materials,
      308−310
  working, 303
Strand woven, 450

Straw bale construction, 127
  applications, 142−152
    BaleHaus, 145f
    design of straw bale buildings, 144−145
    prefabricated panel construction,
      150−152
    in situ straw bale construction, 146−149
    straw bales as insulation, 143−144
    variation of straw bale thermal
      conductivity with density, 143f
  characterisation, 137−142
    acoustic properties, 141
    external render core removed from straw
      bale wall panel, 141f
    fire resistance, 138
    hygrothermal properties and risk of
      decay, 139−140
    physical and mechanical properties,
      137−138
    straw bale panel after fire test, 139f
  Feuillette house, 129f
  future of, 152−153
  LCA, 141−142
  load-bearing straw bale wall, 128f, 130f
  material properties
    CO_2 evolution, 135f
    durability, 132−134
    hygrothermal properties, 131−132
    materials used with straw bales,
      134−137
    plant growth, 130−131
    structure and chemistry of straw,
      131
Straw bales as insulation, 143−144
Stress values, 407
Stress−strain behavior, 460
Structural analysis and design,
    473−474
  Japan Pavilion, 475
  member design, 476−477
Structural elements
  16-mm-thick paper tube, 470f
  arches, 469−473
  beams, 469
  columns, 466−467
  frames, 469−473
  trusses, 469−473
Stumblebloc Mortarless System, 229
Sugar cane bagasse ash (SCBA), 186

Supplementary cementitious materials (SCMs), 159–160, 195
Supply chains, 70–71
Sustainability, 83
Sustainable building materials, 207–208
Sustainable construction materials, 159
  CKD, 160–170
  RHA, 182–189
  SF, 170–173
  WFS, 173–182
Sustainable-scale, 64
  analyzing potential for, 66–74
    life cycle approach, 66–69
      measuring impacts of coir binderless board, 68b
    factors for, 69–74
      systemic factors, 72–74
      technical factors, 69–72
Systemic factors, 72–74
  cultural significance, 72–73
  industrialization and quality control, 74
  localization of resources, materials, and processes, 73
  sophistication of processing, 73–74

T
"Table leg" principle, 24–25
Taq, 16–17, 16f
TCC system. *See* Timber concrete composite system (TCC system)
Technical factors, 69–72
  closed-loop manufacturing, 70
  cost, 72
  disassembly, 69–70
  resilience, 71
  supply chains, 70–71
Temperature, 118
Temporary structures, 473–474
Tensile strength, 433
Tension elements, 415–416
Test standards development, 380–381
TGA. *See* Thermogravimetric analysis (TGA)
Thermal insulation, 127–128, 142
Therme Vals wall structure of Peter Zumthor, 326f
Thermogravimetric analysis (TGA), 118
Thickness direction (ZD), 460

Thin-layer lamination of bamboo strips, 434
Tholoi, 320–321
Three-pinned arch' failure mode, 378–380
Tilt-up timber, 353
Timber concrete composite system (TCC system), 357–358
Timber construction, 5, 11, 16–17, 26–27
Timber plate, 134–136
Timber stakes, 127
Time-dependent behaviour, 345–347
Travertine, 307, 308f
Truck load roadway bridge, 446–447
Trusses, 469–473
Tsai-Wu theory, 437
Turkey, 17, 31f, 33
2×4 stud-wall method, 441–442

U
Ultra-high performance concrete (UHPC), 185–186
Ultrafine palm oil fuel ash (UPOFA), 190
  X-ray diffraction of, 193f
Ultraviolet radiation (UV radiation), 118–119
Uncoursed random rubble masonry, 317–318
Uncoursed square rubble masonry, 317–318, 319f
Unfired clay materials and construction, 255–256, 257t
  agricultural additives, 263–264
  clay-based materials, 251–252, 252f
  industrial additives, 258–263
  small holder fired brick community project, 253f
  structure and properties of clay soils, 252–255
Unworked stones, 310
UPOFA. *See* Ultrafine palm oil fuel ash (UPOFA)
Used foundry sand. *See* Waste foundry sand (WFS)
UV radiation. *See* Ultraviolet radiation (UV radiation)

V
Vapour-permeable materials, 139
Vegetable fibers

Vegetable fibers (*Continued*)
  availability and potential of, 83—84
  characterization for engineering
      applications, 85—95
    macrofibers, 85—89
    nanofibrillated cellulose, 91—93
    Organosolv pulp production and
        properties, 91
    pulp fibers, 89—90
  composites, 95—100
    durability, 100
    engineered composites with hybrid
        reinforcement, 98—100
    pulp-reinforced cement-based
        composites, 96—98
    reinforced with bamboo cellulose fiber,
        95—96
  as reinforcement for brittle matrices,
      84—85
Vegetable lignocellulosic macrofibers,
    84—85
Vernacular, 7—9
  architecture, 8—10, 309
  contemporary design of resort, 9f
  of industrial architecture, 10—11
  materials, 64, 69, 72—74
  nineteenth-century mill buildings, 10f
Vernacular construction. *See also* Straw bale
      construction
  Bam, 3—6
  continuation of vernacular building
      tradition, 7f
  frames, 26—28
  in reinforced concrete, 8f
  San Francisco earthquake and fire,
      19—22
  skeleton-frame construction, 22—26
  solid walls, 26—28
  Srinagar, 13—19
  traditional brick kilns, 9f
Volatized Silica. *See* Silica fume (SF)
Volcanic rocks, 304
Volcanic tuffs, 304—307, 307f

W
w/b ratio. *See* Water-to-binder ratio
    (w/b ratio)
Wall panel compression test, 241

Wall thickness index, 372—374
Walls with diagonal elements, 417
Waste foundry sand (WFS), 173—182
  applications, 180—181
  chemical properties, 179—180, 180t
  effect on properties of mortar and concrete,
      181—182
    durability properties, 182
    strength properties, 181—182
  physical properties, 178—179, 179t
Wastepaper sludge ash (WSA), 262—263,
    263f
Water absorption, 173
Water demand, 171—172
Water sorptivity test, 246
Water-to-binder ratio (w/b ratio), 164, 173
Waterproof asphalt sheets, 446
Weighted reduction parameter, 141
Wet/dry durability test, 245—246
Wet—dry tests, 268—269
WFS. *See* Waste foundry sand (WFS)
Wheat straw
  isopleth, 134f
  isotherms for, 133f
  structure, 132f
Wood cellulose fibers, 93
Wood nature, 342
  orthotropic nature, 343f
  and timber, 343—344
Wooden-frame systems, 321
Workability, 171—172
WSA. *See* Wastepaper sludge ash (WSA)
WUFI, 291—292

X
X-ray diffraction (XRD), 163
XRD. *See* X-ray diffraction (XRD)

Y
Young—Laplace Law, 278

Z
ZD. *See* Thickness direction (ZD)
Zea mays L. See Corn stalk (*Zea mays L.*)
Zenòbito Tower, 305f
Zhao-Zhou Bridge, 40, 40f
Zhaozhou Bridge. *See* Zhao-Zhou Bridge